# SPRINGER TRACTS IN MODERN PHYSICS

Ergebnisse der exakten Naturwissenschaften

Volume 43

Editor: G. Höhler



Springer-Verlag Berlin Heidelberg GmbH 1967

*Manuscripts for publication should be addressed to:*

G. Höhler, Institut für Theoretische Kernphysik der Technischen Hochschule, 75 Karlsruhe, Kaiserstraße 12

*Proofs and all correspondence concerning papers in the process of publication should be addressed to:*

E. A. Niekisch, Kernforschungsanlage Jülich, Arbeitsgruppe Institut für Technische Physik, 517 Jülich, Postfach 365

ISBN 978-3-662-15898-2 ISBN 978-3-540-35553-3 (eBook)
DOI 10.1007/978-3-540-35553-3


Library of Congress Catalog Card Number 25-91 30

Originally published by Springer-Verlag Berlin Heidelberg New York in 1967.

# Recent Developments in Lattice Theory

W. Ludwig

## Contents

# I. Introduction

Since the article on lattice theory in the „Handbuch der Physik" by *G. Leibfried* [55 L 1] and also Born-Huang's book on the „Dynamical theory of crystal lattices" [54 B 1], were published, ten years have been gone. These books contain nearly all the lattice theoretical aspects known up to 1955, or, in other words, they mainly deal with the harmonic theory and central forces or general force constants. In these books there are treated also the general methods and results of lattice theory independent of harmonicity and central forces and they contain already aspects of anharmonicity and noncentral forces.

But since these comprehensive representations have been finished a lot of investigations concerning lattice theoretical questions have been added. These points are essentially a more detailed theory of anharmonic effects in solids, the consideration of non-central forces, especially in form of the shell-model, the discussion of more complicated lattices (molecular crystals), the role of defects in lattice theory and the interaction of phonons with other kinds of radiation or particles. Since the Handbuch article has been published, a new technique was introduced in the theory of crystal lattices, appropriate for dealing with systems of many degrees of freedom: the Green-functions. Their use will be illustrated by some examples. Apart from this there are some concepts which have been known already in 1955 but the significance and usefulness of which have been made clear only later. A number of papers, however, already discuss some of these topics, but nearly all of them are related to special problems of the above mentioned subjects.

In this review we will discuss the main work which has been done since the Handbuch, vol. 7 [refered to as I in the following] and Born-Huang's book [refered to as II] have appeared. It is not possible to mention all the papers that have been published in the recent years and it is a hopeless task to consider all the different models that have been used. Instead we will confine ourselves to a discussion of the general principles and methods, and to a consideration of the most widely used models. We will not claim any completeness but we hope to give an insight in the problems and the results of lattice theory. The list of references is not complete, but from the papers cited, especially the review articles, it is possible to find all the other publications. Further, we will not begin at first principles, but rather give a continuation of the Handbuch article and a summary of recent papers.

For this reason, we will take over most of the nomenclature used in I. Only where there have been adopted different symbols widely in the international publications we will use them. This concerns e.g. the displacements, now mainly called $\boldsymbol{u}$ and the phonon wave number vector, now $\boldsymbol{q}$. For the general force constants or coupling parameters we use $\Phi^{m\,n}_{\substack{\mu\,\nu\\ i\,j}}$, contrary to the writing in the anglo-american papers, because there is no uniformity in different languages, and we think the above writing is the most consistent and unique one.

The importance of the invariance and symmetry relations has been pointed out in several papers [60 L 1, 60 H 1, 65 L 2, 64 L 6, 65 W 1]. They have been known in general already in 1955 [I, II, 50 H 1], but some aspects have been considered only later and so we will discuss them in a comprehensive manner first in chapter II.

Many-body forces have been used in lattice theory mainly since the work of *Löwdin* [48 L 1, 50 L 1, 56 L 1] and *Lundquist* [55 L 2, 57 L 1, 59 L 1] and *Dick* and *Overhauser* [58 D 1], though their first shell-model did not contain many-body forces. Most calculations since then have been done using the shell-model of *Dick* and *Overhauser* and its generalisations. Different aspects will be pointed out in chapter III.

The possible rotational motions in molecular crystals have been discussed only in recent years. We will give some examples in chapter IV.

Anharmonic effects have been discussed already by *Born* and *Brody* [22 B 1], *Schrödinger* [22 S 1] and *Born* and *Blackman* [optical effects 33 B 1, 2] and are reviewed in I and II. More detailed quantum mechanical calculations started in 1958 [58 L 1] and have been discussed in some detail in several papers [61 L 1, 61 M 1, 2, 63 C 2, 64 L 2]. These will be considered in chapter V.

Dynamics of crystals with defects have been investigated first by *Lifshitz* and coworkers [56 L 2, review] and by *Lax, Koster* and *Slater* [54 L 1, 54 K 1, 2]. The number of papers devoted to this problem increased enormously since the calculations of *Montroll* and *Potts* [56 M 1]. More recently some review articles appeared [63 M 1, 64 M 1, 2, 64 L 1, 66 L 1, 66 M 2, 66 N 1] [chapter VI].

The interaction of phonons with other kinds of radiation is also an old problem. We discuss some general aspects of the different types of interaction and show, that they can be represented by correlation functions and Green functions of the atomic displacements [chapter VII] [see also 64 B 1, 66 N 1].

The appendix contains a short compilation of the relations for thermodynamic Green functions as well as a short account on the perturbation methods in connection with Green functions. For phonons most of the relations are very simple.

Prof. *G. Leibfried*, Dr. *W. Biem*, Dr. *K. Dettmann*, Dipl.-Phys. *K. Thoma* and *B. Lengeler* have read parts of the manuscript. I thank them for valuable critizism, suggestions and hints.

## II. Symmetry and Invariance Properties of the Coupling Parameters

### 1. The Potential Energy

We assume a set of interacting particles, the interaction of which is completely described by a potential energy [I, II]. This potential energy is supposed to be a function of the relative positions of the particle sonly, i.e. it is the potential energy of a set of mass-points.* There shall be no hidden variables like internal degrees of freedom of the interacting particles (spin etc.). If the potential energy is given as an analytic function of the ion-coordinates, all the lattice properties of the set can be calculated. For such purposes many models of interaction have been proposed and discussed, mainly using central forces. In recent years also more complicated models (shell-model a.o.) have been treated successfully. But it happens that such models are bad or difficult to handle, especially because our knowledge about the interionic forces is insufficient. Therefore one often uses general coupling-parameters (or force-constants). A second reason for using *general* coupling-parameters is the advantage in *general* considerations. These coupling-parameters (called „cp's" in the following) are defined by an expansion of the potential energy in the vicinity of (not yet specified) equilibrium positions of the ions. It is assumed, that such an expansion exists, and that the physics of lattices can be described completely by restricting the motion of the ions to small relative displacements of the ions (small as compared to the interionic distance). The equilibrium positions can depend on temperature, or external forces and will be specified only later. Before defining the cp's some remarks on the lattice and its description might be useful.

It is often difficult to discuss a finite crystal. Therefore one uses an infinite crystal in many considerations. But since the potential energy of an infinite crystal diverges, one divides this infinite crystal into periodicity volumes simulating in this way a finite crystal. It can be shown that for the discussion of many properties this procedure is correct since surface effects give negligible contributions. This infinite crystal can be described by specifying a matrix $A$, with $\det A \neq 0$ the columns of which define the basis-vectors of the lattice

$$a_j^{(i)} = A_{ji}\,. \tag{1.1}$$

In equilibrium all lattice sites are then given by

$$\boldsymbol{R}^{\boldsymbol{m}} = \sum_i m_i\,\boldsymbol{a}^{(i)} = A\,\boldsymbol{m} \tag{1.2}$$

where $\boldsymbol{m} = \{m_i\}$ is a vector with integer components, where $m_i$ extends from $-\infty$ to $+\infty$. If the lattice is completely described by $A$, we call it Bravais- or primitive lattice. The parallelepiped formed by the $\boldsymbol{a}^{(i)}$ is called unit cell (or elementary cell). Its volume is $|\det A|$. If there is more than one ion in a unit cell, say $s$ ions and if the positions relative to the cell are $\boldsymbol{R}_\mu$, we have the lattice sites

$$\boldsymbol{R}_\mu^{\boldsymbol{m}} = A\,\boldsymbol{m} + \boldsymbol{R}_\mu; \quad \mu = 1, \ldots, s \tag{1.3}$$

* This is essentially the adiabatic approximation of *Born* and *Oppenheimer* [27 B 1, 51 B 1]. We will not discuss here the validity of this approximation. Only in sect. III we will have to mention some aspects of possible non-adiabatic effects. In metals there is a contribution of the „free" electrons to the lattice energy; this interaction can be described by volume forces. It is not difficult, at least in principle, to take into account this contribution in the following treatment of lattice theory. The interacting particles (mass points) will be called ions in what follows; but they might be also atoms or in some cases perhaps molecules as a whole, which are assumed to have no other properties like spin, or other „hidden" variables.

in equilibrium. Such lattices are called non-primitive lattices. The lattices (1.2, 3) are invariant towards translations by multiples of the basis-vectors only. With respect to every Bravais lattice there exists a reciprocal lattice, which is defined by the matrix

$$B = \tilde{A}^{-1} \tag{1.4}$$

or by basis-vectors*

$$\boldsymbol{b}^{(i)} \cdot \boldsymbol{a}^{(j)} = \delta_{ij}\,. \tag{1.5}$$

The reciprocal lattice is the same for all the nonprimitive lattices which have the same basis $A$. Another term we will use is related to the inversion symmetry of the lattice: we call a lattice „parameterfree" if every ion is a center of inversion. That means all the Bravais-lattices are parameterfree, but also some of the non-primitive lattices (e.g. NaCl-Type, CsCl-Type).

All the lattices thus described are called *ideal-infinite-crystal lattices*, if every lattice site is occupied by the corresponding ion. For such an infinite crystal, the translational invariance of the lattice towards translations by multiples of basis-vectors holds (see section 3).

In some cases, surface-effects are not negligible and we are forced to use finite crystals. For such purposes we define an *ideal finite crystal lattice* by assuming that the description of the lattice by (1.1—3) is correct apart from the range of values for $m_i$, which now run from 1 to $N_i$ e.g. In other words, it is assumed, that in ideal finite lattices the lattice spacing is always the same, even at the surface, whereas in real crystals the lattice spacing is slightly different at the surface compared with the interior of the crystal. In such a finite crystal there is no lattice translational invariance (see sections 3 and 5).

All the deviations from these ideal crystals we consider as perturbations. These are e.g. point defects in the crystal (impurities, vacancies, interstitials) or dislocations or stacking faults. In some cases it is even convenient to consider a free surface as a defect of an infinite crystal, thus relating the ideal finite lattice to the ideal infinite lattice.

After specifying the equilibrium positions of the lattice we can give the expansion of the potential energy about these positions, $\boldsymbol{u}_\mu^{\boldsymbol{m}}$ being the displacements of the ions from the equilibrium positions:

$$\Phi\left(\boldsymbol{R}_\mu^{\boldsymbol{m}} + \boldsymbol{u}_\mu^{\boldsymbol{m}}\right) = \Phi_0\left(\boldsymbol{R}_\mu^{\boldsymbol{m}}\right) + \Phi_1 + \Phi_2 + \Phi_3 + \cdots \tag{1.6}$$

with

$$\Phi_1 = \sum_{\boldsymbol{m}\mu i} \Phi_{\mu\, i}^{\boldsymbol{m}}\, u_{\mu\, i}^{\boldsymbol{m}}\,;\; \Phi_2 = \frac{1}{2!} \sum_{\substack{\boldsymbol{m}\mu i\\ \boldsymbol{n}\nu j}} \Phi_{\mu\, i\; \nu\, j}^{\boldsymbol{m}\,\boldsymbol{n}}\, u_{\mu\, i}^{\boldsymbol{m}}\, u_{\nu\, j}^{\boldsymbol{n}}\,;\; \Phi_3 = \frac{1}{3!} \cdots. \tag{1.7}$$

For finite crystals, $\Phi\left(\boldsymbol{R}_\mu^{\boldsymbol{m}} + \boldsymbol{u}_\mu^{\boldsymbol{m}}\right)$ is properly defined. In infinite crystals one considers usually the potential energy of one periodicity volume in order to avoid divergences. In any case, the coupling parameters (cp's)

* Lattice- and reciprocal lattice-vectors are co- and contravariant components of vectors. Since we will not need this distinction in the following, we will not go into details here.

$\Phi^{m\,n\ldots}_{\substack{\mu\,\nu\ldots\\ i\,j\ldots}}$ have a definite meaning: e.g. $-\Phi^{m\,n}_{\substack{\mu\,\nu\\ i\,j}} u^{n}_{\substack{\nu\\ j}}$ is the force, acting on ion $\binom{m}{\mu}$ in $i$-direction, if the ion $\binom{n}{\nu}$ is displaced by $u^{n}_{\substack{\nu\\ j}}$*. The number of index-pairs $\begin{pmatrix} m\\ \mu\\ i\end{pmatrix}, \ldots$, is called the order of the cp. $\Phi^{m}_{\substack{\mu\\ j}}$ vanishes, if the expansion is taken about those positions which make the potential energy a minimum. But these may not be the equilibrium positions, e.g., if there is a finite temperature or if external forces are applied. Then the equilibrium positions can be obtained only by minimizing the Helmholtz free energy $F(V, T) = F(A, \boldsymbol{R}^{\mu}, T)$ with respect to $A$ and $\boldsymbol{R}^{\mu}$. If one starts with the expansion (1.6, 7) using general cp's as the coefficients in (1.7), these cp's have to obey a set of invariance and symmetry relations, which will be discussed in section 2–6.

A first approximation to lattice theory is obtained from (1.6) by dropping terms of higher than 2nd order in the displacements. If one further takes the positions defined by the minimum of the potential energy $\left(\Phi^{m}_{\substack{\mu\\ i}} = 0\right)$ then we call it *harmonic theory*. However, one can account for some anharmonic effects also by restricting to quadratic terms only, if one does not take an expansion about the minimum positions of the potential energy, but an expansion about equilibrium positions determined by minimizing the Helmholtz free energy. In other words, one takes the positions as free parameters which will be calculated only after the free energy has been evaluated. Such a theory we call *quasiharmonic*. If we take into account the terms $\Phi_3$, $\Phi_4$, (and higher ones) explicitly, we speak of a genuine *anharmonic theory*.

## 2. Invariance Relations for Arbitrary Sets of Mass-points

There are relations which must be valid for an arbitrary set of mass-point, not necessarily forming a lattice. These relations can be derived from general invariance requirements, related to homogeneity and isotropy of space. For these relations the lattice has not to be specified, i.e. we need not distinguish between $\boldsymbol{m}$ and $\mu$, we only need a number for the ions. If we can do so, we abbreviate $\binom{m}{\mu}$ by $M$, $\binom{n}{\nu}$ by $N$ etc.

From definition (1.7) it is obvious, that the cp's have to be symmetric towards an interchange of the *index-pairs* $\begin{pmatrix} m\\ \mu\\ i\end{pmatrix} = \binom{M}{i}$, $\begin{pmatrix} n\\ \nu\\ j\end{pmatrix} = \binom{N}{j}$ etc.

$$\Phi^{M\,N}_{i\;j} = \Phi^{N\,M}_{j\;i}\,;\; \Phi^{M\,N\,R}_{i\;j\;k} = \Phi^{N\,R\,M}_{j\;k\;i} = \Phi^{N\,M\,R}_{j\;i\;k} = \cdots \tag{2.1}$$

which holds for both infinite lattices and finite lattices.**

* If the potential energy is well-defined, the cp's can be looked upon as derivatives of $\Phi$ with respect to the positions, e.g. $\Phi^{m}_{\substack{\mu\\ i}} = \partial\Phi \big/ \partial X^{m}_{\substack{\mu\\ i}}$.

** For finite lattices this relation also follows from the definition of the cp's as derivatives (see footnote above).

The potential energy $\Phi$ is a scalar. Therefore it is invariant against any translation or rotation of the coordinate system (or of the crystal), or in other words, it is a function of the relative distances only, or it contains only the internal interaction between the ions.

Let $\Omega$ be a rotation ($\Omega\tilde{\Omega} = 1$) and $\boldsymbol{t}$ a translation, then a vector $\boldsymbol{R}^M + \boldsymbol{u}^M$ is transformed into

$$\boldsymbol{R}^M + \boldsymbol{u}^M \Rightarrow \Omega\,\boldsymbol{R}^M + \Omega\,\boldsymbol{u}^M + \boldsymbol{t} = \boldsymbol{R}^M + \boldsymbol{u}^M + \Delta\,\boldsymbol{u}^M \tag{2.2}$$

with

$$\Delta\,\boldsymbol{u}^M = (\Omega - 1)\,\boldsymbol{R}^M + (\Omega - 1)\,\boldsymbol{u}^M + \boldsymbol{t}\,. \tag{2.3}$$

If we confine ourselves to infinitesimal rotations $\omega$

$$\Omega = 1 + \omega \quad \text{or} \quad \Omega_{ij} = \delta_{ij} + \omega_{ij} \quad \text{with} \quad \omega_{ij} = -\,\omega_{ji} \tag{2.4}$$

we have for infinitesimal operations*

$$\Delta\,\boldsymbol{u}^M = \omega\,\boldsymbol{R}^M + \omega\,\boldsymbol{u}^M + \boldsymbol{t}\,. \tag{2.3a}$$

Now the change of the potential energy, given by the expansion (1.6, 7) under a change $\Delta\,\boldsymbol{u}$ in the displacements, is equal to

$$\Delta\Phi = \sum_{Mi} \Phi_i^M \cdot \Delta u_i^M + \sum \Phi_{i\,j}^{MN} \cdot \Delta u_i^M \cdot u_j^N + \\ + \frac{1}{2!}\sum \Phi_{i\,j\,k}^{MNP} \Delta u_i^M \cdot u_j^N u_k^P + \frac{1}{3!}\sum \Phi_{i\,j\,k\,l}^{MNPR} \Delta u_i^M \cdot u_j^N u_k^P u_l^R\,. \tag{2.5}$$

This expression has to vanish for rotations and translations defined above. Using $\Delta u_i^M = t_i$, i.e. a pure translation of the whole crystal, and remembering that $\Delta\Phi \equiv 0$ for any displacement $u_i^M$, every power of $u_j^N$ has to vanish separately. This leads immediately to the requirements:

$$\sum_M \Phi_i^M = \sum_M \Phi_{i\,j}^{MN} = \sum_M \Phi_{i\,j\,k}^{MNP} = \cdots = 0 \tag{2.6}$$

for every $N, P, \ldots; i, j, k, \ldots$

We call it the condition of translational invariance; it is a consequence of the homogeneity of space.

Now consider an infinitesimal rotation, which can be defined properly only for finite systems, because otherwise the induced displacements $\Delta\,\boldsymbol{u}$ and therefore $\Delta\Phi$ would diverge. Inserting (2.3a) with $\boldsymbol{t} = 0$ into (2.5) we have

$$\Delta\Phi = \sum_{Mij} \Phi_i^M\,\omega_{ij}\,X_j^M + \sum_{Nij}\Big\{\Phi_i^N\,\omega_{ij} + \sum_{Mk} \Phi_{i\,j}^{MN} X_k^M\,\omega_{ik}\Big\}\,u_j^N + \\ + \sum_{\substack{NP\\ ijk}}\Big\{\Phi_{i\,j}^{NP}\,\omega_{ij} + \frac{1}{2}\sum_{Ml} \Phi_{i\,j\,k}^{MNP} X_l^M\,\omega_{il}\Big\}\,u_j^N u_k^P + \cdots \tag{2.7}$$

showing that the cp's of order $r$ and $r-1$ are connected, for every coefficient of the different powers of $\boldsymbol{u}$ has to vanish again separately.

---

* The rotation refers to the (arbitrary) origin of the coordinate system. By adding a translation, a rotation about an arbitrary point can be represented.

But in the general term with $u_{i_2}^{M_2} u_{i_3}^{M_3} \ldots u_{i_\nu}^{M_\nu}$ only that part enters, which is symmetric toward interchange of the pairs $\binom{M_2}{i_2}, \binom{M_3}{i_3}, \ldots$ As the second term in the braces has already this symmetry [see (2.1)], this affects only the first term, giving rise to just $(\nu - 1)$-terms after symmetrization. Having done this, the general term to vanish is

$$\sum_{ij} \omega_{ij} \left\{ \sum_{\lambda=2}^{\nu} \sum_{j_\lambda} \Phi^{M_2 \ldots M_\lambda \ldots M_\nu}_{i_2 \ldots j_\lambda \ldots i_\nu} \delta_{ij_\lambda} \delta_{ji_\lambda} + \sum_{M_1} \Phi^{M_1 M_2 \ldots M_\nu}_{i \; i_2 \ldots i_\nu} X^{M_1}_j \right\} = 0 . \quad (2.8)$$

As $\omega_{ij}$ is arbitrary, but antisymmetric in $i \Leftrightarrow j$, this term vanishes if and only if the term in the braces is symmetric in $i \Leftrightarrow j$. For small $\nu$ we have explicitly

$$\nu = 1: \quad \sum_M \Phi^M_i X^M_j = \quad (2.9\text{a})$$

$$\nu = 2: \quad \Phi^N_i \delta_{jk} + \sum_M \Phi^{MN}_{i\,k} X^M_j = \quad (2.9\text{b})$$

$$\nu = 3: \quad \Phi^{NP}_{i\,l} \delta_{jk} + \Phi^{NP}_{k\,i} \delta_{jl} + \sum_M \Phi^{MNP}_{i\,k\,l} X^M_j = \quad (2.9\text{c})$$

symmetric with respect to $i \Leftrightarrow j$

for every $N, P, \ldots; i, j, k, l, \ldots$

This is the condition of rotational invariance and a consequence of the isotropy of space. [I, II, 60 L 1, 60 H 1, 61 L 1].

Using (2.6) the relations (2.9) can be written in a somewhat different manner, showing that they are independent of the choice of the origin. Adding $-\sum_M \Phi^{MN}_{i\,k} X^N_j \equiv 0$ to (2.9b) we have

$$\nu = 2: \quad \Phi^N_i \delta_{jk} + \sum_M \Phi^{MN}_{i\,k} \left( X^M_j - X^N_j \right) = \text{symmetric in } i \Leftrightarrow j , \quad (2.10)$$

and similarly for $\nu > 2$. This is not possible for the relation (2.9a) with $\nu = 1$. The consequences of this are discussed in sect. 6.*

The relations (2.6) and (2.8, 9) have been derived with the assumption, that space is homogeneous and isotropic. This is valid only, if there are not external forces applied to the ions. However, we can generalize the translational and rotational conditions to the case with external forces applied, provided the forces on the ions have zero resulting force and torque in the equilibrium state** [65 W 1]; if these forces are $F^M_i$, this means

$$\sum_M F^M_i = 0; \quad \sum_M F^M_i X^M_j = \sum_M F^M_j X^M_i . \quad (2.11)$$

Stresses can be represented by such forces. The equilibrium condition then is

$$\Phi^M_i = F^M_i \quad (2.12)$$

---

* The potential energy also may not change by an inversion of the coordinate system. But this gives no general relations (see however section 4).

** This excludes e.g. gravitational forces on the system of ions.

and from (2.11) and (2.12) there follow immediately (2.6a) and (2.9a). In a similar way one can prove the other relations (2.6) and (2.9) again. But if one uses a rotation of the set of ions, one now has to rotate the whole system ions plus applied external forces in such a way that the force acting on an ion remains the same, that means the system ions plus forces has to be embedded in homogeneous and isotropic space. The relations hold for arbitrary sets of mass-points, no matter what the kind and what the range of the *interacting forces of the mass-points* are. However, in infinite lattices the rotational condition (2.9) can be obtained only by extrapolation from the finite case. The usefulness of these relations will be discussed in sect. 6.

## 3. General Invariance Relations in Ideal (Infinite) Lattices

We will turn over to the discussion of lattices, i.e. we must specify again the index $M$ to ${}^{m}_{\mu}$. In these lattices the potential energy and its expansion (1.6, 7) with arbitrary displacements $u_{\mu}^{m}$ has to be invariant against *translations of the displacement pattern* by multiples of the basis-vectors of the lattice (operations of the discrete group of lattice translations) [I, II]; these are the translations

$$\boldsymbol{t}^{\boldsymbol{h}} = A\,\boldsymbol{h}, \quad \boldsymbol{h} \text{ arbitrary integer} \tag{3.1}$$

This means, that

$$\Phi_{\mu\,\nu\,\ldots}^{\boldsymbol{m}\,\boldsymbol{n}\ldots}{}_{i\,j\ldots} = \Phi_{\mu\,\nu\ldots}^{\boldsymbol{m}+\boldsymbol{h},\,\boldsymbol{n}+\boldsymbol{h}\ldots}{}_{i\,j\ldots} \tag{3.2}$$

or the cp's can depend only on the differences $\boldsymbol{n}-\boldsymbol{m}$, $\boldsymbol{p}-\boldsymbol{m}, \ldots$ etc. or in detail for the lowest order cp's

$$\Phi_{\mu\,i}^{\boldsymbol{m}} = \Phi_{\mu\,i}^{\boldsymbol{m}+\boldsymbol{h}} \quad \text{or with} \quad \boldsymbol{h} = -\boldsymbol{m}\colon\; \Phi_{\mu\,i}^{\boldsymbol{m}} = \Phi_{\mu\,i}^{0}, \tag{3.3a}$$

i.e. the first order cp's are independent of the cell-index.

$$\begin{aligned} \Phi_{\mu\,\nu\,i\,j}^{\boldsymbol{m}\,\boldsymbol{n}} &= \Phi_{\mu\,\nu\,i\,j}^{\boldsymbol{m}+\boldsymbol{h}\,\boldsymbol{n}+\boldsymbol{h}} = \Phi_{\mu\,\nu\,i\,j}^{0\,\boldsymbol{n}-\boldsymbol{m}} = \Phi_{\mu\,\nu\,i\,j}^{0\,\boldsymbol{g}} \quad \text{with} \quad \boldsymbol{h} = -\boldsymbol{m} \\ &= \Phi_{\mu\,\nu\,i\,j}^{\boldsymbol{m}-\boldsymbol{n}\,0} = \Phi_{\mu\,\nu\,i\,j}^{-\boldsymbol{g}\,0} \quad \text{with} \quad \boldsymbol{h} = -\boldsymbol{n}, \\ &\qquad\qquad \boldsymbol{g} = \boldsymbol{n}-\boldsymbol{m}\,. \end{aligned} \tag{3.3b}$$

These relations express the *homogeneity of the infinite lattice*. They are no longer valid for finite lattices. However, if the range of the forces is small compared to the dimensions of the lattice, we can define an *interior* region of the lattice, where eqs. (3.3) are valid in a very good approximation. We call a *surface region* of a crystal the region of all the ionic planes lying in a distance from the surface which has the linear dimension of the range of the forces. The rest of the crystal, being not a surface region, is called the interior of the crystal. In defect lattices (e.g. lattices with vacancies) we have to take into account also inner surfaces (see sect. 5). In the interior of the crystal the relations (3.3) hold „nearly" exactly. The problem is more difficult for long-range forces (e.g. Coulomb-forces and forces of external electric fields). Such cases must be investigated in detail.

We will mention shortly the connection of the translational invariances expressed by (2.6) and (3.2). (2.6) states nothing else than the conservation of momentum of the whole crystal, i.e. the momentum $\boldsymbol{P}$ of the center-of-mass-motion of the crystal. On the other hand, (3.2) also refers to a translational invariance, but to a discrete one. Therefore it is often referred to as the conservation of quasi- or pseudo-momentum of the phonons. The phonons have no real momentum, but in interaction processes the quasi- or pseudo-momentum often behaves as a true momentum. Because the pseudo-momentum is related to the discrete reciprocal lattice, it is defined only modulo a reciprocal lattice vector. Now *Süssmann* [56 S 1, 58 S 1] and *Brenig* [55 B 2] have shown, that the difference between the true momentum $\boldsymbol{P}$ of the lattice and the total pseudo-momentum $\Pi = \sum_{q\sigma} n(\boldsymbol{q}\sigma) \cdot \hbar \boldsymbol{q}$ is constant modulo a vector of the reciprocal lattice. Therefore in interaction processes we have to deal with selection rules for the pseudomomentum $\boldsymbol{q} + \boldsymbol{q}' + \boldsymbol{q}'' + \cdots = 2\pi B \boldsymbol{h}$ (reciprocal lattice vector). If a phonon is created, there is also a momentum given to the center-of-mass of the crystal, which in general is not considered. Similar statements hold for the angular momentum. For details we refer to the papers mentioned.

## 4. Symmetry Relations from the Point Group Symmetries of the (Infinite) Lattice

Apart from the general symmetries a lattice is invariant towards a certain finite number of operations which transform the lattice into itself [I, II]. The cp's must show this symmetry also, the number of independent cp's thus being related to the point group symmetry of the crystal. A point group operation can be expressed as a proper or improper rotation $\Omega$. In such a transformation the ionic positions $\boldsymbol{R}_{\mu}^{m}$ change into equivalent positions $\boldsymbol{R}_{\mu'}^{m'}$ with

$$\Omega \boldsymbol{R}_{\mu}^{m} = \boldsymbol{R}_{\mu'}^{m'} \quad \text{or} \quad \sum_{k} \Omega_{ik} X_{\substack{\mu\\k}}^{m} = X_{\substack{\mu'\\i}}^{m'} . \tag{4.1}$$

If $\boldsymbol{R}_{\mu}^{m}$ covers all equivalent ionic positions, so does $\boldsymbol{R}_{\mu'}^{m'}$. The invariance of the potential energy then means

$$\Omega \Phi(\ldots \boldsymbol{R}_{\mu}^{m} + \boldsymbol{u}_{\mu}^{m} \ldots) = \Phi(\ldots \Omega \boldsymbol{R}_{\mu}^{m} + \Omega \boldsymbol{u}_{\mu}^{m} \ldots)$$

$$= \Phi(\ldots \boldsymbol{R}_{\mu'}^{m'} + \Omega \boldsymbol{u}_{\mu}^{m} \ldots) = \Phi(\ldots \boldsymbol{R}_{\mu}^{m} + \boldsymbol{u}_{\mu}^{m} \ldots) .$$

Expanding the last two expressions with respect to $\boldsymbol{u}_{\mu}^{m}$ and comparing the coefficients of equal powers of $\boldsymbol{u}_{\mu}^{m}$ we get

$$\sum_{\substack{i'j'\\ \ldots k'}} \Phi_{\substack{\mu'\\i'}\substack{\nu'\\j'}\ldots\substack{\varkappa'\\k'}}^{m' n' \ldots q'} \Omega_{i'i} \Omega_{j'j} \ldots \Omega_{k'k} = \Phi_{\substack{\mu\\i}\substack{\nu\\j}\ldots\substack{\varkappa\\k}}^{m n \ldots p} , \tag{4.2}$$

where ${}^{m'}_{\mu'}$ are connected with ${}^{m}_{\mu}$ by (4.1). In order to determine the number of independent cp's, one first selects all those symmetry operations, which leave the set $\binom{m\ n\ldots}{\mu\ ,\ \nu\ldots}$ invariant. This gives the relations between the $i, j, \ldots$ components of a cp-tensor with fixed $\binom{m\ \ n}{\mu\ ,\ \nu\ , \ldots}$. Those operations, which transform the points $\binom{m\ \ n}{\mu\ ,\ \nu\ , \ldots}$ into equivalent points $\binom{m'\ \ n'}{\mu'\ ,\ \nu'\ , \ldots}$, transform the corresponding cp's too. Therefore we need to consider only operations of the first kind, which leave the set $\binom{m\ \ n}{\mu\ ,\ \nu\ , \ldots}$ invariant to get the number of independent cp's. With this procedure a number of cp-tensors has been determined and we will refer to the literature for the details (I, 61 L 1].

Only a few essential points shall be discussed. In many lattices, for example in all the Bravais-lattices, the inversion $I$ is a symmetry operation. It is $I\boldsymbol{m} = \boldsymbol{m}' = -\boldsymbol{m}$, and therefore according to (4.2)

$$(-1)^r\, \Phi^{-m\ -n\ \ldots -p}_{\ i\ \ \ \ j\ \ldots\ \ k} = \Phi^{m\ n\ \ldots\ p}_{i\ j\ \ \ \ \ k} \tag{4.3}$$

for the cp's of $r$-th order. Especially,

$$\Phi^{-m}_{\ i} = -\Phi^{m}_{i} \tag{4.3a}$$

and as $\Phi^{m}_{i}$ is independent of $\boldsymbol{m}$ (3.3a), we have

$$\Phi^{m}_{i} = 0 \quad \text{in Bravais-lattices}\,. \tag{4.4}$$

For $r = 2$ we have

$$\Phi^{m\ n}_{i\ j} = \Phi^{-m\ -n}_{\ i\ \ \ \ j}$$

and using (3.3b) with $-\boldsymbol{h} = \boldsymbol{m} + \boldsymbol{n}$

$$\Phi^{-m\ -n}_{\ i\ \ \ \ j} = \Phi^{n\ m}_{i\ j}$$

and therefore

$$\Phi^{m\ n}_{i\ j} = \Phi^{n\ m}_{i\ j} = \Phi^{m\ n}_{j\ i} \tag{4.5}$$

using (2.1).

Thus in ideal infinite Bravais-lattices the cp's of second order are symmetric with respect to the superscripts $\boldsymbol{m} \Leftrightarrow \boldsymbol{n}$ or the subscripts $i \Leftrightarrow j$ alone.

When the number of independent cp's compatible with the symmetry requirements has been determined, one has to look for the conditions of translational invariance (2.6, 9). To satisfy these relations, one has to put further restrictions to the kind and the number of independent cp's. Most of these additional restrictions occur in non-primitive lattices (harmonic and anharmonic) and for the anharmonic cp's of third and higher orders in primitive lattices. (2.6, 9) give, in general, no restrictions for the harmonic cp's in primitive lattices.

All these remarks hold for the ideal infinite lattice. By the same arguments as used in sect. 3, they are also valid in the interior of finite crystals.

## 5. Symmetry Relations Near Surfaces or Defects in Crystals

We restrict ourselves to a very simple example, but it shows at once all the points which have to be regarded in discussing defect lattices. The restrictions imposed on the cp's by the above discussed invariance and symmetry relations are even stronger than in ideal infinite lattices [64 L 1, 65 L 4, 64 L 6]. This is due to the fact that one tries to work with the most simple models in general. All the relations discussed in sect. 2–5 have to be satisfied automatically, if one starts from a known expression for the potential energy. This has to be emphasized strongly. The relations are useful only for general considerations or if one works with cp's as unknown parameters when the potential is not known.

The model we shall discuss is the simple cubic lattice with nearest neighbor interaction only. As "defect" we shall consider a vacancy described by dropping the force constants between one ion and its neighbors (Fig. 2.1) and a free surface described by dropping the force constants between neighboring planes*. We only investigate the relations for the harmonic cp's. The cp-tensor for the ideal lattice or that region of the crystal which is assumed to be ideal is

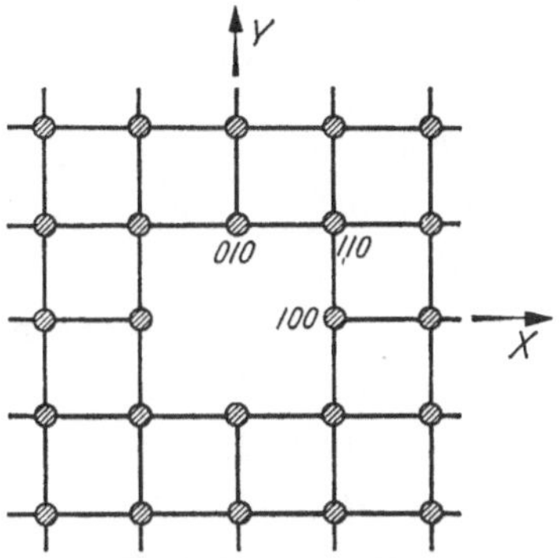


Fig. 2.1. Structure of a simple cubic lattice near a vacancy. The translation symmetry is destroyed

$$\Phi_{i\ j}^{m\ m\pm 100} = -\begin{pmatrix}\alpha & 0 & 0\\ 0 & \beta & 0\\ 0 & 0 & \beta\end{pmatrix} \text{ and cyclic for } \quad (0,\pm 1,0);\ (0,0,\pm 1). \qquad (5.1)$$

$\alpha$ has the meaning of a usual spring constant, whereas $\beta$ is a sort of shear constant, and related to non-central-forces**. If we would assume $\beta = 0$, the lattice would become unstable; for then we could make a (100)-shear without stressing the spring $\alpha$. All invariance and symmetry relations are satisfied with (5.1) in the infinite ideal lattice if we use further

$$\Phi_{i\ j}^{m\ m} = (2\alpha + 4\beta)\begin{pmatrix}1 & 0 & 0\\ 0 & 1 & 0\\ 0 & 0 & 1\end{pmatrix}. \qquad (5.2)$$

If there is a vacancy, say at the origin of the otherwise infinite lattice, the cp-tensors compatible with point group symmetry requirements

* If we work with central forces (springs) between the ions, a vacancy is obtained by cutting the springs between the "vacant ion" and its neighbors. If there are many-body-forces, dropping of some many-body-force-constants implies changes in other many-body-constants!

** Recently *Keating* [66 K 5] claims, that there can be no many-body-forces between nearest neighbors. This is obviously not correct. An example of many-body-forces, which do not vanish for nearest neighbors, is given in sect. 7.

in the vicinity of the vacancy are (with cyclic ones for corresponding other superscripts)

$$\Phi^{0\,100}_{i\;j} = -\begin{pmatrix} \alpha' & 0 & 0 \\ 0 & \beta' & 0 \\ 0 & 0 & \beta' \end{pmatrix}; \quad \Phi^{100\,110}_{i\;\;j} = -\begin{pmatrix} \gamma'' & \delta'' & 0 \\ \vartheta'' & \alpha'' & 0 \\ 0 & 0 & \beta'' \end{pmatrix}$$

$$\Phi^{110\,111}_{i\;\;j} = -\begin{pmatrix} \gamma''' & \varrho''' & \delta''' \\ \varrho''' & \gamma''' & \delta''' \\ \vartheta''' & \vartheta''' & \alpha''' \end{pmatrix}; \quad \text{and so on}, \tag{5.3}$$

where $\alpha'$ and $\beta'$ have to be equal to zero in our model of a vacancy. If we limit the "range" of the defect to third neighbors, the cp's given in (5.3) are the only ones changed. They are no longer diagonal and are even non-symmetric. The restrictions imposed by the translational condition (2.6) are not severe. It can be used to get statements about the cp-tensors $\Phi^{m\,m}_{i\;j}$. But the rotational condition (2.10b) now gives the restrictions for

$$\begin{aligned}
\boldsymbol{m} &= (0,0,0): && \text{none} \\
\boldsymbol{m} &= (1,0,0): && 2\vartheta'' = \beta - \beta' \\
\boldsymbol{m} &= (1,1,0): && 2\vartheta''' = \beta - \beta''; \; -\delta'' = \beta - \gamma'' \\
\boldsymbol{m} &= (1,1,1): && \qquad\qquad -\delta''' = \beta - \gamma'''\,.
\end{aligned} \tag{5.4}$$

For $m \geqq 4$ there are no restrictions under the above assumptions. From the 12 cp's, allowed by symmetry, there are only 7 being independent. Now, sometimes models are used, which specify the cp's even more, e.g. assuming a diagonal form in the subscripts $i, j$.

With $\delta'' = \vartheta'' = \varrho''' = \delta''' = \vartheta''' = 0$ it follows from (5.4)

$$\beta = \beta' = \beta'' = \gamma'' = \gamma''' \tag{5.5}$$

being left with $\alpha'$, $\alpha''$, $\alpha'''$ as the only changes. As $\beta' = 0$ for a vacancy, this means that $\beta = 0$, which makes the lattice unstable. Therefore there is no model for a vacancy with diagonal cp-tensors though an ideal infinite lattice is possible under this assumption. In this model, the simplest assumptions, leading to a consistent vacancy would be

$$\begin{aligned}
&\alpha' = \beta' = 0; \quad \beta'' = \beta; \\
&\vartheta'' = \beta/2; \qquad \delta'' = \gamma'' - \beta; \\
&\varrho''' = \vartheta''' = \delta''' = 0; \\
&\gamma''' = \beta; \\
&\alpha''' = \alpha; \; \alpha'', \gamma'' \quad \text{arbitrary}
\end{aligned} \tag{5.6}$$

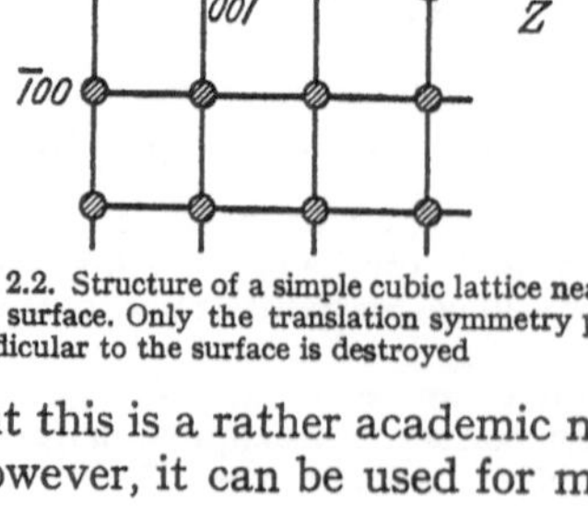

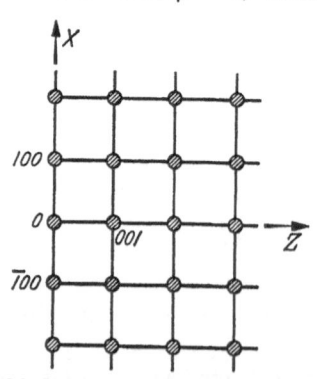

Fig. 2.2. Structure of a simple cubic lattice near a free surface. Only the translation symmetry perpendicular to the surface is destroyed

But this is a rather academic model having no real physical significance. However, it can be used for many qualitative aspects of lattice theory.

A similar model is possible for a free surface. If we limit the "range" of the defect to the surface only, two representative cp-tensors being compatible with symmetry requirements are

$$\Phi_{i\ j}^{000\ \pm 100} = -\begin{pmatrix} \alpha & 0 & \pm\delta \\ 0 & \beta & 0 \\ \mp\delta & 0 & \beta \end{pmatrix}; \quad \Phi_{i\ j}^{000\ 0\pm 10} = -\begin{pmatrix} \beta & 0 & 0 \\ 0 & \alpha & \pm\delta \\ 0 & \mp\delta & \beta \end{pmatrix} \tag{5.7}$$

where it is assumed, that the diagonal elements are those of the interior of the crystal. All other cp-tensors having superscripts $\boldsymbol{m} = [m_1, m_2, m_3]$ with $m_3 > 0$ are unchanged in this model, whereas those with $m_3 < 0$ are zero! Translational invariance (2.6) requires only

$$\Phi_{ij}^{00} = -\begin{pmatrix} 2\alpha + 3\beta & 0 & 0 \\ 0 & 2\alpha + 3\beta & 0 \\ 0 & 0 & \alpha + 4\beta \end{pmatrix}, \tag{5.8}$$

whereas (2.10b) leads to

$$2\delta = \beta\,. \tag{5.9}$$

This is the simplest consistent model of a free surface. The cp-tensor at the surface is again neither diagonal nor symmetric.

By similar arguments it can be shown, that the cp-tensors are never symmetric with respect to the subscripts $i \Leftrightarrow j$ near "defects" in otherwise infinite crystals, whereas they are nondiagonal already in ideal infinite crystals in nearly all other cases apart from this simple cubic model. *The symmetry with respect to the subscripts $i \Leftrightarrow j$ holds only in ideal infinite Bravais-lattices or in the "interior" of finite or defect crystals.* The significance of the distinction between surface and interior region of a crystal becomes obvious only if we calculate macroscopic effects and compare those quantities, which are "surface sensitive" with those which are not.

## 6. Transition to Macroscopic Theory and Consequences of Invariance Relations

The problem we have to deal with is how to get the familiar macroscopic elastic theory from the microscopic lattice theory. This question has been solved essentially by *Born* and *Huang* [II, 50 H 1]; but afterwards there were raised objections to the symmetry of the elastic constants as well as to Born-Huang's derivation [51 L 1, 54 L 2, 55 R 1, 64 L 4] which is connected with these symmetries. The problem has been settled now, showing that the Born-Huang procedure was correct. We will not go into the details of the controverse here, but rather give a straightforward derivation of the macroscopic equations from lattice theory including the generalizations of the Born-Huang procedure which have been done up till now [60 L 1, 60 H 1, 64 L 6, 65 W 1].

The generalizations concern essentially the following. Whereas *Born* and *Huang* consider mainly a crystal with vanishing initial stresses (external forces applied to the surface) in the equilibrium state before an elastic deformation, we will discuss here the case of *finite stresses in the initial equilibrium state, which are produced by applied external forces* $\boldsymbol{F}_\mu^m$ *to the surface ions.* The homogeneous elastic deformation of the crystal is then caused by additionally applied small forces $\boldsymbol{f}_\mu^m$, acting also only on surface ions. The crystal in the initial state then has no longer the point group symmetry of the free crystal but that of the free crystal plus applied forces. E.g. a crystal with cubic symmetry in free state and with applied shear stress has lower than cubic symmetry.

The essential restriction we have made is that the forces are applied only to surface ions (stresses); if we would allow for long range external forces (e.g. electric fields) the theory becomes much more complicated and we cannot eliminate surface effects, as we shall see in the following.

The energy of the system ions plus applied forces is now

$$\Phi_F\left(\boldsymbol{R}_\mu^m\right) = \Phi\left(\boldsymbol{R}_\mu^m\right) - \sum_{m\mu i} F_{\mu\atop i}^m X_{\mu\atop i}^m \tag{6.1}$$

in the equilibrium state, where the $\boldsymbol{R}_\mu^m$ are the equilibrium positions with initially applied forces. The equilibrium condition is the minimum of $\Phi_F$ and therefore [see (2.12)]

$$\Phi_{\mu\atop i}^m = F_{\mu\atop i}^m\,. \tag{6.2}$$

The forces $\boldsymbol{F}_\mu^m$ have to satisfy the equations (2.11).

We seek for the change of the elastic energy of the crystal under a homogeneous deformation, to compare this with the corresponding expression of the macroscopic elastic theory and so to get expressions for the elastic constants.

If the ions suffer a homogeneous displacement from their initial equilibrium positions $X_{\mu\atop i}^m$ this can be expressed by displacements* (I, II)

$$u_{\mu\atop i}^m = u_{\mu\atop i} + \sum_j u_{i|j}\, X_{\mu\atop j}^m\,. \tag{6.3}$$

The $u_{i|j}$ define the deformation and are independent parameters. In non-primitive lattices a relative displacement of the different sub-lattices $\mu$ can be caused by such a deformation, which gives rise to the term $u_{\mu\atop i}$ which is dependent on the $u_{i|j}$ and has to be determined later! The deformation is produced by small additional forces, applied to the surface ions. The forces $\boldsymbol{F}_\mu^m$ are held constant during this deformation.

---

* In macroscopic elastic theory the $u_{i|j}$ are related to the continuously defined displacement field. They are the derivatives of the displacements $u_i$ with respect to the (Lagrange-) coordinates $X_j$ and need not be symmetric in $i \Leftrightarrow j$. We choose here already the same symbol, though it is an adhoc definition here.

The potential energy of the ions per unit volume after this deformation (in the initial state) is obtained by inserting (6.3) into (1.6, 7):

$$\begin{aligned}
\frac{1}{V}\Phi\left(\boldsymbol{R}^{m}_{\mu}+\boldsymbol{u}^{m}_{\mu}\right) &= \frac{1}{V}\Phi_0\left(\boldsymbol{R}^{m}_{\mu}\right)+\frac{1}{V}\sum_{m\mu i}\Phi^{m}_{\underset{i}{\mu}}\,u_{\underset{i}{\mu}}+\frac{1}{V}\sum_{\substack{m\mu\\ ij}}\Phi^{m}_{\underset{i}{\mu}}\,X^{m}_{\underset{j}{\mu}}\,u_{i|j}\\
&+\frac{1}{2V}\sum_{\substack{mn\\ \mu\nu\\ ij}}\Phi^{mn}_{\underset{i}{\mu}\,\underset{j}{\nu}}\,u_{\underset{i}{\mu}}\,u_{\underset{j}{\nu}}+\frac{1}{V}\sum_{\substack{mn\\ \mu\nu\\ ijk}}\Phi^{mn}_{\underset{i}{\mu}\,\underset{j}{\nu}}\,X^{n}_{\underset{k}{\nu}}\,u_{j|k}\,u_{\underset{i}{\mu}}+\\
&+\frac{1}{2V}\sum_{\substack{mn\\ \mu\nu\\ ijkl}}\Phi^{mn}_{\underset{i}{\mu}\,\underset{j}{\nu}}\,X^{m}_{\underset{k}{\mu}}\,X^{n}_{\underset{l}{\nu}}\,u_{i|k}\,u_{j|l}+\\
&+\frac{1}{6V}\sum_{\substack{mnp\\ \mu\nu\varkappa\\ ijk}}\Phi^{mnp}_{\underset{i}{\mu}\,\underset{j}{\nu}\,\underset{k}{\varkappa}}\Big\{u_{\underset{i}{\mu}}\,u_{\underset{j}{\nu}}\,u_{\underset{k}{\varkappa}}+3\sum_{l}X^{p}_{\underset{l}{\varkappa}}\,u_{k|l}\,u_{\underset{i}{\mu}}\,u_{\underset{j}{\nu}}+\\
&+3\sum_{lm}X^{n}_{\underset{l}{\nu}}\,X^{p}_{\underset{m}{\varkappa}}\,u_{j|l}\,u_{k|m}\,u_{\underset{i}{\mu}}+\sum_{lmn}X^{m}_{\underset{l}{\mu}}\,X^{n}_{\underset{m}{\nu}}\,X^{p}_{\underset{n}{\varkappa}}\,u_{i|l}\,u_{j|m}\,u_{k|m}\Big\}+\cdots
\end{aligned}\tag{6.4}$$

It still depends on the relative displacements $u_{\underset{i}{\mu}}$. This can be eliminated by using the equilibrium condition after the small forces $f^{m}_{\mu}$ have been applied:

$$\begin{aligned}
F^{m}_{\underset{i}{\mu}}+f^{m}_{\underset{i}{\mu}} &= \Phi^{m}_{\underset{i}{\mu}}+\sum_{n\nu j}\Phi^{mn}_{\underset{i}{\mu}\,\underset{j}{\nu}}\,u_{\underset{j}{\nu}}+\sum_{\substack{n\nu\\ jk}}\Phi^{mn}_{\underset{i}{\mu}\,\underset{j}{\nu}}\,X^{n}_{\underset{k}{\nu}}\,u_{j|k}+\\
&+\frac{1}{2}\sum_{\substack{np\\ \nu\varkappa\\ jk}}\Phi^{mnp}_{\underset{i}{\mu}\,\underset{j}{\nu}\,\underset{k}{\varkappa}}\Big\{u_{\underset{j}{\nu}}\,u_{\underset{k}{\varkappa}}+2\sum_{l}X^{p}_{\underset{l}{\varkappa}}\,u_{k|l}\,u_{\underset{j}{\nu}}+\sum_{lm}X^{n}_{\underset{l}{\nu}}\,X^{p}_{\underset{m}{\varkappa}}\,u_{j|l}\,u_{k|m}\Big\}\\
&+\cdots
\end{aligned}\tag{6.5}$$

where $F^{m}_{\underset{i}{\mu}}$ and $\Phi^{m}_{\underset{i}{\mu}}$ cancel because of (6.2). Eqs. (6.4) and (6.5) have to be discussed simultaneously. One can see immediately, that if one wants to have the energy (6.4) up to order $r$ in the $u_{i|k}$, one has to solve (6.5) only up to order $(r-1)$ in the $u_{i|k}$. The solution of (6.5) is rather complicated if one takes anharmonic terms into account. It can be done by iteration and we will leave it, dealing now with harmonic theory only.

Before solving (6.5) we will eliminate surface effects as far as possible. This can be done as a result of our assumption of applied forces only to surface ions. Therefore we have [because of (6.2)]

$$\Phi^{m}_{\underset{i}{\mu}}=F^{m}_{\underset{i}{\mu}}=f^{m}_{\underset{i}{\mu}}=0\tag{6.6}$$

for all the $\overset{m}{\mu}$ in the interior.

Let us consider the different expressions occuring in (6.4, 5). It is $V=NV_z$, where $V_z$ is the volume of the unit cell and $N$ the number of cells in the crystal. Then we have

$$\frac{1}{V}\sum_{\mu i}u_{\underset{i}{\mu}}\sum_{m}\Phi^{m}_{\underset{i}{\mu}}\sim\frac{N^{2/3}}{NV_z}\sim\frac{1}{N^{1/3}}\to 0\quad\text{for large } N\tag{6.7}$$

because $\Phi^{m}_{\underset{i}{\mu}}$ is different from zero only for surface ions (order of magnitude $N^{2/3}$). The next term

$$S_{ij} = \frac{1}{V} \sum_{m\mu} \Phi^{m}_{\underset{i}{\mu}} X^{m}_{\underset{j}{\mu}} \tag{6.8}$$

gives contributions only for surface atoms ($N^{2/3}$), but as $X^{m}_{\underset{j}{\mu}}$ extends from one surface to the opposite one, there occur terms with the length $L = aN^{1/3}$ ($a \approx$ lattice constant), independent of the choice of the origin, therefore

$$S_{ij} \sim \frac{N^{2/3} \cdot aN^{1/3}}{N V_z} \sim \frac{1}{a^2}, \quad \text{which is "large"!}$$

Because of the equilibrium condition (6.2) and of (2.11), the $S_{ij}$ are just the stresses in the initial state!

Further we have

$$\hat{C}_{\underset{ij}{\mu\nu}} = -\frac{1}{V} \sum_{mn} \Phi^{mn}_{\underset{i\ j}{\mu\ \nu}} = -\frac{1}{V_z} \sum_{h} \Phi^{0h}_{\underset{ij}{\mu\nu}}, \tag{6.9}$$

because the number of surface terms is small compared to those of the interior and we can use (3.3b) in the interior. The sum over $m$ then only gives a factor $N$. Similarly we obtain

$$\begin{aligned} \hat{C}_{\underset{i,jk}{\mu}} &= -\frac{1}{V} \sum_{\substack{mn\\ \nu}} \Phi^{mn}_{\underset{i\ j}{\mu\ \nu}} X^{n}_{\underset{k}{\nu}} = -\frac{1}{V} \sum_{mn\nu} \Phi^{mn}_{\underset{i\ j}{\mu\ \nu}} \left(X^{n}_{\underset{k}{\nu}} - X^{m}_{\underset{k}{\mu}}\right) \\ &= -\frac{1}{V_z} \sum_{h\nu} \Phi^{0h}_{\underset{ij}{\mu\nu}} \left(X^{h}_{\underset{k}{\nu}} - X^{0}_{\underset{k}{\mu}}\right) \end{aligned} \tag{6.10}$$

using (2.6), in the interior (3.3b) and neglecting again surface contributions. In the last harmonic term

$$\hat{S}_{ik,jl} = \frac{1}{V} \sum_{\substack{mn\\ \mu\nu}} \Phi^{mn}_{\underset{i\ j}{\mu\ \nu}} X^{m}_{\underset{k}{\mu}} X^{n}_{\underset{l}{\nu}} \tag{6.11}$$

elimination of surface effects is not possible.

In (6.5) $f^{m}_{\mu}$ can be neglected, because there are only $N^{2/3}$ terms compared with $N$ terms on the right side of (6.5), being of the same order of magnitude, i.e. it is sufficient to solve (6.5) in the interior of the crystal. The other terms can be expressed by (6.9, 10).

*Symmetries of the coefficients:*

From (2.9a) it follows, that

$$S_{ij} \quad \text{is symmetric with respect to} \quad i \Leftrightarrow j\,. \tag{6.12}$$

From definition,

$\hat{S}_{ij,kl}$ is symmetric toward interchange of the pairs $(ij) \Leftrightarrow (kl)$ (6.13a)

Further, by multiplication of (2.9b) with $X^{n}_{\nu l}$ and summation over $\overset{n}{\nu}$ it follows that

$$\hat{S}_{ij,kl} + S_{il}\,\delta_{jk} = \hat{S}_{ji,kl} + S_{jl}\,\delta_{ik}$$

or

$$\hat{S}_{ij,kl} - S_{jl}\,\delta_{ik} = \hat{S}_{ji,kl} - S_{il}\,\delta_{jk} \tag{6.13b}$$

which is the symmetry relation for an interchange $i \Leftrightarrow j$. From (6.13a, b) there follows a similar relation for $k \Leftrightarrow l$.

$$\hat{C}_{\substack{\mu\nu\\ij}} = \hat{C}_{\substack{\nu\mu\\ji}} \tag{6.14a}$$

from definition. Further with the help of (2.6) and 6.9)

$$\sum_{\nu} \hat{C}_{\substack{\mu\nu\\ij}} = \sum_{\mu} \hat{C}_{\substack{\mu\nu\\ij}} = 0\,. \tag{6.14b}$$

According to (6.10), $\hat{C}_{\substack{\mu\\i,jk}}$ is defined in the "interior" of the crystal, where the $\Phi^{m}_{\substack{\mu\\i}}$ vanish. Therefore we can use (2.9b) or (2.10b) with $\Phi^{m}_{\substack{\mu\\i}} = 0$ and have the relation

$$\hat{C}_{\substack{\mu\\i,jk}} = \hat{C}_{\substack{\mu\\i,kj}}\,,\quad \text{i.e. symmetric in } j \Leftrightarrow k\,. \tag{6.15a}$$

From definition again*

$$\sum_{\mu} \hat{C}_{\substack{\mu\\i,kj}} = 0\,. \tag{6.15b}$$

Using the definitions (6.8—11) and $U = (\Phi - \Phi_0)/V$ we have (including only harmonic terms)

$$U = S_{ij}\,u_{i|j} + \frac{1}{2}\hat{S}_{ij,kl}\,u_{i|k}\,u_{j|l} - \frac{1}{2}\hat{C}_{\substack{\mu\nu\\ij}}\,u_{\substack{\mu\\i}}\,u_{\substack{\nu\\j}} - \hat{C}_{\substack{\mu\\i,jk}}\,u_{\substack{\mu\\i}}\,u_{j|k} \tag{6.16}$$

$$0 = \hat{C}_{\substack{\mu\nu\\ij}}\,u_{\substack{\nu\\j}} + \hat{C}_{\substack{\mu\\i,jk}}\,u_{j|k} \tag{6.17}$$

(using summation convention now). (6.17) is an inhomogeneous equation to be solved for the $u_{\substack{\nu\\j}}$ [I.]. We see immediately, that the homogeneous equation $0 = \hat{C}_{\substack{\mu\nu\\ij}}\,u_{\substack{\nu\\j}}$ has solutions with displacements $u_j$ independent of $\nu$. This follows immediately from (6.14b). Therefore the reciprocal matrix to $\hat{C}_{\substack{\mu\nu\\ij}}$ does not exist, but we can give the solutions of the inhomogeneous equation with another symmetric matrix $R_{\substack{\mu\nu\\ij}}$, which has to be calculated in every definite problem.

Therefore

$$u_{\substack{\mu\\i}} = -R_{\substack{\mu\nu\\ik}}\,\hat{C}_{\substack{\nu\\k,jl}}\,u_{j|l} + u_i\,. \tag{6.18}$$

Here the last term can be dropped as it does not enter (6.16) because of (6.14b, 15b). Now, we first insert (6.17) as a whole into the third term of the right side of (6.16); this term then cancels half the last term. Inserting (6.18) into the remaining half, we finally obtain

$$U = S_{ij}\,u_{i|j} + \frac{1}{2}S_{ij,kl}\,u_{i|j}\,u_{k|l} \tag{6.19}$$

* The $\hat{C}^{\mu}_{i,kj}$ vanish in parameterfree lattices, where *every ion* is a *center of inversion.*

with

$$S_{ij,kl} = \mathring{S}_{ij,kl} + \mathring{C}_{\mu \atop m,ij} R_{\mu \nu \atop mn} \mathring{C}_{\nu \atop n,kl}. \tag{6.20}$$

The first term in (6.20) is surface-sensitive, the second is not. The second term is according to (6.14a, 15a) symmetric with respect to $i \Leftrightarrow j$, $k \Leftrightarrow l$ and the pairs $(ij) \Leftrightarrow (kl)$. Therefore, from (6.13b), $S_{ij,kl}$ has the same symmetries as $\mathring{S}_{ij,kl}$

$$S_{ij,kl} = S_{kl,ij} \tag{6.21a}$$

$$S_{ij,kl} - S_{jl}\,\delta_{ik} = S_{ji,kl} - S_{il}\,\delta_{jk}. \tag{6.21b}$$

We can remove the surface-sensitivity from $\mathring{S}_{ij,kl}$ by introducing a symmetrized "version"

$$\begin{aligned}\mathring{C}_{ik,jl} &= \frac{1}{2}(\mathring{S}_{ij,kl} + \mathring{S}_{il,kj}) = \frac{1}{2V}\sum_{\substack{\mu\nu\\ mn}} \Phi_{\mu\ \nu}^{mn}{}_{j\ l}\left(X_j^{m}{}_{\mu} X_l^{n}{}_{\nu} + X_l^{m}{}_{\mu} X_j^{n}{}_{\nu}\right) = \\ &= -\frac{1}{2V}\sum_{mn\mu\nu} \Phi_{\mu\ \nu}^{mn}{}_{i\ k}\left(X_j^{n}{}_{\nu} - X_j^{m}{}_{\mu}\right)\left(X_l^{n}{}_{\nu} - X_l^{m}{}_{\mu}\right),\end{aligned} \tag{6.22}$$

where use has been made of (2.6). The last term in (6.22) is no longer surface sensitive as it contains only relative distances and the main contribution comes from the interior of the crystal. Therefore we can write

$$\mathring{C}_{ik,jl} = \frac{1}{2}(\mathring{S}_{ij,kl} + \mathring{S}_{il,kj}) = -\frac{1}{2V_z}\sum_{\substack{h\\ \mu\nu}} \Phi_{\mu\nu}^{0h}{}_{ik}\left(X_j^{h}{}_{\nu} - X_j^{0}{}_{\mu}\right)\left(X_l^{h}{}_{\nu} - X_l^{0}{}_{\mu}\right). \tag{6.23}$$

The symmetries of $\mathring{C}_{ik,jl}$ are the following:

$$\mathring{C}_{ik,jl} = \mathring{C}_{ik,lj} \qquad \text{[from (6.23)]} \tag{6.24a}$$

$$\mathring{C}_{ik,jl} = \mathring{C}_{ki,jl} \qquad \text{[from (6.23) with (6.13a)]} \tag{6.24b}$$

$$\mathring{C}_{ik,jl} - S_{jl}\,\delta_{ik} = \mathring{C}_{jl,ik} - S_{ik}\,\delta_{jl} \qquad \text{[from (6.23) with (6.13b)]} \tag{6.24c}$$

Combining this result with (6.20) we have the surface-insensitive quantity

$$\begin{aligned}\bar{S}_{ik,jl} &= \frac{1}{2}(S_{ij,kl} + S_{il,kj}) \\ &= \mathring{C}_{ik,jl} + \frac{1}{2}\mathring{C}_{\mu \atop m,ij} R_{\mu\nu \atop mn} \mathring{C}_{\nu \atop n,kl} + \frac{1}{2}\mathring{C}_{\mu \atop m,il} R_{\mu\nu \atop mn} \mathring{C}_{\nu \atop n,kj},\end{aligned} \tag{6.25}$$

but this does not enter (6.19). The symmetries of $\bar{S}_{ik,jl}$ are those of $\mathring{C}_{ik,jl}$.

Now we must consider the classical macroscopic theory of elasticity. If we perform a homogeneous deformation of the crystal, described by (6.3), this deformation $u_{i|j}$ generally contains rotations of the whole system, not being connected with a change in the energy [see I].
The appropriate quantity to describe true strains, is the tensor of finite strain*

$$\mathscr{V}_{ik} = \frac{1}{2}(u_{i|k} + u_{k|i} + u_{j|i}u_{j|k}). \tag{6.26}$$

* If a vector is transformed according to $X_i' = (\delta_{ij} + u_{i|j})\,X_j$ and a second one according to $Y_i' = (\delta_{ij} + u_{i|j})\,Y_j$, the change in the scalar product is $(X_i'\,Y_i') - (X_i Y_i) = (X_j, (u_{j|k} + u_{k|j} + u_{i|j}u_{i|k})\,Y_k) = 2(X_j, \mathscr{V}_{jk} Y_k)$. The scalar product may not change in a rotation, which is achieved with $\mathscr{V}_{ik} = 0$. So $\mathscr{V}_{jk}$ is the appropriate quantity to describe true strains.

The strain energy therefore can depend only on the quantity $\mathscr{V}_{ik}$ and its expansion (including only harmonic terms) is given by*

$$U = U_0 + C_{ij}\mathscr{V}_{ij} + \frac{1}{2} C_{ij,kl}\mathscr{V}_{ij}\mathscr{V}_{kl} + \cdots. \tag{6.27}$$

The coefficients of this expansion are called the elastic constants of first (stresses), second (usual constants), etc. order. As $\mathscr{V}_{ij}$ is symmetric with respect to $i \Leftrightarrow j$ [from def. (6.26)], so are the $C_{ij,kl}$ [only these symmetric parts enter in (6.27)]. Further they are symmetric with respect to the pairs $(ij) \Leftrightarrow (kl)$ etc., or in other words, they have the complete Voigt-symmetry.

Inserting (6.26) we can give the relation between the $C_{ij,kl}$ and $S_{ij,kl}$:

$$C_{ij} = S_{ij} \tag{6.28a}$$

$$C_{ij,kl} + C_{jl}\,\delta_{ik} = S_{ij,kl} \tag{6.28b}$$

or

$$C_{ij,kl} = S_{ij,kl} - S_{jl}\,\delta_{ik},$$

which agrees in symmetry with that expressed by (6.12, 13, 21).

Now we are in a situation to give the elastic constants as functions of those quantities which can be calculated from lattice theory. From (6.8), (6.2), (2.11) and (6.28) we have the stresses in the initial state

$$C_{ij} = S_{ij} = \frac{1}{V}\sum_{m\mu} \Phi^{m}_{\substack{\mu\\i}} X^{m}_{\substack{\mu\\j}} = \frac{1}{V}\sum_{m\mu} F^{m}_{\substack{\mu\\i}} X^{m}_{\substack{\mu\\j}} \tag{6.29}$$

with the restriction of applying external forces to surface ions only. The elastic constants (of second order) have to be derived from

$$\frac{1}{2}(C_{ij,kl} + C_{il,kj}) = \bar{S}_{ik,jl} - S_{jl}\,\delta_{ik} \tag{6.30}$$

where (6.25) and (6.28) have been used.

These are essentially the so-called the Born-Huang-relations (I, II). *Born* and *Huang* established this relation in order to guarantee the transition from lattice to elastic theory rather than proved it. Also they used the method of long waves (see below). What remains to be done is the solution of (6.30). Such a tensor equation cannot be solved uniquely in general. But the symmetries of the various quantities involved are just necessary and sufficient to give a unique solution of (6.30) for the $C_{ij,kl}$:

Write (6.30) again with $i$ and $j$ interchanged:

$$\frac{1}{2}(C_{ji,kl} + C_{jl,ki}) = \bar{S}_{jk,il} - S_{il}\,\delta_{jk} \tag{6.31}$$

and (6.31) again with $i$ and $k$ interchanged:

$$\frac{1}{2}(C_{jk,il} + C_{jl,ik}) = \bar{S}_{ji,kl} - S_{kl}\,\delta_{ij}. \tag{6.32}$$

* We have discussed here a purely mechanical theory (without temperature). Therefore we need not specify here whether $U$ is the internal energy or the Helmholtz free energy. There is only one sort of elastic constants, as the difference between isothermal and adiabatic constants occurs only if thermodynamics is taken into account.

Now add (6.30) and (6.31) and subtract (6.32), use the various symmetry properties and get

$$C_{ij,kl} = \bar{S}_{ik,jl} + \bar{S}_{jk,il} - \bar{S}_{ij,kl} - S_{jl}\,\delta_{ik} - S_{il}\,\delta_{jk} - S_{kl}\,\delta_{ij} \tag{6.33}$$

or with (6.25)

$$\begin{aligned} C_{ij,kl} = \hat{C}_{ik,jl} + \hat{C}_{jk,il} - \hat{C}_{ij,kl} + \hat{C}_{\substack{\mu\\ m,ij}}\, R_{\substack{\mu\nu\\ mn}}\, \hat{C}_{\substack{\nu\\ n,kl}} - \\ - S_{jl}\,\delta_{ik} - S_{il}\,\delta_{jk} + S_{kl}\,\delta_{ij}\,. \end{aligned} \tag{6.34}$$

From (6.33) or (6.34) $S_{ij,kl}$ can be obtained by adding $S_{jl}\,\delta_{ik}$ [see (6.28b)]. Thus all the quantities needed can be calculated from (surface insensitive) lattice theoretical expressions.

Before discussing the meaning of these results we will give another derivation, using the method of long waves. This is due originally to *Born* in order to avoid surface effects. To eliminate these surface effects, we consider a very large finite crystal (or even an infinite crystal) divided in periodicity volumes of $N$ unit cells. As the external forces are applied to the surface ions only, in the interior of the crystal we have $\Phi^{m}_{\substack{\mu\\ i}} = 0$ because of (6.2) and the equation of motion, derived from (1.6, 7) reads

$$M_\mu\, \ddot{u}^{m}_{\substack{\mu\\ i}} = - \sum_{n\nu j} \Phi^{mn}_{\substack{\mu\nu\\ ij}}\, u^{n}_{\substack{\nu\\ j}} + \text{terms of higher order} \tag{6.35}$$

in the interior of the crystal. The applied external stresses in the initial state $\left(u^{n}_{\substack{\nu\\ j}} = 0\right)$ enter only through the equilibrium positions. For the $\Phi^{mn}_{\substack{\mu\nu\\ ij}}$ we can use the relations of the interior of the crystal (sect. 3, 4.) With the assumed periodicity in the crystal we can solve eq. (6.35) by

$$u^{m}_{\substack{\mu\\ i}} = e_{\substack{\mu\\ i}} \cdot \mathrm{e}^{i\,\boldsymbol{q}\,\boldsymbol{R}^{m}_{\mu}}\,. \tag{6.36}$$

In order to eliminate the relative motions of the different sublattices, we set

$$e_{\substack{\mu\\ i}} = u_{\substack{\mu\\ i}} + u_i \quad \text{with} \quad \sum_\mu M_\mu\, u_{\substack{\mu\\ i}} = 0\,. \tag{6.37}$$

$u_i$ then is the displacement of the center of mass of the unit cell, $u_{\substack{\mu\\ i}}$ are the relative displacements of the sublattices. Inserting (6.36, 37) into (6.35) and expanding with respect to $q_i$ (long waves) we have

$$\begin{aligned} M_\mu\left(\ddot{u}_i + \ddot{u}_{\substack{\mu\\ i}}\right) = &- \sum_{n\nu j} \Phi^{mn}_{\substack{\mu\nu\\ ij}}\, u_j - \sum_{n\nu j} \Phi^{mn}_{\substack{\mu\nu\\ ij}}\, u_{\substack{\nu\\ j}} - \\ &- i \sum_{\substack{n\nu\\ jk}} \Phi^{mn}_{\substack{\mu\nu\\ ij}} \left(X^{n}_{\substack{\nu\\ k}} - X^{m}_{\substack{\mu\\ k}}\right) q_k\, u_j - \\ &- i \sum_{\substack{n\nu\\ jk}} \Phi^{mn}_{\substack{\mu\nu\\ ij}} \left(X^{n}_{\substack{\nu\\ k}} - X^{m}_{\substack{\mu\\ k}}\right) q_k\, u_{\substack{\nu\\ j}} + \end{aligned}$$

$$+\frac{1}{2}\sum_{\substack{n\nu\\jkl}}\Phi^{mn}_{\substack{\mu\nu\\ij}}\left(X^n_{\substack{\nu\\k}}-X^m_{\substack{\mu\\k}}\right)\left(X^n_{\substack{\nu\\l}}-X^m_{\substack{\mu\\l}}\right)q_k q_l\cdot u_j+$$

$$+\frac{1}{2}\sum_{n\nu,jkl}\Phi^{mn}_{\substack{\mu\nu\\ij}}\left(X^n_{\substack{\nu\\k}}-X^m_{\substack{\mu\\k}}\right)\left(X^n_{\substack{\nu\\l}}-X^m_{\substack{\mu\\l}}\right)q_k q_l\cdot u_{\substack{\nu\\j}}\,. \tag{6.38}$$

Summing over $\mu$, we have the equation of motion for the center of mass only. It is $\sum_\mu M_\mu = \varrho V_z$, $\varrho$ being the mass density of the crystal in the initial state! In the summation over $\mu$, the first term on the right side vanishes because of (2.6), the second because of (6.9, 14b), the third because of (6.10, 15b). Using the definitions (6.10) and (6.23), we are left with (now again summation convention)

$$\varrho V_z \ddot{u}_i = -i V_z \hat{C}_{\substack{\nu\\j,ik}} q_k u_{\substack{\nu\\j}} - V_z \hat{C}_{ij,kl} q_k q_l u_j\,, \tag{6.39}$$

where the last term in (6.38) and higher ones can be neglected in the following. The relative motion $u_{\substack{\nu\\j}}$ has to be eliminated. It can be done with the help of (6.38), solved for $\ddot{u}_{\substack{\mu\\i}}$ (the first term vanishes again, see (2.6))

$$M_\mu \ddot{u}_{\substack{\mu\\i}} = -M_\mu \ddot{u}_i + V_z \hat{C}_{\substack{\mu\nu\\ij}} u_{\substack{\nu\\j}} + i V_z \hat{C}_{\substack{\mu\\i,jk}} q_k u_j \tag{6.40}$$

+ higher terms which can be neglected now.

$\ddot{u}_i$ has to be inserted from (6.39), but can be disregarded because we need only terms up to second order in $q_i$ in (6.39).

Now in the limit $q_k \to 0$ we have from (6.40) $M_\mu \ddot{u}_{\substack{\mu\\i}} = V_z \hat{C}_{\substack{\mu\nu\\ij}} u_{\substack{\nu\\j}}$. This must describe just the optical frequencies in the limit $q_k \to 0$ because (6.39) in this limit gives $\ddot{u}_i = 0$, which are the three acoustical limiting frequencies. Because the optical limiting frequencies are high, and we need only the elastical acoustical long wave solution of (6.39), we can neglect the optical waves or in other words, we can use the stationary solutions of (6.40) to get the long wave relation between $u_{\substack{\nu\\j}}$ and $u_j$:

$$\hat{C}_{\substack{\mu\nu\\ij}} u_{\substack{\nu\\j}} = -i \hat{C}_{\substack{\mu\\i,jk}} q_k u_j\,. \tag{6.41}$$

This relation is identical to (6.17) and can be solved by the same method to give

$$u_{\substack{\nu\\j}} = -i R_{\substack{\nu\mu\\ji}} \hat{C}_{\substack{\mu\\i,jk}} q_k u_j\,. \tag{6.42}$$

If we insert this into (6.39) we have finally

$$\varrho \ddot{u}_i = -\left\{\hat{C}_{ik,jl} + \hat{C}_{\substack{\mu\\m,ij}} R_{\substack{\mu\nu\\mn}} \hat{C}_{\substack{\nu\\n,kl}}\right\} q_j q_l u_k\,, \tag{6.43}$$

where some of the summation indices have been interchanged.

Equation (6.43) has to be compared with the elastic equation of motion for continuous media. This can be derived by the usual methods [I, II]

$$\varrho \ddot{u}_i = S_{ij,kl} u_{k|jl} + \cdots;\ u_{k|jl} = \frac{\partial^2 u_k}{\partial X_j \partial X_l} \tag{6.44}$$

or with

$$u_k \sim e^{i\boldsymbol{q}\boldsymbol{R}} \tag{6.45}$$

$$\varrho \ddot{u}_i = -S_{ij,kl}\, q_j q_l u_k \,. \tag{6.46}$$

In (6.43) as well as in (6.46) only those parts of the coefficients enter which are symmetric in $j \Leftrightarrow l$, therefore (6.43) and (6.46) are identical if [$\hat{C}_{ik,jl}$ is symmetric in $j \Leftrightarrow l$]

$$\begin{aligned} \frac{1}{2}(S_{ij,kl} + S_{il,kj}) &= \bar{S}_{ik,jl} \\ &= \hat{C}_{ik,jl} + \frac{1}{2}\hat{C}^{\mu}_{m,ij} R^{\mu\nu}_{mn} \hat{C}^{\nu}_{n,kl} + \frac{1}{2}\hat{C}^{\mu}_{m,il} R^{\mu\nu}_{mn} \hat{C}^{\nu}_{n,kj}, \end{aligned} \tag{6.47}$$

which is identical to (6.25), i.e. both methods give the same result. All the symmetries and relations between the various quantities can therefore be taken from the above discussion. Since only the symmetric part of $S_{ij,kl}$ enters the equation of motion, we can write instead of (6.43) or (6.46)

$$\varrho \ddot{u}_i = -\bar{S}_{ik,jl}\, q_j\, q_l\, u_k = \bar{S}_{ik,jl}\, u_{k|jl} \,. \tag{6.48}$$

This shows clearly that by sound wave measurements one determines the $\bar{S}_{ik,jl}$ rather than the true elastic constants, which must be calculated according to (6.33). If there are stresses in the initial state, these are contained in the $\bar{S}_{ik,jl}$ and can be determined also from them. Using (6.30) and the symmetries of the $C_{ij,kl}$, we obtain e.g.

$$\left.\begin{aligned} S_{12} &= \bar{S}_{ii,12} - \bar{S}_{21,ii} \\ S_{23} &= \bar{S}_{ii,23} - \bar{S}_{32,ii}\,, \\ S_{31} &= \bar{S}_{ii,31} - \bar{S}_{13,ii} \end{aligned}\right. \qquad i \text{ arbitrary} \tag{6.49a}$$

$$\begin{aligned} S_{11} - S_{33} &= \bar{S}_{33,11} - \bar{S}_{11,33} \\ S_{22} - S_{11} &= \bar{S}_{11,22} - \bar{S}_{22,11} \\ S_{33} - S_{22} &= \bar{S}_{22,33} - \bar{S}_{33,22}\,. \end{aligned} \tag{6.49b}$$

These equations allow to determine all the stresses except for an isotropic homogeneous pressure, defined by

$$p = -\frac{1}{3}(S_{11} + S_{22} + S_{33}). \tag{6.50}$$

The eq. (6.49b) can be written in the alternative form

$$S_{11} = -p + \frac{1}{3}(\bar{S}_{22,11} + \bar{S}_{33,11} - \bar{S}_{11,22} - \bar{S}_{11,33}) \tag{6.49c}$$

and cyclic for $S_{22}$, $S_{33}$.

Also in static measurements of the elastic constants one has to specify exactly the experimental situation and to look for the parameters which describe the experimental situation appropriately. Then one can determine the corresponding constants and calculate from them the elastic constants [I, II, 60 L 1, 61 L 1, 65 W 1].

Recently, *Laval* again has raised objections against lattice theory of elastic constants [64 L 4]. His objections are based on a severe misunderstanding of the (many body-) potential energy. He denies that in a lattice without spins* (or other hidden variables) the potential energy is a function of all the mutual distances of the ions. Therefore he does not believe the condition of rotational invariance. Because of the obvious confusion of different things in Laval's paper, we will not go into details here.

By comparing the results for the $C_{ij,kl}$ with the corresponding lattice theoretical expressions (6.34) we can get a knowledge about the ionic force constants, the cp's. But this works only, if we would have purely mechanical behaviour. However, finite temperatures (or even $T = 0$ where we have zero point vibrations) mean, that we have to take into account thermodynamics. Therefore one does not measure these „purely mechanical" constants, but rather temperature dependent elastic constants (adiabatic or isothermal) and one must eliminate the temperature effects to obtain information about the true mechanical ionic cp's, defined by the properties of the interaction potential of the ions only (chapter V).

The significance of the conditions of rotational invariance can be seen from another effect which is related to surface phenomena. In finite elastic media there exists the possibility of surface waves (Rayleigh waves). These waves propagate in a direction lying in the surface and have exponentially decreasing amplitudes perpendicular to the surface.

To treat these waves, we have to solve eq. (6.44)** [see (6.28b)]

$$\varrho \ddot{u}_i = S_{ij,kl}\, u_{k|jl} = C_{ij,kl}\, u_{k|jl} \tag{6.51}$$

with the boundary condition of a free surface, i.e. the forces on this surface have to vanish. If we consider a semi-infinite crystal with plane surface $X_3 = 0$, the stresses corresponding to the small forces $f_\mu^m$ on this surface have to be zero (see page 16):

$$S_{13} = S_{23} = S_{33} = 0 \quad \text{at} \quad X_3 = 0\,. \tag{6.52}$$

Instead of solving (6.51) with the boundary condition (6.52) we can also discuss the solution of

$$\varrho \ddot{u}_i = C_{ij,kl|j}\, u_{k|l} + C_{ij,kl}\, u_{k|jl} \tag{6.53}$$

where

$$C_{ij,kl|j} = \sum_j \frac{\partial C_{ij,kl}}{\partial X_j}$$

and this is even the more general equation of motion, valid for non-homogeneous elastic media (varying elastic constants). It can be derived just as the usual equation of motion (6.44), with the assumption that

* Of course, there is a condition for the rotational invariance also, if spin-interactions are taken into account. These relations are more complicated than (2.9) and involve all the parameters which describe the interactions (potential derivatives, spin-interaction-parameters etc.). On deriving elastic constants which take into account spin interactions, such relations have to be considered.

** We assume vanishing initial stresses now! There may be the possibility of surface waves at an initially stressed crystal too, but this case we will not consider here.

the $S_{ij,kl}$ (or $C_{ij,kl}$ if $S_{ij}=0$) depend on the position. If the elastic constants can be described by a step-function at the plane $X_3=0$, jumping from zero to the value of the interior of the crystal, it can be shown, that (6.53) is equivalent to (6.51) plus (6.52). With $S_{ij}=0$ the derived constants $S_{ij,kl|j}$ or

$$C_{ij,kl|j} \text{ are symmetric with respect to } k \Leftrightarrow l. \tag{6.54}$$

Now let us look for the corresponding lattice expression. For simplicity, we discuss only Bravais-lattices; the others can be treated in a similar way. The equation of motion reads now

$$M \ddot{u}_i^m = - \sum_{nk} \Phi_{i\ k}^{mn} u_k^n \tag{6.55}$$

[here $\Phi_i^m = 0$ everywhere because of the assumption $S_{ij}=0$.] To discuss this equation for long waves, we introduce a slowly varying displacement field (I, II), i.e. neighboring ions shall differ in the displacement by a very small amount only. The variation is to be small over the range of the interionic forces. Then

$$u_k^n = u_k^m + u_{k|l}^m \left(X_l^n - X_l^m\right) + \frac{1}{2} u_{k|jl}^m \left(X_j^n - X_j^m\right)\left(X_l^n - X_l^m\right) \tag{6.56}$$

and introducing this into (6.55), using (2.6), dividing by $V_z$ and dropping $m$ on the now continuously defined function $u_j(\boldsymbol{R})$ we have

$$\begin{aligned} \varrho \ddot{u}_i(\boldsymbol{R}) = &- \frac{1}{V_z} \sum_{jl} \sum_{n} \Phi_{i\ k}^{mn} \left(X_l^n - X_l^m\right) u_{k|l} - \\ &- \frac{1}{2V_z} \sum_{jkl} \sum_{n} \Phi_{i\ k}^{mn} \left(X_j^n - X_j^m\right)\left(X_l^n - X_l^m\right) u_{k|jl} . \end{aligned} \tag{6.57}$$

*In infinite homogeneous lattices* or *in the interior of finite lattices* the first term on the right side vanishes because of (3.3) and (4.3). The condition (2.9b) or (2.10b) has not been used here. The second term is independent on $m$ in these cases and gives the usual elastic constants (see the discussion above).

*In inhomogeneous media*, e.g. at the surface, the relations (3.3) and (4.3) do not hold in the surface region. Therefore the first term in (6.57) does not vanish. It corresponds exactly to the first term in (6.53) with symmetry (6.54). Therefore the first term in (6.57) must be symmetric in $k \Leftrightarrow l$ too, to get the elastic limit from lattice theory. This is just guaranteed by the rotational condition (2.9b) or (2.10b), with $\Phi_i^m = 0$ in this case.

The second term has to be identified with the elastic constants. In inhomogeneous media it depends on $m$, the position in the crystal. But this may happen also in elastic theory and can be discussed in the usual way (see above). The first term also depends on $m$; but if the inhomogeneities are limited to the immediate surface, it is just equivalent to using (6.51) plus (6.52). A step-function in the elastic constants corresponds to a case where the first term in (6.57) is different from zero only for $m$-values in the surface.

We have stressed the importance of the rotational condition because often surface-models have been used which violated the condition (2.9b, 10b). Then, in the limit of long waves, lattice theory cannot agree with elastic theory.

Some remarks concerning the third rotational condition (2.9c) shall finish this section. In a purely harmonic theory one would have to satisfy also (2.9c) with $\Phi^{mnp}_{i\,j\,k} = 0$ for all $\boldsymbol{m}, \boldsymbol{n}, \boldsymbol{p}$. This is possible only with

$$\Phi^{mn}_{i\,j} = \varphi^{mn}\,\delta_{ij}\,, \tag{6.58}$$

as one can see immediately by considering the possible combinations of indices.

From (6.23) then (Bravais-lattices!)

$$\hat{C}_{ik,jl} = -\frac{1}{2V_z}\sum_{h}\varphi^{0h}X^h_j X^h_l\,\delta_{ik} = \varrho_{jl}\,\delta_{ik} \tag{6.59}$$

and from (6.24c)

$$(\varrho_{jl} - S_{jl})\,\delta_{ik} = (\varrho_{ik} - S_{ik})\,\delta_{jl} \tag{6.60}$$

which can be solved again only with

$$\varrho_{jl} - S_{jl} = \alpha\,\delta_{jl}\,. \tag{6.61}$$

Inserting this result into (6.34) we have

$$C_{ij,kl} = \alpha(\delta_{ik}\,\delta_{jl} + \delta_{jk}\,\delta_{il} - \delta_{ij}\,\delta_{kl})\,. \tag{6.62}$$

Now the stability of the crystal demands the shear moduli and the compressibility to be positive (I, II). Eq. (6.62) expresses the elastic constants of an isotropic medium with the shear moduli

$$C_{12,12} = \frac{1}{2}(C_{11,11} - C_{11,22}) = \alpha \tag{6.63a}$$

and the compressibility $\varkappa$

$$1/\varkappa = \frac{1}{3}(C_{11,11} + 2C_{11,22}) = -\frac{1}{3}\alpha\,, \tag{6.63b}$$

thus either (6.63a) or (6.63b) is negative and therefore the crystal is unstable. We arrive at the conclusion, that a purely harmonic theory is not possible.

Nevertheless, the familiar harmonic theory may be used as a very good approximation. This can be seen by using a central force model between nearest neighbors only. If

$$\Phi = \frac{1}{2}\sum_{mn}\varphi\left(\boldsymbol{R}^m - \boldsymbol{R}^n\right) \tag{6.64}$$

we have ($a$: nearest neighbor distance)

$$\Phi^{mn}_{i\,j} \sim \text{ terms with } \varphi'' + \text{ terms with with } \varphi'/a \tag{6.65a}$$

$$\Phi^{mnp}_{i\,j\,k} \sim \text{terms with } \varphi''' + \text{terms with } \varphi''/a + \text{ terms with } \varphi'/a^2\,. \tag{6.65b}$$

If now $\varphi''' = 0$ (no genuine anharmonicity) the forces caused by anharmonicity are $\varphi''/a \cdot (\delta a)^2$, whereas the harmonic forces are $\varphi'' \cdot \delta a$, $\delta a$ being the mean change in lattice constant. We have to compare $\varphi''$ with $\varphi'' \cdot \delta a/a$ where $\delta a/a$ is comparable with the thermal expansion, i.e. $10^{-2}$ to $10^{-3}$. This shows, that the harmonic approximation is rather good and that it is sufficient to satisfy (2.9b), but to neglect (2.9c) in a purely harmonic theory. Further, it turns out, that the genuine anharmonicity ($\varphi'''$) is actually larger than the "necessary" anharmonicity $\varphi''/a$ by one order of magnitude. At any case, the „good old" harmonic theory is an appropriate starting point for lattice theoretical investigations and there are no discrepancies between the elastic limit of lattice theory and the classical theory of elasticity.

# III. Different Models for the Potential Energy in Crystals

## 7. General Remarks on Many-body Potentials

Whereas up to 1956 mainly central-forces and volume-forces have been considered in investigations of the interionic potentials, in recent years the discussion of many-body-potentials has become fruitful. In many crystals there are effects which can be explained only by the use of such many-body-potentials. For example, recent measurements of the phonon-dispersion-curves in alkali halides [60 W 1, 63 W 1, 62 B 2, 3, 7] cannot be understood by taking into account central-forces only. The most striking effect, however, seems to be the deviation from the Cauchy relation for the elastic constants. In parameterfree cubic crystals (each ion is a center of inversion!) it can be derived, using central-forces only,

$$c_{44} - c_{12} = \frac{27\,s}{V_z}\,\gamma^2\,k\,T \tag{7.1}$$

for temperatures $T$ above the Debye-temperature [I, 61 L 1], $\gamma$ is the Grüneisen constant, $c_{44}$ and $c_{12}$ are the adiabatic elastic constants. Eq. (7.1) has been derived in a certain approximation, which affects however neither the sign nor the temperature dependence. In alkali-halides, where volume forces can be neglected, therefore $c_{44} > c_{12}$ and if one extrapolates (7.1) linearly to $T = 0$, $c_{44}$ and $c_{12}$ should be equal. This is satisfied in none of the alkali-halides, where rather different situations might occur. Thus, in LiF there is indeed $c_{44} > c_{12}$, but in the extrapolation it is $c_{44} \neq c_{12}$. On the other hand, in NaJ, the extrapolation gives fairly well $c_{44} = c_{12}$, but with increasing temperature it is $c_{12} > c_{44}$! This leads to the conclusion, that an essential part of the interionic forces in ionic crystals has non-central character.

There are no sufficient measurements for van der Waals-crystals to decide whether there are non-central-forces or not. In metals there are volume-forces (i.e. the electronic contribution to the energy, depending on the volume of the elementary cell mainly) which give already deviations from Cauchy's relations and which can be regarded as a special kind of non-central forces [I]. In homopolar crystals one has to expect a rather large contribution of non-central-interactions.

We would avoid the use of non-central-potentials, if we would work with general cp's. These must be fitted from experiments; that is, we have to restrict the number of independent cp's to the number of experiments which allow a unique fitting. Then it is often more convenient to use given potentials and to determine from experiment the constants of the potential only. In many cases this leads to fewer constants being needed for fitting.

To give an idea of non-central-forces we will first derive some expressions for the potential energy from simple models. As is known a second order perturbation treatment of two neutral atoms leads to the famous van-der-Waals-forces [39 M 1]. The conjecture is, that a third order perturbation method would give rise to a three-body-potential; we will show this to be true. The Schrödinger-perturbation-theory gives up to third order (for the ground state):

$$E = E_0 + W_{00} - \sum_{\alpha \neq 0} \frac{W_{0\alpha} W_{\alpha 0}}{E_\alpha - E_0} + \sum_{\substack{\alpha \neq 0 \\ \gamma \neq 0}} \frac{W_{0\alpha} W_{\alpha\gamma} W_{\gamma 0}}{(E_\alpha - E_0)(E_\gamma - E_0)} - W_{00} \sum_{\alpha \neq 0} \frac{W_{0\alpha} W_{\alpha 0}}{(E_\alpha - E_0)^2} . \tag{7.2}$$

$W$ is the perturbation, $E_0$ the unperturbed ground state energy and $\alpha$ is an excited state.

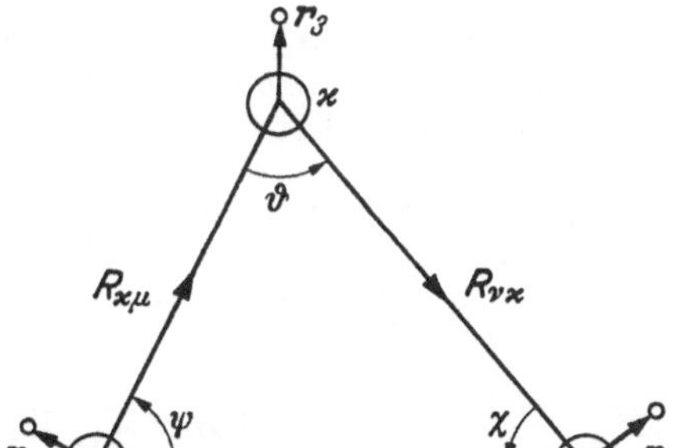

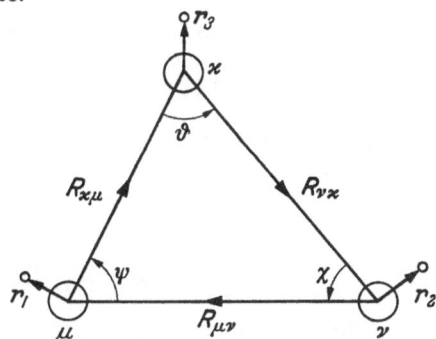

Fig. 3.1. Positions of three interacting atoms

Consider the model given in Fig. 3.1. If the charge* of the core is $Q_\mu$, the Hamiltonian of the system is

$$\mathscr{H} = \mathscr{H}_0 + W; \quad \mathscr{H}_0 = h_\mu(1) + h_\nu(2) + h_\varkappa(3)$$
$$W = W_{\mu\nu}(12) + W_{\nu\varkappa}(23) + W_{\varkappa\mu}(31) \tag{7.3}$$

$h(1)$ is the Hamiltonian for the atom $\mu$, electron 1 and satisfies

$$h_\mu(1)\, \varphi_\mu^{(\alpha_1)}(1) = \varepsilon_\mu^{(\alpha_1)}\, \varphi_\mu^{(\alpha_1)}(1) \tag{7.4}$$

* The calculation is reasonable only for $Q_\mu = Q =$ one elementary charge, but in order to distinguish between the atoms, we write $Q_\mu$.

$\alpha_1$ denotes the state of the atom. The interaction is given by

$$W_{\mu\nu}(12) = Q_\mu Q_\nu \left\{\frac{1}{R^{\mu\nu}} - \frac{1}{R^{\mu 2}} - \frac{1}{R^{\nu 1}} + \frac{1}{R^{12}}\right\}, \quad \text{cyclic} \tag{7.5}$$

with

$$\boldsymbol{R}^{\mu 2} = \boldsymbol{R}^{\mu\nu} - \boldsymbol{r}_2; \quad \boldsymbol{R}^{\nu 1} = -\boldsymbol{R}^{\mu\nu} - \boldsymbol{r}_1; \quad \boldsymbol{R}^{12} = \boldsymbol{R}^{\mu\nu} - \boldsymbol{r}_2 + \boldsymbol{r}_1 .$$

We are interested only in the interaction at relatively large distances of the atoms and can expand (7.5) wich respect to $r_1/R^{\mu\nu}$, $r_2/R^{\mu\nu}$, .... Then

$$\begin{aligned} W_{\mu\nu}(12) &\approx \frac{Q_\mu Q_\nu}{(R^{\mu\nu})^3} g_{\mu\nu}(12) + \frac{3 Q_\mu Q_\nu}{2(R^{\mu\nu})^4} h_{\mu\nu}(12) \\ g_{\mu\nu}(12) &= (\boldsymbol{r}_1 \boldsymbol{r}_2) - 3(\boldsymbol{n}_{\mu\nu}\boldsymbol{r}_1)(\boldsymbol{n}_{\mu\nu}\boldsymbol{r}_2) \\ h_{\mu\nu}(12) &= 5(\boldsymbol{n}_{\mu\nu}\boldsymbol{r}_1)(\boldsymbol{n}_{\mu\nu}\boldsymbol{r}_2)(\boldsymbol{n}_{\mu\nu}, \boldsymbol{r}_1 - \boldsymbol{r}_2) - 2(\boldsymbol{r}_1\boldsymbol{r}_2)(\boldsymbol{n}_{\mu\nu}, \boldsymbol{r}_1 - \boldsymbol{r}_2) \\ &\quad - r_1^2(\boldsymbol{n}_{\mu\nu}\boldsymbol{r}_2) + r_2^2(\boldsymbol{n}_{\mu\nu}\boldsymbol{r}_1) \end{aligned} \tag{7.6}$$

$\boldsymbol{n}_{\mu\nu} = \boldsymbol{R}^{\mu\nu}/R^{\mu\nu}$ = unit vector in the direction $\nu \to \mu$. This has to be inserted into (7.2), where $\alpha$ has to be replaced by the set of quantum numbers $\alpha_1 \alpha_2 \alpha_3$. It is easily seen that $W_{00}$ vanishes. Further we have e.g.

$$\langle 0 \mid W_{\mu\nu}(12) \mid \alpha_1\alpha_2\alpha_3 \rangle = \langle 0 \mid W_{\mu\nu}(12) \mid \alpha_1\alpha_2 \rangle \, \delta_{0\alpha_3}, \ldots . \tag{7.7}$$

This factor $\delta_{0\alpha_3}$, in connection with corresponding factors at the other terms, lets in $\sum_\alpha W_{0\alpha} W_{\alpha 0}/(E_\alpha - E_0)$ only enter terms of the form

$$\sum_{\alpha_1\alpha_2} \frac{\langle 0 \mid W_{\mu\nu}(12) \mid \alpha_1\alpha_2 \rangle \langle \alpha_1\alpha_2 \mid W_{\mu\nu}(12) \mid 0 \rangle}{(\varepsilon_\mu^{\alpha_1} - \varepsilon_\mu^0 + \varepsilon_\nu^{\alpha_2} - \varepsilon_\nu^0)} .$$

There are contributions only for $\alpha_1 \neq 0$ and $\alpha_2 \neq 0$. Now, if $\varepsilon_\mu^{\alpha_1} - \varepsilon_\mu^0$ is nearly the same for every $\alpha_1$ or if the transition elements are small compared to only *one* large element, we can replace $\varepsilon_\mu^{\alpha_1} - \varepsilon_\mu^0$ by the ionization energy $I_\mu$ and are left with*

$$\frac{1}{I_\mu + I_\nu} \sum_{\alpha_1\alpha_2} \langle 0 \mid W_{\mu\nu} \mid \alpha_1\alpha_2 \rangle \langle \alpha_1\alpha_2 \mid W_{\mu\nu} \mid 0 \rangle = \frac{1}{I_\mu + I_\nu} \langle 0 \mid W_{\mu\nu}^2 \mid 0 \rangle .$$

By the same arguments it can be shown, that the only term, which enters the third order perturbation term, is

$$\frac{4(I_\mu + I_\nu + I_\varkappa)}{(I_\mu + I_\nu)(I_\nu + I_\varkappa)(I_\varkappa + I_\mu)} \langle 0 \mid W_{\mu\nu} W_{\nu\varkappa} W_{\varkappa\mu} \mid 0 \rangle .$$

With the spherically symmetric distribution of the ground state the calculation of the remaining matrix-elements is easy. They are

$$\begin{aligned} \langle 0 \mid g_{\mu\nu}^2(12) \mid 0 \rangle &= \frac{2}{3} \overline{r_\mu^2}\, \overline{r_\nu^2}, \quad \text{cyclic} \\ \langle 0 \mid h_{\mu\nu}^2(12) \mid 0 \rangle &= \frac{4}{9} \overline{r_\mu^2\, r_\nu^2\, (r_\mu^2 + r_\nu^2)}, \quad \text{cyclic} \\ \langle 0 \mid g_{\mu\nu}(12)\, g_{\nu\varkappa}(23)\, g_{\varkappa\mu}(31) \mid 0 \rangle &= \frac{1}{36} \overline{r_\mu^2}\, \overline{r_\nu^2}\, \overline{r_\varkappa^2} \times \\ &\quad \times \{1 - 3\cos 2\chi - 3\cos 2\vartheta - 3\cos 2\psi\} \end{aligned} \tag{7.8}$$

---

* This holds, if all the excited states have an energy comparable with $I_\mu$ (see 39 M 1).

with

$$\overline{r_\mu^2} = \int \varphi_\mu^{*0}(1)\, r_1^2\, \varphi_\mu^0(1)\, d\tau_1\,, \quad \text{etc.,} \tag{7.9}$$

$$\cos\chi = -(\boldsymbol{n}_{\mu\nu}, \boldsymbol{n}_{\nu\varkappa}); \quad \cos\vartheta = -(\boldsymbol{n}_{\nu\varkappa}\, \boldsymbol{n}_{\varkappa\mu}); \quad \cos\psi = -(\boldsymbol{n}_{\varkappa\mu}\, \boldsymbol{n}_{\mu\nu})\,.$$

The matrix elements $\langle 0 \mid g_{\mu\nu}(12)\, h_{\mu\nu}(12) \mid 0\rangle$ vanish again. As the final result, we have

$$\begin{aligned} E = E_0 &- \frac{2}{3}\, \frac{Q_\mu^2\, Q_\nu^2\, \overline{r_\mu^2}\, \overline{r_\nu^2}}{I_\mu + I_\nu} \cdot \frac{1}{(R^{\mu\nu})^6} + \text{cyclic terms} \\ &- \frac{Q_\mu^2\, Q_\nu^2\, \overline{r_\mu^2\, r_\nu^2\, (r_\mu^2 + r_\nu^2)}}{I_\mu + I_\nu} \cdot \frac{1}{(R^{\mu\nu})^8} + \text{cyclic terms} \\ &+ \frac{1}{9}\, \frac{Q_\mu^2\, Q_\nu^2\, Q_\varkappa^2\, (I_\mu + I_\nu + I_\varkappa)\, \overline{r_\mu^2}\, \overline{r_\nu^2}\, \overline{r_\varkappa^2}}{(I_\mu + I_\nu)\,(I_\nu + I_\varkappa)\,(I_\varkappa + I_\mu)} \cdot \frac{1 - 3\cos 2\gamma - 3\cos 2\vartheta - 3\cos 2\psi}{(R^{\mu\nu}\, R^{\nu\varkappa}\, R^{\varkappa\mu})^3}\,. \end{aligned} \tag{7.10}$$

The first term is the usual van-der-Waals-interaction, the second is a central-force-correction to the van-der-Waals-term, already given by *Margenau* [39 M 1]. The last term represents a genuine three-body-interaction*. The $\overline{r_\mu^2}, \ldots$ can be related, at least approximately, to the

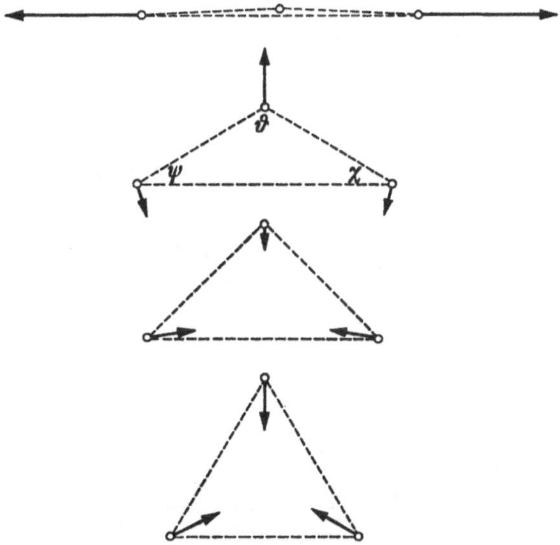


Fig. 3.2. Mutual forces on one ion, exerted from the other two, according to a three-body van der Waals-interaction (7.10)

static polarizability of the ions, but as we are interested only in the position dependent part, we will not go into those details. The three-body-term can be, in a certain sense, attractive or repulsive. In Fig. 3.2

* It should be mentioned, that this term can be calculated semiclassically too. One starts with a Thomson-Modell for the three atoms, calculates the zero-point energy as a function of the three distances (including the interaction of course) and expands with respect to the distances. This gives just the same term as in (7.10). The term can be derived also with a variational procedure.

there are shown the mutual forces for certain arrangements of the atoms. The arrows indicate the force on an atom, exerted from the other two, if these other two are kept fixed. It is seen, that these forces essentially are of angle-bending character. For some cases it is convenient to replace the double angles in (7.10) by the angles themselves.

Using

$$\boldsymbol{R}^{\mu\nu} + \boldsymbol{R}^{\nu\varkappa} + \boldsymbol{R}^{\varkappa\mu} = 0 \tag{7.11}$$

after a somewhat lengthy calculation we have for the three-body-term

$$\Phi^{(3)} = \frac{Q_\mu^2 Q_\nu^2 Q_\varkappa^2 (I_\mu + I_\nu + I_\varkappa)\, \overline{r_\mu^2}\, \overline{r_\nu^2}\, \overline{r_\varkappa^2}}{9 (I_\mu + I_\nu)(I_\nu + I_\varkappa)(I_\varkappa + I_\mu)} \times \\ \times \left\{ \frac{1}{(R^{\mu\nu} R^{\nu\varkappa} R^{\varkappa\mu})^3} - \frac{3(\boldsymbol{n}_{\mu\nu}\,\boldsymbol{n}_{\nu\varkappa})}{(R^{\mu\nu} R^{\nu\varkappa})^4 R^{\varkappa\mu}} - \right. \\ \left. - \frac{3(\boldsymbol{n}_{\nu\varkappa}\,\boldsymbol{n}_{\varkappa\mu})}{R^{\mu\nu}(R^{\nu\varkappa} R^{\varkappa\mu})^4} - \frac{3(\boldsymbol{n}_{\varkappa\mu}\,\boldsymbol{n}_{\mu\nu})}{(R^{\mu\nu} R^{\varkappa\mu})^4 R^{\nu\varkappa}} \right\}. \tag{7.12}$$

In forming the potential energy of the whole lattice, we have to sum up over all pairs of atoms, triple pairs etc., giving after some changes in indices, in the usual way the two-body-potential*.

$$\Phi^{(2)} = - \sum_{\substack{m n \\ \mu\nu}}{}'' \frac{Q_\mu^2 Q_\nu^2}{3(I_\mu + I_\nu)} \left\{ \frac{\overline{r_\mu^2}\, \overline{r_\nu^2}}{(R_{\mu\ \nu}^{m\ n})^6} + \frac{3}{2} \frac{\overline{r_\mu^2\, r_\nu^2\, (r_\mu^2 + r_\nu^2)}}{(R_{\mu\ \nu}^{m\ n})^8} + \cdots \right\} \tag{7.13a}$$

and additionally a three-body-potential*

$$\Phi^{(3)} = \frac{1}{54} \sum_{\substack{m n p \\ \mu\nu\varkappa}}{}''' \frac{Q_\mu^2 Q_\nu^2 Q_\varkappa^2 (I_\mu + I_\nu + I_\varkappa)\, \overline{r_\mu^2}\, \overline{r_\nu^2}\, \overline{r_\varkappa^2}}{(I_\mu + I_\nu)(I_\nu + I_\varkappa)(I_\varkappa + I_\mu)} \times \\ \times \left\{ \frac{1}{(R_{\mu\ \nu}^{m\ n} R_{\nu\ \varkappa}^{n\ p} R_{\varkappa\ \mu}^{p\ m})^3} - \frac{9(\boldsymbol{R}_{\mu\ \nu}^{m\ n}\, \boldsymbol{R}_{\nu\ \varkappa}^{n\ p})}{(R_{\mu\ \nu}^{m\ n} R_{\nu\ \varkappa}^{n\ p})^5 R_{\varkappa\ \mu}^{p\ m}} \right\}. \tag{7.13b}$$

With the assumption, that all the interacting ions are of the same kind, we can rewrite (7.13a, b) into

$$\Phi^{(2)} = -\frac{3}{8} \alpha^2 I \sum_{\substack{m n \\ \mu\nu}}{}'' \frac{1}{(R_{\mu\ \nu}^{m\ n})^6} + \cdots \\ \Phi^{(3)} = \frac{3}{32} \alpha^3 I \sum_{\substack{m n p \\ \mu\nu\varkappa}}{}''' \frac{1 + 3\cos\chi \cos\vartheta \cos\psi}{(R_{\mu\ \nu}^{m\ n} R_{\nu\ \varkappa}^{n\ p} R_{\varkappa\ \mu}^{p\ m})^3}. \tag{7.13c}$$

$I$ is the ionization-energy of the ion, $\alpha = 2 Q^2 \overline{r^2} / 3 I$ its polarizability.

$\chi$, $\vartheta$, $\psi$ are the angles defined above. In van-der-Waals-crystals, this three-body-potential could be the "largest" many-body-contribution. Higher order terms in (7.2) as well as in (7.6) would give rise to other three-body- as well as to four-body, ... forces. If the electron-shells of different ions overlap, this overlap might lead to another kind of three-body-potentials, and this effect might be more significant than (7.13b)

* The primes on the summation sign indicate, that terms with $\boldsymbol{R}_{\mu\ \nu}^{m\ n} = 0$ have to be omitted. From this expression it can be seen, that many-body-forces do not vanish between nearest neighbors. See footnote page 11.

when the overlap is large. Overlap effects have been investigated by *Abrahamson* et al. [61 A 1] and *Jansen* [62 J 1, 64 J 1, 65 L 1, 66 J 1].

Though the model (7.13b) is a rather simple three-body-potential, no one seems to have used it in calculations of lattice-properties, apart from the energy per cell. The reason is, that there are „two-center-sums"

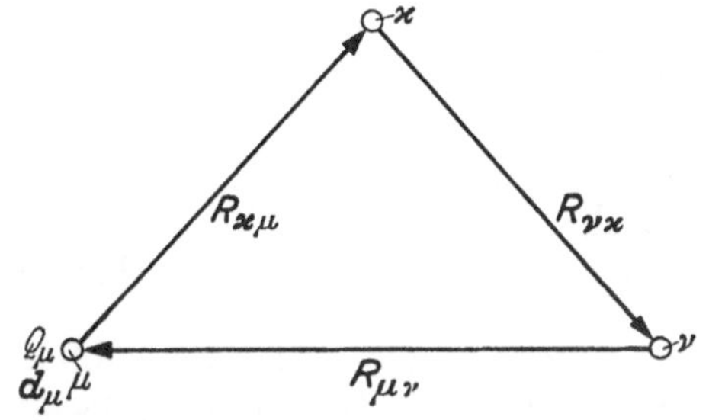


Fig. 3.3. Definition of parameters for three interacting polarizable ions

involved, which can be estimated only roughly. Where they have been calculated the numbers are very questionable [49 A 1, 51 A 1]. Similar potentials or even more complicated ones have been derived in some papers [49 A 1, 51 A 1, 55 J 2, 55 S 1, 56 M 2, . . .].

In ionic crystals the main part of the interaction is due to the electrostatic forces between point charges or between polarizable ions and charges and so on. Also this interaction leads to many-body-forces. Let us consider again three ions with effective charges $Q_\mu$ (Fig. 3.3). Let these ions have induced dipole-moments $\boldsymbol{d}^\mu$, quadrupole-moments $\boldsymbol{q}^\mu$ etc. The interaction-energy between two of these ions then is

$$\begin{aligned} E_{\nu\varkappa} = \frac{Q_\nu Q_\varkappa}{R^{\nu\varkappa}} &- \frac{1}{(R^{\nu\varkappa})^3}\{Q_\nu(\boldsymbol{R}^{\nu\varkappa}\,\boldsymbol{d}^\nu) + Q_\varkappa\,(\boldsymbol{R}^{\varkappa\nu}\,\boldsymbol{d}^\varkappa)\} \\ &- \frac{3}{2\,(R^{\nu\varkappa})^5}\sum_{ij}\{Q_\nu\,X_i^{\nu\varkappa}\,q_{ij}^\varkappa\,X_j^{\nu\varkappa} + Q_\varkappa\,X_i^{\nu\varkappa}\,q_{ij}^\nu\,X_j^{\nu\varkappa}\} \\ &- \frac{1}{(R^{\nu\varkappa})^5}\{3\,(\boldsymbol{R}^{\nu\varkappa}\,\boldsymbol{d}^\nu)\,(\boldsymbol{R}^{\nu\varkappa}\,\boldsymbol{d}^\varkappa) - (R^{\nu\varkappa})^2\,(\boldsymbol{d}^\nu\,\boldsymbol{d}^\varkappa)\}\,. \end{aligned} \tag{7.14}$$

If the dipole-moments, quadrupole-moments etc. are induced by electrostatic fields $\boldsymbol{E}$, we have*

$$\boldsymbol{d}^\nu = \chi_\nu \cdot \boldsymbol{E}(\boldsymbol{R}^\nu)\,; \quad 3q_{ik}^\nu = -\,\beta_\nu\,E_{i|k}(\boldsymbol{R}^\nu) \quad \text{with} \quad E_{i|k} = \frac{\partial E_i}{\partial X_k}\,. \tag{7.15}$$

$\chi_\nu$ is the polarizability of the ion $\nu$, $\beta_\nu$ a „quadrupolarizability". If this field $\boldsymbol{E}$ originates from an ion $\mu$ in $\boldsymbol{R}^\mu$ we have

$$\boldsymbol{E}(\boldsymbol{R}^\nu) = -\,Q_\mu\,\frac{\boldsymbol{R}^{\mu\nu}}{(R^{\mu\nu})^3}$$

---

* Strictly, $\chi_\nu$ is a tensor of 2nd rank, $\beta_\nu$ a tensor of 4th rank. For discussing the main features of many-body-interactions, it is sufficient to take $\chi_\nu$ and $\beta_\nu$ as numbers (special diagonal tensors). The generalization to more complicated cases is obvious.

and therefore for the interaction energy between two ions in an electrostatic field produced by another ion $\mu$ only:

$$\begin{aligned} W^{(\mu)}_{\nu\varkappa} = {} & \frac{Q_\nu Q_\varkappa}{R^{\nu\varkappa}} + Q_\mu Q_\nu \chi_\nu \frac{(\boldsymbol{R}^{\mu\nu} \boldsymbol{R}^{\nu\varkappa})}{(R^{\mu\nu} R^{\nu\varkappa})^3} + Q_\mu Q_\varkappa \chi_\varkappa \frac{(\boldsymbol{R}^{\nu\varkappa} \boldsymbol{R}^{\varkappa\mu})}{(R^{\nu\varkappa} R^{\varkappa\mu})^3} - \\ & - \frac{Q_\mu Q_\nu \beta_\varkappa}{2(R^{\nu\varkappa} R^{\varkappa\mu})^5} \{3(\boldsymbol{R}^{\nu\varkappa} \boldsymbol{R}^{\varkappa\mu})^2 - (R^{\nu\varkappa} R^{\varkappa\mu})^2\} - \\ & - \frac{Q_\mu Q_\varkappa \beta_\nu}{2(R^{\mu\nu} R^{\nu\varkappa})^5} \{3(\boldsymbol{R}^{\mu\nu} \boldsymbol{R}^{\nu\varkappa})^2 - (R^{\mu\nu} R^{\nu\varkappa})^2\} + \\ & + \frac{Q_\mu Q_\lambda \chi_\nu \chi_\varkappa}{(R_{\mu\nu} R_{\nu\varkappa} R_{\varkappa\lambda})^3} \left\{3 \frac{(\boldsymbol{R}^{\mu\nu} \boldsymbol{R}^{\nu\varkappa})(\boldsymbol{R}^{\nu\varkappa} \boldsymbol{R}^{\varkappa\lambda})}{(R^{\nu\varkappa})^2} - (\boldsymbol{R}^{\mu\nu} \boldsymbol{R}^{\varkappa\lambda})\right\} \end{aligned} \tag{7.16}$$

where we have allowed for another fourth ion which polarizes the ion $\varkappa$ (in the last term).

The first term represents the usual Coulomb-central-force-interaction; the next four terms are dependent on the positions of three ions, therefore containing three-body-forces. They represent the interaction between point charges and dipoles, quadrupoles, ..., whereas the last term represents a dipole-dipole interaction and contains even four-body-potentials. But (7.16) does not describe the whole interaction because we have to take into account the mutual interactions and fields. This leads essentially to a symmetrization of (7.16), and results in a potential energy for the lattice:

$$\Phi = \Phi^{(2)} + \Phi^{(3)} + \Phi^{(4)} + \cdots$$

$$\Phi^{(2)} = \frac{1}{2} \sum_{\substack{m\,n \\ \mu\,\nu}}'' \frac{Q_\mu Q_\nu}{R^{m\,n}_{\mu\,\nu}} \tag{7.17a}$$

$$\begin{aligned} \Phi^{(3)} = \sum_{\substack{m\,n\,p \\ \mu\,\nu\,\varkappa}}''' \Bigg\{ & Q_\mu Q_\nu \chi_\nu \frac{(\boldsymbol{R}^{m\,n}_{\mu\,\nu} \boldsymbol{R}^{n\,p}_{\nu\,\varkappa})}{(R^{m\,n}_{\mu\,\nu} R^{n\,p}_{\nu\,\varkappa})^3} - \frac{1}{2} Q_\varkappa Q_\mu \beta_\nu \times \\ & \times \frac{3(\boldsymbol{R}^{m\,n}_{\mu\,\nu} \boldsymbol{R}^{n\,p}_{\nu\,\varkappa})^2 - (R^{m\,n}_{\mu\,\nu} R^{n\,p}_{\nu\,\varkappa})^2}{(R^{m\,n}_{\mu\,\nu} R^{n\,p}_{\nu\,\varkappa})^5} \Bigg\} \end{aligned} \tag{7.17b}$$

$$\begin{aligned} \Phi^{(4)} = {} & \frac{1}{2} \sum'''' \frac{Q_\mu Q_\lambda \chi_\nu \chi_\varkappa}{(R^{m\,n}_{\mu\,\nu} R^{n\,p}_{\nu\,\varkappa} R^{p\,h}_{\varkappa\,\lambda})^3} \times \\ & \times \left\{3 \frac{(\boldsymbol{R}^{m\,n}_{\mu\,\nu} \boldsymbol{R}^{n\,p}_{\nu\,\varkappa})(\boldsymbol{R}^{n\,p}_{\nu\,\varkappa} \boldsymbol{R}^{p\,h}_{\varkappa\,\lambda})}{(R^{n\,p}_{\nu\,\varkappa})^2} - (\boldsymbol{R}^{m\,n}_{\mu\,\nu} \boldsymbol{R}^{p\,h}_{\varkappa\,\lambda})\right\}. \end{aligned} \tag{7.17c}$$

Essentially these expressions have been used by *Herpin* [53 H 1] for the calculation of the elastic constants in ionic crystals*; especially the deviations of the Cauchy- relations have been considered. However, *Herpin* does not use these simple classical expressions, but rather a quantum mechanical calculation of the above discussed interaction, similar to the derivation given for van-der-Waals-crystals on page 27. The depen-

* Eq. (7.17) as well as Herpin's equations do *not* describe *completely* the effects of the polarization in the interaction. In any case it is supposed, that we can look upon the ions as *point* charges in a homogeneous electrostatic field.

dence on the relative positions of the ions in Herpin's result is the same as given in (7.17), apart from Herpin's neglect of (7.17c). It can be shown that in parameterfree lattices (7.17b) gives no contribution to the energy per unit cell, and even to the elastic constants only the second (quadrupole-) term in (7.17b) gives a contribution.

Another approach to the interionic potential was given by *Löwdin* [48 L 1, 56 L 1]. He started quantum mechanical calculations by using one-electron-wave-functions for the ions, which are connected rigidly to the ions and which are not deformable. The antisymmetrized product of these one-electron-functions is then used for calculating the interionic forces. The antisymmetry requirement causes certain characteristic deviations from the point charge model. The investigation by *Löwdin* seems to be the most comprehensive one up till now. But it has certain disadvantages. The expressions are very complex and difficult to interpret. In a series of papers, *Lundquist* has tried to express the results of *Löwdin* by rather simple potential-functions [52 L 1, 55 L 2, 57 L 1, 59 L 1]. The main point is again the occurrence of three-body-potentials. Besides the usual two-body-potentials *Lundquist* finds that Löwdin's results can be represented by three-body-potentials of the form [57 L 1]*:

$$\Phi^{(3)} = e \sum_{\substack{m\,n\,p \\ \mu\,\nu\,\varkappa}} \left\{ f_\nu\,(R^{n\,p}_{\nu\,\varkappa}) \cdot \frac{Q_\mu}{R^{m\,n}_{\mu\,\nu}} - h_\nu (R^{n\,p}_{\nu\,\varkappa}) \cdot \frac{Q_\mu (\boldsymbol{R}^{n\,p}_{\nu\,\varkappa}, \boldsymbol{R}^{m\,n}_{\mu\,\nu})}{R^{n\,p}_{\nu\,\varkappa}\,(R^{m\,n}_{\mu\,\nu})^3} \right\}. \quad (7.18)$$

Here $f_\nu$ and $h_\nu$ are relatively complicated functions, which can be expressed by overlap-, Coulomb-, . . . integrals of the one-electron-wave-functions. *Lundquist* made it plausible, that the potential (7.18) also contains the polarization forces of *Herpin* (or at least parts of them). This can also be seen by an immediate comparison between the first term in (7.17b) and the second term in (7.18). On the other hand, (7.18) contains also overlap-effects, which are not included in (7.17). This short account shows already the complexity of three-body-forces in crystals. For a very consistent theory all the different terms have to be considered and that includes the knowledge of all the different parameters occuring in the potentials, as polarizabilities, ,,quadrupolarizabilities'', overlap-, Coulomb-integrals and so on. As these are unknown in general, it is often better to work with general expressions for the potential and to fit the parameters by a comparison with experiments, or to work with simple models, like the shell model e.g. This shall be discussed in the next section.

It has been shown, that all the different models used (like the shell-model [58 D 1, 59 C 1, 63 D 1] or the polarization-models of the russian school mainly [57 M 1, 61 T 1, 61 M 4, 62 C 1]) and also the above mentioned papers of *Herpin* and *Lundquist*, contain essentially the same physical ideas on the interionic forces [61 L 2, 65 L 3]. Also the results are in a nice agreement, at least qualitatively. There are some quantitative differences which arise mainly from the lack of knowledge of the

---

* In (7.18) we have adapted our writing of the symbols; but it is essentially the equ. (5) in [57 L 1].

different ionic parameters. This is also the main reason for the discrepancies between theory and experiment.

Let us discuss some general aspects of all these three-body-potentials. All these can be written additively by potentials of the general form*

$$\Phi^{(3)} = \frac{1}{3!} \sum_{\substack{mnp \\ \mu\nu\varkappa}} \overset{1}{\varphi}_{\mu\nu} \left(R^{mn}_{\mu\,\nu}\right) \overset{2}{\varphi}_{\nu\varkappa} \left(R^{np}_{\nu\,\varkappa}\right) \overset{3}{\varphi}_{\varkappa\mu} \left(R^{pm}_{\varkappa\,\mu}\right). \tag{7.19}$$

Here $\overset{1}{\varphi}_{\mu\nu}, \overset{2}{\varphi}_{\nu\varkappa}, \overset{3}{\varphi}_{\varkappa\mu}$ are arbitrary functions restricted only by the requirement that the sum converges. Further, *one* of the functions $\overset{1}{\varphi}, \overset{2}{\varphi}, \overset{3}{\varphi}$ can be a constant. E.g. this is the case for the potentials in (7.17b) and (7.18). It is always possible to eliminate „angle-dependent" parts $\left(\boldsymbol{R}^{mn}_{\mu\,\nu}, \boldsymbol{R}^{np}_{\nu\,\varkappa}\right)$ by (7.11) and to write also the „angle-dependent" potentials in the form (7.19), though it might be inconvenient in some cases. Of course, the eq. (7.11) for the arguments of the functions has to be satisfied. The functions $\overset{1}{\varphi}_{\mu\nu}, \ldots$ must even not be symmetric with respect to the subscripts $\mu\nu$. This can be seen also with the examples given in eqs. (7.17b) and (7.18).

Some general considerations can be done, starting with (7.19). In order to have a more or less complete model, we have to add the two-body-forces

$$\Phi^{(2)} = \frac{1}{2} \sum_{\substack{mn \\ \mu\nu}} \varphi_{\mu\nu} \left(R^{mn}_{\mu\,\nu}\right). \tag{7.20}$$

We discuss the sum of these two potentials (7.19, 20). If there are different kinds of three-body-potentials, we have only to replace the resulting expressions by the corresponding sums of three-body-potentials, because they are additive if they are of the form (7.19). What we will investigate mainly are the elastic constants. This illustrates already the general procedure. Other quantities can be calculated in just a similar way.

For this purpose, we have to calculate all the quantities discussed in sect. 6. The first order cp's are the derivatives of (7.19, 20) with respect to $X^{m}_{\underset{i}{\mu}} = X^{M}_{i}$ **. To have a simpler writing, we introduce the operator

$$O = \frac{1}{R} \frac{\partial}{\partial R}. \tag{7.21a}$$

* We define $\overset{1}{\varphi}_{\mu\nu}(0) = \overset{2}{\varphi}_{\nu\varkappa}(0) = \overset{3}{\varphi}_{\varkappa\mu}(0) \equiv 0$ and also all the derivatives with zero argument. This is to have no difficulties in the summation. (7.19) is not the most general form of three-body-potentials. Other forms can be thought of, for oxample $\{A + (R^{mn} R^{np})^{\nu}\}^{-1}$, $A$ and $\nu$ being constants. But it seems, that in physically reasonable models, which can be gotten by perturbation treatment, the potentials have the form (7.19).

** If possible, we take again $^{m}_{\mu}$ together as $M$, in order to save space. It is no difficulty for the reader, to remark where $M$ means $^{m}_{\mu}$.

If we write $O\varphi_{\mu\nu}$, this symbol means, that the operator $O$ acts on the function *just* behind $O$ and only on this function, and the differentiation is with respect to the argument of $\varphi_{\mu\nu}$, therefore

$$O\,\varphi_{\mu\nu}(R^{MN})\,\chi_{\nu\varkappa}(R^{NP}) = \left[\frac{1}{R^{MN}}\frac{\partial\varphi_{\mu\nu}(R^{MN})}{\partial R^{MN}}\right]\cdot\chi_{\nu\varkappa}(R^{NP})\,. \quad (7.21\text{b})$$

A simple calculation then leads to the expression*:

$$\frac{1}{V}\sum_M \Phi_i^M X_j^M = \frac{1}{2V}\sum_{MN} X_j^M X_i^{MN}\{O\,\varphi_{\mu\nu}(R^{MN}) + O\,\varphi_{\nu\mu}(R^{MN})\} +$$
$$+\frac{1}{3!\,V}\sum_{MNP} X_j^M X_i^{MN}\Big\{O\,\overset{1}{\varphi}_{\mu\nu}\,\overset{2}{\varphi}_{\nu\varkappa}\,\overset{3}{\varphi}_{\varkappa\mu} + O\,\overset{1}{\varphi}_{\nu\mu}\,\overset{2}{\varphi}_{\mu\varkappa}\,\overset{3}{\varphi}_{\varkappa\nu} +$$
$$+ \text{cyclic terms in the supercripts } 1, 2, 3\Big\}\,.$$

Now this expression can be written down once again with $M$ and $N$ (and also $\mu$ and $\nu$ of course) interchanged. Then nothing happens in the braces. We can add the two expressions and take the halfth to have

$$\frac{1}{V}\sum_M \Phi_i^M X_j^M = \frac{1}{2\cdot 2V}\sum_{MN} X_i^{MN} X_j^{MN}\{O\,\varphi_{\mu\nu} + O\,\varphi_{\nu\mu}\} +$$
$$+\frac{1}{2\cdot 3!\,V}\sum_{MNP} X_i^{MN} X_j^{MN}\Big\{O\,\overset{1}{\varphi}_{\mu\nu}\,\overset{2}{\varphi}_{\nu\varkappa}\,\overset{3}{\varphi}_{\varkappa\mu} +$$
$$+ O\,\overset{1}{\varphi}_{\nu\mu}\,\overset{2}{\varphi}_{\mu\varkappa}\,\overset{3}{\varphi}_{\varkappa\nu} + \text{cyclic terms}\Big\}\,.$$

The braces are symmetric with respect to $\mu \Leftrightarrow \nu$, $M \Leftrightarrow N$, and so are the $X_i^{MN}$. Therefore we can interchange the indices in the second halfth of terms to have with (6.8)

$$S_{ij} = S_{ij}^{(2)} + S_{ij}^{(3)}$$
$$S_{ij}^{(2)} = \frac{1}{2V}\sum_{MN} X_i^{MN} X_j^{MN}\cdot O\,\varphi_{\mu\nu}(R^{MN})$$
$$S_{ij}^{(3)} = \frac{1}{3!\,V}\sum_{MNP} X_i^{MN} X_j^{MN}\Big\{O\,\overset{1}{\varphi}_{\mu\nu}\,\overset{2}{\varphi}_{\nu\varkappa}\,\overset{3}{\varphi}_{\varkappa\mu} + \overset{1}{\varphi}_{\varkappa\mu}\,O\,\overset{2}{\varphi}_{\mu\nu}\,\overset{3}{\varphi}_{\nu\varkappa} + \quad (7.22)$$
$$+ \overset{1}{\varphi}_{\nu\varkappa}\,\overset{2}{\varphi}_{\varkappa\mu}\,O\,\overset{3}{\varphi}_{\mu\nu}\Big\}\,.$$

For vanishing stresses $S_{ij}$ in the initial state, the sum of $S_{ij}^{(2)} + S_{ij}^{(3)}$ has to vanish. This is the equilibrium condition, or it is equivalent with it, resp. This holds for finite lattices, but can be extrapolated to infinite lattices (see sect. 6).

On the other hand, the usual equilibrium for infinite lattices is the minimum of the lattice energy per unit cell (see detailed discussion in I), with respect to variations of the basis-vectors $A_{ij}$ and cell vectors $X_j^\mu$:

$$\frac{\partial\varepsilon_z}{\partial A_{ij}} = 0\,; \quad \frac{\partial\varepsilon_z}{\partial X_j^\mu} = 0\,. \quad (7.23)$$

* Argument of $\overset{1}{\varphi}_{\mu\nu}$ is always $R^{MN}$, of $\overset{1}{\varphi}_{\nu\varkappa}$ always $R^{NP}$, etc.

The energy per cell can be gotten from (7.19, 20) by dropping the sum over $\boldsymbol{m}$ (not over $\mu$). This gives just the energy of the cell $\boldsymbol{m}$. But in infinite homogeneous lattices this must be independent of $\boldsymbol{m}$. In the moment, to have a better comparison with (7.22), we will keep the index $\boldsymbol{m}$. A simple, but rather lengthy calculation gives

$$\frac{\partial \varepsilon_z(\boldsymbol{m})}{\partial A_{jk}} = \frac{1}{2} \sum_{\substack{\boldsymbol{n} \\ \mu\nu}} (\boldsymbol{m} - \boldsymbol{n})_k \underset{j}{X}{}^{\boldsymbol{m}\boldsymbol{n}}_{\mu\nu} \Big\{ O\, \varphi_{\mu\nu}\Big(R^{\boldsymbol{m}\boldsymbol{n}}_{\mu\nu}\Big) + \\ + \frac{1}{3} \sum_{\boldsymbol{p}\varkappa} \Big[ O\, \overset{1}{\varphi}_{\mu\nu} \cdot \overset{2}{\varphi}_{\nu\varkappa} \cdot \overset{3}{\varphi}_{\varkappa\mu} + \text{two cyclic terms} \Big] \Big\} \tag{7.24}$$

$$\frac{\partial \varepsilon_z(\boldsymbol{m})}{\partial \underset{j}{X}_{\mu}} = \frac{1}{2} \sum_{\boldsymbol{n}\nu} \underset{j}{X}{}^{\boldsymbol{m}\boldsymbol{n}}_{\mu\nu} \Big\{ O\, \varphi_{\mu\nu}\Big(R^{\boldsymbol{m}\boldsymbol{n}}_{\mu\nu}\Big) + \frac{1}{3} \sum_{\boldsymbol{p}\varkappa} \Big[ O\, \overset{1}{\varphi}_{\mu\nu}\, \overset{2}{\varphi}_{\nu\varkappa}\, \overset{3}{\varphi}_{\varkappa\mu} + \text{cyclic terms} \Big] \Big\} - \\ - \frac{1}{2} \sum_{\boldsymbol{n}\nu} \underset{j}{X}{}^{\boldsymbol{m}\boldsymbol{n}}_{\mu\nu} \Big\{ O\, \varphi_{\nu\mu}\Big(R^{\boldsymbol{m}\boldsymbol{n}}_{\mu\nu}\Big) + \frac{1}{3} \sum_{\boldsymbol{p}\varkappa} \Big[ O\, \overset{1}{\varphi}_{\nu\mu}\, \overset{2}{\varphi}_{\mu\varkappa}\, \overset{3}{\varphi}_{\varkappa\nu} + \text{cyclic terms} \Big] \Big\}.$$

We will show, that if (7.24) is satisfied, (7.22) is too. Therefore form

$$\sum_{k} A_{ik} \frac{\partial \varepsilon_z(\boldsymbol{m})}{\partial A_{jk}} + \sum_{\mu} \underset{i}{X}_{\mu} \frac{\partial \varepsilon_z(\boldsymbol{m})}{\partial \underset{j}{X}_{\mu}} \tag{7.25}$$

$$= \frac{1}{2} \sum_{\substack{\boldsymbol{n} \\ \mu\nu}} \underset{i}{X}{}^{\boldsymbol{m}\boldsymbol{n}}_{\mu\nu} \underset{j}{X}{}^{\boldsymbol{m}\boldsymbol{n}}_{\mu\nu} \Big\{ O\, \varphi_{\mu\nu} + \frac{1}{3} \sum_{\boldsymbol{p}\varkappa} \Big[ O\, \overset{1}{\varphi}_{\mu\nu}\, \overset{2}{\varphi}_{\nu\varkappa}\, \overset{3}{\varphi}_{\varkappa\mu} + \text{cyclic terms} \Big] \Big\}.$$

In the procedure, some changes in the summation indices $\boldsymbol{n}$, $\boldsymbol{p}$, $\mu$, $\nu$, $\varkappa$ have been done. (7.25) vanishes in the case of zero initial stresses because of (7.24). If there are stresses in the initial state, (7.25) does not vanish for $\boldsymbol{m}$ being in the surface of the crystals. $\varepsilon_z(\boldsymbol{m})$ depends on the position of the cell in the crystal. Then, by summing (7.25) over $\boldsymbol{m}$ and dividing by the volume $V$ of the crystal, we just get the stresses (7.22), which originate from the surface contribution in (7.25) as well as in (7.22). Thus all these conditions are consistent with each other. Of course, the minimalization of $\varepsilon_z$ in the infinite crystal determines the $3(s+1)$ data of the *unit cell in infinite crystals* (3 lengths of basis-vectors, 3 angles between them and $3(s-1)$ components for relative positions in the cell). The corresponding relation in the finite crystal is $\underset{i}{\Phi}{}^{\boldsymbol{m}}_{\mu} = 0$ for all $\boldsymbol{m}$, $\mu$, $i$ or the minimum of $\varepsilon_z(\boldsymbol{m})$ for all $\boldsymbol{m}$. This defines the positions of *all the atoms in finite crystals* (apart from a translation or rotation of the whole system). This is discussed in detail in I.

The equilibrium being specified, it is a straightforward calculation to get the Born-Huang- tensors $\hat{C}_{ik,jl}$ and the elastic constants $C_{ij,kl}$. We will give the relations for parameterfree crystals here $\left(\hat{C}^{\mu}_{ijk} = 0\right)$. The non-primitive terms can be handled in the same way. Using (6.22), (7.22) and the fact that each ion is a center of inversion in parameterfree lattices we obtain with some lengthy transformations

$$\hat{C}_{ij,kl} = \delta_{ij} \cdot S_{kl} + \frac{1}{2V} \sum_{MN} X_i^{MN} X_j^{MN} X_k^{MN} X_l^{MN} O^2 \varphi_{\mu\nu} (R^{MN}) +$$

$$+ \frac{1}{3!\,V} \sum_{MNP} X_i^{MN} X_j^{MN} X_k^{MN} X_l^{MN} \times$$

$$\times \left\{ O^2 \overset{1}{\varphi}_{\mu\nu} \cdot \overset{2}{\varphi}_{\nu\varkappa} \cdot \overset{3}{\varphi}_{\varkappa\mu} + \text{cyclic terms} \right\} +$$

$$+ \frac{1}{2 \cdot 3!\,V} \sum_{MNP} X_i^{MN} X_j^{MP} \left\{ X_k^{MN} X_l^{MP} + X_l^{MN} X_k^{MP} \right\} \times$$

$$\times \left\{ O \overset{1}{\varphi}_{\mu\nu} \overset{2}{\varphi}_{\nu\varkappa} O \overset{3}{\varphi}_{\varkappa\mu} + O \overset{1}{\varphi}_{\nu\mu} O \overset{2}{\varphi}_{\mu\varkappa} \overset{3}{\varphi}_{\varkappa\nu} + \text{cyclic terms} \right\}. \qquad (7.26)$$

Because the sums are no longer surface-sensitive (compare sect. 6), we can replace $M = \binom{\boldsymbol{m}}{\mu}$ by $\binom{\boldsymbol{o}}{\mu}$, the summation over $\boldsymbol{m}$ giving $N$, which cancels the factor $N$ in $V = N V_z$.

The most important fact is the occurence of the term $\delta_{ij} S_{kl}$ in (7.26). It vanishes with zero stress in the initial state. But if $S_{kl} \neq 0$ it can be seen from (6.34) that it does not enter the elastic constants because it is just canceled out by the stress-terms in (6.34). Further, from (6.24c) it follows, that all terms in (7.26) apart from $\delta_{ij} S_{kl}$ must be symmetric with respect to $i \Leftrightarrow j$, $k \Leftrightarrow l$, and $(ij) \Leftrightarrow (kl)$. This can be seen by direct inspection. Thus, with (6.34) we have the elastic constants

$$C_{ij,kl} = \frac{1}{2V} \sum_{MN} X_i^{MN} X_j^{MN} X_k^{MN} X_l^{MN} \cdot O^2 \varphi_{\mu\nu} (R^{MN}) +$$

$$+ \frac{1}{3!\,V} \sum_{MNP} X_i^{MN} X_j^{MN} X_k^{MN} X_l^{MN} \left\{ O^2 \overset{1}{\varphi}_{\mu\nu} \overset{2}{\varphi}_{\nu\varkappa} \overset{3}{\varphi}_{\varkappa\mu} + \text{cyclic terms} \right\} +$$

$$+ \frac{1}{3!\,V} \sum_{MNP} X_i^{MN} X_j^{MN} X_k^{MP} X_l^{MP} \times$$

$$\times \left\{ O \overset{1}{\varphi}_{\mu\nu} \overset{2}{\varphi}_{\nu\varkappa} O \varphi_{\varkappa\mu} + O \overset{1}{\varphi}_{\nu\mu} O \overset{2}{\varphi}_{\mu\varkappa} \overset{3}{\varphi}_{\varkappa\nu} + \text{cyclic terms} \right\}. \qquad (7.27)$$

They have just the correct symmetries. They do not contain the stresses of the initial state explicitly. But, if stresses are present, the equilibrium positions $\boldsymbol{R}_{\mu\,\nu}^{mn}$, at which the derivatives have to be taken, are those which are defined by the initially applied stresses, and thus the elastic constants (7.27) depend implicitly on the initial stresses!

It should be mentioned here, that this method is essentially that which was used by *Bauer* [55 B 4] to prove the Born-Huang-conditions. He represents the potential energy by a sum of arbitrary two-, three-, four-, . . . body-potentials and investigates the equilibrium conditions and the expressions of the elastic constants, ending with similar conclusions as we have stated in the simple example of three-body-potentials only.

The most interesting quantity is the deviation from the Cauchy-relation. Only the last term in (7.27) enters, the first two being totally symmetric in $i, j, k, l$. We consider the simplest case of cubic crystals only. Because of cubic symmetry we have (in parameterfree lattices)

$$\sum_{\text{all } N \text{ with equal distance to } M} X_i^{MN} X_j^{MN} \times \text{function of distances only} \tag{7.28}$$

$$= \frac{1}{3}\,\delta_{ij} \sum_{\text{all } N \text{ with equal distance to } M} (R^{MN})^2 \times \text{function of distances only}$$

thus (with Voigts notation)

$$C_{11,22} - C_{12,12} = c_{12} - c_{44} = \frac{1}{9} \cdot \frac{1}{3!\,V} \sum_{MNP} (R^{MN})^2\,(R^{MP})^2 \times \tag{7.29}$$

$$\times \left\{ O\overset{1}{\varphi}_{\mu\nu}\,\overset{2}{\varphi}_{\nu\varkappa}\,O\,\overset{3}{\varphi}_{\varkappa\mu} + O\overset{1}{\varphi}_{\nu\mu}\,O\overset{2}{\varphi}_{\mu\varkappa}\,\overset{3}{\varphi}_{\varkappa\mu} + \text{cyclic terms} \right\}.$$

With this formula the difference between $c_{12}$ and $c_{44}$ can be calculated, if the potential is known. Using for example the potential (7.13b) for equal ions, we arrive after some simple transformations at

$$c_{12} - c_{44} = \frac{(Q^2\,\overline{r^2})^3}{24\,I^2 V_z} \sum_{np} \left\{ \frac{1}{(R^n R^{np} R^p)^3} - \frac{5}{2}\,\frac{R^p}{(R^n R^{np})^5} + \frac{23}{3}\,\frac{1}{(R^n)^5\,(R^{np})^3\,R^p} \right\}. \tag{7.30}$$

This should be the deviation from Cauchy's relation for van-der-Waals-crystals (at least one of the contributions). The main difficulty is the evaluation of the sums, which are two-center-sums. They have not been tabulated up till now; a rough estimate, using (7.13b, 7.30), shows, that the deviation* $c_{12} - c_{44}$ in van der Waals-crystals should be positive and in order of magnitude between $10^7$ dyn/cm$^2$ (Ne) and $3 \cdot 10^8$ dyn/cm$^2$ (Xe). The corresponding elastic constants are about $2 \cdot 10^{10}$ dyn/cm$^2$. So the values for $c_{12} - c_{44}$ seem to be rather small. Probably there are other three-body-effects (overlap) which make the difference larger.

Similar calculations can be done for the other potentials. If these are such, that one of the functions $\overset{1}{\varphi}, \overset{2}{\varphi}, \overset{3}{\varphi}$ is a constant, it can easily be seen, that only one-center sums occur, which in most cases are tabulated.

We will give here some of the results that have been gotten using different models for the potential in ionic crystals. Table 1 shows the theoretical values for $c_{12} - c_{44}$ from different authors in comparison to the experimental data (extrapolated linearly from $T >$ Debye-temperature to $T = 0$, i.e. harmonic theory values). The results of *Löwdin* and *Lundquist*, who use the same model essentially give approximately (within a factor of two) good values and the correct sign, whereas Herpin's results give the wrong sign. As Herpin's values are related to a charge-quadrupole-interaction mainly, while Löwdin-Lundquist's model contains also quantum-mechanical effects (overlap) we may conclude

* This is for the elastic constants in harmonic approximation, i.e. the „high temperature" values above Debye-temperature linearly extrapolated to $T = 0$.

that this last interaction gives the main contribution to the deviation from two-body-interactions in ionic crystals.

Also the generalized shell-model (next section) contains a part corresponding to charge-multipole interaction and a part related to overlap effects [63 D 1]. Therefore this should be quite useful in calculating the difference $c_{12} - c_{44}$. On the other hand, the models of

Table 1. $c_{12} - c_{44}$ in $10^{11}$ dyn cm$^{-2}$

| Author | LiF | LiCl | NaCl | KCl | KBr | KJ |
|---|---|---|---|---|---|---|
| experimental | −0.95 | −0.43 | −0.16 to −0.36 | −0.12 | −0.04 | −0.13 |
| *Löwdin* [50 L 1] | −1.938 | −0.516 | −0.543 | −0.347 | | |
| *Lundquist* [52 L 1] | | | −0.483 | | | |
| *Lundquist* [55 L 2] | | | −0.456 | | | |
| *Herpin* [53 H 1] | | | +0.083 | +0.017 | +0.023 | +0.032 |
| *Dick* [63 D 1] | +0.326 | +0.309 | +0.154 | −0.233 | −0.118 | −0.014 |

*Lundquist* and the shell-model, and also some other models (*Lax* [61 L 7, 65 L 3], *Tolpygo, Mashkevich* [59 M 1, 61 T 1]) are nearly equivalent. This has been shown by *Lundquist* [61 L 2] and follows also by the physics underlying these models. All these models, for example, lead to an effective ionic charge $e^*$ in the Szigeti-relation

$$9 \frac{\varepsilon_0 - \varepsilon_\infty}{(\varepsilon_\infty + 2)^2} = \frac{4\pi N (Z e^*)^2}{\mu\, \omega_0^2}\,, \tag{7.31}$$

which relates the dielectric constants ($\varepsilon_0$ static; $\varepsilon_\infty$ high frequency) to the maximum absorption frequency $\omega_0$ in the infrared ($\mu$: reduced mass in a lattice with two components; $N$: number of unit cells per unit volume). But also other types of potentials would give rise to an effective charge, in the same order of magnitude. Therefore, from this behavior one cannot decide which is the most appropriate model.

The use of expressions for the potential has certain advantages. Once the potential is given, we can calculate all the physical quantities wanted in a straightforward manner. Contrary, in the shell model, one has to eliminate the shell-coordinates, which is somewhat tedious in some cases. On the other hand, the shell-model is very illustrative and shows the physical aspects underlying the interaction concepts. Therefore we will discuss it in some detail in the following.

## 8. The Shell-model

The shell-model, first introduced by *Dick* and *Overhauser* [58 D 1], was thought to describe the polarization effects of the ions, mainly in alkali-halides. It has been used also for other crystals more or less successfully. The applications of the model concern mostly the description of the optical properties, as infrared absorption and related phenomena. We will not give a complete description of the different applications, but only discuss the model and the parameters involved.

The essential point is the division of the ion into a core and a shell, both with an effective charge the sum of which is equal to the ionic charge. The mass of the ions is concentrated in the core, assuming it as a mass-point, the shell is supposed to have the electronic or nearly

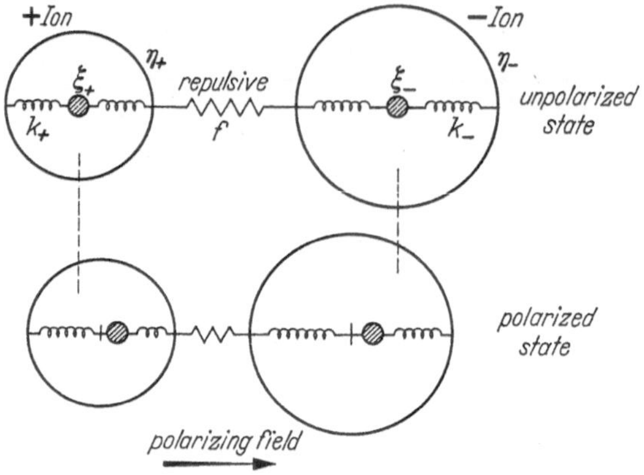


Fig. 3.4. Definition of parameters and constants in the simplest form of the shell-model

zero mass. The forces between core and shell and neighboring shells are represented by springs. Fig. 3.4 shows the simplest model. An electric field polarizes the ions, i.e. core and shell are displaced against each other. In this simplest model the shell is supposed to be rigid, and neighboring shells may not overlap; then there is no deformation and change in overlap during polarization or elastic distortion. The polarizability is related to the spring-constants $k$. A similar model is due to *Hanlon* and *Lawson* [59 H 1]; *Havinga* [60 H 2] compared these two models and improved them somewhat.

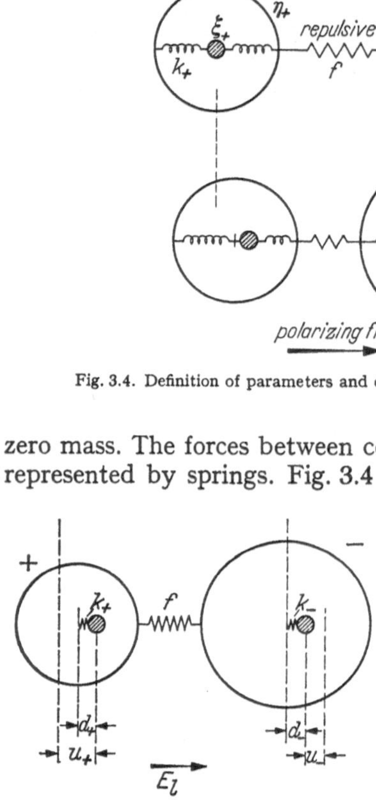


Fig. 3.5. Displacements of cores and shells in an electric field.

To obtain the relation between the shell-model-constants and the optical and elastic data of the crystals, we have to write down the equations of motion and to solve them by eliminating the shell-coordinates. Fig. 3.5 explains the choice of the coordinates. The springs $k$ in Fig. 3.4 have been replaced by a spring which connects the core and the center of the shell. If the time dependence of the motion is chosen to be $\exp\{-i\omega t\}$, the equations of motion are for the cores:

$$
\begin{aligned}
-M_+\,\omega^2\,\boldsymbol{u}_+ &= k_+ d_+ + \xi_+\,eE_l \\
-M_-\,\omega^2\,\boldsymbol{u}_- &= k_- d_- + \xi_-\,eE_l
\end{aligned}
\tag{8.1}
$$

and for the shells:

$$0 = -k_+ d_+ + f(u_- - u_+ + d_- - d_+) + \eta_+ eE_l$$
$$0 = -k_- d_- + f(u_+ - u_- + d_+ - d_-) + \eta_- eE_l \tag{8.2}$$

because the shell is assumed as massless. $\xi_\pm$, $\eta_\pm$ are the fractional charges of cores or shells, resp. $E_l$ is the local field in the crystal, which is related to the external applied field by

$$E_l = E_0 + \frac{4\pi}{3} \cdot P,$$

$P$ being the polarization. Eq. (8.2) can be used to eliminate the shell coordinates. From (8.1) and (8.2) we obtain

$$\mu\omega^2 (u_+ - u_-) = f(u_+ - u_- + d_+ - d_-) + Q_+ eE_l \tag{8.3}$$

$\mu$ is the reduced mass, $Q_+ = -Q_- = \xi_+ + \eta_+ = -\xi_- - \eta_-$ is the charge of the ions. The dipole moment per ion pair is

$$P = eQ_+(u_+ - u_-) + \eta_+ d_+ + \eta_- d_- \tag{8.4}$$

and therefore the polarizability per pair of ions, $P = \alpha E_l$,

$$\alpha = e^2 \frac{\xi_+^2/k_+ + \xi_-^2/k_- + Q_+^2/f - \mu\omega^2 \{\eta_+^2/fk_+ + \eta_-^2/fk_- + (\eta_+ + \eta_-)^2/k_+k_-\}}{1 - \mu\omega^2\{1/f + 1/k_+ + 1/k_-\}}. \tag{8.5}$$

The equation of motion for the relative displacements then is

$$-\mu\omega^2(u_+ - u_-) = -\frac{f}{1 + f \cdot \frac{k_+ + k_-}{k_+ k_-}}(u_+ - u_-) -$$
$$-\left\{Q_+ + f\frac{k_- \eta_+ + k_+ \eta_-}{f(k_+ + k_-) + k_+ k_-}\right\} eE_l. \tag{8.6}$$

If $k_\pm \Rightarrow \infty$ (rigid ions), the formulae become identical with those of the point-ion-model [e.g. I].

Eqs. (8.4–6) can be used for a determination of dielectric constants (static and high frequency) and the infrared dispersion, if the shell-constants are known. Or the constants can be obtained by a comparison with these macroscopic quantities. Three possibilities have been used:

i) Consider a free ion. The equation of motion for the shell is

$$\eta_\pm eE - k_\pm d_\pm = -\eta_\pm m\omega^2 d_\pm \tag{8.7}$$

$E = E_0 \exp(i\omega t)$ is the applied field, $m$ the mass of one electron in the shell. It is

$$P = \eta_\pm ed_\pm = \frac{(\eta_\pm e)^2}{k_\pm - \eta_\pm m\omega^2} E = \alpha_\pm^f \cdot E, \tag{8.8}$$

where $\alpha_\pm^f$ is the polarizability of the free ions. For small frequencies $(\omega \to 0)$

$$\alpha_\pm^f = (\eta_\pm e)^2/k_\pm; \tag{8.9}$$

this gives the charges of the shells from the free ion polarizability, if $k_\pm$ is known. *Dick* and *Overhauser* used (8.9) with some additional assumptions to determine $\eta_\pm$.

ii) In the ultraviolet region, the ion cores can be looked upon as fixed. The equation of the anion-shell is

$$\eta - eE_l - k_- d_- - f(d_- - d_+) = -m\omega^2 d_- \eta_- . \qquad (8.10)$$

The anions are responsible for the ultraviolet absorption. Neglecting $|d_+| \ll d_-|$, which is allowed in Li- and Na-halides, we obtain for the polarizability

$$\alpha_-^\infty = e^2 \eta_- d_- /(eE_l) = \frac{\eta_-^2}{k_- + f - m\,\omega^2\,\eta_-} . \qquad (8.11)$$

The dielectric constant, which is related to the absorption, satisfies for high-frequencies

$$\alpha_- \approx \alpha_+ + \alpha_- = \frac{3}{4\pi N} \cdot \frac{\varepsilon_\infty - 1}{\varepsilon_\infty + 2} , \quad \alpha_+ \ll \alpha_- , \qquad (8.12)$$

where $N$ is the number of ion pairs per unit volume. The absorption is maximal for $\varepsilon_\infty \to \infty$, i.e.

$$\alpha_- \approx 3/(4\pi N) .$$

Inserting this into (8.11) we have with (8.9)

$$\eta_-^2 \left\{\frac{4\pi}{3} N e^2 - e^2 / \alpha_-^f\right\} + \eta_- m\omega^2 - f = 0 . \qquad (8.13)$$

This gives a determination of $\eta_-$, provided $\alpha_-^f$ (free ions) and $f$ (Born-Mayer-potential, elastic constants) are known. Then also through (8.9) $k_-$ can be determined. Only the larger values of the roots of (8.13) are reasonable, because the smaller ones would contradict the assumptions in neglecting $d_+$ in (8.10).

iii) The static polarizibility $(\omega \to 0)$ of the crystal is, according to (8.5)

$$\alpha_0 = e^2 \{\xi_+^2/k_+ + \xi_-^2/k_- + Q_+^2/f\} . \qquad (8.5\text{a})$$

Since $\alpha_+ \ll \alpha_-$ (Li-, Na-halides), also $\xi_+^2/k_+ \ll \xi_-^2/k_-$. If $f$ is known sufficiently well (Born-Mayer-potential), and if we use again (8.9)

$$\alpha_0 = \alpha_-^f (1 + Q_+/\eta_-)^2 + e^2 Q_+^2/f \qquad (8.14)$$

from which $\eta_-$ can be obtained.

Table 2. *Shell-constants for halide-ions*

| Ion | $\alpha_-^f$ (in $10^{-24}$ cm$^3$) (*Pauling*) | $\eta$- according to (8.9) | (8.13) | (8.14) | $k_-$ (in $10^6$ dyn/cm) according to (8.9) | (8.13) | (8.14) |
|---|---|---|---|---|---|---|---|
| $F^-$ | 1.04 | 4.8 | 0.9 | 3.0 | 5.11 | 0.18 | 2.00 |
| $Cl^-$ | 3.66 | 8.7 | 4.3 | 2.1 | 4.77 | 1.17 | 0.28 |
| $Br^-$ | 4.77 | 9.9 | 3.8 | 2.7 | 4.74 | 0.70 | 0.35 |
| $J^-$ | 7.10 | 11.3 | 4.0 | 3.0 | 4.15 | 0.52 | 0.29 |

Table 2 shows some values for $\eta_-$, determined according to the three procedures. The agreement between the methods is not very good. The reason might be, that some of the above assumptions fail; on the other hand it might be necessary to generalize the simple shell model; this has been done in some respects.

*Cochran* [59 C 1] generalized the shell model by introducing interactions (in form of springs) between cores and shells of neighboring ions as sketched in Fig. 3.6. There are three additional parameters which must be fitted by comparison with experimental data. This model however, has so many parameters, that after fitting the experimental results all quantities should be described very well. The equations of motion are

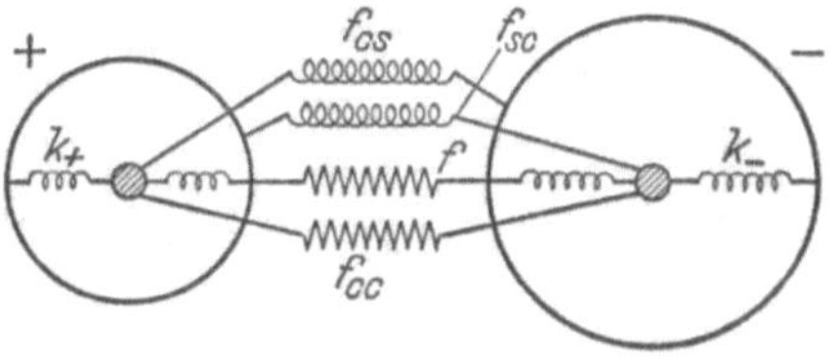


Fig. 3.6. Springs in the generalized shell-model

more complicated than those given by (8.1, 2), but they can be solved in a similar way. This model has been used widely for a calculation of the dispersion curves of alkali-halides and even homopolar crystals, and it describes the curves in a good approximation. Further generalizations concern the consideration of core- and shell-interactions between next-nearest neighbors and farther ones. The number of parameters increases so that nearly every experiment can be described sufficiently well by taking further neighbor interactions into account.

All these models contain effective constants, after the shell-coordinates have been eliminated. The effective constants, which describe the motions of cores (ionic displacements for the center of mass) are essentially due to central-force-interactions; this holds also for the polarization effects included. Thus from these models no deviations from the Cauchy relations for the elastic constants result (in lattices, where every ion is a center of inversion, harmonic theory). To obtain deviations from the Cauchy relations, we need many-body-potentials. Many-body-interactions can be included in the shell-model via "exchange charges", already introduced in *Dick* and *Overhausers* first paper [58 D 1] (fig. 3.7). This model has been used by *Dick* [63 D 1] to calculate the elastic constants of the alkali-halides and their deviations from the Cauchy relations ($c_{12} - c_{44} \neq 0$), in the harmonic approximation. Probably the exchange charges are not very essential for the optical effects, but for the Cauchy relations they are important, because in the framework of the shell-model they are the only interactions of many-body-character.

The exchange-charges are a consequence of the overlap between the shells of neighboring ions and the resulting exchange integral. If the ions are displaced, the charge distribution and the exchange integral alters. This leads to an effective polarization called "exchange charge polarization". If the ions come closer together, besides the electrostatic attraction there is a repulsion due to Pauli's principle, since because of the overlap electrons have to go into other states, or charges have to be brought away from the overlap region. These charges are supposed to be distributed homogeneously over the ions. The removal of the charges is described by small additional point charges in the overlap region, specially localized on the line connecting neighboring ion cores, at

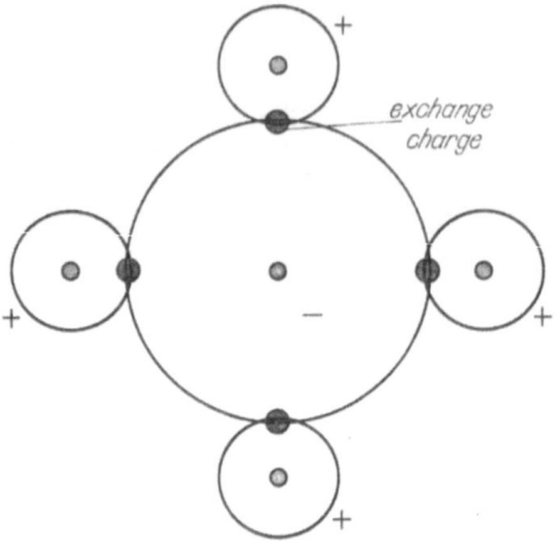


Fig. 3.7. Position of the exchange charges in the shell-model with overlapping shells

distance $r_+$ from the positive ion core in equilibrium state*. $r_+$ is the positive ion radius. In a strained crystal, the distance is

$$r_{q+} = r_+ \cdot a/a_0 \tag{8.15}$$

where $a$ is the strained, $a_0$ the unstrained nearest neighbor distance. The magnitude of the exchange charge is supposed to be

$$q = \frac{a}{\gamma e} B \, \mathrm{e}^{-a/\varrho} \tag{8.16}$$

where $B$ and $\varrho$ are the constants of the Born-Mayer-interaction

$$\varphi(r) = B \, \mathrm{e}^{-r/\varrho} .$$

$a$ is the nearest-neighbor-distance, $e$ the electronic charge and $\gamma$ a parameter to be fitted. Thus the repulsion of the exchange effect is related to the Born-Mayer-repulsion, which seems to be a plausible assumption. In order to preserve the over-all neutrality of the lattice, the ions (positive or negative) have modified charges

$$\pm e - \frac{1}{2} \sum_i q_i \tag{8.17}$$

where the sum extends over all the neighboring exchange charges $q_i$ of one ion.

The exchange charges change their magnitude (8.16) and position (8.15), when the crystal is strained. The parameters magnitude and position depend on the positions of the neighboring ions. Therefore the

* This implies, that neighboring positive and negative ions overlap. If next-nearest neighbors overlap, one has to choose other parameters.

interactions between exchange charges depend on the positions of at least three ions, thus leading to a many-body-character of the interactions and to deviations from the Cauchy relation. *Dick* [63 D 1] has calculated the elastic constants according to this model; the calculation is lengthy and tedious. We give some of his results in Table 1 for the difference $c_{12} - c_{44}$. The agreement with the experimental data is not very good, but not worse than other calculations. In Dick's paper it looks like a better agreement, because he compares with the experimental values of $c_{12} - c_{44}$ for room temperature; but these values contain anharmonic effects, which also produce deviations from the Cauchy relations. One should compare a harmonic theory with harmonic experimental values. These can be gotten by an appropriate extrapolation for the elastic constants [see 61 L 1 and chapter V]. The experimental values of Table 1 are the extrapolated (harmonic) ones. The table shows, that none of the models is sufficient, and that probably a number of different many-body-interactions are present; different signs for $c_{12} - c_{44}$ ($\gtrless 0$) occur experimentally also for some Rb- and Cs-halides. On the other hand, the many-body-effects discussed by the different authors are not completely independent from each other. To a certain extent, the models contain a similar physical background.

The shell-model could be improved even more by allowing the shells to deform during a displacement of the ions. This effect has been indicated also by *Dick* and *Overhauser*, and leads to a further many-body-interaction, similar to the interaction mentioned on page 31 and discussed by *Herpin* [53 H 1]. If at all, it is difficult to find the effect which is mostly responsible for the difference $c_{12} - c_{44}$ in alkali-halides. The effect of deformable shells is also investigated in a recent paper by *Schröder*; he can describe the dispersion curves of NaJ and KBr very well with a few parameters only [66 S 1].

The question arises whether the shell-model can be justified from a quantum-mechanical point of view. Such a foundation is indeed possible, but, of course, involves a number of assumptions and approximations. Most of the quantum-mechanical derivation has been done by *Tolpygo* and *Mashkevitch* [50 T 1, 56 M 3, 57 M 1, 59 M 1, 61 T 1, 61 M 4]. Starting with the complete Hamiltonian of the crystal, the problem is to bring the resulting lattice energy into a form, which allows for a power expansion in the ionic displacement and dipole components. From this then equations of motion for the ionic and dipole-coordinates can be derived, which have the form of the shell-model-equations. Of course, the adiabatic approximation of *Born* and *Oppenheimer* is involved in such a procedure. A similar treatment has been used already by *Yamashita* and *Kurosawa* [55 Y 1]. Later on *Cowley* [62 C 1] brought the work of the russian school into a simpler form, and derived formal expressions for the elastic constants and dielectric constants, but without the use of exchange charges (i.e. the Cauchy relations hold in this procedure).

As already mentioned, the shell-model was developed for an use in alkali-halide crystals, and has been proved to be successful in homopolar crystals, too, though it seems quite sure, that many-body-interactions play a far greater role in homopolar crystals than it can be expressed with the shell model. The situation is even more difficult in metals, where the energy of the normal vibrations is comparable with the energy of electronic excitations.

## 9. Interactions in Metals

The situation of interacting forces is most difficult in metals. The main point is the contribution of the free or nearly free electrons to the interactions. The first step has been done by *K. Fuchs* [35 F 1, 36 F 1],

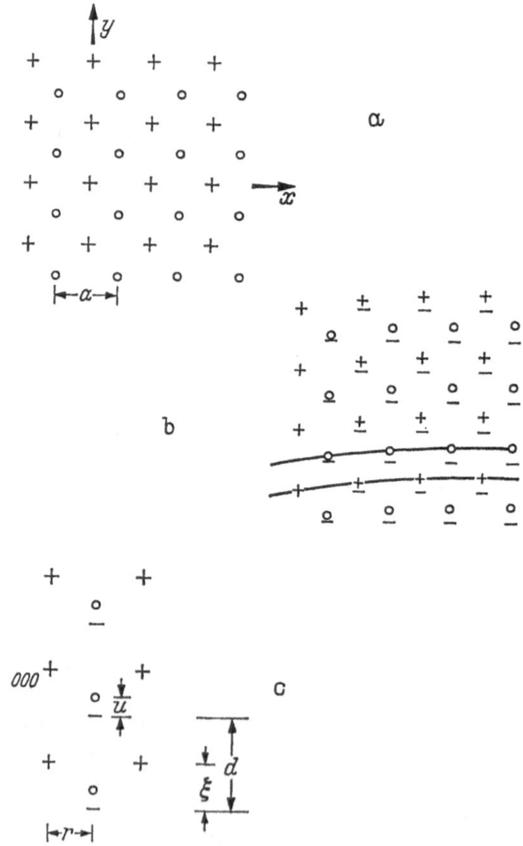


Fig. 3.8a — c. Positions of ions in a b.c.c. lattice. a) in equilibrium, + : in drawing plane, ○ : at distance $a/2$ before or behind the plane; b) with a transverse sound wave in $x$-direction, —: equilibrium positions, ○, + : instantaneous positions; c) definition of displacement parameters

who calculated the shear moduli (long wave part of transverse dispersion curves) with the assumption, that the electrostatic interaction of the positive ions in the sea of negative electrons gives the main contribution. This should be justified well in lattices, where the interionic distance is "large" compared to the ionic radii (alkali-metals).

The most promising recent work is that of *Toya* [53 T 1, 55 T 1, 58 T 1, 61 T 2, 65 T 1] and of *Vosko* et al. [65 V 1]. These authors started with a self-consistent treatment of the Hamiltonian, containing the ionic displacements as parameters, and using an expansion with respect to these parameters. But all the work is done in the adiabatic approximation. This might be questioned in metals. *Chester* [61 C 1] however conjectured that the adiabatic approximation should be justified for all the electrons, which are not close to the Fermi energy. This would not affect the cohesive energy and ionic interactions very much. But its influence should be larger for energies near the Fermi energy, where the electron-phonon-interaction might become essential. An effect had been predicted by *Kohn* [59 K 1]; as a consequence of this interaction together with the interaction of the conduction electrons, the group velocity of the phonons should have a logarithmic singularity at

$$\boldsymbol{q} = 2\boldsymbol{k}_{\mathrm{F}} + 2\pi B\boldsymbol{h}\,. \tag{9.1}$$

$\boldsymbol{q}$ is the phonon wave vector, $\boldsymbol{k}_{\mathrm{F}}$ the Fermi-momentum and $2\pi B\boldsymbol{h}$ a reciprocal lattice vector. *Woll* and *Kohn* [62 W 1], *Taylor* [63 T 1] and others then have done quantitative calculations of this effect. There is also some experimental evidence for the Kohn-effect in lead [61 B 2].

We will indicate the calculation of dispersion curves for pure transverse modes and afterwards discuss the extension to other cases. Suppose a cubic-body-centered lattice with a transverse wave in the $X$-direction (Fig. 3.8). We assume that the negative charge is distributed homogeneously over the lattice, whereas the positive point ions move in this charge distribution. When one plane of ions, say the plane $\frac{a}{2}\,(1h_2h_3)$ is displaced rigidly by $u_2^1$, there is a force on the ion 000, which can be described by the force of a plane of dipoles. This force can be calculated according to method of *Madelung* [I, II]. Having calculated the effective forces for every displaced plane, we can give the equation of motion for the transverse wave described above:

$$M\,\ddot{u}_2^{000} = -\sum_{h_2h_3}\left\{\Phi_{2\;2}^{0\,1\,h_2h_3}\,u_2^{1\,h_2h_3} + \Phi_{2\;2}^{0\,2\,h_2h_3}\,u_2^{2\,h_2h_3} + \cdots\right\}. \tag{9.2}$$

In this case, $u_2^{1\,h_2h_3}, \ldots$ are independent of $h_2$, $h_3$ and introducing

$$\sum_{h_2h_3}\Phi_{2\;2}^{0\,1\,h_2h_3} = \Phi_{22}^{01}, \ldots, \tag{9.3}$$

using this and the solution of (9.2) for the transverse mode with $\boldsymbol{q} = (q, 0, 0)$,

$$u_2^{h_1h_2h_3} = e_2 \exp\{iqX^{h_1} - i\omega t\} \tag{9.4}$$

we obtain the squared frequencies

$$M\omega^2 = \sum_{h_1} \Phi_{2\,2}^{0\,h_1} \exp(iqX^{h_1}) = \sum_{h_1} \Phi_{2\,2}^{0\,h_1} \cos q\,X^{h_1}$$
$$= -2 \sum_{h_1} \Phi_{2\,2}^{0\,h_1} \sin^2 (qX^{h_1}/2) \tag{9.5}$$

because (sect. 2)

$$\sum_{h_1} \Phi_{2\,2}^{0\,h_1} = \sum_{h_1 h_2 h_3} \Phi_{2\quad 2}^{0\,h_1 h_2 h_3} = 0\,.$$

The problem is only to calculate the effective force-constants (9.3) with the help of Madelung's sums. Of course, when doing this and using it all other interactions, e.g. between the ion cores, have been neglected (see below). The electrostatic potential energy at a positive charge at (000) of a line of dipoles at distance $r$ (Fig. 3.8c) is given by [I]:

$$\Psi(\xi, r) = -\frac{2e^2}{d} \sum_{\nu=0}^{\infty} K_0\left(2\pi\,|\nu|\,\frac{r}{d}\right) e^{i2\pi\nu\frac{\xi}{d}}\left\{1 - e^{-i2\pi\nu\frac{u}{d}}\right\}. \tag{9.6}$$

$d$ is the distance of positive charges on the line, $\xi$ the distance of the first lattice point on the line from zero (Fig. 3.8c). $K_0$ is a modified Hankel-function, decreasing essentially exponentially with increasing $r$. The second derivative of (9.6) at $u = 0$ gives the effective force constants, when the sum is taken over all the lines forming one of the rigidly displaced planes:

$$\Phi_{i\,i}^{\text{eff}} = \sum_{\text{all lines of a plane}} + \frac{8\pi^2 e^2}{d^3} \sum_{\nu \neq 0} \nu^2 e^{i2\pi\nu\xi/d} \cdot K_0\left(2\pi\,|\nu|\,\frac{r}{d}\right). \tag{9.7}$$

Table 3 contains some of these effective force-constants for a transverse motion.

Table 3. *Effective force-constants for transverse modes, electrostatic interaction only, in units $e^2/a^3$*

| | 1. neighbor-plane | 2. neighbor-plane | 3. neighbor-plane |
|---|---|---|---|
| b.c.c. | | | |
| $q = q\,(1, 0, 0)$ | —2.088 | +0.164 | —0.006 |
| $q = \frac{q}{\sqrt{2}}\,(1, 1, 0)$; $e = (0, 0, 1)$ | —0.756 | +0.0038 | —0.000005 |
| ; $e = \frac{1}{\sqrt{2}}\,(\bar{1}, 1, 0)$ | —0.109 | +0.0024 | —0.0003 |
| $q = \frac{q}{\sqrt{3}}\,(1, 1, 1)$ | —2.795 | —1.055 | +0.719 |
| f.c.c. | | | |
| $q = q\,(1, 0, 0)$ | —2.023 | +0.032 | —0.0004 |
| $q = \frac{q}{\sqrt{2}}\,(1, 1, 0)$; $e = (0, 0, 1)$ | —8.384 | +1,411 | —0.141 |
| ; $e = \frac{1}{\sqrt{2}}\,(\bar{1}, 1, 0)$ | —2.011 | +0.418 | —0.010 |
| $q = \frac{q}{\sqrt{3}}\,(1, 1, 1)$ | —0.569 | —0.0016 | +0.00003 |

We cannot expect, that the dispersion curves (fig. 3.9) are represented very well because we have neglected all other interactions. In alkali-metals the neglect might be justified to a large extent; but in noble metals the ion cores are relatively large compared to the neighbor-distance, so that the interactions (repulsion) between the ion-cores come into play. In principle, this repulsion can be represented by spring-constants, which can be fitted to experimental data.

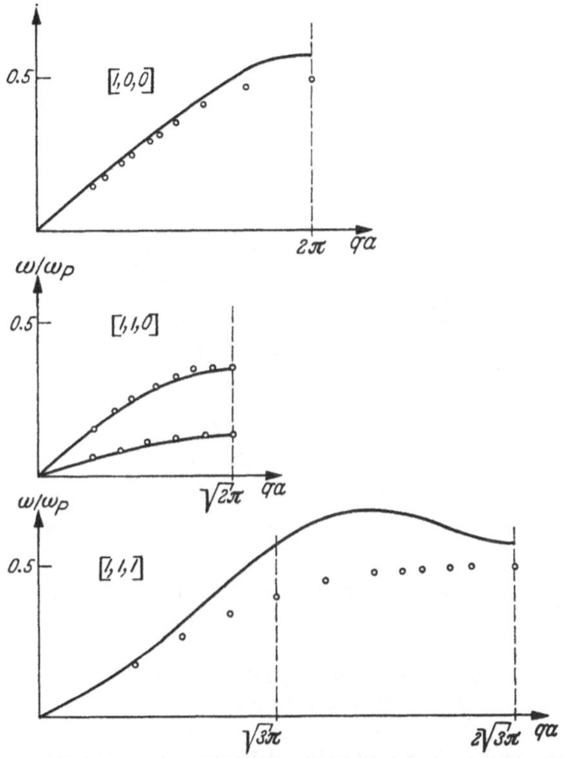


Fig. 3.9. Transverse dispersion curves for Na. Full lines according to the simple model (9.5—7). The agreement with experimental values (circles) according to [63 W 3] is quite good, which is not necessarily a merit of the model. Nevertheless, for transverse motions the electrostatic interaction should give the main contribution, apart from the [111]-direction perhaps, where the distance between neighboring atomic planes is very small. $\omega_p = (4\pi e^2/MV_z)^{1/2} = 4.52 \cdot 10^{13}$ sec$^{-1}$ for Na

The situation is extremely more difficult for the longitudinal modes or the longitudinal components of the modes. This has been investigated recently by *Toya* and *Vosko* et al. *Toya* starts with a Hamiltonian of the whole system, which contains the Coulomb-interactions of the electrons, the Coulomb-interactions between ions and electrons, and those between the ions and the overlap- (exchange) repulsion between the ions. The latter one is assumed to have the Born-Mayer-form $\exp(-r/\varrho)$. The ion coordinates are written as equilibrium positions plus small displacements $\boldsymbol{u}^m$ from them. The $\boldsymbol{u}^m$ are represented by the normal mode representation. Eigenfunctions and eigenvalues of the Hamiltonian are determined in a self-consistent way, but with a number of restrictions on the form of the wave-functions (orthogonalized plane waves). The energy is expanded up to the quadratic terms in the displacements, and from its dependence on the wave-number introduced via the normal-mode-representation, the dispersion-curves can be found. The

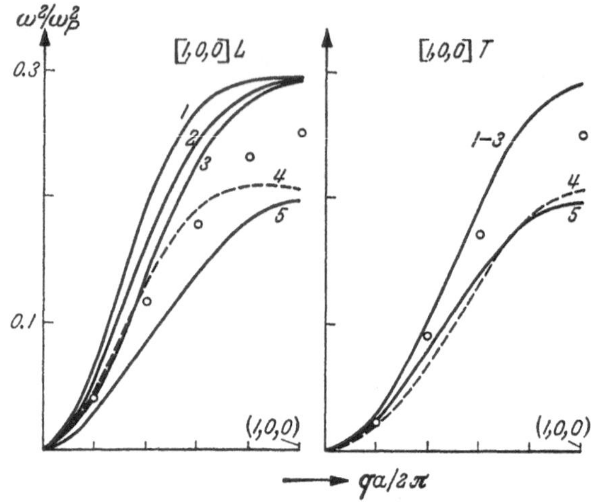


Fig. 3.10a

Fig. 3.10 a—c. Comparison of different models for the dispersion curves of Na. Experimental values according to [63 W 3], theoretical curves from [65 V 1]; 1—5 indicate different models, which differ mainly in the calculation of the electron-ion-interaction matrix elements and in the electron-screening-effect on the ion-ion-interaction. For $\omega_p$ se Fig. 3.9

calculations are somewhat lengthy; for the transverse modes the results are not so different from the simple calculations discussed above. The main advantage lies in the knowledge we get for the longitudinal modes.

*Vosko* et al. improved Toya's method especially by dropping the restrictions on the form of wave-functions. They allowed for general electronic wave-functions. In numerical calculations they give a comparison of the results obtained using different wave functions. Differences occur for large wave-numbers. Fig. 3.10 shows a comparison with experimental results of *Brockhouse* et al. [62 B 3] for Sodium. There is no preference for one of the special models for the wave-function.

We should mention a number of other papers, which are devoted to the same problem. *Sham* [65 S 1] calculated the dispersion curves of Na taking into account the conduction electrons and the electron-phonon-scattering with a pseudopotential in a Hartree-Fock-procedure. In a similar way a number of calculations by *Harrison* [63 H 1, 64 H 1, 65 H 2] are done for several metals. A very simple method giving surprisingly good results, has been used by *Krebs* [65 K 1] and *Mahesh* and *Dayal* [66 M 1]. They used arbitrary force-constants for the repulsive interaction between neighbors, or next-nearest neighbors and represented the electronic contribution by an interaction which results from a pseudopotential assumption. The parameters are fitted to the elastic constants and to the maximum frequency. The dispersion curves calculated then agree well with experimental ones.

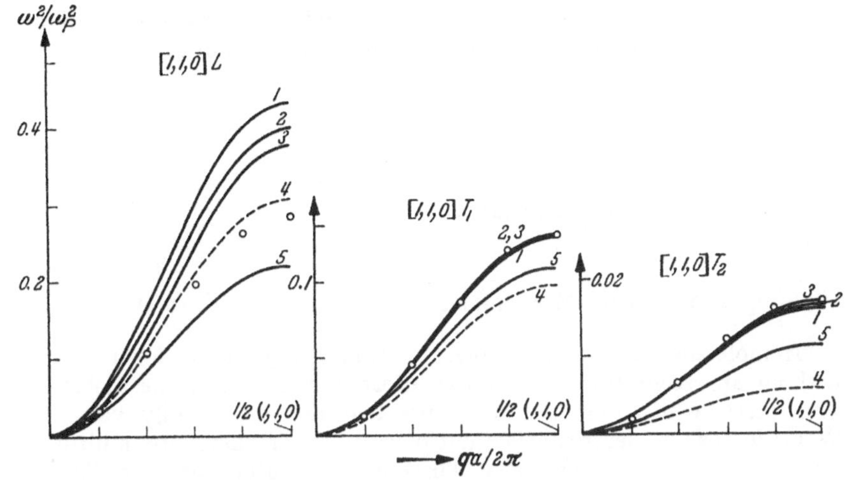


Fig. 3.10 b

As mentioned above, all these calculations are done using the adiabatic approximation. Though the feeling is, that this is sufficient (apart from Kohn-effect etc.), it is still an open question, how far the validity

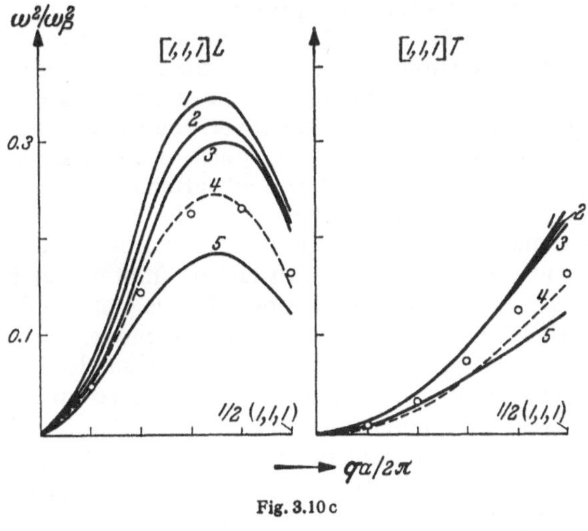


Fig. 3.10 c

of the adiabatic assumption is correct. Especially, if temperature effects are included (anharmonic effects) it is even more uncertain whether non-adiabatic contributions play a role. Most of the measurements have been done at finite temperatures. The results are compared with harmonic calculations (no temperature-effect included). From a comparison, force constants are fitted. Such a fit can be done sometimes only by including $10^{th}$ or higher neighbor-interactions. Statements of this kind should be doubted, because in a comparison between results at finite temperatures* and harmonic calculations differences should occur by means of anharmonic effects, which can mislead to an assumption of higher neighbor interactions than are present in reality.

# IV. Dynamics of Molecular Crystals

## 10. Simple Models for Molecular Motions in Perfect Crystals

In molecular crystals the molecules are put together to form a lattice without an essential change of the internal structure of the molecules. The forces, which hold the molecules together, are mainly van- der-Waals-forces or electrostatic multipole-forces, if there are multipoles associated with the molecules. Therefore the intermolecular distance is relatively large compared with the distance of the atoms in the molecules. Typical examples of such crystals are organic substances, e.g. a solid methane-crystal. These crystals are of course non-primitive crystals, or even non-parameterfree** crystals except in few cases.

In principle, the ionic motions in molecular crystals can be described in the same way as in other non-primitive lattices. But as the intramolecular binding energy of the atoms is large compared with the potential energy between the different molecules, the frequencies of the internal vibrations are large as compared with the frequencies connected with relative motions of the molecules. Moreover, the frequencies of the internal vibration will not be different very much from those of the internal vibrations of the free molecules. These vibrations will not be considered in the following, as we are interested only in lattice effects.

Those motions which are essentially modified by the arrangement of the molecules in crystals are the translations (motions of the center of mass of the molecules) and the rotations (because of the finite moment of inertia of the molecules). The translational modes can be supposed to be similar to those of ionic lattices. Rotational modes do not occur in ionic crystals (zero moment of inertia, as is supposed always). We must distinguish between two possible cases:

1. The molecules rotate freely or nearly freely (as in a gas e.g.), every orientation is possible, the angle of rotation is "large". Probably there is no strong correlation between the motions of different molecules.

---

* Even measurements at $T = 0$ contain anharmonic effects via the zero-point-motion, whereas the harmonic theory does not include this.

** In parameterfree crystals (see page 4) *every* ion must be a centre of inversion!

2. There exist characteristic equilibrium positions for the orientation of a molecule, which can be described by a minimum of a potential for the orientation of the molecules. If the rotational energy of the molecules is small (e.g. low temperatures), the molecules will oscillate or better librate about the equilibrium position of orientation.

In the following we will discuss only the second case, assuming small (infinitesimal) angles of libration. This corresponds to the case of small displacements as starting point for the theory of lattice vibrations. The theory of these small librations has been developed in a number of papers [63 B 1, 2; 64 C 1]. Our outline is based on these papers.

With the neglect of the internal degrees of freedom of the molecules, and under the assumption of small (infinitesimal) translations and librations of the molecules we can immediately give a set of "phenomenological" equations for the motions of the rigid *molecules:*

$$M \ddot{u}_i^{m} = -\sum_{nj} \Phi^{mn}_{\substack{v\,v\\ i\,j}} \cdot u_j^{n} - \sum_{nj} \Phi^{mn}_{\substack{v\,\omega\\ i\,j}} \cdot \omega_j^{n} \tag{10.1a}$$

$$I_i \ddot{\omega}_i^{m} = -\sum_{nj} \Phi^{mn}_{\substack{\omega\,v\\ i\,j}} \cdot u_j^{n} - \sum_{nj} \Phi^{mn}_{\substack{\omega\,\omega\\ i\,j}} \cdot \omega_j^{n}\,. \tag{10.1b}$$

Here we have assumed, that the principle axes of inertia of the equal molecules in equilibrium have the same orientation. $\boldsymbol{m}$ numbers the cell of the Bravais-lattice, $i$ the coordinate direction, as usual. $M$ is the total mass of a molecule, $I_i$ the moment of inertia connected with a rotation about an axis in direction $i$. We have supposed, that the tensor of the moment of inertia is diagonal, $I_{ij} = I_i \cdot \delta_{ij}$, which means, that we have chosen as the directions of the coordinate axes the directions of the principal axes of the tensor of the moment of inertia*.

Then $u_i^{m}$ are the displacements of the centers of mass in these directions and $\omega_i^{m}$ are the angles of rotation about these axes. As we have restricted the angles to infinitesimal ones, librations about different axes do commute!**

The coupling parameters describe the interaction between translational motions ($\Phi\, v v$), between librational motions ($\Phi\, \omega\omega$) and the coupling between both ($\Phi\, v\omega$, $\Phi\, \omega v$). Eqs. (10.1) can be derived from the equation of motion for all ions of a molecular crystal by an appropriate transformation [§ 11, 63 B 1]. Of course, the coupling constants have to satisfy certain symmetry and invariance requirements. As the potential energy belonging to (10.1) is a homogenous quadratic function in $u_i^{m}$, $\omega_i^{m}$, it follows immediately

$$\Phi^{mn}_{\substack{v\,v\\ i\,j}} = \Phi^{nm}_{\substack{v\,v\\ j\,i}}\;;\quad \Phi^{mn}_{\substack{v\,\omega\\ i\,j}} = \Phi^{nm}_{\substack{\omega\,v\\ j\,i}}\;;\quad \Phi^{mn}_{\substack{\omega\,\omega\\ i\,j}} = \Phi^{nm}_{\substack{\omega\,\omega\\ j\,i}}\,. \tag{10.2}$$

* This is not necessary, but it seems, that it is the most convenient way of describing the motions and solving the problem. The axes refer to the equilibrium orientation of the molecules.

** This theory is valid for small libration angles only, as has to be emphasized again. If the angles are large, or the molecules rotate nearly freely (as is the case perhaps in solid methane), the theory has to be improved, which is rather more involved. A case, in which the above theory fails, is the solid $H_2$-crystal.

Further, if we displace all the molecules of the crystal by the same amount $a_j$ no force is exerted on the crystal (in homogeneous space). Therefore it follows

$$\sum_n \Phi^{m\,n}_{\substack{v\\i}\,\substack{v\\j}} = \sum_m \Phi^{m\,n}_{\substack{v\\i}\,\substack{v\\j}} = 0; \quad \sum_n \Phi^{m\,n}_{\substack{\omega\\i}\,\substack{v\\j}} = \sum_m \Phi^{m\,n}_{\substack{v\\i}\,\substack{\omega\\j}} = 0\,. \tag{10.3}$$

Use has been made of (10.2). No relation for $\Phi^{m\,n}_{\substack{\omega\\i}\,\substack{\omega\\j}}$ results. The most important relation is that which results from the lattice structure of the crystal. Translation by multiples of basis-vectors does not change the displacement-pattern of the lattice. Therefore

$$\Phi^{m\,n}_{\substack{\alpha\\i}\,\substack{\beta\\j}} = \Phi^{m+h\;n+h}_{\substack{\alpha\\i}\;\;\substack{\beta\\j}} = \Phi^{0\;n-m}_{\substack{\alpha\\i}\,\substack{\beta\\j}} \quad \text{for } \alpha, \beta = v, \omega\,. \tag{10.4}$$

Further relations will be discussed in § 12. Also the point-group-symmetries give rise to relations. E.g., if the center of mass of every molecule is a center of inversion, we have

$$\Phi^{-m-n}_{\substack{v\\i}\;\;\substack{v\\j}} = \Phi^{m\,n}_{\substack{v\\i}\,\substack{v\\j}}; \quad \Phi^{-m-n}_{\substack{v\\i}\;\;\substack{\omega\\j}} = -\,\Phi^{m\,n}_{\substack{v\\i}\,\substack{\omega\\j}}; \quad \Phi^{-m-n}_{\substack{\omega\\i}\;\;\substack{\omega\\j}} = \Phi^{m\,n}_{\substack{\omega\\i}\,\substack{\omega\\j}}\,. \tag{10.5}$$

The minus-sign in the second relation results from the fact, that a displacement-vector changes sign under inversion, a rotation-vector (pseudovector) does not.

We will discuss the principle effects which might occur in molecular lattices with the simple example of an infinite linear arrangement of molecules (linear chain of molecules, fig. 4.1). This model shows already all the features of more realistic ones and is very simple to handle. A special property of it is the occurrence of another invariance relation, which is strongly connected with this one-dimensional model and which

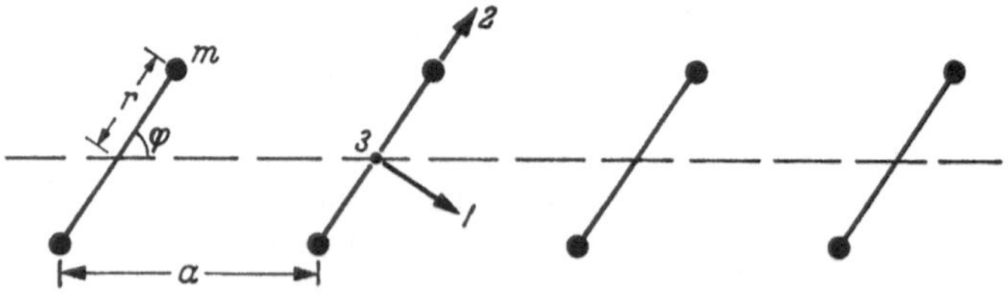


Fig. 4.1. Linear lattice of identical dumbbell molecules

is *not valid in three-dimensional models* (§ 12). For, if we rotate all the molecules by the same angle about the axis of the chain, through the centers of mass, no force is exerted on the molecules. Such an infinitesimal rotation can be represented by rotations about the molecules principal axes in the form

$$\omega_1 = \tau \sin\varphi; \quad \omega_2 = \tau \cos\varphi; \quad \omega_3 = 0\,, \tag{10.6}$$

$\tau$ being an arbitrary infinitesimal angle. Introducing this into (10.1), we have the relations

$$\sum_n \left\{\Phi\begin{smallmatrix} m & n \\ \omega & \omega \\ i & 1\end{smallmatrix} \sin\varphi + \Phi\begin{smallmatrix} m & n \\ \omega & \omega \\ i & 2\end{smallmatrix} \cos\varphi\right\} = 0\,, \tag{10.7a}$$

$$\sum_n \left\{\Phi\begin{smallmatrix} m & n \\ v & \omega \\ i & 1\end{smallmatrix} \sin\varphi + \Phi\begin{smallmatrix} m & n \\ v & \omega \\ i & 2\end{smallmatrix} \cos\varphi\right\} = 0 \quad \text{for all } i,\, m\,.$$

These relations simplify even more, if the moment of inertia about the 2-axis vanishes (dumbbell-molecules), for then the $\Phi\begin{smallmatrix} m & n \\ \omega & \omega \\ i & 2\end{smallmatrix}$ and $\Phi\begin{smallmatrix} m & n \\ v & \omega \\ i & 2\end{smallmatrix}$ vanish (or if they vanish by another reason.) Then simply

$$\sum_n \Phi\begin{smallmatrix} m & n \\ \omega & \omega \\ i & 1\end{smallmatrix} = 0;\; \sum_n \Phi\begin{smallmatrix} m & n \\ v & \omega \\ i & 1\end{smallmatrix} = 0\,. \tag{10.7b}$$

We discuss the dumbbell-model $I_2 = 0$ and neglect further the translational motions in the 3-direction. The latter serves only to simplify the calculations and does not affect the physics of the problem appreciably. Satisfying the relations (10.2, 3, 5, 7), as the inversion is a symmetry operation, we have for the remaining force constants:

$$\alpha = -\Phi\begin{smallmatrix} 0 & 1 \\ v & v \\ 1 & 1\end{smallmatrix} = \frac{1}{2}\Phi\begin{smallmatrix} 0 & 0 \\ v & v \\ 1 & 1\end{smallmatrix};\quad \beta = -\Phi\begin{smallmatrix} 0 & 1 \\ v & v \\ 2 & 2\end{smallmatrix} = \frac{1}{2}\Phi\begin{smallmatrix} 0 & 0 \\ v & v \\ 2 & 2\end{smallmatrix};$$

$$\gamma = -\Phi\begin{smallmatrix} 0 & 1 \\ v & v \\ 1 & 2\end{smallmatrix} = \frac{1}{2}\Phi\begin{smallmatrix} 0 & 0 \\ v & v \\ 1 & 2\end{smallmatrix};$$

$$\delta = -\Phi\begin{smallmatrix} 0 & 1 \\ v & \omega \\ 1 & 1\end{smallmatrix} = \Phi\begin{smallmatrix} 0 & 1 \\ \omega & v \\ 1 & 1\end{smallmatrix};\quad \zeta = -\Phi\begin{smallmatrix} 0 & 1 \\ v & \omega \\ 1 & 3\end{smallmatrix} = \Phi\begin{smallmatrix} 0 & 1 \\ \omega & v \\ 3 & 1\end{smallmatrix};\quad \vartheta = -\Phi\begin{smallmatrix} 0 & 1 \\ v & \omega \\ 2 & 1\end{smallmatrix} = \Phi\begin{smallmatrix} 0 & 1 \\ \omega & v \\ 1 & 2\end{smallmatrix};$$

$$\varkappa = -\Phi\begin{smallmatrix} 0 & 1 \\ v & \omega \\ 2 & 3\end{smallmatrix} = \Phi\begin{smallmatrix} 0 & 1 \\ \omega & v \\ 3 & 2\end{smallmatrix};\quad \lambda = -\Phi\begin{smallmatrix} 0 & 1 \\ \omega & \omega \\ 1 & 1\end{smallmatrix} = \frac{1}{2}\Phi\begin{smallmatrix} 0 & 0 \\ \omega & \omega \\ 1 & 1\end{smallmatrix};$$

$$\nu = -\Phi\begin{smallmatrix} 0 & 1 \\ \omega & \omega \\ 1 & 3\end{smallmatrix} = \frac{1}{2}\Phi\begin{smallmatrix} 0 & 0 \\ \omega & \omega \\ 1 & 3\end{smallmatrix};\quad \mu = -\Phi\begin{smallmatrix} 0 & 1 \\ \omega & \omega \\ 3 & 3\end{smallmatrix}$$

$$2\mu + 4\mu' = \Phi\begin{smallmatrix} 0 & 0 \\ \omega & \omega \\ 3 & 3\end{smallmatrix}. \tag{10.8}$$

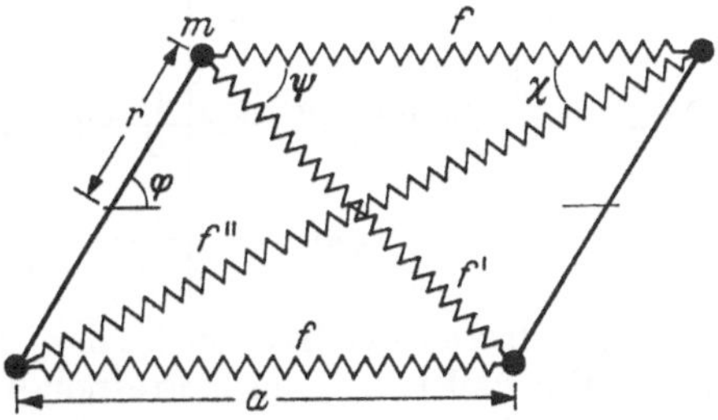


Fig. 4.2. Definition of springs and geometrical data for a linear lattice of dumbbell molecules

Under certain conditions some of them can be simplified further. E.g. if $\varphi = \pi/2$, reflection at a plane through the centers of mass, perpendicular to the dumpbell-axis, is a symmetry operation. This implies

$$\gamma = \delta = \zeta = 0\,. \tag{10.9}$$

Another simplification results from the assumption of central forces between the ions of the molecules (Fig. 4.2). With

$$\operatorname{tg}\psi = \frac{2r\sin\varphi}{a - 2r\cos\varphi}; \quad \operatorname{tg}\chi = \frac{2r\sin\varphi}{a + 2r\cos\varphi} \tag{10.10a}$$

we have

$$\delta = \vartheta = \lambda = \nu = 0;$$

$$\begin{aligned}
\alpha &= 2f\sin^2\varphi + f'\sin^2(\varphi+\psi) + f''\sin^2(\varphi-\chi)\\
\beta &= 2f\cos^2\varphi + f'\cos^2(\varphi+\psi) + f''\cos^2(\varphi-\chi)\\
\gamma &= 2f\sin\varphi\cos\varphi + f'\sin(\varphi+\psi)\cos(\varphi+\psi) + f''\sin(\varphi-\psi)\cos(\varphi-\psi)\\
\zeta &= r[f'\sin^2(\varphi+\chi) - f''\sin^2(\varphi-\chi)]\\
\varkappa &= r[f'\sin(\varphi+\chi)\cos(\varphi+\chi) - f''\sin(\varphi-\chi)\cos(\varphi-\chi)]\\
\mu &= r^2[2f\sin^2\varphi - f'\sin^2(\varphi+\chi) - f''\sin^2(\varphi-\chi)]\\
\mu' &= r^2[f'\sin^2(\varphi+\chi) + f''\sin^2(\varphi-\chi)]\,.
\end{aligned} \tag{10.10b}$$

Because of (10.4), the solution of (10.1) is given by*

$$\begin{aligned}
u_i^m &= \frac{e_i}{\sqrt{M}}\cdot e^{iqam - i\omega t}\\
\omega_i^m &= \frac{\varepsilon_i}{\sqrt{I_i}}\cdot e^{iqam - i\omega t}
\end{aligned} \tag{10.11}$$

$e_i$ and $\varepsilon_i$ are the polarization vectors for "translational" and "librational" modes of the lattice. In the above example it is

$$M = 2m; \quad I_1 = I_3 = 2mr^2 = Mr^2 = I; \quad I_2 = 0\,. \tag{10.12}$$

Inserting this as well as (10.10, 11) into (10.1) we have the secular equation:

$$0 = \begin{Bmatrix}
\frac{4\alpha}{M}\sin^2 qa/2 - \omega^2 & \frac{4\gamma}{M}\sin^2 qa/2 & -i\frac{2\delta}{\sqrt{MI}}\sin qa & -i\frac{2\zeta}{\sqrt{MI}}\sin qa\\
\frac{4\gamma}{M}\sin^2 qa/2 & \frac{4\beta}{M}\sin^2 qa/2 - \omega^2 & -i\frac{2\vartheta}{\sqrt{MI}}\sin qa & -i\frac{2\varkappa}{\sqrt{MI}}\sin qa\\
i\frac{2\delta}{\sqrt{MI}}\sin qa & i\frac{2\vartheta}{\sqrt{MI}}\sin qa & \frac{4\lambda}{I}\sin^2 qa/2 - \omega^2 & \frac{4\nu}{I}\sin^2 qa/2\\
i\frac{2\zeta}{\sqrt{MI}}\sin qa & i\frac{2\varkappa}{\sqrt{MI}}\sin qa & \frac{4\nu}{I}\sin^2 qa/2 & \frac{4\mu'}{I} + \frac{4\mu}{I}\sin^2 qa/2 - \omega^2
\end{Bmatrix} \times \begin{Bmatrix} e_1/\sqrt{M}\\ e_2/\sqrt{M}\\ \varepsilon_1/\sqrt{I}\\ \varepsilon_3/\sqrt{I}\end{Bmatrix}. \tag{10.13}$$

In the limit with $q \to 0$, we have three frequencies $\omega = 0$, two corresponding to translations of the molecules, one corresponding to a libration (about the 1-axis). This, of course, must be so. The fourth frequency $\omega^2 = 4\mu'/I$ corresponds to a libration about the 3-axis; direct inspection shows, that this libration cannot have zero frequency in the limit $q \to 0$ (for finite $\mu'$).

* The frequency $\omega$ may not be confused with the libration angle $\omega_i^m$. But $\omega_i^m$ has always indices $m$, $n$ $i$, $j$ etc., whereas the frequency $\omega$ has no index or the wave number $q$ as an index.

The libration mode with $\omega(q \to 0) = 0$ is uninteresting and even vanishing in a central-force model. Therefore we will drop it from further considerations. We assume *central forces* and start with the case $\varphi = \pi/2$.

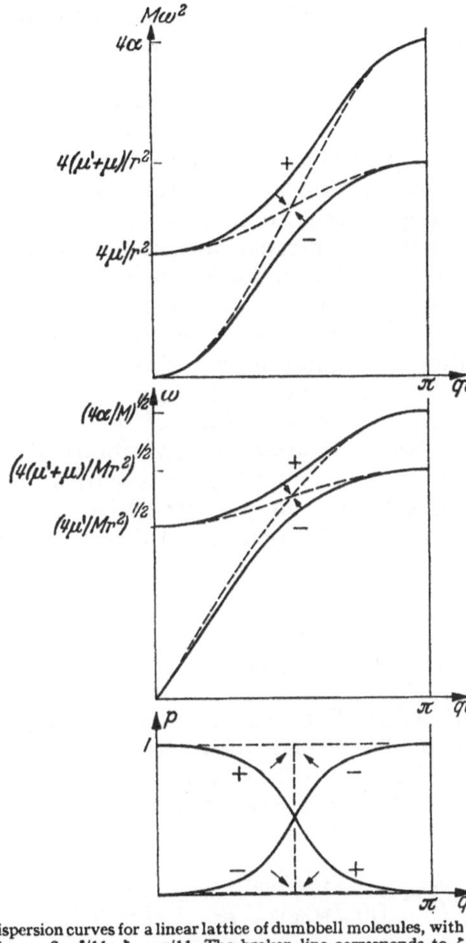


Fig. 4.3 a

Fig. 4.3 b

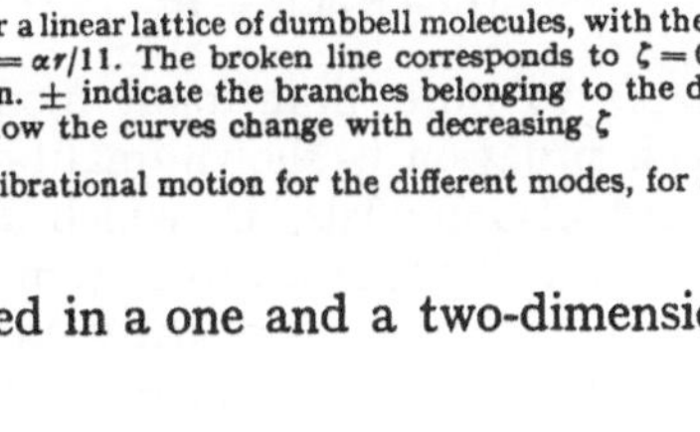


Fig. 4.4

Fig. 4.3a and b. Dispersion curves for a linear lattice of dumbbell molecules, with the following choice of parameters: $\mu' = 4\alpha r^2/11$; $\mu = 3\alpha r^2/11$; $\zeta = \alpha r/11$. The broken line corresponds to $\zeta = 0$, i.e. no coupling between librational and translational motion. $\pm$ indicate the branches belonging to the different signs of the root in equ. (10.15). The arrows indicate how the curves change with decreasing $\zeta$

Fig. 4.4. Relative contribution to librational motion for the different modes, for a linear lattice of dumbbell molecules. See Fig. 4.3

Then (10.13) is separated in a one and a two-dimensional problem with the solutions:

$$M\omega^2 = 4\alpha \sin^2 qa/2;$$

$$M\omega^2 = 2\mu'/r^2 + 2(\beta + \mu/r^2)\sin^2 qa/2 \pm$$

$$\pm 2\sqrt{[\mu'/r^2 + (\mu/r^2 - \beta)\sin^2 qa/2]^2 + \frac{4\varkappa^2}{r^2}\sin^2 qa}\,. \qquad (10.14)$$

If one inserts the parameters from (10.10b), one can see immediately that one of the squared frequencies becomes negative for small $q$. That means, that the model with central forces and $\varphi = \pi/2$ is a labile equilibrium, every small displacement in the „2-direction" enlarges the displacement. If one wants to use this model, one has to impose the constraint, that motion in the „2-direction" is impossible. On the other hand, if one would discuss (10.14) without this constraint, one finds, that libration and vibration in „2-direction" are separated and that there is small influence on each other. Therefore the use of the constraint seems to be allowed in a qualitative discussion of what happens in molecular crystals, especially because linear lattices in this sense do not exist in nature. This case is discussed in [63 B 1], though it is not mentioned there that the lattice is instable without the constraint. Fig. 4.3 (broken line) shows the dispersion curves for $a/r = 4$ and $f = f'$ or $f = 4f'$ resp. with constraint: no motion in „2-direction".

The situation changes, if we consider the case $\varphi \neq \pi/2$; then it is possible to have a stable configuration, if $f$, $f'$, $f''$ satisfy a certain condition depending on $\varphi$. If $\varphi$ is near $\pi/4$, e.g., the stability demands essentially that $f$ is in order of magnitude of or larger than $f' + f''$. But the interaction between libration and „2-vibration" is also small and we impose the same constraint ($\beta = \gamma = \varkappa = 0$) in order to make a comparison between $\varphi = \pi/2$ and $\varphi \neq \pi/2$. With $\zeta \neq 0$ there is no separation of libration and vibration in „1-direction [63 B 1]. The frequencies are then given by

$$M\omega^2 = 2\mu'/r^2 + 2(\alpha + \mu/r^2)\sin^2 qa/2$$
$$\pm 2\sqrt{[\mu'/r^2 + (\mu/r^2 - \alpha)\sin^2 qa/2]^2 + \frac{4\zeta^2}{r^2}\sin^2 qa} \tag{10.15}$$

and shown in Fig. 4.3 (full line). As is to be expected, the crossing dispersion curves split and the modes change their character from vibrational type to librational type from $q = 0$ to $q = \pi/a$ or vice versa. As a measure of the librational part of two interacting modes (librational and vibrational) we may consider the quantity

$$p(\omega_i) = \frac{[\sqrt{I_i/M}\,\omega_i]^2}{[\sqrt{I_i/M}\,\omega_i]^2 + u_i^2} = \frac{\varepsilon_i^2}{\varepsilon_i^2 + e_i^2}\,. \tag{10.16}$$

$p(\omega_i) = 1$ corresponds to pure librational modes, $p(\omega_i) = 0$ to pure vibrational ones. The situation is shown qualitatively in Fig. 4.4.

We can construct also models, in which there is no interaction between different modes. Suppose a three-atomic "dumbbell"-molecule with a large mass $m_1 \gg m$ in the midth between the two masses $m$. The coupling shall be (or nearly be) the same as in the above case. Then

$$M = m_1 + 2m \approx m_1\,;\; I = 2m\,r^2 \approx M \cdot \frac{2m}{m_1} r^2 = M \cdot r_{\text{eff}}^2\,. \tag{10.17}$$

The frequencies of the libration modes are raised by a factor $r/r_{\text{eff}}$; no splitting occurs because there is no crossing of dispersion curves.

As a simple example of a dimolecular linear lattice we consider the model given in Fig. 4.5. In order to describe the molecules in one lattice cell, we have to introduce further indices $\mu$ on the coupling parameters $\overset{m\,n}{\underset{i\ \ j}{\Phi^{\mu\ \nu}_{v\ \omega}}}$ etc. Apart from this the procedure is just as before. Owing to the fact that there are more than one molecule in the cell,

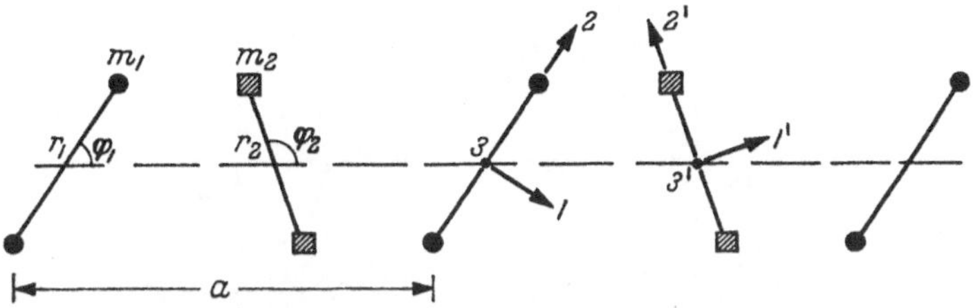


Fig. 4.5. Dimolecular linear lattice of dumbbell molecules

we have acoustical *and* optical branches as well for vibrational modes as for librational ones. All kinds of interactions are possible: vibrational acoustical with librational acoustical, vibrational optical with librational acoustical and optical and so on. What happens in detail depends on the magnitude of force constants, masses and moments of inertia.

In a one-molecular lattice we have chosen the coordinates to be parallel to the principal axes of the molecule. If we do a corresponding choice in a dimolecular lattice we have to introduce different coordinate systems for different molecules in the lattice-cell*. Let $K^{\mu}_{ij}$ be the (orthogonal) transformation between a space-coordinate-system (say the axes of the lattice e. g.) and the molecules own coordinates. Then, of course we have to replace the conditions of translational invariance (10.3) by

$$\sum_{\substack{n\nu\\ j}} \overset{m\,n}{\underset{i\ \ j}{\Phi^{\mu\ \nu}_{v\ v}}} K^{\nu}_{jk} = 0\,; \quad \sum_{\substack{n\nu\\ j}} \overset{m\,n}{\underset{i\ \ j}{\Phi^{\mu\ \nu}_{\omega\ v}}} K^{\nu}_{jk} = 0\,, \tag{10.18}$$

if the coupling-constants are defined in the coordinate-systems of the molecules. Only if all the principal axes of the molecules have the same orientation (10.18) simplifies to (10.3). However, (10.4) remains unchanged, whereas point group symmetry-relations have to be changed in some cases too. With this description, the modes of vibration and libration in dimolecular crystals refer to the molecule's system of axes.

The simplest model (Fig. 4.6), using central forces between the ions of the dumbbell-molecules, leads to a six-dimensional secular equation. In the one-molecular case the "2-vibration" does not interact

* As already mentioned this is not a necessary choice, but only convenient, at least for molecules with one vanishing principal moment of inertia (dumbbell-molecules, see § 11).

with other modes under reasonable assumptions for the forces. However, in a two-molecular lattice the "optical 2-vibrations" have higher frequencies than the acoustical one and may well interact with librational modes. Thus the neglect of the "2-vibration" seems not to be justified.

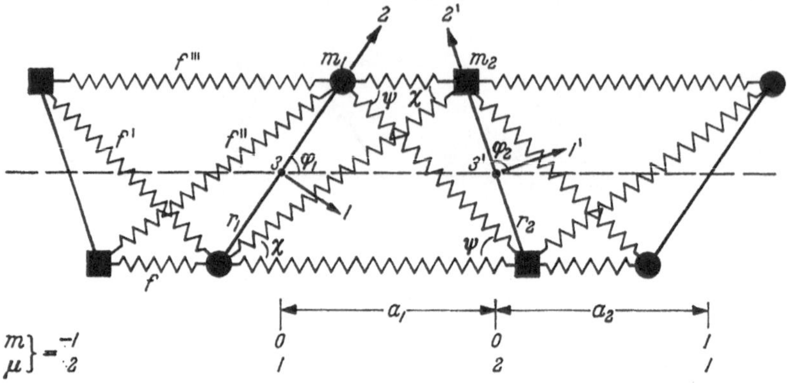


Fig. 4.6. Definition of force-constants and parameters for a dimolecular lattice of dumbbell molecules

We assume $a_1 = a_2 = a/2$, to save the inversion-symmetry (at the centers of mass) and further $r_1 = r_2$. The different coupling-constants which enter are the following ones.

*v — v-coupling:*

$$\alpha = -\Phi\begin{smallmatrix}00\\12\\11\end{smallmatrix} = -\Phi\begin{smallmatrix}01\\21\\11\end{smallmatrix};\quad 2\alpha' = \Phi\begin{smallmatrix}00\\11\\11\end{smallmatrix};\quad 2\alpha'' = \Phi\begin{smallmatrix}00\\22\\11\end{smallmatrix};$$

$$\beta = -\Phi\begin{smallmatrix}00\\12\\22\end{smallmatrix} = -\Phi\begin{smallmatrix}01\\21\\22\end{smallmatrix};\quad 2\beta' = \Phi\begin{smallmatrix}00\\22\\22\end{smallmatrix};\quad 2\beta'' = \Phi\begin{smallmatrix}00\\11\\22\end{smallmatrix};$$

$$\gamma = -\Phi\begin{smallmatrix}00\\12\\12\end{smallmatrix} = -\Phi\begin{smallmatrix}01\\21\\21\end{smallmatrix};\quad \gamma' = -\Phi\begin{smallmatrix}00\\12\\21\end{smallmatrix} = -\Phi\begin{smallmatrix}01\\21\\12\end{smallmatrix};$$

$$2\gamma'' = \Phi\begin{smallmatrix}00\\11\\12\end{smallmatrix} = \Phi\begin{smallmatrix}00\\11\\21\end{smallmatrix};\qquad 2\gamma''' = \Phi\begin{smallmatrix}00\\22\\12\end{smallmatrix} = \Phi\begin{smallmatrix}00\\22\\21\end{smallmatrix}. \tag{10.19}$$

*v — ω-coupling:*

$$r\zeta = -\Phi\begin{smallmatrix}00\\12\\13\end{smallmatrix} = \Phi\begin{smallmatrix}10\\12\\13\end{smallmatrix};\quad r\zeta' = -\Phi\begin{smallmatrix}01\\21\\13\end{smallmatrix} = \Phi\begin{smallmatrix}00\\21\\13\end{smallmatrix};$$

$$r\varkappa = -\Phi\begin{smallmatrix}00\\12\\23\end{smallmatrix} = \Phi\begin{smallmatrix}10\\12\\23\end{smallmatrix};\quad r\varkappa' = -\Phi\begin{smallmatrix}01\\21\\23\end{smallmatrix} = \Phi\begin{smallmatrix}00\\21\\23\end{smallmatrix};$$

$$0 = \Phi\begin{smallmatrix}00\\11\\13\end{smallmatrix} = \Phi\begin{smallmatrix}00\\11\\23\end{smallmatrix} = \Phi\begin{smallmatrix}00\\22\\13\end{smallmatrix} = \Phi\begin{smallmatrix}00\\22\\23\end{smallmatrix}.$$

*ω — ω-coupling:*

$$r^2\mu = -\Phi\begin{smallmatrix}00\\12\\33\end{smallmatrix} = -\Phi\begin{smallmatrix}01\\21\\33\end{smallmatrix};\quad 2\mu' r^2 = \Phi\begin{smallmatrix}00\\11\\33\end{smallmatrix};\quad 2\mu'' r^2 = \Phi\begin{smallmatrix}00\\22\\33\end{smallmatrix}.$$

Fig. 4.7, 8 show dispersion curves for such a dimolecular "linear" lattice. They differ only in the choice of the $\mu'$-coupling-constant, which describes essentially the librational coupling. The values of constants chosen are given in the legend of the figures, where we have put

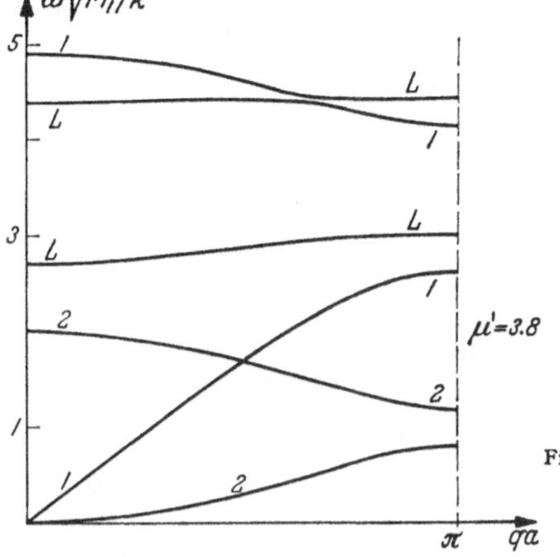


Fig. 4.7 a

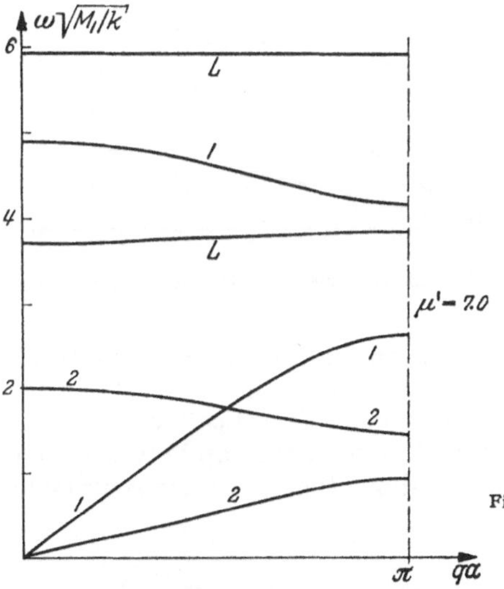


Fig. 4.7 b

Fig. 4.7a and b. Dispersion curves for a dimolecular lattice of dumbbell molecules; "1" means translational motion in 1-direction, "2" mainly translational in 2 direction, "$L$" means mainly librational motion. The parameters are the following: $\alpha = 3.4k$; $\beta = 0.6k$; $\varkappa = -0.92k$; $\gamma = 0.3k$; $\zeta = 0.3k$; $\mu = 0.6k$; and in a) $\mu' = 3{,}8k$; in b) $\mu' = 7{,}0k$. $k$ is an arbitrary unit. The masses are chosen as $M_1 = 2.5M_2$

$$\alpha' = \alpha'' = \alpha\,; \quad \beta' = \beta'' = \beta\,; \quad \gamma = \gamma' = \gamma'' = \gamma'''\,;$$
$$\zeta = \zeta'\,; \quad \varkappa = \varkappa'\,; \quad \mu' = \mu''$$

for simplicity. An even simpler choice would be $\varphi_1 = \varphi_2 = \pi/2$ and central forces between the ions. Then the vibration in "1-direction" is again separated and

$$\gamma = \zeta = 0; \quad \mu' = \mu'' = \mu + 4f' \cos^2 \psi .$$

As the figures indicate, by appropriate choice of the parameters (force-constants, masses, moments of inertia) all kinds of relative position of the dispersion curves can be obtained. It might be, that

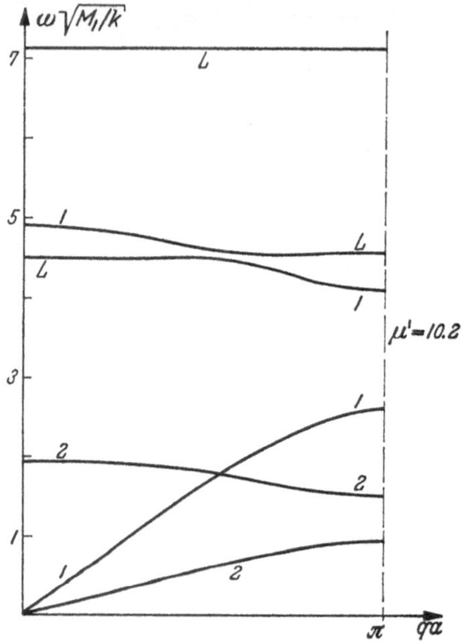


Fig. 4.8. The same as Fig. 4.7, but with $\mu' = 10.2k$

translational and librational acoustical branches interact, and as well for the optical ones. But it might also be that the translational branch interacts with the librational "acoustical" one. It can happen, that all the "librational" branches lie in the gap between acoustical and optical translational modes or above the optical translational modes.

## 11. General Derivation of the Equation of Motion for Molecular Crystals

In the last section we have set up the equation of motion phenomenologically and have considered simple examples. We will now derive these equations from the general equation of motion for *harmonic lattices*, and discuss the approximations involved [63 B 1].

We use the following notations: $\boldsymbol{m}$ is the number of the unit cell of the molecular lattice, $\mu$ is the molecule in the cell, $\alpha$ denotes the ions (atoms) in the molecule $\mu$; thus $\alpha$ and $\mu$ together replace the index $\mu$ of chapter II. $\alpha$ runs over $1 \ldots r_\mu$ ($r_\mu$ ions in molecule $\mu$), $\mu$ over $1 \ldots s$. There are $\sum_{\mu=1}^{s} r_\mu$ ions in the unit cell of the lattice.

For the positions we take the following descriptions: $X_i^{m}$ is the position of one point of cell $\boldsymbol{m}$; $X_i^{\mu}$ is the position of the center of mass of molecule $\mu$ relative to the cell $\left(\text{to } X_i^{m}\right)$ and $X_{\alpha i}^{\mu}$ is the position of ion $\alpha$ in molecule $\mu$ relative to the center of mass of $\mu$. Then the position of one ion is in equilibrium:

$$\overset{m}{X_{\alpha i}^{\mu}} = \overset{m}{X_i} + X_i^{\mu} + X_{\alpha i}^{\mu}\,. \tag{11.1}$$

The displacement from the equilibrium position is $\overset{m}{U_{\alpha i}^{\mu}}$. Let $m_\alpha^\mu$ be the mass of ion $\alpha$ in molecule $\mu$, then

$$M_\mu = \sum_\alpha m_\alpha^\mu \tag{11.2}$$

is the mass of molecule $\mu$. Further it follows from the definition of the center of mass

$$\sum_\alpha m_\alpha^\mu X_{\alpha i}^{\mu} = 0 \quad \text{for} \quad \text{every} \quad \mu, i\,, \tag{11.3a}$$

and for the displacement of the center of mass of molecule $\mu$

$$U_{\mu i}^{m} = \frac{1}{M_\mu} \sum_\alpha m_\alpha^\mu \overset{m}{U_{\alpha i}^{\mu}}\,. \tag{11.3b}$$

Now it is convenient, at least in some cases, to introduce for every molecule a coordinate system with the principal axes of inertia of the molecule as coordinate axes $\left(Y_{\alpha i}^{\mu}\right)$. The transformation* from the "space coordinate system" $X_{\alpha i}^{\mu}$ to the $Y_{\alpha i}^{\mu}$ is given by an orthogonal transformation

$$Y_{\alpha i}^{\mu} = \sum_j K_{ij}^{\mu} X_{\alpha j}^{\mu}; \quad \overset{m}{V_{\alpha i}^{\mu}} = \sum_j K_{ij}^{\mu} \overset{m}{U_{\alpha j}^{\mu}}; \quad \sum_j K_{ij}^{\mu} K_{kj}^{\mu} = \delta_{ik} \tag{11.4a}$$

and of course

$$\sum_\alpha m_\alpha^\mu Y_{\alpha i}^{\mu} = 0\,. \tag{11.4b}$$

The essential step is now the introduction of coordinates, which replace the $3N \sum_{\mu=1}^{s} r_\mu$ diplacement-coordinates and which describe the

* If one does not want to take a molecular coordinate system, but rather a space coordinate system, one can drop the index $\mu$ on $K_{ij}^{\mu}$ or even take $K_{ij}^{\mu} = \delta_{ij}$ for every $\mu$. With this convention all the following formulas can be taken in every case. See footnote on page 59.

translations of the center of mass of the molecules ($\alpha = 1$), the (infinitesimal) rotations or librations of the rigid molecules ($\alpha = 2$) and the internal vibrations of the molecules (normal modes of internal degrees of freedom, $\alpha = 3 \ldots$)*. The new displacements are denoted $\overset{m}{\underset{i}{t^\mu_\alpha}}$ with $\overset{m}{U^\mu_i} = \overset{m}{\underset{i}{t^\mu_1}}$ as the center of mass translation and $\overset{m}{\omega^\mu_i} = \overset{m}{\underset{i}{t^\mu_2}}$ as the infinitesimal angle of libration.

Then

$$\overset{m}{\underset{i}{V^\mu_\alpha}} = \sum_{\beta k} \underset{ik}{\tilde{L}^{\alpha\beta}}(\mu)\, \overset{m}{\underset{i}{t^\mu_\beta}} = \overset{m}{U^\mu_i} + \sum_k \tilde{L}^{\alpha\,2}_{i\,k}(\mu)\, \overset{m}{\underset{k}{\omega^\mu}} + \sum_{\beta=3,k}^{r_\mu} \tilde{L}^{\alpha\beta}_{i\,k}(\mu)\, \overset{m}{\underset{k}{t^\mu_\beta}} \tag{11.5}$$

with**

$$\tilde{L}^{\alpha\,1}_{i\,k}(\mu) = \delta_{ik} \tag{11.5a}$$

$$\tilde{L}^{\alpha\,2}_{i\,k}(\mu) = \sum_j \varepsilon_{ikj}\, \underset{j}{Y^\mu_\alpha} \tag{11.5b}$$

and further

$$\sum_\alpha m^\mu_\alpha \tilde{L}^{\alpha\,1}_{i\,k}(\mu) = M^\mu\, \delta_{ik} \tag{11.5c}$$

$$\sum_\alpha m^\mu_\alpha \tilde{L}^{\alpha\beta}_{i\,k}(\mu) = 0 \quad \text{for} \quad \beta \geqq 2\,. \tag{11.5d}$$

$L$ is *not* an *orthogonal* transformation ($\tilde{L} \neq L^{-1}$). By some re-definition it might be possible, to have an orthogonal transformation. But there is no need for it in the following. The inverse transformation $L^{-1}$ has to satisfy certain conditions:

$$\sum_\alpha L^{-1}{}^{\alpha\beta}_{i\,k}(\mu) = \delta_{ik} \tag{11.6a}$$

$$\sum_\alpha L^{-1}{}^{\alpha\beta}_{i\,k}(\mu) = 0 \quad \text{for} \quad \beta \geqq 2 \tag{11.6b}$$

$$\sum_{\substack{\beta \geqq 3\\ \alpha,\,k}} L^{-1}{}^{\alpha\beta}_{i\,k}\, \tilde{L}^{\alpha\beta}_{j\,k} = (r_\mu - 2)\, \delta_{ik} \quad \text{(trivial)} \tag{11.6c}$$

$$\sum_\alpha \left\{ L^{-1}{}^{\alpha\beta}_{j\,i}\, \underset{k}{Y^\mu_\alpha} - L^{-1}{}^{\alpha\beta}_{k\,i}\, \underset{j}{Y^\mu_\alpha} \right\} = \varepsilon_{ijk}\, \delta_{2\beta} \tag{11.6d}$$

or with (11.4a)

$$\sum_{\alpha,\,l} \left\{ L^{-1}{}^{\alpha\beta}_{j\,i}\, K^\mu_{kl}\, \underset{l}{X^\mu_\alpha} - L^{-1}{}^{\alpha\beta}_{k\,i}\, K^\mu_{jl}\, \underset{l}{X^\mu_\alpha} \right\} = \varepsilon_{ijk}\, \delta_{2\beta}\,.$$

These conditions follow from the definition

$$\sum_{\alpha i} L^{\alpha\beta}_{k\,i}\, L^{-1}{}^{\alpha\beta'}_{i\,k'} = \sum_{\alpha i} \tilde{L}^{\alpha\beta}_{i\,k}\, L^{-1}{}^{\alpha\beta'}_{i\,k'} = \delta_{\beta\beta'}\, \delta_{kk'}$$

together with (11.5a, b).

---

* Sometimes it might be more convenient, to use $\alpha = 3$ for all the internal degrees of freedom and to have lower indices $i$, $k = 1, 2, \ldots, 3r_\mu - 6$ for these degrees of freedom ($3r_\mu - 5$ in case of one vanishing moment of inertia of the molecule $\mu$). In the following the choice is meaningless.

** The other components of the transformation $\tilde{L}$ need not to be specified in the following. $\tilde{L}$ is the transposed of a matrix $L$, $\varepsilon_{ijk}$ the total antisymmetric tensor of rank 3. If $I^\mu_k$ vanishes for some $k$ (in dumbbell-molecules), the index $k$ for $\beta = 2$ takes only 2 values, and also $\overset{m}{\omega^\mu_k}$ has only two components, and so for $\tilde{L}^{\alpha\,2}_{i\,k}$ too.

Introducing tensor notation

$$\boldsymbol{U} = \overset{m}{\underset{i}{U^{\mu}_{\alpha}}}\,; \quad \boldsymbol{\Phi} = \overset{m\,n}{\underset{i\;j}{\Phi^{\mu\;\nu}_{\alpha\;\beta}}}\,;$$

$$\mathbf{m} = \delta^{mn}\,\delta_{\mu\nu}\,\delta_{\alpha\beta}\,\delta_{ij}\cdot m^{\mu}_{\alpha} \tag{11.7a}$$

$$\mathbf{K} = \delta^{mn}\,\delta_{\mu\nu}\,\delta_{\alpha\beta}\,K^{\mu}_{ik}\,; \quad \widetilde{\mathbf{L}} = \delta^{mn}\,\delta_{\mu\nu}\,\tilde{L}^{\alpha\beta}_{i\,k}(\mu)$$

we have

$$\boldsymbol{V} = \mathbf{K}\,\boldsymbol{U} \quad \text{and} \quad \mathbf{K}\,\mathbf{m}\,\widetilde{\mathbf{K}} = \mathbf{m}\,,$$

because the diagonal mass-tensor does not change in an orthogonal transformation. After the above transformations we have for the kinetic and potential energy

$$E_{\text{kin}} = \frac{1}{2}\,(\dot{\boldsymbol{t}}, \mathbf{M}\,\dot{\boldsymbol{t}})\,; \quad E_{\text{pot}} = \frac{1}{2}\,(\boldsymbol{t}, \boldsymbol{\Psi}, \boldsymbol{t})\,;$$

with (11.7b)

$$\mathbf{M} = \mathbf{L}\,\mathbf{m}\,\widetilde{\mathbf{L}}\,; \quad \boldsymbol{\Psi} = \mathbf{L}\,\mathbf{K}\,\boldsymbol{\Phi}\,\widetilde{\mathbf{K}}\,\widetilde{\mathbf{L}}\,.$$

The components of the mass-tensor **M** are the following*:

$$\overset{m\,n}{\underset{i\;k}{M^{\mu\;\nu}_{1\;1}}} = \delta^{mn}\,\delta_{\mu\nu}\,\delta_{ik}\,M_{\mu} \tag{11.8a}$$

$$\overset{m\,n}{\underset{i\;k}{M^{\mu\;\nu}_{2\;2}}} = \delta^{mn}\,\delta_{\mu\nu}\,I^{\mu}_{ik} = \delta^{mn}\,\delta_{\mu\nu}\,\delta_{ik}\,I^{\mu}_{i} \tag{11.8b}$$

with the tensor of the moment of inertia

$$I^{\mu}_{ik} = \sum_{\alpha} m^{\mu}_{\alpha}\Big[\delta_{ik}\sum_{l} \underset{l}{Y^{\mu}_{\alpha}}\,\underset{l}{Y^{\mu}_{\alpha}} - \underset{i}{Y^{\mu}_{\alpha}}\,\underset{k}{Y^{\mu}_{\alpha}}\Big]\,;$$

$$\overset{m\,n}{\underset{i\;k}{M^{\mu\;\nu}_{1\;\beta}}} = 0 \quad \text{if} \quad \beta \geqq 2\,, \tag{11.8c}$$

because of $\sum_{\alpha} m^{\mu}_{\alpha}\,\tilde{L}^{\alpha\beta}_{k\,l}(\mu) = 0$ if $\beta \geqq 2$, which follows from (11.3b).

For the terms having indices $2\,\beta$ with $\beta \geqq 3$ we write down the corresponding kinetic energy

$$\sum_{\substack{\beta\geqq 3\\ m\,n\\ \mu\;\nu\\ i\;k}} \overset{m}{\underset{i}{\dot{\omega}^{\mu}}}\,\overset{m\,n}{\underset{i\;k}{M^{\mu\;\nu}_{2\;\beta}}}\,\overset{.n}{\underset{k}{t^{\nu}_{\beta}}} = \sum_{\alpha,l} m^{\mu}_{\alpha}\,\underset{l}{Y^{\mu}_{\alpha}}\Big[\overset{m}{\boldsymbol{W}^{\mu}_{\alpha}} \times \overset{m}{\dot{\boldsymbol{\omega}}^{\mu}}\Big]_{l}$$

with (11.8d)

$$\overset{m}{\underset{j}{W^{\mu}_{\alpha}}} = \sum_{\beta\geqq 3,\,k} \tilde{L}^{\alpha\beta}_{j\,k}(\mu)\,\overset{.m}{\underset{k}{t^{\mu}_{\beta}}}\,.$$

This term represents the Coriolis-energy connected with the internal vibrations due to the libration of the molecules. It will be neglected (see p. 67). The rest of the kinetic energy is that of the internal motion of the ions and does not contribute to the equation for translational

* $I^{\mu}_{ik}$ is diagonal if one chooses the principal axes of the molecules as the coordinate system. Otherwise $I^{\mu}_{ik}$ is not diagonal.

vibration and libration, if one neglects the coupling between internal motions and the other ones.

The potential energy reads after performing the transformation

$$
\begin{aligned}
E_{\text{pot}} = {} & \frac{1}{2} \sum_{\substack{mn\\ \mu\nu\\ ij}} \overset{mn}{\underset{i\,j}{\Psi^{\mu\,\nu}_{1\,1}}}\, \overset{m}{\underset{i}{U^{\mu}}}\, \overset{n}{\underset{j}{U^{\nu}}} + \sum \overset{mn}{\underset{i\,j}{\Psi^{\mu\,\nu}_{1\,2}}}\, \overset{m}{\underset{i}{U^{\mu}}}\, \overset{n}{\underset{j}{\omega^{\nu}}} + \\
& + \frac{1}{2} \sum \overset{mn}{\underset{i\,j}{\Psi^{\mu\,\nu}_{2\,2}}}\, \overset{m}{\underset{i}{\omega^{\mu}}}\, \overset{n}{\underset{j}{\omega^{\nu}}} + \underbrace{\sum_{\beta \geqq 3} \overset{mn}{\underset{i\,j}{\Psi^{\mu\,\nu}_{1\,\beta}}}\, \overset{m}{\underset{i}{U^{\mu}}}\, \overset{n}{\underset{j}{t^{\nu}_{\beta}}}}_{E_{vi}} + \qquad (11.9) \\
& + \underbrace{\sum_{\beta \geqq 3} \overset{mn}{\underset{i\,j}{\Psi^{\mu\,\nu}_{2\,\beta}}}\, \overset{m}{\underset{i}{\omega^{\mu}}}\, \overset{n}{\underset{j}{t^{\nu}_{\beta}}}}_{E_{\omega i}} + \underbrace{\sum_{\alpha,\beta \geqq 3} \overset{mn}{\underset{i\,j}{\Psi^{\mu\,\nu}_{\alpha\,\beta}}}\, \overset{m}{\underset{i}{t^{\mu}_{\alpha}}}\, \overset{n}{\underset{j}{t^{\nu}_{\beta}}}}_{E_{ii}} .
\end{aligned}
$$

If one neglects the coupling $E_{vi}$, $E_{\omega i}$, the equation of motion can be easily derived:

$$
\begin{aligned}
M_{\mu}\, \overset{m}{\underset{i}{\ddot{U}^{\mu}}} &= - \sum_{n\nu j} \overset{mn}{\underset{i\,j}{\Psi^{\mu\,\nu}_{1\,1}}}\, \overset{n}{\underset{j}{U^{\nu}}} - \sum_{n\nu j} \overset{mn}{\underset{i\,j}{\Psi^{\mu\,\nu}_{1\,2}}}\, \overset{n}{\underset{j}{\omega^{\nu}}} - \cdots \\
I^{\mu}_{ik}\, \overset{m}{\underset{k}{\ddot{\omega}^{\mu}}} &= - \sum_{n\nu j} \overset{mn}{\underset{i\,j}{\Psi^{\mu\,\nu}_{2\,1}}}\, \overset{n}{\underset{j}{U^{\nu}}} - \sum_{n\nu j} \overset{mn}{\underset{i\,j}{\Psi^{\mu\,\nu}_{2\,2}}}\, \overset{n}{\underset{j}{\omega^{\nu}}} - \cdots .
\end{aligned} \qquad (11.10)
$$

This is equivalent to (10.1), if one replaces 11 by $v\,v$, 12 by $v\,\omega$ and 22 by $\omega\,\omega$. The molecular $\Psi$ can be calculated from the ionic $\Phi$, but one can also use the $\Psi$ as new coupling parameters to be determined by symmetry considerations.

If one uses coordinate systems, in which all the $I^{\mu}_{ij}$ are diagonal ($K^{\mu}_{ij} \neq \delta_{ij}$), the solution of (11.10) is simple and can be done as illustrated in § 10. But then all the displacements and librations as well as the coupling parameters $\overset{mn}{\underset{i\,j}{\Psi^{\mu\,\nu}_{\alpha\,\beta}}}$ refer to the coordinate system of principal axes of the molecules, being different for each molecule of a cell in general. But it is possible to use a coordinate system in which the $I^{\mu}_{ik}$ are not diagonal. The solution is then possible again, of course. But, e. g., if there are two kinds of dumbbell-molecules in the lattice with different orientation, the determinant of the tensor of inertia vanishes for both kinds of molecules. Then it seems to be more cumbersome to use a unique coordinate system for all the molecules, but rather a system for each molecule which is oriented parallel to the principal axes. In principle one can use the system one likes.

We will discuss the approximations involved in the derivation of (11.9) from the ionic equation of motion. The main point is the limitation to infinitesimal $\overset{m}{\underset{i}{\omega^{\mu}}}$, that is to an equation of motion which is linear in the

$\overset{m}{\omega}{}^{\mu}_{i}$. This implies that there are small angles of libration. Only in this case rotations about different axes commute. As soon as the angles become larger (for example if free rotations are possible) the theory breaks down. This is already the case in an anharmonic theory, the equation of motion being quadratic in $\overset{m}{\omega}{}^{\mu}_{i}$. This cannot be described in the above theory with infinitesimal, commutable rotations!

The molecules have been considered as rigid. This implies, that the molecules are in the ground-state of the internal vibrations, their atoms having small zero point amplitudes, so that the moments of inertia can be calculated with the equilibrium positions of the ions. The frequencies of the internal vibrations have to be large compared to the frequencies of translational vibrations and librations. No overlap may occur. The same argument holds for the neglect of the Coriolis-energy (11.8d). This contains the angular momentum of the internal vibrations, which vanishes in the ground-state of the internal vibrations.

The other approximation concerns the neglect of the potential energy of interaction $E_{vi}$, $E_{\omega i}$ in (11.9, 10). In [63 B 1] it has been shown, that the contribution from this interaction term to the frequencies is small. But it seems to us, that this approximation, compared to the other ones before, is not as good as those! However, the equations of motion (11.10) should describe very well the translational vibrations and librations of molecular crystals which allow for small libration angles only.

## 12. Some Relations for Coupling-parameters and Derived Expressions in Molecular Crystals

As for the usual coupling matrices $\overset{mn}{\Phi}{}^{\mu\,\nu}_{i\,j}$ also for the $\overset{mn}{\Psi}\begin{smallmatrix}\mu & \nu\\ \alpha & \beta\\ i & j\end{smallmatrix}$ general invariance relations can be derived. In principle, this is possible by transforming the relations of chapter I to the new coordinates. But then one has to know $L$ *and* $L^{-1}$. In general it is simpler to derive the invariance relations for molecular lattices separately.

The most obvious relation is of course the interchange of index pairs:

$$\overset{mn}{\Psi}\begin{smallmatrix}\mu & \nu\\ \alpha & \beta\\ i & j\end{smallmatrix} = \overset{nm}{\Psi}\begin{smallmatrix}\nu & \mu\\ \beta & \alpha\\ j & i\end{smallmatrix} \tag{12.1}$$

following directly from definition.

An arbitrary equal translation of all the centers of mass of the molecules may not cause any force on the molecules. Therefore it follows immediately

$$\sum_{n\nu j} \overset{mn}{\Psi}\begin{smallmatrix}\mu & \nu\\ \alpha & 1\\ i & j\end{smallmatrix} K^{\nu}_{jk} = \sum_{m\mu j} K^{\mu}_{jk} \overset{mn}{\Psi}\begin{smallmatrix}\mu & \nu\\ 1 & \beta\\ j & i\end{smallmatrix} = 0\,. \tag{12.2}$$

The transformation matrices **K** occur, because in non-primitive molecular lattices the $\Psi$ refer to different coordinate systems $\left(\underset{i}{Y^{\mu}_{\alpha}}\right)$ having different axes with respect to a common translation. With the same argument it can be derived for infinitesimal rigid rotation of the whole crystal*

$$\sum_{m\mu l}\left\{\overset{m}{\underset{k}{X^{\mu}}}\,\underset{lj}{K^{\mu}}-\overset{m}{\underset{j}{X^{\mu}}}\,\underset{lk}{K^{\mu}}\right\}\overset{mn}{\underset{l\;\;i}{\Psi^{\mu\;\nu}_{1\;\alpha}}}=\sum_{\substack{m\mu l\\ rs}}\underset{rk}{K^{\mu}}\,\underset{sj}{K^{\mu}}\,\varepsilon_{rsl}\,\overset{mn}{\underset{l\;\;i}{\Psi^{\mu\;\nu}_{2\;\alpha}}}$$

or (12.3a)

$$\sum_{n\nu l}\overset{mn}{\underset{i\;\;l}{\Psi^{\mu\;\nu}_{\alpha\;1}}}\left\{\underset{lj}{K^{\nu}}\,\overset{n}{\underset{k}{X^{\nu}}}-\underset{lk}{K^{\nu}}\,\overset{n}{\underset{j}{X^{\nu}}}\right\}=\sum_{\substack{n\nu l\\ rs}}\overset{mn}{\underset{i\;\;l}{\Psi^{\mu\;\nu}_{\alpha\;2}}}\,\varepsilon_{lrs}\,\underset{rk}{K^{\nu}}\,\underset{sj}{K^{\nu}}\,.$$

In this rotation every molecule remains in its position relative to its own coordinate system $\underset{i}{Y^{\mu}_{\alpha}}$. $\overset{m}{\underset{i}{X^{\mu}}}$ refers to a space coordinate system for the centers of mass of the molecules! If all the molecules have the same directions for the principle axis of inertia, we can choose $K^{\mu}_{rk}=\delta_{rk}$ for all $\mu$ and we have

$$\sum_{n\nu}\left\{\overset{mn}{\underset{i\;\;j}{\Psi^{\mu\;\nu}_{\alpha\;1}}}\,\overset{n}{X^{\nu}_{k}}-\overset{mn}{\Psi^{\mu\;\nu}_{i\;\;k}}\,\overset{n}{X^{\nu}_{j}}\right\}=\sum_{n\nu}\overset{mn}{\underset{i\;\;l}{\Psi^{\mu\;\nu}_{\alpha\;2}}}\,\varepsilon_{lkj}$$

or (12.3b)

$$\sum_{\substack{n\nu\\ jk}}\overset{mn}{\underset{i\;\;j}{\Psi^{\mu\;\nu}_{\alpha\;1}}}\,\overset{n}{X^{\nu}_{k}}\,\varepsilon_{lkj}=\sum_{n\nu}\overset{mn}{\underset{i\;\;l}{\Psi^{\mu\;\nu}_{\alpha\;2}}}\,.$$

A further essential relation, which allows the solution of (11.10) is the invariance of the displacement pattern in a translation of the lattice by multiples of the basis-vectors:

$$\overset{mn}{\underset{i\;\;j}{\Psi^{\mu\;\nu}_{\alpha\;\beta}}}=\overset{m+h\;n+h}{\underset{i\;\;\;\;\;\;j}{\Psi^{\mu\;\;\;\;\nu}_{\alpha\;\;\;\;\beta}}}\qquad\text{for every } \boldsymbol{h}\,. \tag{12.4}$$

This is referred to as the conservation of quasimomentum or, for the librational modes, the conservation of quasi-angular momentum. Of course, it only holds in infinite crystals or in the "interior" of finite crystals (see sect. 6).

Apart from these general relations a lot of statements about the c.p'.s can be derived from the point group symmetries of the special molecular lattices. For simple cases this has been done in some detail: *Hexamethylenetetramine* $(CH_2)_6{\cdot}N_4$ [63 B 2, 64 C 1] and *Adamantan*

---

* Anharmonic effects are excluded (see sect. 11). (12.3) is not stated correctly in [63 B 1]. It can be obtained in a straightforward way by starting with the rotational condition for the $\boldsymbol{\Phi}$ and using the $L$-transformation with properties (11.5, 11.6).

$(CH_2)_6 \cdot (CH)_4$. Both molecules have essentially the same shape. The six $CH_2$-groups form an octahedron containing inside of it a tetrahedron of the four N- or CH-groups. The tetrahedron is arranged in such a way that every N- or CH-group is bonded to three $CH_2$-groups (Fig. 4.9). If one assumes the intermolecular forces to be acting mainly between the outer $CH_2$-groups, the coupling between the molecules has the complete cubic ($O_h$) symmetry, otherwise it has tetrahedral symmetry. The tetrahedral symmetry is strongly related to the positions of the H-ions in the different groups. Hexamethylenetetramine has cubic-body-centered crystal structure and Adamantan cubic-face-centered i. e. both are primitive molecular lattices (no $\mu$, $\nu$-indices). *Biem* has discussed the coupling-matrices up to neighbors of fourth order with the assumption of complete cubic and tetrahedral symmetry of the coupling. The inversion is a symmetry operation only in the first case. In symmetry operations translations transform as polar vectors, librations (rotations) as axial (pseudo) vectors. We will not go into the details of calculating the independent elements of coupling-matrices. It can be done just as it is explained in [I], e. g. if the inversion is a symmetry operation it holds ($\alpha, \beta = 1, 2$)

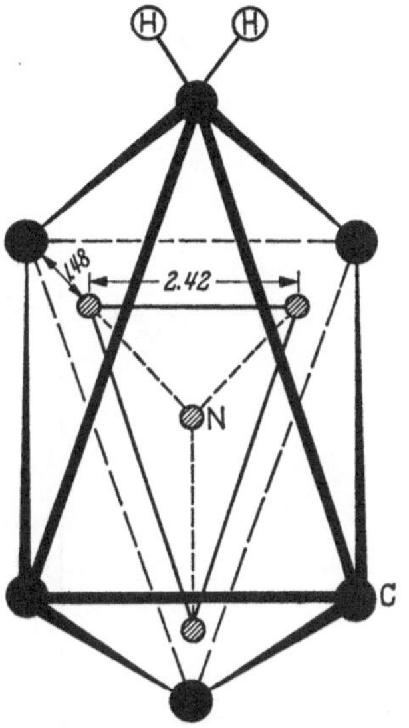


Fig. 4.9. The Hexamethylenetetramine-molecule, $(CH_2)_6 \cdot N_4$ schematically. The C-atoms form an octahedron, the N-atoms a tetrahedron inside the octahedron. The H-atoms are connected to the C-atoms indicated only for one C-atom. If the intermolecular forces are determined by the C-atoms, there is octahedral symmetry, otherwise (and most probably) there is only tetrahedral symmetry

$$\Psi^{0\,h}_{\substack{\alpha\alpha \\ i\,j}} = \Psi^{h\,0}_{\substack{\alpha\alpha \\ i\,j}}; \quad \Psi^{0\,h}_{\substack{\alpha\beta \\ i\,j}} = -\Psi^{h\,0}_{\substack{\alpha\beta \\ i\,j}} \quad \text{if} \quad \alpha \neq \beta\,. \tag{12.5}$$

The minus-sign is related to the axial character of the libration-displacement. Direct inspection gives the following independent matrices. Equivalent matrices can be gotten by a suitable transformation. We have put

$$\Psi^{0\,h}_{\substack{\alpha\beta \\ i\,j}} = \begin{Bmatrix} \Psi^{0\,h}_{\substack{11 \\ i\,j}} & \Psi^{0\,h}_{\substack{12 \\ i\,j}} \\ \Psi^{0\,h}_{\substack{21 \\ i\,j}} & \Psi^{0\,h}_{\substack{22 \\ i\,j}} \end{Bmatrix}. \tag{12.6}$$

It is

$$\Psi^{0\,0}_{\substack{\alpha\beta \\ i\,j}} = \overset{0}{A_\alpha}\, \delta_{\alpha\beta}\, \delta_{ij}\,. \tag{12.7}$$

Further we have

1) tetrahedral symmetry of the coupling:

$$-\Psi^{0\,100}_{\substack{\alpha\beta\\ i\;j}} = \left\{\begin{matrix} A_1 & 0 & 0 & 0 & 0 & 0 \\ 0 & B_1 & 0 & 0 & a_1 & b_1 \\ 0 & 0 & B_1 & 0 & -b_1 & -a_1 \\ 0 & 0 & 0 & \alpha_1 & 0 & 0 \\ 0 & a_1 & b_1 & 0 & \beta_1 & 0 \\ 0 & -b_1 & -a_1 & 0 & 0 & \beta_1 \end{matrix}\right\};$$

6 independent parameters

(12.8)

$$-\Psi^{0\,110}_{\substack{\alpha\beta\\ i\;j}} = \left\{\begin{matrix} B_2 & C_2 & D_2 & a_2 & b_2 & -d_2 \\ C_2 & B_2 & D_2 & -b_2 & -a_2 & d_2 \\ -D_2 & -D_2 & A_2 & c_2 & -c_2 & 0 \\ a_2 & -b_2 & -c_2 & \beta_2 & \gamma_2 & \delta_2 \\ b_2 & -a_2 & c_2 & \gamma_2 & \beta_2 & \delta_2 \\ d_2 & -d_2 & 0 & -\delta_2 & -\delta_2 & \alpha_2 \end{matrix}\right\}.$$

12 independent parameters

$$-\Psi^{0\,111}_{\substack{\alpha\beta\\ i\;j}} = \left\{\begin{matrix} A_3 & C_3 & C_3 & 0 & a_3 & -a_3 \\ C_3 & A_3 & C_3 & -a_3 & 0 & a_3 \\ C_3 & C_3 & A_3 & a_3 & -a_3 & 0 \\ 0 & a_3' & -a_3' & \alpha_3 & \gamma_3 & \gamma_3 \\ -a_3' & 0 & a_3' & \gamma_3 & \alpha_3 & \gamma_3 \\ a_3' & -a_3' & 0 & \gamma_3 & \gamma_3 & \alpha_3 \end{matrix}\right\};$$

6 independent parameters

$$-\Psi^{0\,211}_{\substack{\alpha\beta\\ i\;j}} = \left\{\begin{matrix} A_4 & D_4 & D_4 & 0 & c_4 & -c_4 \\ D_4 & B_4 & C_4 & d_4 & a_4 & b_4 \\ D_4 & C_4 & B_4 & -d_4 & b_4 & -a_4 \\ 0 & -d_4' & d_4' & \alpha_4 & \delta_4 & \delta_4 \\ -c_4' & -a_4' & b_4' & \delta_4 & \beta_4 & \gamma_4 \\ c_4' & -b_4' & a_4' & \delta_4 & \gamma_4 & \beta_4 \end{matrix}\right\}.$$

16 independent parameters

2) octahedral symmetry of the coupling:

Now we have additionally to the former symmetries the inversion symmetry (12.5). Therefore it is in case*

100:

$$a_1 = 0$$

5 independent parameters

110:

$$a_2 = b_2 = 0$$

$$D_2 = \delta_2 = 0$$

8 independent parameters

---

* $c_2$ and $d_2$ are not zero, as stated in [63 B 2]. Therefore translations are not independent from the librations for nearest neighbor coupling. For the case of nearest neighbor in Adamantan the conditions of translational and rotational invariance read: $A_1^0 = -4(A_2 + 2B_2)$; $4(c_2 + d_2)\,a = A_2^0 + 4(\alpha_2 + 2\beta_2)$, $a$: lattice constant. Recent measurements indicate, that the cubic-face-centered structure of Adamantan is stable only at high temperatures; at low temperatures a tetragonal structure is stable. For this case the coupling-parameters are discussed by *Stockmeyer* [65 S 9].

111:

$$a_3 = a_3'$$

5 independent parameters

211:

$$a_4 = a_4'; \quad b_4 = b_4'$$
$$c_4 = c_4'; \quad d_4 = d_4' \qquad (12.9)$$

12 independent parameters

After specifying the parameters, the conditions (12.2) and (12.3) have to be satisfied. In this case (12.2) relates $A_1^0$ of (12.7) with the other c.p'.s, whereas (12.3) gives a certain relation for the mixed coefficients $\Psi_{12\,ij}^{0h}$! With these c.p'.s the dispersion curves can be calculated, at least in principle and for some symmetry directions of propagation (100-, 110-, 111-direction). Because of (12.2) there are three modes with frequency $\omega \sim q \to 0$ with wave-vector $q \to 0$. The three other modes have librational character for $\omega \to 0$ and finite frequency. For larger $\omega$ translational (vibrational) and librational modes have more interaction and not a "pure" character. We will not give details of a calculation, but refer to [63 B 2, 64 C 2]. Fig. 4.10 is taken from [64 C 2] having been calculated by a machine with parameters taken from experimental values of the elastic constants and Raman-frequencies of Hexamethylenetetramine. The librational frequencies are in order of magnitude of $\omega \approx 10^{13}\,\text{sec}^{-1}$, whereas the internal vibrations of molecules have frequencies (angular) in the order $4.10^{13}$ to $10^{15}\,\text{sec}^{-1}$, in general.

The calculation of elastic and caloric data is more involved as in Bravais-lattices, because of the many degrees of freedom. Only in very special cases it can be simplified. As an example, we will consider the

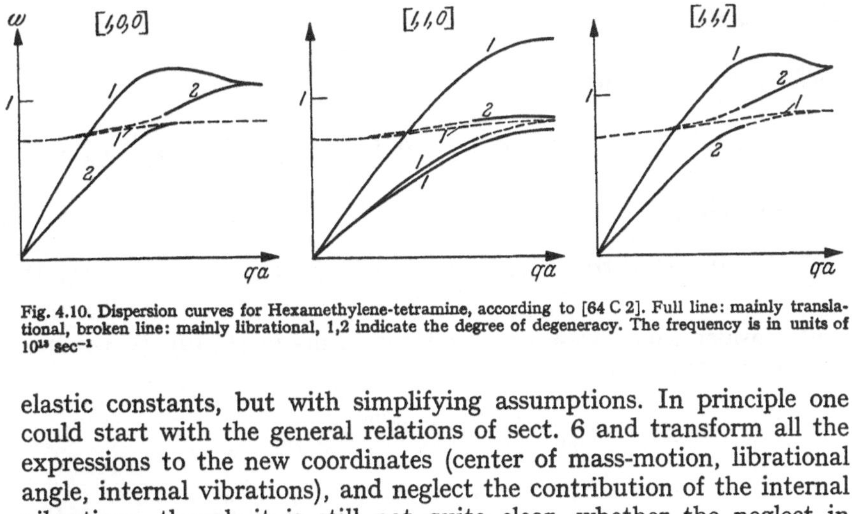


Fig. 4.10. Dispersion curves for Hexamethylene-tetramine, according to [64 C 2]. Full line: mainly translational, broken line: mainly librational, 1,2 indicate the degree of degeneracy. The frequency is in units of $10^{13}\,\text{sec}^{-1}$

elastic constants, but with simplifying assumptions. In principle one could start with the general relations of sect. 6 and transform all the expressions to the new coordinates (center of mass-motion, librational angle, internal vibrations), and neglect the contribution of the internal vibrations, though it is still not quite clear, whether the neglect in calculations of elastic constants is justified. If the internal degrees of freedom can be omitted, in many cases it is easier to start with (11.10) and to take the long wave limit of it. In more complicated cases (no inversion symmetry, non-primitive molecular lattices) it seems to be more appropriate to proceed as in section 6, which is correct for every lattice build up by point-ions.

Of course, the Cauchy-relations (I) will not be valid in any molecular lattice. For them to be satisfied, *each ion* (not only the centers of mass of the molecules) has to be a center of inversion symmetry. Molecular lattices which fulfill this requirement seem not to be existing. So what we expect is at least a violation of Cauchy's relation, e. g. $c_{12} = c_{44}$ for cubic crystals, even if all the forces are central forces.

The calculations that have been done up till now [64 C 1; 66 D 1] use the assumption, that it holds

$$\sum_h \Psi_{\substack{11\\ij}}^{0h} X_k^h = 0\,. \tag{12.10}$$

Obviously this is true in the following cases: *Every ion* of the lattice (not only the molecules' center of masses) is a center of inversion, or rather, the interatomic forces have this symmetry, — or *all the forces* are central forces. Apart from this (12.10) is satisfied in some special cases, which can be seen from the coupling matrices (12.8).

With this restriction the calculation of the elastic constants is simple and we will consider this simple model as an example in the following. The molecular crystal is supposed further to be a "primitive" molecular crystal. These assumptions hold e. g. for Hexamethylenetetramine or Adamantan, if the intermolecular forces have octahedral symmetry.

We introduce slowly varying "displacement" fields

$$\begin{aligned}
u_j^m &= u_j(\boldsymbol{R})\,; \quad \boldsymbol{R} = \boldsymbol{R}^m; \qquad\qquad |k = \frac{\partial}{\partial X_k}\\
u_j^n &= u_i(\boldsymbol{R} + \boldsymbol{R}^n - \boldsymbol{R}^m)\\
&= u_j(\boldsymbol{R}) + \sum_k \left(X_k^n - X_k^m\right) u_{j|k} + \frac{1}{2}\sum_{k,l}\left(X_k^n - X_k^m\right)\left(X_l^n - X_l^m\right) u_{j|kl} + \cdots\\
\omega_j^m &= \omega_j(\boldsymbol{R})\\
\omega_j^n &= \omega_j(\boldsymbol{R} + \boldsymbol{R}^n - \boldsymbol{R}^m) = \omega_j(\boldsymbol{R}) + \sum_k \left(X_k^n - X_k^m\right)\omega_{j|k} + \cdots
\end{aligned} \tag{12.11}$$

and consider only lowest non-vanishing terms in (11.10), ($\mu, \nu$-indices can be omitted now).

$$\begin{aligned}
M\,\ddot{u}_i = &- \sum_h \Psi_{\substack{11\\ij}}^{0h} u_j - \sum_h \Psi_{\substack{11\\ij}}^{0h} X_k^h u_{j|k} -\\
&- \frac{1}{2}\sum_h \Psi_{\substack{11\\ij}}^{0h} X_k^h X_l^h u_{j|kl} - \sum_h \Psi_{\substack{12\\ij}}^{0h} \omega_j -\\
&- \sum_h \Psi_{\substack{12\\ij}}^{0h} X_k^h \omega_{j|k} - \cdots
\end{aligned} \tag{12.12a}$$

$$I_{ik}\,\ddot{\omega}_k = -\sum_{\boldsymbol{h}} \Psi\overset{0\boldsymbol{h}}{\underset{ij}{21}}\, u_j - \sum_{\boldsymbol{h}} \Psi\overset{0\boldsymbol{h}}{\underset{ij}{21}}\, \overset{\boldsymbol{h}}{\underset{k}{X}}\, u_{j|k} - \\ - \sum_{\boldsymbol{h}} \Psi\overset{0\boldsymbol{h}}{\underset{ij}{22}}\, \omega_j - \cdots \tag{12.12b}$$

where $\boldsymbol{n} - \boldsymbol{m} = \boldsymbol{h}$; $\boldsymbol{R}^n - \boldsymbol{R}^m = \boldsymbol{R}^h$ (see 12.4). A lot of terms vanish: In (12.12a) the first because of (12.2) with $\overset{\nu}{\underset{jk}{K}} = \delta_{jk}$ (only one sort of molecules and orientation), the second because of (12.10), the fourth because of (12.2) and

$$\sum_{\boldsymbol{h}} \Psi\overset{0\boldsymbol{h}}{\underset{ij}{12}} = \sum_{\boldsymbol{h}} \Psi\overset{-\boldsymbol{h}0}{\underset{ij}{12}} = \sum_{\boldsymbol{h}} \Psi\overset{\boldsymbol{h}0}{\underset{ij}{12}} = 0\,,$$

and in (12.12b) the first term because of (12.2).

We introduce $\varrho = M/V_z$,

$$\hat{C}_{ij,kl} = -\frac{1}{2V_z}\sum_{\boldsymbol{h}} \Psi\overset{0\boldsymbol{h}}{\underset{ij}{11}}\, \overset{\boldsymbol{h}}{\underset{k}{X}}\, \overset{\boldsymbol{h}}{\underset{l}{X}}\,,$$

$$\hat{C}_{i,jk} = -\frac{1}{V_z}\sum_{\boldsymbol{h}} \Psi\overset{0\boldsymbol{h}}{\underset{ij}{22}}\, \overset{\boldsymbol{h}}{\underset{k}{X}}\,, \tag{12.13}$$

$$\hat{C}_{ij} = -\frac{1}{V_z}\sum_{\boldsymbol{h}} \Psi\overset{0\boldsymbol{h}}{\underset{ij}{22}}\,; \quad (\hat{C}^{-1})_{ij} = R_{ij}\,.$$

Because of (12.3) and (12.4) the following symmetry relations hold:

$$\hat{C}_{ij,kl} = \hat{C}_{ji,kl} = \hat{C}_{ij,lk}\,, \tag{12.14a}$$

$$\hat{C}_{i,jk} - \hat{C}_{i,kj} = \sum_l \hat{C}_{il}\,\varepsilon_{lkj}\,, \tag{12.14b}$$

$$\hat{C}_{ij} = \hat{C}_{ji}\,; \quad R_{ij} = R_{ji}\,. \tag{12.14c}$$

Then from (12.12) we have

$$\varrho\,\ddot{u}_i = \hat{C}_{ij,kl}\,u_{j|kl} - \hat{C}_{m,ik}\,\omega_{m|k} \tag{12.15a}$$

$$\frac{1}{V_z} I_{ik}\,\ddot{\omega}_k = \hat{C}_{i,jl}\,u_{j|l} + \hat{C}_{im}\,\omega_m\,. \tag{12.15b}$$

$\omega_m$ has to be eliminated from the second equation and to be inserted into (12.15a). The term $I_{ik}\,\ddot{\omega}_k$ can be neglected with the same argument as in sect. 6, page 21: In the limit of very long waves $u_{j|l}$ and $\omega_{m|k}$ vanish, therefore also $\ddot{u}_i$ which describes the acoustical waves in this limit. $\ddot{\omega}_k \sim \hat{C}_{im}\,\omega_m$ must describe the librational modes with non-vanishing frequencies in this limit. These frequencies are "high" compared with the elastical acoustical ones, which are of sole interest here. We can neglect librational waves here and use stationary solutions of (12.15b), which are

$$\omega_m = -(\hat{C}^{-1})_{mn}\,\hat{C}_{n,jl}\,u_{j|l} = -R_{mn}\,\hat{C}_{n,jl}\,u_{j|l}\,. \tag{12.16}$$

$\hat{C}^{-1}_{ij}$ does exist always, because it can be seen in the limit of $u_{j|l} = 0$ (long wave limit) that $\hat{C}_{im}$ is connected with the frequencies of the librational modes of this limit, which are nonzero (see page 57/62). Inserting (12.16) into (12.15a) we have

$$\varrho\,\ddot{u}_i = \{\hat{C}_{ij|kl} + \hat{C}_{m,ik}\,R_{mn}\,\hat{C}_{n,jl}\}\,u_{j|kl} \tag{12.17}$$

which is identical in form to (6.43). This is to be expected because a molecular lattice is a special form of a non-primitive lattice, where internal motions of the molecules have been neglected. (12.17) has to be compared with the usual elastic equation of motion (6.44)

$$\varrho\, \ddot{u}_i = S_{ik,jl}\, u_{j|kl} \tag{12.18}$$

in our case of a stress-free reference state. In (12.18) only that part of $S_{ik,jl}$ enters which is symmetric in $k$ and $l$. Therefere the symmetric parts in $k$ and $l$ of (12.17) and (12.18) have to be identical. This gives

$$\begin{aligned} S_{ik,jl} + S_{il,jk} = 2\hat{C}_{ij,kl} + \hat{C}_{m,ik} R_{mn} \hat{C}_{n,jl} + \\ + \hat{C}_{m,il} R_{mn} \hat{C}_{n,jk}\,. \end{aligned} \tag{12.19}$$

In this way one obtains a statement only for the combination $S_{ik,jl} + S_{il,jk}$. The $\hat{C}_{m,ik}$ do not have the usual symmetries (see 12.14b) and so do the $S_{ik,jl}$. Therefore (12.19) cannot be solved for the $S_{ik,jl}$. If the above procedure would be correct, in sound wave experiments one would measure just (12.19). But (12.19) includes certain assumptions, which cannot be justified in general:

(i) Primitive molecular crystals (examples exist, see above).

(ii) The intermolecular forces have inversion symmetry, or *all the forces* are central forces. It is questionable, whether this is satisfied in any molecular crystal. In some models at least it can be justified, condition (12.10).

(iii) The contribution of the internal degrees of freedom of the molecules to the elastic constants can be neglected. This assumption seems to be most restrictive for *all* molecular crystals, and it seems questionable, whether it can be justified in any case. Probably this neglect is related to the deviation from symmetry in (12.19)*.

Because of all these reasons, one should be cautious when using (12.19). The same expression has been given in [66 D 1], using a somewhat different derivation, and it has been identified with the elastic constants, which might be questioned.

Experimental values of the elastic constants exist for Hexamethylenetetramine [58 H 1]. For Adamantan there are no experimental values; a rough estimate can be done using van der Waals-central-forces between molecules, the strength being of the same order of magnitude as in Hexamethylenetetramine. So we have

| Hexamethylenetetramine (experimental) | | Adamantan (estimated) | |
|---|---|---|---|
| $c_{11}$ | 1.643 | 0.86 | $10^{11}$ dyn cm$^{-2}$ |
| $c_{12}$ | 0.433 | 0.43 | $10^{11}$ |
| $c_{44}$ | 0.515 | 0.43 | $10^{11}$ |
| $c_{44} - c_{12}$ | 0.082 | ? | $10^{11}$ |
| $a$ | 7.02 | 9.43 | $10^{-8}$ cm |

* One could argue, that in a molecular lattice the ions cannot be looked upon as mass-points. But one could use a model lattice with strong springs between the point-ions of a molecule, and very weak springs between the different molecules; then the elastic constants must have the correct symmetries, which they do not have according to [66 D 1], the procedure being sketched above.

The difference $c_{44} - c_{12}$ is not large. The measurements have been done at 20° C, so that anharmonic effects are possible, and could explain the difference quite well. Measurements over the entire range of temperature in principle can settle this question (sect. 15). Apart from this, all the points (i) (ii) (iii) can be used as an explanation.

The force-constants can be estimated using (12.19). We have (neglecting the last term)

| Hexamethylenetetramine (experimental) | Adamantan (estimated) |
|---|---|
| $c_{11} = \frac{1}{a}(2A_3 + 4A_1)$ | $c_{11} = \frac{1}{a} 4B_2$ |
| $c_{12} = \frac{1}{a}(4C_3 - 2A_3 - 4B_1)$ | $c_{12} = \frac{1}{a}(4C_2 - 2A_2 - 2B_2)$ |
| $c_{44} = \frac{1}{a}(2A_3 + 4B_1)$ | $c_{44} = \frac{1}{a}(2A_2 + 2B_2)$ |

Assuming $B_1$ small, we find

| Hexamethylenetetramine | Adamantan |
|---|---|
| $A_3 = 1.81 \cdot 10^3$ dyn cm$^{-1}$ | $B_2 = 2.03 \cdot 10^3$ dyn cm$^{-1}$ |
| $C_3 = 1.67 \cdot 10^3$ dyn cm$^{-1}$ | $C_2 = 2.03 \cdot 10^3$ dyn cm$^{-1}$ |
| $A_1 = 1.98 \cdot 10^3$ dyn cm$^{-1}$ | $A_2 = ?$ |
| $B_1 < 0.1 \cdot 10^3$ dyn cm$^{-1}$ | |

In a central force model for Hexamethylenetetramine one would expect $3A_3 = 3C_3 = f_3$; $A_1 = f_1$; $B_1 = 0$. The deviations are very small and can be explained as well by anharmonicities as by many-body-forces as by tetrahedral symmetry of otherwise central-forces. The proportion $f_1/f_3 \approx 0.38$ is compatible with the assumption of essentially van der Waals-forces between the molecules.

# V. Anharmonic Effects in Thermodynamical Properties

## 13. Mechanical Discussion of a Simple Linear Model

The harmonic theory of the thermodynamical properties of crystal lattices does not account for a number of effects such as thermal expansion, the temperature dependence and the difference between isothermal and adiabatic elastic constants, the difference between the specific heat at constant volume and constant pressure and a lot of other effects [I, II]. These effects are strongly related to anharmonic terms in the expansion of the potential energy with respect to the displacements (1.6, 7). Before discussing the general quantum mechanical treatment of such an anharmonic system of oscillators, we will describe a simple model, which at least shows most features which occur.

We consider a linear lattice with two types of ions in it, and generalize this model in a certain respect beyond the usual one:

i) We discuss a finite lattice in order to show the influence of surfaces, if such an effect is present at all.

ii) We include electrical effects in order to obtain statements about pyroelectric, piezoelectric, etc. constants; that means we assume the ions to have charges, and we assume the presence of an external electric field. In order to have a consistent linear model, — the linear Poisson equation has to be satisfied — we cannot represent the ionic charges by point charges, but only by charged plates (Fig. 5.1), the dimension of which can be assumed of the order of magnitude of a lattice constant (area $b^2$).

(iii) The linear lattice indicated in Fig. 5.1. has a finite dipolemoment $Nea_1$. In general the lattices in equilibrium do not have a finite dipolemoment. Space-charges in the vicinity of the surface will be build up to compensate such a dipole-moment. The space charges originate (a) from the electrons of the crystal, if it has a small but finite conductivity, which allows the electrons to move to the surface, or (b) from electrons taken from the surrounding medium. To describe this effect, we allow the surfaces to have additional charges $\pm Q$. In our model we assume these charges to be located at the ends of the lattice. For a complete compensation it has to be

$$Q = \frac{a_1}{a - a_2/N}\, e \approx \frac{a_1}{a}\, e\,. \tag{13.1}$$

with $N \gg 1$; terms with $O(1/N)$ will be dropped in the following.

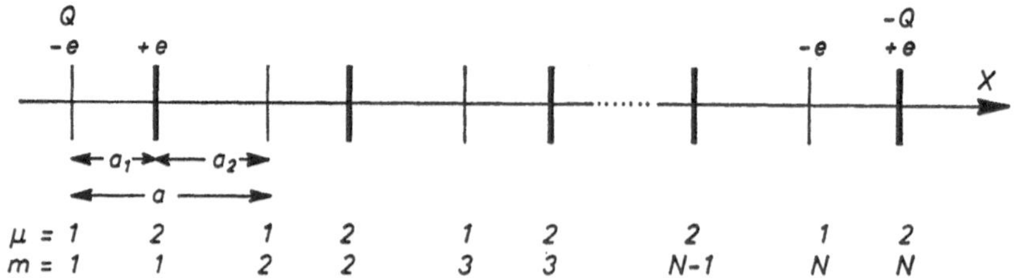


Fig. 5.1. Linear chain of condensor plates, each having a charge $\pm e$. The end plates have additional surface charges $\pm Q$. Equilibrium configuration

This model is essentially due to *K. Thoma* [67 T 1]; *Ninio* [62 N 2] discusses a similar model, but he looses some statements by a symmetrization of the energy with respect to charges ($\pm e$) and force-constants ($f_1, f_2$), the justification of which is not given. The above model without charges and electric field has been discussed in [I] and [64 L 2].

The equilibrium positions of the ions are denoted by $X_\mu^m$, $m = 1 \ldots N$, $\mu = 1, 2$. It is

$$X_2^m - X_1^m = a_1; \quad X_1^{m+1} - X_2^m = a_2\,. \tag{13.2}$$

The length (volume) of the cell is $a = a_1 + a_2$. The masses and charges of the particles are $M_1, M_2$ or $e_1 = -e$, $e_2 = e$, resp. The mechanical potential between the particles is assumed to be

$$\begin{aligned} &\varphi_1\left(a_1 + \overset{m}{u_2} - \overset{m}{u_1}\right) \quad \text{between} \quad \overset{m}{X_1} \text{ and } \overset{m}{X_2} \\ &\varphi_2\left(a_2 + \overset{m+1}{u_1} - \overset{m}{u_2}\right) \quad \text{between} \quad \overset{m}{X_2} \text{ and } \overset{m+1}{X_1}. \end{aligned} \tag{13.3}$$

Here $\overset{m}{u_\mu}$ are the displacements from the equilibrium positions. Only "small" displacements are considered in the following, so that an expansion with respect to the $\overset{m}{u_\mu}$ is reasonable. Two kinds of external

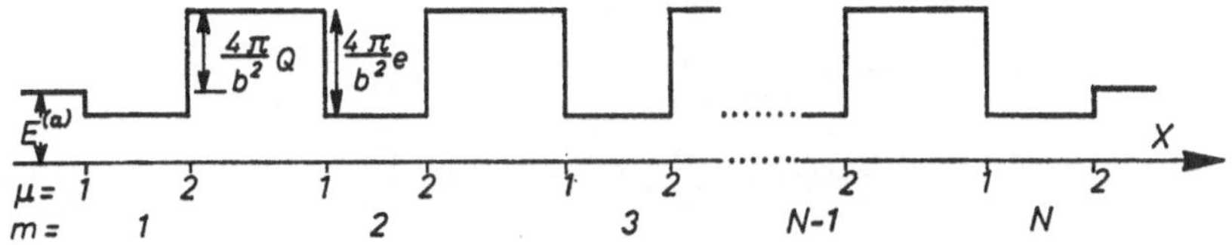


Fig. 5.2. The electric field for the model of fig. 5.1, shown in the equilibrium configuration

forces will be considered: Mechanical surface forces $\overset{m}{f_\mu}$ with $\overset{1}{f_1} = -F$ and $\overset{N}{F_2} = F$, all other $\overset{m}{f_\mu} = 0$ (these surface forces represent stresses $\sigma = F/b^2$, where $b^2$ is the area of the plates) and an external electric field $E^{(a)}$, which allows us to define electrical properties (piezoelectric constants etc.) in the following.

The electrical interaction is more complicated than the mechanical one. The total electric field consists of three contributions:

i) The internal field of the charges $\pm e$. It is zero outside the lattice and changes discontinuously by $\pm 4\pi e/b^2$ at the site of the ions. The total lattice is electrically neutral.

ii) The field of the surface charges $\pm Q$, which is equal to $4\pi Q/b^2$ in the interior of the lattice and zero outside.

iii) The external field $E^{(a)}$.

The total field is (Fig. 5.2)

$$\begin{aligned} E(X) = E^{(a)} + \frac{4\pi}{b^2} \sum_{m\mu} e_\mu\, \theta\left(X - \overset{m}{X_\mu} - \overset{m}{u_\mu}\right) + \\ + (4\pi Q/b^2)\left\{\theta\left(X - \overset{1}{X_1} - \overset{1}{u_1}\right) - \theta\left(X - \overset{N}{X_2} - \overset{N}{u_2}\right)\right\}; \end{aligned} \tag{13.4}$$

$$\theta(X) = \begin{cases} 1 & \text{if } X > 0 \\ 0 & \text{if } X < 0. \end{cases}$$

If we average this field in the interior of the crystal, we have in the interior of the crystal (in the equilibrium state $\overset{m}{u_\mu} = 0$),

$$\overline{E} = E^{(a)} + \frac{4\pi}{b^2} Q - \frac{4\pi}{b^2} \cdot \frac{a_1}{a} e \tag{13.5}$$

and in the special case that there is no total dipole-moment (13.1)

$$\overline{E} = E^{(a)}. \tag{13.5a}$$

The electrostatic potential of the field (13.4) is given by

$$
\begin{aligned}
V_E(X) = {} & -E^{(a)}X - \frac{2\pi}{b^2}\sum_{m\mu} e_\mu \left|X - \overset{m}{X}_\mu - \overset{m}{u}_\mu\right| - \\
& - \frac{2\pi}{b^2} Q\left\{\left|X - \overset{1}{X}_1 - \overset{1}{u}_1\right| - \left|X - \overset{N}{X}_2 - \overset{N}{u}_2\right|\right\} + \\
& + \frac{2\pi}{b^2} N(e a_1 - Q a) + \frac{2\pi}{b^2}\left\{\sum_{m\mu} e_\mu \overset{m}{u}_\mu - Q\left(\overset{N}{u}_2 - \overset{1}{u}_1\right)\right\}
\end{aligned}
\tag{13.6}
$$

apart from terms $O(1/N)$.

The electrostatic energy of this system can be divided into three contributions: the interaction energy of the charges, which is the Coulomb-interaction for this linear system

$$
\begin{aligned}
\Phi_C = {} & (2\pi e^2/b^2)\sum_m \left(\overset{m}{X}_2 - \overset{m}{X}_1 + \overset{m}{u}_2 - \overset{m}{u}_1\right) - \\
& - (4\pi Q e/b^2)\sum_m \left(\overset{m}{X}_2 - \overset{m}{X}_1 + \overset{m}{u}_2 - \overset{m}{u}_1\right) + \\
& + (2\pi Q^2/b^2)\left(\overset{N}{X}_2 - \overset{1}{X}_1 + \overset{N}{u}_2 - \overset{1}{u}_1\right),
\end{aligned}
\tag{13.7}
$$

the contribution, which is linear in the applied field,

$$
\begin{aligned}
\Phi_E & = - E^{(a)}\left\{\sum_{m\mu} e_\mu \left(\overset{m}{X}_\mu + \overset{m}{u}_\mu\right) - Q\left(\overset{N}{X}_2 + \overset{N}{u}_2 - \overset{1}{X}_1 - \overset{1}{u}_1\right)\right\} \\
& = - E^{(a)}\mathcal{M}
\end{aligned}
\tag{13.8}
$$

where $\mathcal{M}$ is the total dipole-moment of the lattice. The third contribution is quadratic in $E^{(a)}$ and represents the field energy, which is irrelevant in the following considerations.

The potential of the external mechanical forces is

$$
\Phi_K = - F\left(\overset{N}{X}_2 + \overset{N}{u}_2 - \overset{1}{X}_1 - \overset{1}{u}_1\right) \tag{13.9}
$$

and that of the mechanical interaction

$$
\Phi_M = \sum_m \left\{\varphi_1\left(a_1 + \overset{m}{u}_2 - \overset{m}{u}_1\right) + \varphi_2\left(a_2 + \overset{m+1}{u}_1 - \overset{m}{u}_2\right)\right\}. \tag{13.10}
$$

Terms of the order $O(1/N)$ have been neglected always. The total interaction is then

$$
\Phi = \Phi_M + \Phi_C + \Phi_K + \Phi_E\,. \tag{13.11}
$$

A common displacement of the whole system in the homogeneous field $E^{(a)}$ does not change the interaction energy of the system. This can be seen immediately. The internal interaction is represented by the first two terms in (13.11). We make the usual assumption, that it is reasonable to expand $\Phi_M$ with respect to the $\overset{m}{u}_\mu$, and that this expansion converges sufficiently well. With

$$
\begin{aligned}
& \varphi_0 = \varphi_1(a_1) + \varphi_2(a_2) + (2\pi/b^2)\,(e^2 a_1 - 2Qea_1 + Q^2 a) \\
& d_1 = \varphi_1'(a_1) + (2\pi/b^2)\,(Q-e)^2; \quad d_2 = \varphi_2'(a_2) + 2\pi Q^2/b^2; \\
& f_1 = \varphi_1''(a_1); \quad f_2 = \varphi_2''(a_2); \quad g_1 = \varphi_1'''(a_1); \ldots
\end{aligned}
\tag{13.12}
$$

we obtain

$$\Phi_M + \Phi_C = N\varphi_0 + \tag{13.13}$$
$$+ \sum_{m=1}^{N} \left\{ d_1\left(\overset{m}{u^2} - \overset{m}{u^1}\right) + \frac{1}{2} f_1 \left(\overset{m}{u^2} - \overset{m}{u^1}\right)^2 + \frac{1}{6} g_1 \left(\overset{m}{u^2} - \overset{m}{u^1}\right)^3 + \cdots \right\} +$$
$$+ \sum_{m=1}^{N-1} \left\{ d_2\left(\overset{m+1}{u^1} - \overset{m}{u^2}\right) + \frac{1}{2} f_2 \left(\overset{m+1}{u^1} - \overset{m}{u^2}\right)^2 + \frac{1}{6} g_2 \left(\overset{m+1}{u^1} - \overset{m}{u^2}\right)^3 + \cdots \right\}.$$

In the case of vanishing external forces, the equilibrium positions are defined by the minimum of the potential energy $\Phi_M + \Phi_C$, therefore

$$d_1(a_1) = d_2(a_2) = 0\,. \tag{13.14}$$

This defines $a_1$ and $a_2$.

Under the influence of the external force $F$ and the homogeneous field $E^{(a)}$ the crystal will suffer a homogeneous deformation, which can be described by

$$\overset{m}{u^\mu} = v\overset{m}{X^\mu} + v^\mu, \quad v a = \Delta a\,, \tag{13.15}$$

with $v$ and $v^\mu$ independent of the position. $v$ is the linear strain tensor, $v^\mu$ describes the individual displacements of the two sublattices. Because of the translation symmetry, the energy may depend only on the relative displacements

$$w = v^2 - v^1\,. \tag{13.15a}$$

Substituting (13.15) into (13.13) and taking into account $\Phi_K$ and $\Phi_E$ we obtain

$$\begin{aligned}\Phi/N = {} & \varphi_0 + \frac{1}{2} f_1 (v a_1 + w)^2 + \frac{1}{6} g_1 (v a_1 + w)^3 + \cdots \\ & + \frac{1}{2} f_2 (v a_2 - w)^2 + \frac{1}{6} g_2 (v a_2 - w)^3 + \cdots \\ & - F a (1 + v) + E^{(a)} Q a (1 + v) - \\ & - E^{(a)} e a_1 (1 + v) - E^{(a)} e w\,.\end{aligned} \tag{13.16}$$

Always terms $O(1/N)$ have been neglected. To allow for a comparison with the threedimensional case and with the macroscopic elastic theory, we define stress and polarization

$$\sigma = F/b^2, \quad P = \mathcal{M}/N V_{z0} \tag{13.17}$$

where $V_{z0} = a \cdot b^2$ corresponds to the volume of a unit cell. The polarization of the equilibrium state is

$$P_0 = (e a_1 - Q a)/V_{z0}\,. \tag{13.18}$$

Inserting (13.17) and (13.18) into (13.16) we have

$$\begin{aligned}\Phi/N V_{z0} = {} & \varphi_0/V_{z0} + E^{(a)}(Q a - e a_1)/V_{z0} - F/b^2 + \\ & + \frac{1}{V_{z0}} \left\{ \frac{1}{2} f_1 (v a_1 + w)^2 + \frac{1}{2} f_2 (v a_2 - w)^2 + \cdots \right\} - \\ & - v(\sigma + E^{(a)} P_0) - E^{(a)} e w / V_{z0}\,.\end{aligned} \tag{13.19}$$

If we consider $v$ and $w$ as independent variables, the conjugated ones are $\sigma + E^{(a)} P_0$ and $E^{(a)}$. $\sigma + E^{(a)} P_0$ represents a generalized stress, which includes the Maxwellian stress of the charged system. $\Phi/N V_{z0}$

is the energy density of the system, related to the undeformed volume $V_{z0}$. The deformation of the lattice generates a change in the polarization of the lattice, which can be found as

$$\Delta P = v(ea_1 - Qa)/V_{z0} + ew/V_{z0} = vP_0 + ew/V_{z0}, \tag{13.20}$$

where the change is again related to the undeformed volume. It contains two contributions: one originates from the homogeneous strain of the whole lattice, the other originates from the relative displacement of opposite charges.

The equilibrium position with external force and field present is

$$\frac{\partial}{\partial v}(\Phi/NV_{z0}) = \frac{\partial}{\partial w}(\Phi/NV_{z0}) = 0. \tag{13.21}$$

This gives

$$\begin{aligned} V_{z0}(\sigma + E^{(a)}P_0) &= f_1 a_1(va_1 + w) + \frac{1}{2}g_1 a_1(va_1 + w)^2 + \cdots \\ &\quad + f_2 a_2(va_2 - w) + \frac{1}{2}g_2 a_2(va_2 - w)^2 + \cdots \\ eE^{(a)} &= f_1(va_1 + w) + \frac{1}{2}g_1(va_1 + w)^2 + \cdots \\ &\quad - f_2(va_2 - w) - \frac{1}{2}g_2(va_2 - w)^2 + \cdots. \end{aligned} \tag{13.22}$$

In this equation all the quantities are properly defined. $\sigma$ is essentially the external mechanical force, $E^{(a)}$ the given external field. $v$ can be measured by the change in length of the crystal, $w$ by the change in polarization, due to external forces. It is a matter of convenience, which of the quantities are considered as the independent ones, and which are the dependent ones. For theoretical purposes it is useful, to take $v$ and $E^{(a)}$ as independent variables. A solution of (13.22) with respect to $\sigma + E^{(a)}P_0$ and $w$ gives, up to the same order in anharmonicity ($g_1, g_2$)

$$\begin{aligned} \sigma + E^{(a)}P_0 &= c_{vv}\cdot v + c_{vE}\cdot E^{(a)} + \frac{1}{2}c_{vvv}\cdot v^2 + \\ &\quad + c_{vvE}\cdot vE^{(a)} + \frac{1}{2}c_{vEE}\cdot E^{(a)2} + \cdots \end{aligned} \tag{13.23a}$$

$$\begin{aligned} ew/V_{z0} &= -c_{vE}\cdot v - c_{EE}E^{(a)} - \frac{1}{2}c_{vvE}\cdot v^2 - \\ &\quad - c_{vEE}\cdot vE^{(a)} - \frac{1}{2}c_{EEE}\cdot E^{(a)2} + \cdots \end{aligned} \tag{13.23b}$$

with

$$\begin{aligned} c_{vv} &= \frac{a^2 f_1 f_2}{V_{z0}(f_1 + f_2)}; & c_{vvv} &= \frac{a^3(g_1 f_2^3 + g_2 f_1^3)}{V_{z0}(f_1 + f_2)^3}; \\ c_{vE} &= -\frac{e_1 a_1 f_1 + e_2 a_2 f_2}{V_{z0}(f_1 + f_2)}; & c_{vvE} &= -\frac{a^2(e_1 g_1 f_2^2 + e_2 g_2 f_1^2)}{V_{z0}(f_1 + f_2)^3}; \\ c_{EE} &= -\frac{e_1^2 + e_2^2}{2V_{z0}(f_1 + f_2)}; & c_{vEE} &= \frac{a(e_1^2 g_1 f_2 + e_2^2 g_2 f_1)}{V_{z0}(f_1 + f_2)^3}; \\ c_{EEE} &= -\frac{e_1^3 g_1 + e_2^3 g_2}{V_{z0}(f_1 + f_2)^3}. \end{aligned} \tag{13.23c}$$

If we want to have $\sigma$ and $\Delta P$, we obtain from (13.23) with (13.20)

$$\sigma = c_{vv} \cdot v + (c_{vE} - P_0) \cdot E^{(a)} + \cdots \tag{13.24a}$$

$$\Delta P = -(c_{vE} - P_0) \cdot v + c_{EE} \cdot E^{(a)} + \cdots . \tag{13.24b}$$

The coefficients, which enter in (13.23) are symmetric against an interchange of particle "1" with particle "2", whereas for the coefficients $c_{vE} - P_0$, which enter in (13.24) this does not hold, because of the unsymmetry involved in $P_0$. However, if there is no permanent polarization $P_0$ in the equilibrium state without external forces, i.e. if equation (13.1) is satisfied, this unsymmetry vanishes again. In an experimental situation, one will have $\sigma$, $\Delta P$, $v$ and $E^{(a)}$ as variables, therefore (13.24) will have to be used for an interpretation of static measurements. The coefficients, which enter, are the elastic modulus at constant field $c_{vv}$, the piezoelectric moduls $v_{vE} - P_0$, the susceptibility at constant strain $v_{EE}$, the third order elastic modulus* $c_{vvv}$, the electrostrictive modulus $c_{vEE}$ and so on. The third order constants are essentially related to anharmonic effects. The coefficients $c_{vE}$, $c_{vvE}$ and $c_{EEE}$ vanish in crystals with inversion symmetry ($e_1 = -e_2$, $a_1 = a_2$, $f_1 = f_2$, $g_1 = g_2$).

But it has to be emphasized, that (13.24) only describes static measurements. If one wants to interpret dynamical measurements (sound wave measurements) a thorough investigation of the equation of motion is necessary. For this we refer to *K. Thoma* [67 T 1].

Before giving a more general investigation of anharmonic effects, we use this simple model to show the influence of temperature on the coefficients defined above, restricting ourselves to the classical limit. This already gives the essential features.

## 14. Thermodynamical Discussion of a Simple Linear Model

The thermodynamic quantities can be derived from the partition function $Z$ which contains as independent parameters the temperature $T$, the mechanical force $F$ and the electric field $E^{(a)}$ or, as discussed in section 13, the generalized force $F_t = F + b^2 E^{(a)} P_0$ instead of $F$. As we are interested only in the classical limit for this simple model, the partition function is ($\beta = 1/kT$)

$$Z(T, F_t, E^{(a)}) = \int e^{-\beta E_{\text{kin}}} \cdot dp_\mu^m \cdot \int e^{-\beta \Phi} \, du_\mu^m . \tag{14.1}$$

The corresponding thermodynamic function is the free enthalpy (Gibbs free energy)

$$G(T, F_t, E^{(a)}) = -\frac{1}{\beta} \ln Z(T, F_t, E^{(a)}) . \tag{14.2}$$

* $v$ is the so called infinitesimal strain (tensor). In the linear case the relation between finite $\mathscr{V}$ and infinitesimal strain is simply $\mathscr{V} = v + v^2/2$. The elastic moduli are in a more consistent way defined as the coefficients in the expansion with respect to $\mathscr{V}$. In the linear case it is simple to relate the various coefficients. We will not give the relations here (see also sect. 6)!

Starting with the expansion (13.13), where the expansion is taken from the equilibrium positions defined by (13.14), i.e. the minimum of the potential energy without external forces, we obtain

$$\begin{aligned}\Phi = N\varphi_0 - NaF_t + \\ + \sum_{m=1}^{N} \left\{\bar{d}_1 \xi_1^m + \frac{1}{2} f_1 \left(\xi_1^m\right)^2 + \frac{1}{6} g_1 \left(\xi_1^m\right)^3 + \frac{1}{24} h_1 \left(\xi_1^m\right)^4 \cdots\right\} + \\ + \sum_{m=1}^{N-1} \left\{\bar{d}_2 \xi_2^m + \frac{1}{2} f_2 \left(\xi_2^m\right)^2 + \frac{1}{6} g_2 \left(\xi_2^m\right)^3 + \frac{1}{24} h_2 \left(\xi_2^m\right)^4 \cdots\right\}.\end{aligned} \tag{14.3}$$

Here the following abbreviations have been used:

$$F_t = F + b^2 E^{(a)} P_0 = F + eE^{(a)} a_1/a - QE^{(a)}, \tag{14.4a}$$

$$\bar{d}_1 = -F_t - eE^{(a)} a_2/a; \quad \bar{d}_2 = -F_t + eE^{(a)} a_1/a, \tag{14.4b}$$

$$\xi_1^m = u_2^m - u_1^m; \quad \xi_2^m = u_1^{m+1} - u_2^m. \tag{14.4c}$$

With the help of (14.2) and (14.3) it can be shown easily, that the derivatives of the enthalpy with respect to the external forces determine the length of the lattice and the relative displacements of the two sublattices. It is

$$-\left(\frac{\partial G}{\partial F_t}\right)_{T, E^{(a)}} = N(a + \Delta a), \tag{14.5a}$$

$$-\left(\frac{\partial G}{\partial E^{(a)}}\right)_{T, F_t} = +New, \tag{14.5b}$$

where $\Delta a$ and $w$ are defined in (13.15), but now they include the (homogeneous) temperature displacements, thus being functions of $T$, $F$ and $E^{(a)}$. In (14.5) the kinetic part of (14.1) does not enter and can be neglected. A further quantity of interest is the specific heat at constant pressure (force) and constant field,

$$C_p = -T\left(\frac{\partial^2 G}{\partial T^2}\right)_{F, E^{(a)}}. \tag{14.6}$$

We seek for an expansion of (14.1) with respect to $F$ and $E^{(a)}$. Further we assume that the anharmonic effects (related to $g_1$, $g_2$, $h_1$ and $h_2$) are "small" compared to the harmonic effects ($f_1$, $f_2$), and that it is sufficient to consider the lowest order anharmonic theory. As we will see, in some averages terms linear in $g_1$, $g_2$ vanish, the first anharmonic effects being due to $g_1^2$, $g_2^2$, $h_1$ and $h_2$. In a reasonable expansion of the above kind we must expect effects of $g^2$ to be in the same order of magnitude as $h$ (decreasing terms with increasing order of terms). Therefore we take into account all terms up to $g^2$ and $h$ and neglect terms with $g^3$, $gh$, $h^2$ and higher ones [I, II, 61 L 1, 64 L 2, et al.]*. How many terms in $F$ and $E^{(a)}$ we have to consider, depends on the order of the effect we wish to

* It has to be emphasized, that if one takes into account $g^3$, $gh$ etc. one has to take also higher order terms in (14.3). It can be seen easily, that the next terms in (14.3) give rise to effects of the same order of magnitude as $g^3$, $gh$ etc. Otherwise the theory is not consistent.

discuss. We expand up to third order in $F$ and $E^{(a)}$, giving rise to quadratic effects in $va$ and $w$. With

$$\Phi\left(\xi_1^m\right) = \bar{d}_1 \xi_1^m + \frac{1}{2} f_1 \left(\xi_1^m\right)^2 + \frac{1}{6} g_1 \left(\xi_1^m\right)^3 + \frac{1}{24} h_1 \left(\xi_1^m\right)^4$$

we have $\left(\mathrm{d} u_1^m \Rightarrow \mathrm{d}\xi_1^m\right)$

$$\ln \int \exp\left\{-\beta \sum_m \Phi\left(\xi_1^m\right)\right\} \mathrm{d}\xi_1^m = N \ln \int \exp\{-\beta \Phi(\xi)\}\, \mathrm{d}\xi$$

$$= N \ln \int \exp\{-\beta f_1 \xi^2/2\}\, \mathrm{d}\xi +$$

$$+ N \ln \left\langle \exp\left[-\beta \bar{d}_1 \xi - \frac{1}{6} \beta g_1 \xi^3 - \frac{1}{24} \beta h_1 \xi^4\right]\right\rangle_{\mathrm{AV}}$$

where the average has to be calculated with the harmonic distribution $\exp\{-\beta f_1 \xi^2/2\}$. Expanding the last expression with respect to $\bar{d}$, $g$, $h$ up to the order discussed above, the averages are easily obtained. The complete free enthalpy then is*

$$\frac{1}{N} G(T, F_t, E^{(a)}) = \varphi_0 - aF_t - \frac{1}{\beta} \ln (2\pi \sqrt{M_1 M_2}/\beta) -$$

$$- \frac{1}{\beta} \ln (2\pi/\beta \sqrt{f_1 f_2}) - \frac{1}{24\beta^2} \{5g_1^2/f_1^3 - 3h_1/f_1^2 + 5g_2^2/f_2^3 - 3h_2/f_2^2\} -$$

$$- \frac{\bar{d}_1 g_1}{2\beta f_1^2} - \frac{\bar{d}_2 g_2}{2\beta f_2^2} - \frac{\bar{d}_1^2}{2}\left\{\frac{1}{f_1} - \frac{h_1}{2\beta f_1^3} + \frac{g_1^2}{\beta f_1^4}\right\} -$$

$$- \frac{\bar{d}_2^2}{2}\left\{\frac{1}{f_2} - \frac{h_2}{2\beta f_2^3} + \frac{g_2^2}{\beta f_2^4}\right\} - \frac{\bar{d}_1^3 g_1}{6 f_1^3} - \frac{\bar{d}_2^3 g_2}{6 f_2^3}\,. \tag{14.7}$$

According to (14.5) we have from (14.7)**

$$\Delta a = - \frac{kT}{2}\left\{\frac{g_1}{f_1^2} + \frac{g_2}{f_2^2}\right\} - \frac{\bar{d}_1}{f_1} - \frac{\bar{d}_2}{f_2} - \frac{\bar{d}_1^2 g_1}{2 f_1^3} - \frac{\bar{d}_2^2 g_2}{2 f_2^3} - \tag{14.8a}$$

$$- \frac{kT \bar{d}_1}{f_1^4} (g_1^2 - h_1 f_1/2) - \frac{kT \bar{d}_2}{f_2^4} (g_2^2 - h_2 f_2/2)\,;$$

$$w = - \frac{kT}{2a}\left\{\frac{a_2 g_1}{f_1^2} - \frac{a_1 g_2}{f_2^2}\right\} - \frac{\bar{d}_1 a_2}{f_1 a} + \frac{\bar{d}_2 a_1}{f_2 a} - \frac{\bar{d}_1^2 a_2 g_1}{2 a f_1^3} + \tag{14.8b}$$

$$+ \frac{\bar{d}_2^2 a_1 g_2}{2 a f_2^3} - \frac{kT a_2 \bar{d}_1}{a f_1^4} (g_1^2 - h_1 f_1/2) + \frac{kT a_1 \bar{d}_2}{a f_2^4} (g_2^2 - h_2 f_2/2)\,.$$

With zero external forces, $\bar{d}_1 = \bar{d}_2 = 0$, these equations give the temperature effect on the lattice constant (thermal expansion) and on the relative displacements of the sublattices (pyroelectric effect). We will first discuss these two effects.

* In the following, all terms of $O(1/N)$ are neglected, as we are concerned with large systems ($N$) only.

** The surface charges $Q$ may change (cf. eq. 13.1) when $F_t$, $E^{(a)}$ or $T$ vary. Therefore the change $\partial Q/\partial F_t$, etc. has to be taken into account in the derivatives (14.5, 6). This gives additional terms in (14.7, 8) and the following equations, especially in (14.9, 14). Since the additional terms are rather lengthy, they have been dropped in the following. The principle features of all the phenomena can be seen without these terms. For the additional terms see *K. Thoma* [67 T 1].

$\Delta a$ is simply the change in lattice constant due to temperature, so that

$$a(T) = a - \frac{kT}{2}\left\{\frac{g_1}{f_1^2} + \frac{g_2}{f_2^2}\right\}. \tag{14.9}$$

In the classical limit the lattice constant is proportional to temperature. Qualitatively this is found experimentally; deviations from this behavior can be explained by higher order anharmonicities. Because the thermal expansion is positive in the classical region, $g_1$ and $g_2$ have to be negative (or the resulting expression in (14.9)). This is in accordance with the potentials in crystals. All the potentials which have been investigated, have negative third derivatives ($g_1$, $g_2$) at nearest neighbor distance. This is known already for a long time *(Grüneisen!)*. The thermal expansion is the temperature derivative of the relative change of $a$, therefore

$$\alpha(T) = \frac{1}{a}\frac{\mathrm{d}a(T)}{\mathrm{d}T} = -\frac{k}{2a}\left\{\frac{g_1}{f_1^2} + \frac{g_2}{f_2^2}\right\}, \tag{14.10}$$

being a constant. Higher order anharmonicities lead to a temperature dependence of $\alpha(T)$ in the classical limit. $k$ is the classical specific heat per degree of freedom. Therefore we might conclude, that $\alpha(T)$ is proportional to the specific heat. In sect. 15 we will see this to be true, but only in a certain approximation and with certain restrictions.

The pyroelectric effect is related to a change in the polarization. The total dipole-moment of the linear lattice in a homogeneously strained state is, according to (13.8, 13.15)

$$\mathcal{M} = N\{e(a_1 + v a_1) - Q(a + va) + ew\}. \tag{14.11}$$

The polarization is the dipole-moment divided by the volume of the lattice $NV_z = Nab^2(1+v)$, therefore

$$P = \frac{ea_1 - Qa}{ab^2} + \frac{ew}{ab^2(1+v)}. \tag{14.12}$$

As we are interested only in the lowest order effect we can neglect $v$ in the denominator of (14.12). The additional surface charges have the effect of making the total polarization of the crystal zero, as discussed in the introduction of sect. 13. If $T = 0$, $Q$ will be equal to $ea_1/a$ (13.1). At finite temperature we then have

$$Q(T) = ea_1/a + (e/a)\cdot w(T)$$

or if the temperature is raised from $T_0$ to $T_0 + \Delta T$, we have

$$\Delta Q = +\frac{e}{a}\cdot\left(\frac{\partial w}{\partial T}\right)_{\sigma,E}\cdot\Delta T \tag{14.13}$$

where the derivative here is at constant pressure ($F = \sigma b^2$) and field. $\Delta Q$ is the amount of charge, which is added at one surface and lost at the other, if the temperature changes by $\Delta T$, to make the polarization zero again*.

$$\pi_{\sigma E} = \frac{e}{ab^2}\left(\frac{\partial w}{\partial T}\right)_{\sigma E} = -\frac{e}{ab^2}\cdot\frac{k}{2a}\left\{\frac{a_2 g_1}{f_1^2} - \frac{a_1 g_2}{f_2^2}\right\} \tag{14.14}$$

is the pyroelectric constant at constant pressure and field; $\pi_{\sigma E} - \pi_{vE}$ is sometimes called the secondary effect [58 S 2]. It vanishes in crystals with inversion symmetry ($f_1 = f_2$, etc.). The temperature dependence of $\pi_{\sigma E}$ is the same as for $\alpha(T)$. The sign of the effect depends on the relative magnitudes of $g_1$ and $g_2$, and the other quantities.

Sometimes the pyroelectric constant at constant volume and field, $\pi_{vE}$ (effect of the first kind) is used. It is

$$\mathrm{d}w = \left(\frac{\partial w}{\partial T}\right)_{vE}\mathrm{d}T + \left(\frac{\partial w}{\partial E}\right)_{vT}\mathrm{d}E + \left(\frac{\partial w}{\partial v}\right)_{TE}\mathrm{d}v$$

and from this with $F \sim \sigma = \text{const.}$, $E^{(a)} = \text{const.}$

$$\pi_{\sigma E} = \pi_{vE} + \frac{e}{a^2 b}\cdot\left(\frac{\partial w}{\partial v}\right)_{TE}\cdot\left(\frac{\partial v}{\partial T}\right)_{\sigma E} = \pi_{vE} - c_{vE}\cdot\alpha(T) \tag{14.15}$$

we obtain the "true" constant

$$\pi_{vE} = -\frac{ek}{2ab^2(f_1 + f_2)}\left\{\frac{g_1}{f_1} - \frac{g_2}{f_2}\right\}, \tag{14.16}$$

which vanishes, of course, in lattices with inversion symmetry.

If the thermal equilibrium is specified by $a(T)$ and $w(T)$, we can give the deviations from this equilibrium by external mechanical forces $F$ and external fields $E^{(a)}$. We describe their effect by a homogeneous strain $v$ and an additional relative displacement $\Delta w$. They are defined by

$$\begin{aligned} a(v, T) &= (1 + v)\, a(T)\,, \\ \Delta w &= w - w(T) \end{aligned} \tag{14.17}$$

and can be obtained immediately from (14.8). Using (14.8a), we would have $v \cdot a(T)$ as a function of $F_t$ and $E^{(a)}$; but we can invert the resulting equation to have

$$F_t/b^2 = \sigma + P_0 \cdot E^{(a)}$$

as a function of $v$ and $E^{(a)}$. This gives

$$\begin{aligned} \sigma + P_0 E^{(a)} = c_{vv}(T)\cdot v + c_{vE}(T)\cdot E^{(a)} + \frac{1}{2}c_{vvv}\cdot v^2 + \\ + c_{vvE}\cdot vE^{(a)} + \frac{1}{2}c_{vEE}\cdot E^{(a)2} + \cdots \end{aligned} \tag{14.18a}$$

* This charge transfer can be measured experimentally. If a crystal is cooled very rapidly by bringing it from room-temperature to nitrogen-temperature, the formation of a finite polarization can be observed. The charges need a certain time, to compensate the dipole-moment. But in most experiments the temperature changes so slowly that charges can move and compensate the dipole moment during the experiment.

with*

$$c_{vv}(T) = \frac{a(T)\cdot f_1 f_2}{b^2(f_1+f_2)}\left\{1 - \frac{kT}{f_1+f_2}\left[\frac{f_2}{f_1^3}(g_1^2 - h_1 f_1/2) + \frac{f_1}{f_2^3}(g_2^2 - h_2 f_2/2)\right]\right\}$$

$$\begin{aligned} c_{vE}(T) &= \frac{e(a_1 f_1 - a_2 f_2)}{a b^2 (f_1+f_2)}\left\{1 - \frac{akT}{(a_1 f_1 - a_2 f_2)(f_1+f_2)}\times\right. \\ &\quad \times \left.\left[\frac{f_2}{f_1^2}(g_1^2 - h_1 f_1/2) - \frac{f_1}{f_2^2}(g_2^2 - h_2 f_2/2)\right]\right\} \end{aligned} \tag{14.19}$$

$$\begin{aligned} c_{EE}(T) &= \frac{-e^2}{a(T)\cdot b^2(f_1+f_2)}\left\{1 + \frac{kT}{f_1+f_2}\times\right. \\ &\quad \times \left.\left[\frac{1}{f_1^2}(g_1^2 - h_1 f_1/2) + \frac{1}{f_2^2}(g_2^2 - h_2 f_2/2)\right]\right\}. \end{aligned}$$

$c_{vvv}$, $c_{vvE}$, $c_{vEE}$ and $c_{EEE}$ are identical with those given in (13.23c). In our approximation this is trivial and only a check for the calculation.

From (14.8b) we obtain $\Delta w$, primarily as a function of $F_t$ and $E^{(a)}$, but after substituting $F_t$ according to (14.18a), we have $\Delta w$ as a function of $v$ and $E^{(a)}$:

$$\begin{aligned} \frac{e}{V_z(T)}\Delta w &= -c_{vE}(T)\cdot v - c_{EE}(T)\cdot E^{(a)} - \frac{1}{2}c_{vvE}\cdot v^2 - \\ &\quad - c_{vEE}\cdot vE^{(a)} - \frac{1}{2}c_{EEE}\cdot E^{(a)2} + \cdots . \end{aligned} \tag{14.18b}$$

Comparing (14.18) with (13.23) we see, that the two expansions are in complete agreement, as it must be. The gain is the temperature dependence of the second order constants $c_{vv}$, $c_{vE}$, $c_{EE}$. At high temperatures they are changing linearly with temperature. The temperature dependence of the third order constants would correspond to higher order anharmonicities.

We have started from the free enthalpy, but finally we finished with an expansion, using $v$ and $E^{(a)}$ as independent variables. The corresponding thermodynamical function would be the free energy $F(T, v, E^{(a)})$. It would be given by

$$\begin{aligned} \frac{1}{NV_z(T)}F(T, v, E^{(a)}) &= f_0(T) + \frac{1}{2}c_{vv}\cdot v^2 + c_{vE}\cdot vE^{(a)} + \\ &\quad + \frac{1}{2}c_{EE}\cdot E^{(a)2} + \frac{1}{6}c_{vvv}\cdot v^3 + \frac{1}{2}c_{vvE}\cdot v^2 E^{(a)} + \\ &\quad + \frac{1}{2}c_{vEE}\cdot vE^{(a)2} + \frac{1}{6}c_{EEE}\cdot E^{(a)3} + \cdots . \end{aligned} \tag{14.20}$$

No linear terms in $v$ and $E^{(a)}$ occur because of our definition of the equilibrium state (no stress, no field). Sometimes, especially in three-dimensional lattices, it is more convenient to use the finite strains as the

* All the force-constants $f_1$, $f_2$, $g_1$, $g_2$, $h_1$, $h_2$ in (14.19) refer to the minimum of the potential energy, whereas the cell parameter, $a_1(T)$, $a_2(T)$, $a(T)$ contain the temperature dependence up to the first anharmonic order, i.e. $g_1$, $g_2$.

expansion parameters. The relation is very simple in the linear case, $\mathcal{V} = v + v^2/2$. Then we have

$$\frac{1}{N V_s(T)} \cdot F(T, \mathcal{V}, E^{(a)}) = f_0(T) + \frac{1}{2} c_{vv} \cdot \mathcal{V}^2 + c_{vE} \mathcal{V} E^{(a)} + \\ + \frac{1}{2} c_{EE} \cdot E^{(a)} + \frac{1}{6} (c_{vvv} - 3 c_{vv}) \mathcal{V}^3 + \\ + \frac{1}{2} (c_{vvE} - c_{vE}) \mathcal{V}^2 E^{(a)} + \frac{1}{2} c_{vEE} \mathcal{V} E^{(a)2} + \frac{1}{6} c_{EEE} \cdot E^{(a)3} . \tag{14.20a}$$

From (13.15) we have for non-vanishing polarization $\overline{E} = E^{(a)} - 4\pi P$; here $\overline{E}$ is the *macroscopic* electric field in the interior of the crystal, whereas $E^{(a)}$ corresponds to the displacement field $D$ in the interior. Therefore (13.23, 24; 14.8, 18, 20) are expansions with respect to $v$ and $D$, which can be converted into an expansion with respect to $v$ and $\overline{E}$. Thus we have the isothermal elastic modulus* at constant $D$, the piezoelectric constant and the static isothermal dielectric susceptibility** at constant strain

$$c_{11}^{\text{is}, D} = c_{vv}; \quad e_{11} = -\frac{c_{vE}}{1 + 4\pi c_{EE}}; \quad \chi_{11} = -\frac{c_{EE}}{1 + 4\pi c_{EE}} . \tag{14.21}$$

For the piezoelectric coefficient at constant stress it follows

$$g_{11} = d_{11}/(1 + 4\pi \chi_{11}^{\sigma}) = \left(\frac{\partial v}{\partial D}\right)_{\sigma} = -\frac{e(a_1 f_1 - a_2 f_2)}{a^2 f_1 f_2} \left\{1 - \frac{k T f_1 f_2}{a_1 f_1 - a_2 f_2} \times \right. \\ \left. \times \left[\frac{a_2}{f_1^4} (g_1^2 - h_1 f_1/2) - \frac{a_1}{f_2^4} (g_2^2 - h_2 f_2/2)\right]\right\} . \tag{14.22}$$

and the susceptibility at constant stress, using (14.8)

$$\chi_{11}^{\sigma}/(1 + 4\pi \chi_{11}^{\sigma}) = \frac{e^2 (a_1^2 f_1 + a_2^2 f_2)}{V_s(T) \cdot a^2 f_1 f_2} \left\{1 + \frac{k T f_1 f_2}{a_1^2 f_1 + a_2^2 f_2} \times \right. \\ \left. \times \left[\frac{a_2^2}{f_1^4} (g_1^2 - h_1 f_1/2) + \frac{a_1^2}{f_2^4} (g_2^2 - h_2 f_2/2)\right]\right\} . \tag{14.23}$$

From the next order coefficients we will mention the isothermal elastic constant of third order,

$$c_{111}^{\text{is}} = c_{vvv} - 3 c_{vv} \\ = \frac{a^3}{V_s (f_1 + f_2)^3} \{(g_1 - 3 f_1/a) f_2^3 + (g_2 - 3 f_2/a) f_1^3 - 6 f_1^2 f_2^2/a\} \tag{14.24}$$

and the electrostrictive constant, which measures the quadratic effect of the electric field on the stress at constant volume, $c_{vEE}$.

Another group of coefficients is related to temperature derivatives of the Gibbs or Helmholtz free energy. This is the entropy $S$ and especially

* We use the Voigt-notation, which has only 1-indices in the linear case.

** According to the model, this expression represents only the "rigid-ion" contribution to the susceptibility. Ionic polarizabilities and other contributions have to be added in an appropriate way, if they are present.

the specific heat. According to (14.6) we obtain from (14.7) the specific heat at constant pressure (stress) and field $E^{(a)} = D$

$$C_{pD} = 2Nk\left\{1 + \frac{kT}{24}\left[5g_1^2/f_1^3 + 5g_2^2/f_2^3 - 3h_1/f_1^2 - 3h_2/f_2^2\right]\right\}, \qquad (14.25)$$

which means a linear increase or decrease with temperature above the Dulong-Petit-value $2Nk$. The difference between the specific heat at constant pressure and constant volume is

$$C_{pE} - C_{vE} = T\left(\frac{\partial S}{\partial v}\right)_{TE} \cdot \left(\frac{\partial v}{\partial T}\right)_{\sigma E} = T\alpha(T) \cdot \left(\frac{\partial S}{\partial v}\right)_{TE}$$

and

$$\left(\frac{\partial S}{\partial v}\right)_{TE} = NV_z \cdot c_{vv} \cdot \alpha(T)\,,$$

thus

$$\begin{aligned} C_{pE} - C_{vE} &= NV_z \cdot c_{vv} \cdot T \cdot \alpha^2(T) \\ &= 2Nk \cdot \frac{kT f_1 f_2}{8(f_1 + f_2)} (g_1/f_1^2 + g_2/f_2^2)^2\,. \end{aligned} \qquad (14.26)$$

Similarly the difference in the specific heat at constant field $E$ and constant electric displacement $D = E^{(a)} = \bar{E} + 4\pi P$ can be calculated. From

$$C_{pE} - C_{pD} = T\left(\frac{\partial S}{\partial D}\right)_{\sigma T}\left(\frac{\partial D}{\partial T}\right)_{E\sigma}$$

and

$$\left(\frac{\partial S}{\partial D}\right)_{\sigma T} = \frac{NV_z}{1 + 4\pi\chi_{11}} \cdot \pi_{\sigma E}\,; \quad \left(\frac{\partial D}{\partial T}\right)_{E\sigma} = 4\pi \cdot \pi_{\sigma E}$$

we have

$$C_{pE} - C_{pD} = \frac{4\pi V_z N T}{1 + 4\pi\chi_{11}} (\pi_{\sigma E})^2\,. \qquad (14.27)$$

Further we give the difference between isothermal and adiabatic elastic constants,

$$\begin{aligned} c_{11}^{\mathrm{ad}} - c_{11}^{\mathrm{is}} &= \frac{NV_z T}{C_{VE}} [c_{11}^{\mathrm{is}} \cdot \alpha(T)]^2 \\ &= \frac{kT}{8V_z} \cdot \frac{a^2 f_1^2 f_2^2}{(f_1 + f_2)^2} \{g^1/f_1^2 + g_2/f_2^2\}^2\,, \end{aligned} \qquad (14.28)$$

and between the elastic moduli at constant field $E$ and constant electric displacement $D$:

$$c_{11}^{\mathrm{is},D} - c_{11}^{\mathrm{is},E} = \frac{4\pi}{1 + 4\pi\chi_{11}} (e_{11})^2 = \frac{4\pi}{1 + 4\pi c_{EE}} (c_{vE})^2. \qquad (14.29)$$

We will not mention all the possible relations which can be derived further. The only purpose of this simple example is to show some general features and the influence of anharmonicity on the different coefficients of physical interest.

Three types of coefficients can be distinguished; the largest group contains those effects which vanish in a harmonic theory, and which are determined essentially by third order anharmonicity ($g_1, g_2$). There are two types of effects. The first type shows a temperature dependence in our approximation (14.9, 14, 16): thermal expansion $a(T)$ and pyroelectric moment $ew(T)/V_z$. If the anharmonicity vanishes, then also the differences between certain coefficients are zero; i.e. between those the difference of which is essentially determined by $\alpha(T)$ or $\pi_{\sigma E}(T)$. The difference between specific heats at constant volume and constant pressure (14.26), between specific heats at constant field $E$ and displacement $D$ (14.27), and between isothermal and adiabatic elastic constants (14.28) vanishes.

The next type of coefficients is not temperature dependent in our approximation, but is also determined by anharmonicity. These are the third order elastic constants (14.24), the electrostrictive constants, etc.

The harmonic theory (only $f_1$, $f_2$ non-zero) makes statements about the third type of effects, being essentially the *quadratic* coefficients in the expansion of the free energy with respect to small changes (in temperature $T$, in stress $\sigma$ or strain $v$ and in electric field $E$) from equilibrium: the specific heat (14.25), the elastic constants (14.21), the piezoelectric constants (14.22) and the dielectric susceptibility (14.23). Considering anharmonic effects in these coefficients, we find a temperature dependence proportional to $T$ (in the classical limit), which is connected with second order anharmonicity $g_1^2$, $g_2^2$, $h_1$, $h_2$.

All these statements hold for the lowest order anharmonic approximation. The next order would give rise to one higher power in the temperature dependence, as can easily be seen. For example, the pyroelectric moment would involve a term with $T^2$, thus $\pi_{\sigma E}$ being proportional to $T$, or the third order elastic constants would show a linear temperature dependence and so on. This is correct in the classical limit. At lower temperatures quantum mechanical corrections have to be considered. In a rough way one can say, that in every place $kT$ has to be replaced by $\varepsilon(\omega, T)$, averaged over the spectrum of phonon frequencies $\omega$. But as we will see in the next sections, this procedure gives at most a very crude estimate and the true behavior can be quite different. It might even happen, that ecrtain coefficients are not monotonic functions of temperature.

Emphasis has to be put on the fact, that it is very essential to take the surface charges into account, and to describe the experimental situation in an appropriate way.

We will conclude this section with a few historical remarks. The influence of anharmonicity on thermodynamic properties has been first discussed by *Born* and *Brody* [21 B 1] and *Schrödinger* [22 S 1] and by *Grüneisen* [12 G 1] for the thermal expansion. During the following years, mainly *Born* and his collaborators have made a large number of investigations devoted to this problem. This is summarized essentially in [II]. Most of the problems dealt with are related to mechanical and electrical properties of the lattice. A comprehensive quantum-mechanical expansion of the anharmonic free energy of the crystal seems to have not been given up to 1958 [58 L 1; 61 L 1]. Such an expansion is necessary mainly for calculating caloric quantities (see next section), whereas mechanical quantities can be obtained in a simpler way. In recent years, similar or slightly different methods have been discussed [61 M 1—3]. Other investigations use thermodynamical Green functions as a powerful tool [63 C 2 and the other papers, discussed in sect. 17]. But in phonon-calculations its main purpose is to show general relations. In every definite calculation one again uses a perturbation expansion, being equivalent to the quantum-mechanical expansion of the free-energy.

## 15. Thermal Expansion. Temperature Dependence of Elastic Constants

The calculations of the preceding section have to be generalized to threedimensional lattices, and allowance should be made for quantum effects, i.e. the theory has to be extended to temperatures below the Debye-temperature. Now, quantum effects can be handled most simply by using an infinite crystal, divided in periodicity volumes of $N$ unit cells [I, II, 40 K 1]. Thus it is convenient to work with the Helmholtz free energy $F$, taking volume $V = N V_z$ and temperature $T$ as independent variables. The connection between using $F$ with variable $V$ and free enthalpy $G$ with forces as the independent variables has been discussed in some detail in [I]. This procedure is appropriate for the discussion of mechanical properties (lattice constant, elastic constants) and the specific heat; if electric fields are present, the Gibbs free enthalpy is the more appropriate function [see sect. 14, 61 L 1].

In sect. 14 we have seen, that there are always $g^2/f$- and $h$-terms which enter the various coefficients and which lead to the same temperature dependence. Thus, if we start with the potential energy (1.6), we have to use a perturbation treatment which contains the anharmonicities $\Phi_3$ up to quadratic order, $\Phi_4$ up to linear terms. Further, $\Phi_3$ is of odd order in the displacements $u_{\mu}^{m}{}_{i}$, and therefore it is also of odd order if we transform the displacements in the usual way to phonon creation and annihilation operators. Such terms give no contribution in a first order perturbation treatment (expectation value zero). Therefore we can formally write for the Helmholtz free energy

$$F = \Phi_0 + F_2(\Phi_2) + F_3(\Phi_3^2) + F_4(\Phi_4) + F_1(\Phi_1^2) + \cdots \qquad (15.1)$$

$F_2(\Phi_2)$ is the harmonic vibrational part of the free energy.

In a harmonic theory the equilibrium positions of the ions are those defined by the minimum of the potential energy, independent of temperature [I, II, 61 L 1]. This however is not expected to be true if anharmonic effects are considered. Sect. 14 shows that there is a thermal expansion, i.e. temperature dependent equilibrium positions. Therefore it is convenient, to use the following procedure: the expansion (1.6) is performed by starting with unknown equilibrium positions $\boldsymbol{R}_{\mu}^{m}$, which are looked upon as parameters to be determined later by the equilibrium condition, which is the minimum of the free energy*. Thus all the different terms in (15.1) depend on these parameters, and therefore contain a certain anharmonic effect via the (temperature dependent) equilibrium positions. If this expansion is now limited to the quadratic terms** or to

$$F_{qh} = \Phi_0 + F_2(\Phi_2) \tag{15.2}$$

it is called the quasiharmonic approximation, because it is formally identical to the harmonic theory, but the coupling-parameters depend on the parameters $\boldsymbol{R}^{m}_{\mu}$.

In calculating the equilibrium positions and the elastic constants, one has to take derivatives with respect to strains or to the components of the matrix $A$ of basis vectors etc. These enter the free energy only via the c.p'.s. Derivation of $\Phi_3$ with respect to strain gives $\Phi_4$-coefficients, of $\Phi_4$ gives $\Phi_5$ etc. and changes the order of anharmonicity. Because we have neglected higher order anharmonicities, we have to neglect these higher order terms too. We arrive at the statement: All properties which are related to *strain-derivatives of the free energy* can be obtained from the *quasiharmonic approximation*, if we are interested in lowest order anharmonic effects only. In higher approximations, always two *further orders of magnitude have to be included.*

The quasiharmonic theory starts with the Helmholtz free energy

$$F_{\text{qh}} = \Phi_0\left(\boldsymbol{R}^{m}_{\mu}\right) + kT \sum_{\boldsymbol{q}\sigma} \ln\{2\sinh[\hbar\omega(\boldsymbol{q}\sigma)/2kT]\}\,, \tag{15.3}$$

which is the harmonic free energy, apart from the fact, that $\Phi_0$ and the frequencies $\omega(\boldsymbol{q}\sigma)$ now depend on the position-parameters $\boldsymbol{R}^{m}_{\mu}$. The frequencies are related to the c.p.'s of second order by the wellknown relation

$$\omega^2(\boldsymbol{q}\sigma)\, e^{\mu}_{i}(\boldsymbol{q}\sigma) = \sum_{\nu j} t^{\mu\nu}_{ij}(\boldsymbol{q})\, e^{\nu}_{j}(\boldsymbol{q}\sigma) \tag{15.4}$$

with***

$$t^{\mu\nu}_{ij}(\boldsymbol{q}) = \sum_{\boldsymbol{h}} \frac{\Phi^{0\,\boldsymbol{h}}_{\substack{\mu\nu\\ ij}}}{\sqrt{M_\mu M_\nu}}\, e^{i\boldsymbol{q}A\boldsymbol{h}} = t^{*\nu\mu}_{ji}(\boldsymbol{q}) \tag{15.4a}$$

$e^{\mu}_{i}(\boldsymbol{q}\sigma)$ are the polarization-vectors of the lattice waves with wave vector $\boldsymbol{q}$ and polarization state $\sigma$. We choose [see I]

$$e^{\mu}_{i}(-\boldsymbol{q}\sigma) = -e^{*\mu}_{i}(\boldsymbol{q}\sigma)\,. \tag{15.4b}$$

* If there are no stresses in the equilibrium state. Otherwise the derivatives of the free energy with respect to strains have to be equal to the stresses.

** If we take an expansion about arbitrary positions, the $\Phi_1$-term does not vanish in general. But the deviations of the expansion positions from the positions of minimal potential energy are of the order of magnitude of thermal expansion, which is expected to be proportional to $\Phi_3$. Thus $\Phi_1$ can be considered just as $\Phi_3$ and can be disregarded in a quasiharmonic theory (see 61 L 1 and sect. 14).

*** Often the exponent is chosen in the form $\exp\{i\boldsymbol{q}(A\boldsymbol{h}+\boldsymbol{R}_\nu-\boldsymbol{R}_\mu)\}$ but this means only a phase-factor in the polarization vectors $\boldsymbol{e}^{\mu}$; in the following it is convenient to use the above definition.

Now we can determine the equilibrium positions, by taking the derivatives of $F_{qh}$ with respect to $A_{ik}$ and $X_{\mu}^{i}$, giving the cell-data and the relative positions of the ions in the cell as functions of temperature. Instead of $A_{ij}$ we will use the thermal deformation of the cell as the describing parameter, i.e. we introduce the tensor of finite thermal strain $\mathscr{V}_{ij}^{\text{th}}$ to describe the temperature variation of the cell-data. The derivative of $F_{qh}$ with respect to $\mathscr{V}_{ij}^{\text{th}}$ has to be zero in case of vanishing stresses in the equilibrium state,

$$\frac{\partial F_{qh}}{\partial \mathscr{V}_{ij}^{\text{th}}} = 0 = \frac{\partial \Phi_0}{\partial \mathscr{V}_{ij}^{\text{th}}} + \sum_{\boldsymbol{q}\sigma} \frac{\partial \hbar \omega(\boldsymbol{q}\sigma)}{\partial \mathscr{V}_{ij}^{\text{th}}} [\bar{n}(\boldsymbol{q}\sigma) + 1/2] \tag{15.5}$$

since only $\omega(\boldsymbol{q}\sigma)$ depends on the parameters $\boldsymbol{R}_{\mu}^{m}$ or $\mathscr{V}_{ij}^{\text{th}}$. $\bar{n}(\boldsymbol{q}\sigma)$ is the mean thermal occupation number of an oscillator $\boldsymbol{q}\sigma$:

$$\bar{n} = \{\exp(\hbar\omega/kT) - 1\}^{-1}.$$

If we introduce the mean thermal energy $\varepsilon(\omega, T) = \hbar\omega(\bar{n} + 1/2)$ of an oscillator, we have ($\omega = \omega(\boldsymbol{q}\sigma)$!)

$$\begin{aligned} &\frac{\partial \Phi_0}{\partial \mathscr{V}_{ij}^{\text{th}}} + \sum_{\boldsymbol{q}\sigma} \frac{1}{\omega(\boldsymbol{q}\sigma)} \cdot \frac{\partial \omega(\boldsymbol{q}\sigma)}{\partial \mathscr{V}_{ij}^{\text{th}}} \varepsilon(\omega, T) \\ &\quad = \frac{\partial \Phi_0}{\partial \mathscr{V}_{ij}^{\text{th}}} + \frac{1}{2} \sum_{\boldsymbol{q}\sigma} \frac{\partial \ln \omega^2(\boldsymbol{q}\sigma)}{\partial \mathscr{V}_{ij}^{\text{th}}} \varepsilon(\omega, T) = 0 . \end{aligned} \tag{15.5a}$$

The change of $\Phi_0$ with thermal strain can be related to the elastic constants. The same deformation which can be produced by a mechanical force can be produced by an equivalent thermal force; we can use the relation (6.27) as describing the difference in the potential energy, once taken at the minimum positions and once taken at the positions after introducing thermal strain. Therefore with zero stresses

$$\frac{\partial \Phi_0}{\partial \mathscr{V}_{ij}^{\text{th}}} = N V_z \tilde{C}_{ij,kl} \mathscr{V}_{kl}^{\text{th}} \tag{15.6}$$

with the elastic constants $\tilde{C}_{ij,kl}$ of the harmonic theory*.

$$\gamma_{ij}(\boldsymbol{q}\sigma) = -\frac{1}{6} \frac{\partial \ln \omega^2(\boldsymbol{q}\sigma)}{\partial \mathscr{V}_{ij}^{\text{th}}} \tag{15.7}$$

is a generalized Grüneisen-constant. It depends in this definition on the frequencies or on the special mode. In a cubic crystal, neglecting the dependence on $\boldsymbol{q}\sigma$ it is identical with the usual definition:

$$\gamma(\omega) = -\frac{a_0}{6} \cdot \frac{\partial \ln \omega^2}{\partial a} = -\frac{V_z}{2} \frac{\partial \ln \omega^2}{\partial V_z} = -\frac{\partial \ln \omega}{\partial \ln V_z} = -\frac{\partial \ln \theta}{\partial \ln V_z} \tag{15.8}$$

if $\omega$ is replaced by $\bar{\omega}$, which is proportional to the Debye-temperature in Debye's approximation.

* All quantities which are related to harmonic theory (minimum of the potential energy) are denoted by a tilde (in sect. 15, 16!).

With (15.6, 7) we have from (15.5), solving for $\mathscr{V}_{ij}^{\text{th}}$ with*

$$\sum_{jl} \tilde{C}_{ik,jl}\tilde{S}_{jl,rs} = \delta_{ir}\delta_{ks}$$

$$\mathscr{V}_{kl}^{\text{th}} = \frac{3}{N V_z} \sum_{ij} \tilde{S}_{kl,ij} \sum_{q\sigma} \gamma_{ij}(\boldsymbol{q}\sigma) \cdot \varepsilon(\omega, T) \tag{15.9}$$

and the thermal expansion coefficient

$$\alpha_{kl}^{\text{th}} = \frac{\partial \mathscr{V}_{kl}^{\text{th}}}{\partial T} = \frac{3}{N V_z} \sum_{ij} \tilde{S}_{kl,ij} \sum_{q\sigma} \gamma_{ij}(\boldsymbol{q}\sigma) \frac{\partial \varepsilon(\omega, T)}{\partial T}. \tag{15.10}$$

In cubic crystals we have $\gamma_{ij} = \gamma \cdot \delta_{ij}$ and

$$\sum_{ij} \tilde{S}_{11,ij} = \tilde{S}_{11,11} + \tilde{S}_{11,22} + \tilde{S}_{11,33} = \frac{\varkappa}{3}$$

($\varkappa$: compressibility) and so with $\mathscr{V}_{kl}^{\text{th}} = \mathscr{V}^{\text{th}} \cdot \delta_{kl}$

$$\mathscr{V}^{\text{th}} = \frac{\varkappa}{N V_z} \sum_{q\sigma} \gamma(\boldsymbol{q}\sigma) \cdot \varepsilon(\omega, T) \tag{15.9a}$$

and a corresponding relation for (15.10).

At high temperatures, the deformation of the lattice is proportional to $T$, the expansion coefficient is constant. This was already stated with our simple model in sect. 14. At low temperatures ($T \to 0$) the lattice parameter takes constant values, being different from the harmonic values because of the interaction of anharmonicities with the zero-point-vibrations. The expansion coefficient becomes zero. For intermediate temperatures the behavior is more complicated. If we are interested only in a rough overall-estimate of the temperature dependence, we may use the following procedure [61 L 1, 43 B 2]: the Grüneisenconstant is replaced by a certain mean value $\bar{\gamma}_{ij} = \frac{1}{3N} \sum_{q\sigma} \gamma_{ij}(\boldsymbol{q}\sigma)$; the remaining sum $\sum_{q\sigma} \varepsilon(\omega, T) = U(T)$ is just the internal energy of the system of oscillators. If this approximation is justified, the temperature dependence of lattice parameters is given by the internal energy, which is proportional to $T^4$ at low temperatures. The expansion coefficient is then proportional to the specific heat of the harmonic theory, vanishing as $T^3$ with $T \to 0$ (Fig. 5.3). However, in some cases this approximation is too bad. Let us consider (15.10) at low temperatures. Then we have

$$\partial\varepsilon/\partial T = k\left(\frac{\hbar\omega}{kT}\right)^2 \exp\{-\hbar\omega/kT\}$$

and the essential quantity is

$$k \sum_{q\sigma} \gamma_{ij}(\boldsymbol{q}\sigma) \left(\frac{\hbar\omega}{kT}\right)^2 \mathrm{e}^{-\hbar\omega/kT} \tag{15.11a}$$

compared with

$$k \sum_{q\sigma} \gamma_{ij}(\boldsymbol{q}\sigma) \tag{15.11b}$$

---

* $\tilde{S}_{jl,rs}$ may not be confused with the quantities $S_{ij,kl}$, $\hat{S}_{ij,kl}$, $\bar{S}_{ij,kl}$ and $S_{ij}$, defined in sect. 6 and at the end of this section.

at high temperatures. In (15.11a) only that part of $\gamma_{ij}(\boldsymbol{q}\sigma)$ enters, which is related to small frequencies ($\hbar\omega \lesssim kT$), whereas at high temperatures only the sum $\sum \gamma_{ij}(\boldsymbol{q}\sigma)$ is of interest. As the thermal expansion is known to be positive at high temperatures, this sum has to be positive.

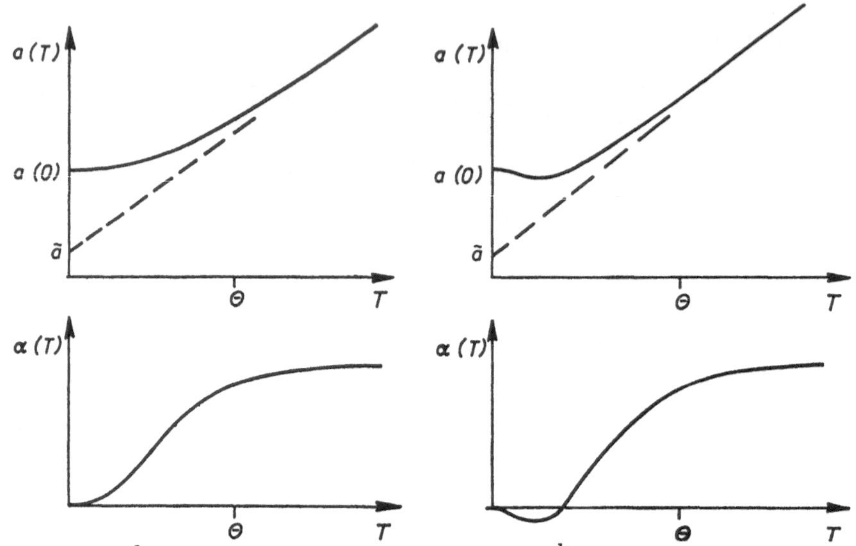


a b

Fig. 5.3. a and b. Qualitative behavior of lattice constant and thermal expansion as a function of temperature. Extrapolation of the high temperature behavior to $T = 0$ gives the harmonic values. a) "normal" behavior b) at low temperatures, $\alpha(T)$ becomes negative for a certain range of temperature

Table 4. *Given are substances with negative thermal expansion coefficients $\gamma$ (volume expansion) below $T(\gamma = 0)$*

| substance | structure | $T(\gamma = 0)$ (° K) | minimum at (° K) | $\Theta_D$ (° K) |
|---|---|---|---|---|
| ZnS | cubic | 63 | | 330 |
| InSb | cubic | 55 | | 202 |
| Si | cubic | 120 | | 653 |
| Ge | cubic | 33 | | 375 |
| $Be_3Al_2(SiO_3)_6$ | hexagonal | 262 | | |
| $H_2O$ | hexagonal | 63 | 36 | 192 |
| $D_2O$ | hexagonal | 63 | 36 | 185 |
| $CaCO_3$ | rhombic | 103 | | |
| U | rhombic | 42 | | 200 |

This does not exclude the case, that for certain low frequency values $\gamma_{ij}(\boldsymbol{q}\sigma)$ is negative and thus making (15.11a) and the thermal expansion coefficient negative. Whether this happens or not depends on the interaction potential of the ions and on the crystal structure. It might even happen that $\alpha_{ij}^{\text{th}}$ is negative for some $i, j$, and positive for others at low temperatures. The dependence on the interaction-potential and the crystal structure has been discussed by *Blackman* [55 B 1, 57 B 1, 2,

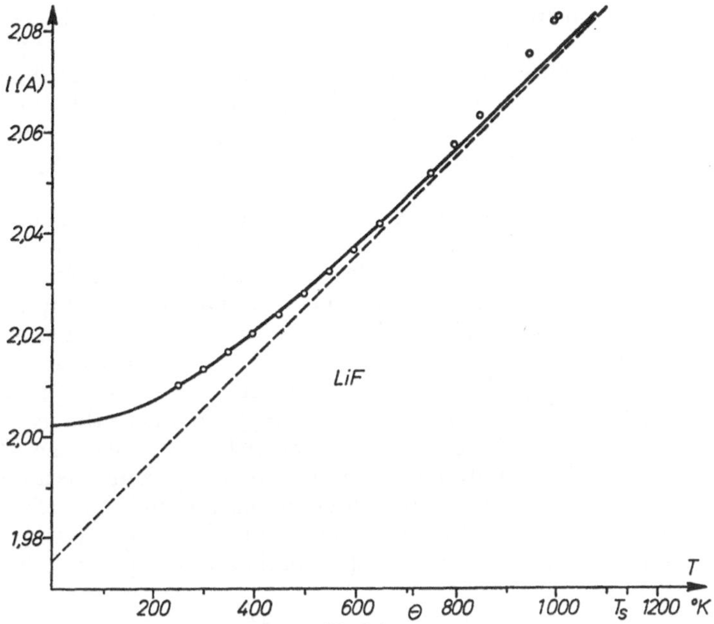


Fig. 5.4 a

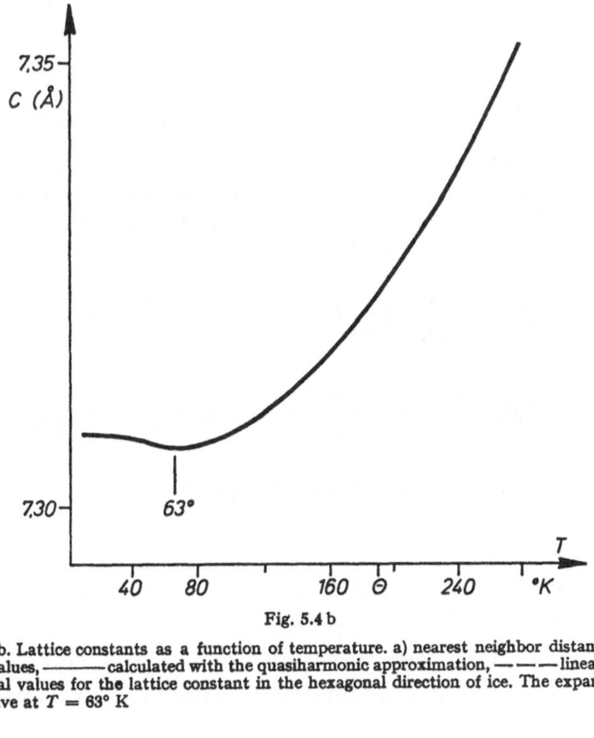


Fig. 5.4 b

Fig. 5.4 a and b. Lattice constants as a function of temperature. a) nearest neighbor distance for LiF, • • • experimental values, ——— calculated with the quasiharmonic approximation, — — — linear extrapolation. b) Experimental values for the lattice constant in the hexagonal direction of ice. The expansion coefficient becomes negative at $T = 63°$ K

58 B 1, 59 B 1] in some detail. He finds that in fact the expansion coefficient is negative in some ionic lattices, mainly of Zincblende-type. This is confirmed by measurements of the thermal expansion of ionic and some other hexagonal crystals (tabl. 4). In any case, the thermal expansion coefficient should be positive at high temperatures ($T \gtrsim \theta$, Debye-temperature) and take a constant value. If at high temperatures ($T > \theta$) the lattice parameter does not vary proportional to $T$ or $\alpha(T)$ is not a constant, this is due to higher order anharmonic effects or to the influence of point defects and other defects in the lattice.

However, in most of the simple lattices (parameterfree cubic lattices) the description by a mean Grüneisen-constant or by an only slightly varying constant is sufficient (Fig. 5.4). In this case often further approximations are made. $\gamma_{ij}(\boldsymbol{q}\sigma)$ is not only replaced by an average value, but the average is performed in a very special way. Instead of averaging $\ln \omega^2(\boldsymbol{q}\sigma)$ it is replaced by $\ln\langle\omega^2(\boldsymbol{q}\sigma)\rangle$, where the average is taken over the squared frequencies only [61 L 1]. It has been shown, that this procedure is quite good in most cases, especially because only derivatives of this average value enter in the thermal expansion. The reason for this replacement is the simple evaluation of the average $\langle\omega^2\rangle$ which can be evaluated from (15.4) ($s$: number of ions per unit cell):

$$\langle\omega^2\rangle = \overline{\omega^2} = \frac{1}{3s}\sum_{\mu i}\frac{\Phi^{00}_{\substack{\mu\mu\\ ii}}}{M_\mu}\,. \tag{15.12}$$

With this approximation we have e.g.

$$\mathscr{V}^{\text{th}}_{kl} = -\frac{1}{2NV_z}\sum_{ij}\tilde{S}_{kl,ij}\frac{\partial\ln\overline{\omega^2}}{\partial\mathscr{V}^{\text{th}}_{ij}}\cdot U(T) \tag{15.13}$$

which can be used at high temperatures ($T > \theta$) and for the limiting value $T = 0$, and in many cases also in the intermediate region.

Another expression for $\gamma_{ij}(\boldsymbol{q}\sigma)$ can be derived by performing the differentiation with respect to $\mathscr{V}_{ij}$.

$$\frac{\partial\omega^2}{\partial\mathscr{V}_{kl}} = \sum_{\mu\nu,ij}\frac{\partial}{\partial\mathscr{V}_{kl}}\left\{t_{\substack{\mu\nu\\ ij}}\,e^*_{\substack{\mu\\ i}}\,e_{\substack{\nu\\ j}}\right\} = \sum_{\substack{\mu\nu\\ ij}}\frac{\partial t^{\mu\nu}_{ij}}{\partial\mathscr{V}_{kl}}\,e^*_{\substack{\mu\\ i}}\,e_{\substack{\nu\\ j}}$$

because

$$\sum_{\mu\nu,ij} t_{\substack{\mu\nu\\ ij}}\left\{e^*_{\substack{\mu\\ i}}\frac{\partial e^\nu_j}{\partial\mathscr{V}_{kl}} + \frac{\partial e^{*\mu}_i}{\partial\mathscr{V}_{kl}}\,e_{\substack{\nu\\ j}}\right\} = \omega^2(q\sigma)\,\frac{\partial}{\partial\mathscr{V}_{kl}}\left(e^*_{\substack{\mu\\ i}}\,e_{\substack{\nu\\ j}}\right) = 0$$

since the $e_{\substack{\mu\\ i}}$ are orthonormalized and complete.

The derivative of $t_{\substack{\mu\nu\\ ij}}$ with respect to $\mathscr{V}_{kl}$ can be calculated then simply again. $\boldsymbol{q}A\boldsymbol{h}$ does not depend on the deformation, for as the lattice $A$ changes, so the reciprocal lattice $B$ does, and therefore the product $\boldsymbol{q}A\boldsymbol{h} = 2\pi(B\boldsymbol{m}, A\boldsymbol{h})/N^{1/3}$ is unaffected by a homogeneous deformation. The first derivative of $\Phi^{0\boldsymbol{h}}_{\substack{\mu\nu\\ ij}}$ with respect to $\mathscr{V}_{kl}$ is equal to the first derivative with respect to the infinitesimal strain $v_{kl}$, if we are concerned with linear terms only. Then we have

$$\frac{\partial t_{ij}^{\mu\nu}}{\partial \mathscr{V}_{kl}} = \frac{1}{(M_\mu M_\nu)^{1/2}} \sum_{hg,\varkappa} \Phi^{0\,h\,g}_{\substack{\mu\,\nu\,\varkappa\\ i\,j\,k}} X^g_{\substack{\varkappa\\ l}} \, \mathrm{e}^{i\boldsymbol{q}A\boldsymbol{h}}$$

and

$$\gamma_{kl}(\boldsymbol{q}\sigma) = \frac{1}{6} \sum_{\substack{hg\\ \mu\nu\varkappa,ij}} \frac{\Phi^{0\,h\,g}_{\substack{\mu\,\nu\,\varkappa\\ i\,j\,k}}}{(M_\mu M_\nu)^{1/2}} X^g_{\substack{\varkappa\\ l}} \frac{e^{*\mu}_i(\boldsymbol{q}\sigma)\, e^\nu_j(\boldsymbol{q}\sigma)}{\omega^2(\boldsymbol{q}\sigma)} \, \mathrm{e}^{i\boldsymbol{q}A\boldsymbol{h}} \tag{15.14}$$

Inserting (15.14) into (15.9), we obtain

$$\mathscr{V}^{\mathrm{th}}_{kl} = -\frac{1}{2NV_z} \cdot \sum_{ij} \tilde{S}_{kl,ij} \sum_{\substack{hg\\ \mu\nu\varkappa,rs}} \frac{\Phi^{0\,h\,g}_{\substack{\mu\,\nu\,\varkappa\\ r\,s\,i}}}{(M_\mu M_\nu)^{1/2}} \cdot X^g_{\substack{\varkappa\\ j}} \times$$

$$\times \sum_{\boldsymbol{q}\sigma} \frac{e^{*\mu}_r(\boldsymbol{q}\sigma)\, e^\nu_s(\boldsymbol{q}\sigma)}{\omega^2(\boldsymbol{q}\sigma)} \, \mathrm{e}^{i\boldsymbol{q}A\boldsymbol{h}} \cdot \varepsilon(\omega_{\boldsymbol{q}\sigma}, T)$$

Using the Fourier-coefficients of the thermodynamic Green-function defined in the Appendix (31.25) we obtain with the harmonic Hamiltonian

$$\varepsilon(\omega_{\boldsymbol{q}\sigma}, T) = \frac{1}{2} \hbar\omega(\boldsymbol{q}\sigma) \cdot \sum_\nu a_\nu(\boldsymbol{q}\sigma) \, .$$

The Green-function in the Appendix is defined with the operator $b_{\boldsymbol{q}\sigma} - b^+_{-\boldsymbol{q}\sigma}$ and its adjoint one. The Green-function corresponding to the displacements can be obtained by multiplying with the eigenvectors of the harmonic Hamiltonian (see 31.2, 3). We define the Fourier coefficients by

$$G^{m\,n}_{\substack{\mu\,\nu\\ i\,j}}(i\omega_\lambda) = \frac{i}{N\sqrt{M_\mu M_\nu}} \sum_{\boldsymbol{q}\sigma} \frac{e^{*\mu}_i(\boldsymbol{q}\sigma)\, e^\nu_j(\boldsymbol{q}\sigma)}{\omega(\boldsymbol{q}\sigma)} \, \mathrm{e}^{i\boldsymbol{q}(\boldsymbol{R}^n - \boldsymbol{R}^m)} \cdot a_\lambda(\boldsymbol{q}\sigma)$$

with $\omega_\lambda = 2\pi\lambda/\beta\hbar$ (see 31.26). With these last two equations we have

$$\mathscr{V}^{\mathrm{th}}_{kl} = -\frac{1}{4V_z} \sum_{ij} \tilde{S}_{kl,ij} \sum_\lambda \sum_{\substack{hg\\ \mu\nu\varkappa,rs}} \frac{\hbar}{i} G^{0\,h}_{\substack{\mu\,\nu\\ r\,s}}(i\omega_\lambda)\, \Phi^{0\,h\,g}_{\substack{\mu\,\nu\,\varkappa\\ r\,s\,i}} X^g_{\substack{\varkappa\\ j}} \, . \tag{15.14a}$$

This form has been given by *Cowley* [63 C 2]; it is completely equivalent, but there is no real gain, because all the difficulties are now in the calculation of the Green-function, which has to be known explicitly for quantitative purposes. Which of these formulas is appropriate in the calculation of thermal expansion is a matter of convenience and depends partly on the special model under consideration. We will not go into the details of different models, which have been discussed, but only refer to the corresponding papers [61 L 1, 65 B 1, 62 M 5]. The simplest case of a monatomic linear lattice gives

$$M\omega^2 = 4f\sin^2 qa/2$$

and

$$\frac{\partial \ln\omega^2}{\partial \mathscr{V}} = \frac{1}{f}\frac{\partial f}{\partial \mathscr{V}} = \frac{\tilde{g}\,\tilde{a}}{\tilde{f}}$$

because it is

$$f(a) = \tilde{f} + \tilde{g}\,\delta a; \quad \mathscr{V} = \delta a/\tilde{a} + \frac{1}{2}(\delta a/\tilde{a})^2$$

($a$: nearest neighbor distance; $\delta a$: variation of $a$) and therefore with $V_z = ab^2$; $\tilde{C} = \tilde{f}\tilde{a}/b^2$ [see sect. 14]

$$\mathscr{V}^{\text{th}} = -\frac{\tilde{g}}{2\tilde{f}^2\tilde{a}} \cdot \frac{U(T)}{N} \tag{15.15}$$

which corresponds completely to (14.9) and extends it to low temperatures. The volume of the unit cell in this case is half the volume of the model in sect. 14, if we put $f_1 = f_2, \ldots$!

The equilibrium positions of the anharmonic theory as functions of temperature being specified, all the further quantities of interest can be calculated. The *(isothermal) elastic constants* are essentially the second derivatives of the free energy with respect to mechanical strains, related to the equilibrium positions at a given temperature $T$. A similar procedure as that which leads to equ. (15.5) now gives

$$\begin{aligned} N V_z C^{\text{is}}_{ij,kl} = {} & \frac{\partial^2 \Phi_0}{\partial \mathscr{V}_{ij}\,\partial \mathscr{V}_{kl}} + \frac{1}{2}\sum_{q\sigma} \frac{\partial^2 \ln \omega^2(q\sigma)}{\partial \mathscr{V}_{ij}\,\partial \mathscr{V}_{kl}} \cdot \varepsilon(\omega, T) + \\ & + \frac{1}{4}\sum_{q\sigma} \frac{\partial \ln \omega^2(q\sigma)}{\partial \mathscr{V}_{ij}} \cdot \frac{\partial \ln \omega^2(q\sigma)}{\partial \mathscr{V}_{kl}} \left[\varepsilon(\omega, T) - T\frac{\partial \varepsilon(\omega, T)}{\partial T}\right]. \end{aligned} \tag{15.16}$$

The general features of this expression are the same as for the thermal expansion. At high temperatures, the elastic constants vary linearly with temperature. But a few remarks concerning the first term in (15.16) have to be added. Whereas the logarithmic derivatives are directly proportional to anharmonicities ($\Phi_3^2$, $\Phi_4$), the first term contains harmonic contributions ($\Phi_2$) and through the (temperature dependent) equilibrium positions also anharmonic contributions ($\Phi_3$). Therefore in (15.16) we can take the anharmonic c.p.'s as defined by the minimum of the potential energy, deviations being of higher order. But we want to relate the first term in (15.16) to the elastic constants of the harmonic theory, therefore we have to calculate it at the temperature dependent equilibrium positions and expand this with respect to the thermal expansion. As the lattice data vary (in a certain approximation, see above) as the internal energy of the lattice, this expansion gives a further term which is proportional to the internal energy, or proportional to $T$ at high temperatures. Formally this first term can be gotten in the following way: $(1/NV_z)\,\partial^2\Phi_0/\partial\mathscr{V}_{ij}\,\partial\mathscr{V}_{kl}$ is formed in the same way as the harmonic elastic constants, but with temperature dependent positions and therefore also c.p.'s. The temperature dependent positions can be described by a thermal strain from the initial harmonic equilibrium positions. So one has to calculate the expression for arbitrary equilibrium positions and expand this expression with respect to the thermal deviations from the harmonic equilibrium positions (minimum of potential energy). We can write

$$\frac{1}{NV_z} \cdot \frac{\partial^2 \Phi_0}{\partial \mathscr{V}_{ij}\,\partial \mathscr{V}_{kl}} = \tilde{C}_{ij,kl} + \tilde{B}_{ij,kl,mn}\,\mathscr{V}^{\text{th}}_{mn}\,. \tag{15.17}$$

Here, of course, the $\tilde{C}_{ij,kl}$ are the elastic constants of the harmonic theory (sect. 6). But the $\tilde{B}_{ij,kl,mn}$ are not the elastic constants of the third order, as one would perhaps suppose from an extension of (6.27). This would be so in a linear theory in the strains only, if the strains are additive quantities. But anharmonic terms cause an essential non-linear theory (third order terms) and the finite strains are non-additive quantities. If we introduce a higher order Grüneisen-constant by

$$\Gamma_{ij,kl}(\boldsymbol{q}\sigma) = -\frac{1}{18}\frac{\partial^2 \ln \omega^2(\boldsymbol{q}\sigma)}{\partial \mathscr{V}_{ij}\,\partial \mathscr{V}_{kl}} \tag{15.18}$$

we have with (15.9), (15.16) and (15.17)

$$\begin{aligned} C^{\text{is}}_{ij,kl} = \tilde{C}_{ij,kl} &- \\ &- \frac{9}{N V_z}\sum_{\boldsymbol{q}\sigma}\Big\{\Gamma_{ij,kl}(\boldsymbol{q}\sigma) - \gamma_{ij}(\boldsymbol{q}\sigma)\,\gamma_{kl}(\boldsymbol{q}\sigma) - \\ &- \frac{1}{3}\sum_{\substack{mn\\rs}} \tilde{B}_{ij,kl,mn}\,\tilde{S}_{mn,rs}\,\gamma_{rs}(\boldsymbol{q}\sigma)\Big\}\cdot \varepsilon(\omega, T) - \\ &- \frac{9}{N V_z}\cdot\sum_{\boldsymbol{q}\sigma}\gamma_{ij}(\boldsymbol{q}\sigma)\,\gamma_{kl}(\boldsymbol{q}\sigma)\cdot T\,\frac{\partial\varepsilon(\omega, T)}{\partial T}\,. \end{aligned} \tag{15.19}$$

The main difficulty of a discussion of this expression lies in the calculation of the different coefficients. We will illustrate this with the simple example of a monatomic linear lattice.

Suppose that there is given a function of the squared positions of the ions, i.e. $G\left[\left(\boldsymbol{R}^m_\mu\right)^2\right]$. If there is a deformation of the lattice, we have [sect. 6, I, II]

$$\boldsymbol{R}^2 - \tilde{\boldsymbol{R}}^2 = 2\sum_{ij}(\tilde{X}_i \mathscr{V}_{ij} \tilde{X}_j)\,.$$

With the operator

$$O = \frac{1}{R}\frac{\partial}{\partial R} = 2\frac{\partial}{\partial R^2}$$

we have the expansion of $G(R^2)$ with respect to the finite strain components

$$G(R^2) = G(\tilde{R}^2) + (OG)\,\tilde{X}_i\,\tilde{X}_j\,\mathscr{V}_{ij} + \frac{1}{2}(O^2G)\,\tilde{X}_i\,\tilde{X}_j\,\tilde{X}_k\,\tilde{X}_l\cdot\mathscr{V}_{ij}\,\mathscr{V}_{kl} + \cdots \tag{15.20}$$

With this expansion all the strain derivatives of quantities above can be calculated simply. Every lattice potential energy and frequencies, etc., if they depend on the position, can be written as functions of squared distances. This holds also for many-body-potentials. In the case of linear lattice we have now

$$\frac{1}{N}\Phi_0 = \frac{1}{2N}\sum_{m\neq 0}\varphi(R^m) = \varphi(a)$$

and

$$\ln\omega^2 = \ln f(a) + \cdots$$

if we restrict the interaction to nearest neighbors. From this we have

$$\frac{\partial^2 \varphi(a)}{\partial \mathscr{V}^2} = a^4 (O^2 \varphi) = a^2 \cdot f(a) - a \cdot d(a)$$

$$\frac{\partial \ln \omega^2}{\partial \mathscr{V}} = a^2 (O f)/f = \frac{a}{f} g(a) = -6\gamma$$

$$\frac{\partial^2 \ln \omega^2}{\partial \mathscr{V}^2} = \frac{a^4}{f} O^2 f - \frac{a^4}{f^2} (O f)^2 = \frac{a^2}{f} h - \frac{a}{f} g - \frac{a^2}{f^2} g^2 = -18\Gamma$$

$$d = \frac{\partial \varphi}{\partial a}; \quad f = \frac{\partial^2 \varphi}{\partial a^2}; \quad g = \frac{\partial^3 \varphi}{\partial a^3} = \frac{\partial f}{\partial a}; \quad h = \frac{\partial^2 f}{\partial a^2}.$$

We can calculate the coefficients. It is

$$\frac{1}{N V_z} \cdot \frac{\partial^2 \Phi_0}{\partial \mathscr{V}^2} = \frac{1}{a} \frac{\partial^2 \varphi(a)}{\partial \mathscr{V}^2} = a f - d.$$

Expansion with respect to $\mathscr{V}^{\mathrm{th}}$ gives in the same way

$$a f = \tilde{a}\tilde{f} + \tilde{f}\tilde{a} \cdot \mathscr{V}^{\mathrm{th}} + \tilde{a}^2 \tilde{g} \mathscr{V}^{\mathrm{th}}$$

$$d = \tilde{d} + \tilde{f}\tilde{a} \cdot \mathscr{V}^{\mathrm{th}}$$

where the tilde denotes the harmonic equilibrium positions, defined by the minimum of the potential energy. Therefore $\tilde{d} = 0$ and we have

$$\tilde{C} = \tilde{a}\tilde{f}; \quad \tilde{B} = \tilde{g}\tilde{a}^2; \quad \gamma = -\frac{\tilde{g}\tilde{a}}{6\tilde{f}}; \quad \Gamma = \frac{\tilde{a}^2}{18\tilde{f}^2} (\tilde{g}^2 - \tilde{h}\tilde{f} + \tilde{g}\tilde{f}/\tilde{a}).$$

All quantities are independent of $\boldsymbol{q}\sigma$ in this simple model and we have

$$C^{\mathrm{is}} = \tilde{a}\tilde{f} \left\{ 1 - \frac{U(T)}{4\tilde{f}^3 N} (3\tilde{g}^2 - 2\tilde{h}\tilde{f} + 2\tilde{g}\tilde{f}/\tilde{a}) - \frac{T C_V}{4\tilde{f}^3 N} \tilde{g}^2 \right\} \tag{15.21}$$

$C_V = \partial U / \partial T$ is the specific heat in the harmonic approximation. In the high temperature limit (15.21) agrees with (14.25, 26), if we put $f_1 = f_2$, etc., introduce $a(T)$ as given by (14.9) or (15.15), and divide by $b^2$ in order to get the usual dimension of elastic constants. (Note, that the lattice constant now is half of that in sect. 14) (see page 98).

This simple example shows the general features of the temperature dependence of the elastic constants. They vary linearly with $T$ at high temperatures ($T > \theta$), and they reach a constant value at $T = 0$, which is given by the interaction between anharmonicity and zero-point vibrations of the lattice. These zero temperature constants are not the elastic constants of the harmonic theory. Rather, *the harmonic values of the elastic constants can be obtained from the linear extrapolation of the high temperature behavior* ($T > \theta$) *to* $T = 0$. *Only these linearly extrapolated elastic constants are the true harmonic constants and only these should be taken for a calculation of the harmonic (second order) coupling parameters.* Otherwise one would obtain temperature dependent coupling parameters, which cannot be defined in a unique way. If one would work with temperature dependent c.p.'s, these would be different if taken from different experiments (see below).

A general discussion of (15.16) or (15.19), especially for intermediate temperatures, is very cumbersome. All those phenomena as discussed in connection with the thermal expansion might happen. It is possible,

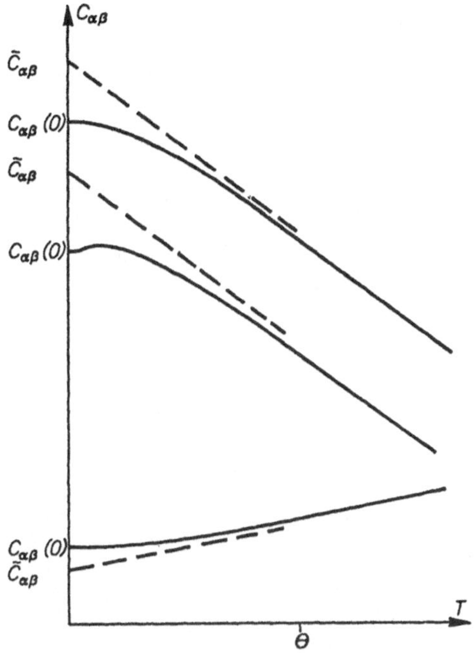


Fig. 5.5. Possible qualitative behavior of the elastic constants as functions of temperature. Linear extrapolation from high temperatures to $T = 0$ gives the harmonic values

that there is a slight increase of elastic constants, when temperature is raised, and at higher temperatures only the general decrease of elastic constants with temperature begins (Fig. 5.5). However, as far as measurements are available, the isothermal elastic constants show a monotonic behavior, in nearly all cases a monotonic decrease with increasing temperature. This behavior can be described in general by the approximation mentioned for the thermal expansion: The logarithmic derivatives of the squared frequencies are replaced by the derivatives of the mean squared frequencies $\overline{\omega^2}$, so that

$$\gamma_{ij} = -\frac{1}{6}\cdot\frac{\partial \ln \overline{\omega^2}}{\partial \mathscr{V}_{ij}}; \quad \Gamma_{ij,kl} = -\frac{1}{18}\frac{\partial^2 \ln \overline{\omega^2}}{\partial \mathscr{V}_{ij}\,\partial \mathscr{V}_{kl}}. \tag{15.22}$$

This procedure has been used in [61 L 1] to discuss the temperature dependence of the elastic constants in a number of cases. The agreement between theory and experiment is quite good. Fig. 5.6 shows as an example the temperature dependence of the elastic constants of LiF, NaJ and Ag. In this review we will not go into the details of the different

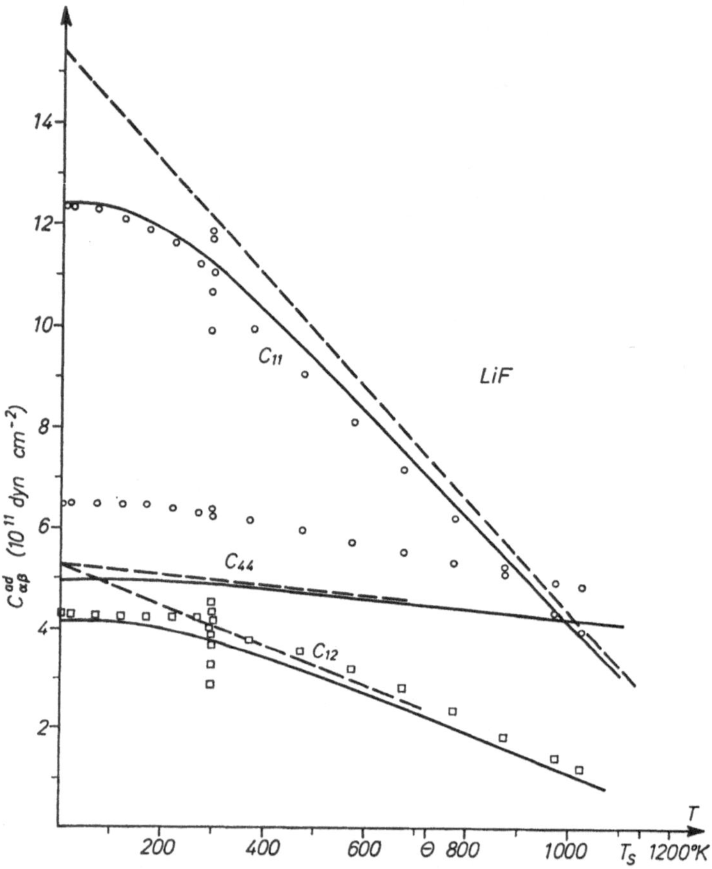


Fig. 5.6a. Adiabatic elastic constants of LiF as functions of temperature. ∘ ∘, □ □ experimental values, ——— according to the quasiharmonic approximations, with electrostatic and Born-Mayer-forces only [61 L 1], — — — linear extrapolation

models which have been investigated and mainly refer to the original papers [61 L 1, 58 L 2].

But the measured constants in general are not the isothermal, but the adiabatic elastic constants. These can be calculated simply, if the isothermal constants and the thermal expansion is known [I, II]. The difference between adiabatic and isothermal constants is given by

$$C^{ad}_{ij,kl} - C^{is}_{ij,kl} = N V_z \cdot \frac{T}{C_V} \sum_{mn,rs} C^{is}_{ij,mn} C^{is}_{kl,rs} \alpha^{th}_{mn} \alpha^{th}_{rs}, \qquad (15.23)$$

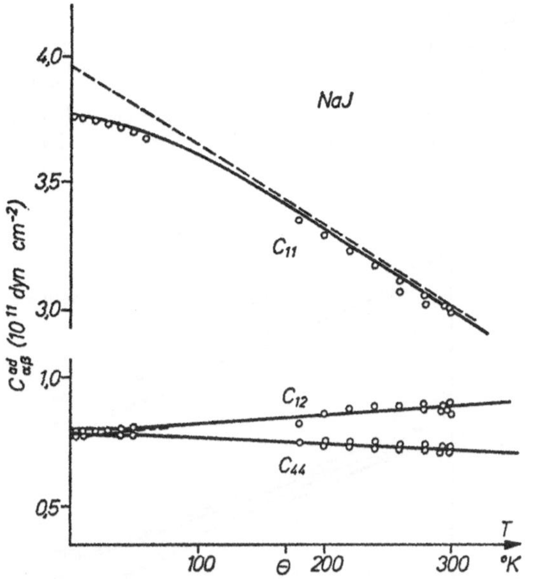


Fig. 5.6b. Elastic constants of NaJ. ——— according to the quasiharmonic approximation, but fitted to the experimental values ∘ ∘ ∘; $c_{12}$ is larger than $c_{44}$, an indication for the presence of many-body-forces

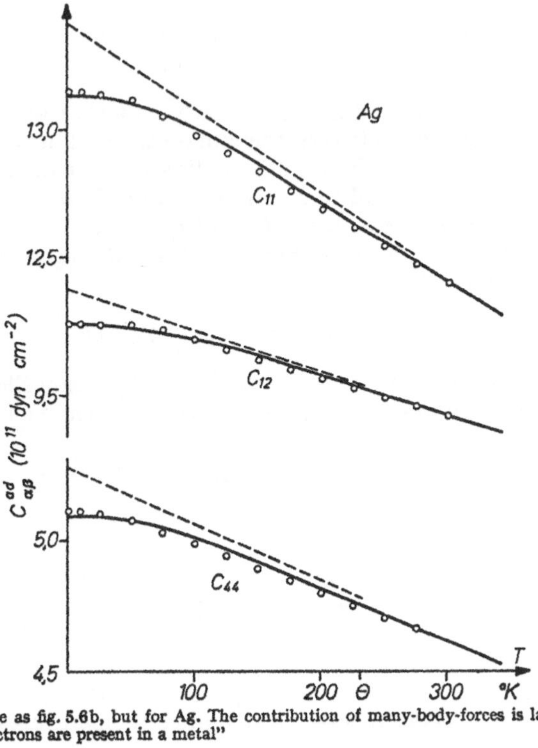


Fig. 5.6c. The same as fig. 5.6b, but for Ag. The contribution of many-body-forces is large, owing to the fact, that "free electrons are present in a metal"

where the elastic constants on the right hand side of (15.23) can be replaced by the harmonic values in our approximation. Inserting (15.10) into (15.23) we have

$$C^{\text{ad}}_{ij,kl} - C^{\text{is}}_{ij,kl} = \frac{9T}{N V_z C_V} \sum_{\mathbf{q}\sigma,\mathbf{q}'\sigma'} \gamma_{ij}(\mathbf{q}\sigma)\, \gamma_{kl}(\mathbf{q}'\sigma') \frac{\partial \varepsilon(\omega, T)}{\partial T} \cdot \frac{\partial \varepsilon(\omega', T)}{\partial T} \,. \tag{15.24}$$

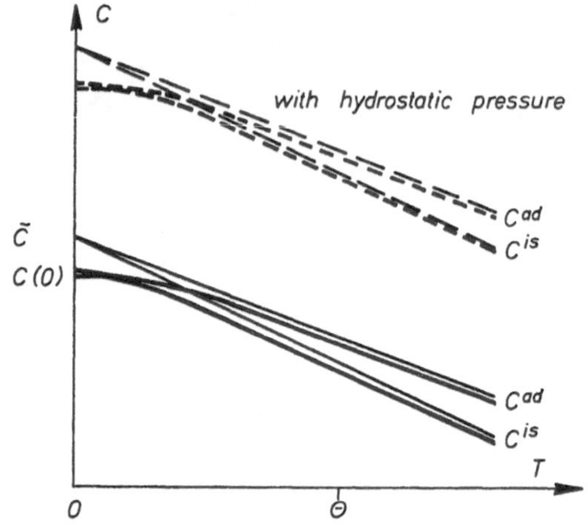


Fig. 5.7. Qualitative temperature dependence of adiabatic and isothermal elastic constants. The effect of a hydrostatic pressure is also indicated

If introduced into (15.19), this term nearly cancels the second term in (15.19). The difference between the adiabatic and isothermal elastic constants vanishes at $T = 0$, which is wellknown. At high temperatures the difference is proportional to $T$, and to obtain *the true elastic constants in the harmonic theory an extrapolation of the adiabatic as well as of the isothermal constants can be used* (Fig. 5.7). With the approximation used in (15.22) we have instead of (15.24)

$$C^{\text{ad}}_{ij,kl} - C^{\text{is}}_{ij,kl} = \frac{9TC_V}{N V_z} \cdot \gamma_{ij}\, \gamma_{kl} \tag{15.25}$$

and in this approximation the last term in (15.19) is cancelled completely (this is exact for the monatomic linear lattice!). Then we have

$$C^{\text{ad}}_{ij,kl} = \tilde{C}_{ij,kl} - \frac{9U(T)}{N V_z} \cdot \left\{ \Gamma_{ij,kl} - \gamma_{ij}\, \gamma_{kl} - \frac{1}{3} \sum_{\substack{mn\\ rs}} \tilde{B}_{ij,kl,mn}\, \tilde{S}_{mn,rs} \cdot \gamma_{rs} \right\} \tag{15.26}$$

the adiabatic elastic constants are directly proportional to the internal energy. This approximation seems to be quite good and can be used for the representation of the temperature dependence of the elastic constants in nearly all cases.

The limitations of this first order anharmonic theory can be seen by a comparison with experiments. If at higher temperatures ($T \gg \theta$) the elastic constants do not vary linearly with temperature, we must conclude, that higher order anharmonic effects are essential or even that there are other influences in the lattice (defects). Thus in the vicinity of the melting point of lattices this theory is not sufficient.

Some general remarks should be added concerning general statements on the elastic constants. In cubic crystals, we have $\gamma_{ij} = \gamma \cdot \delta_{ij}$. Then the difference between adiabatic and isothermal constants vanishes for $c_{44}$ and for $c_{11} - c_{12}$ (*Voigt's* notation) in agreement with the general statement, that adiabatic and isothermal shear moduli are equal. For the other cubic elastic constants, $c_{11}$ and $c_{12}$ as well as for most of the constants in other crystal systems, the adiabatic constants are larger than the isothermal ones.

In a harmonic crystal with central forces only and inversion symmetry, the Cauchy relations (e.g. $c_{12} = c_{44}$ in cubic crystals) should be satisfied. Now $c_{12} = C^{\mathrm{ad}}_{11,22}$ contains a term with $\gamma_{11} \cdot \gamma_{22} = \gamma^2$, whereas $c_{44} = C^{\mathrm{ad}}_{12,12}$ contains $\gamma_{12} \cdot \gamma_{12}$ which is zero in cubic crystals. Therefore the Cauchy-relations cannot be satisfied, even at $T = 0$, and even if all the other restrictions hold. Cauchy's relations can be satisfied only accidentally. As the difference $c_{12} - c_{44}$ is temperature dependent, it might happen that Cauchy's relations are satisfied at higher temperatures and not at $T = 0$, or contrary. The details depend very sensitively on the interacting potential energy of the ions.

If only central forces are present in parameterfree crystals (every ion is a center of inversion), then the elastic constants of the harmonic approximation $\tilde{C}_{ij,kl}$ are totally symmetric in the indices. $\Gamma_{ij,kl}$ can be written as

$$\Gamma_{ij,kl} = -\frac{1}{18} \cdot \frac{\partial^2 \ln \omega^2}{\partial \mathscr{V}_{ij}\, \partial \mathscr{V}_{kl}} = -\frac{1}{18\omega^2} \cdot \frac{\partial^2 \omega^2}{\partial \mathscr{V}_{ij}\, \partial \mathscr{V}_{kl}} + 2\gamma_{ij}\, \gamma_{kl} \,.$$

The first term of the right hand side then is totally symmetric in the indices. This can be seen immediately from (15.20), because central forces can always be represented by functions $G[(\boldsymbol{R}^m)^2]$. The only term which is not totally symmetric is that containing the factor $\gamma_{ij} \cdot \gamma_{kl}$. This gives for the difference (in cubic crystals)

$$C^{\mathrm{ad}}_{12,12} - C^{\mathrm{ad}}_{11,22} = c^{\mathrm{ad}}_{44} - c^{\mathrm{ad}}_{12} = \frac{9\,U(T)}{N V_z} \cdot \gamma^2 > 0 \quad \text{with} \quad \gamma_{ij} = \gamma \delta_{ij} \,. \tag{15.27}$$

This statement is not restricted to our approximation of replacing $\gamma_{ij}(\boldsymbol{q}\sigma)$ by mean values, as can be seen from the original formula. Thus in cubic crystals, if there are only central forces present, $c^{\mathrm{ad}}_{12}$ should be smaller than $c^{\mathrm{ad}}_{44}$. If this is not so, this is a direct hint for the presence of many body forces in the lattice. On the other hand, if (15.27) is satisfied, this would *not* exclude the presence of many body forces.

Apart from the temperature-dependence of the elastic constants due to anharmonic effects, there is another source of temperature variation. The thermal energy of a (free) electron gas is proportional to $T^2$. When deforming the lattice, this energy of the electron gas changes and therefore gives a contribution to the elastic constants, and thermal expansion.

In the usual electron gas approximation this contribution varies as $T^2$, for the whole range of temperatures. It should be possible to observe this effect mainly at low temperatures ($T \to 0$), just as the electronic contribution to the specific heat (see sect. 16). At higher temperatures the effect of anharmonicity is probably larger. Of course, the electronic contribution is present in metals only. Calculations have been made by *Bernstein* [63 B 4, 65 B 3].

Some remarks concerning the measurements may be added. In sound wave measurements of the elastic constants the adiabatic constants are measured. This was assumed always and can be proved in a certain sense from general considerations using energy and momentum conservation [61 L 5, see also 67 G 1]. From these measurements, if done for various temperatures, by a linear extrapolation one can calculate the constants of the harmonic theory and the coupling parameters (see page 100). By static measurements, the isothermal constants will be obtained, and a similar extrapolation can be done.

On the other hand, phonon dispersion curves are measured by X-ray- and especially by neutron-diffraction at crystals. The long wave dispersion is generally assumed to give the elastic constants. But then one has to ask, whether these are adiabatic or isothermal elastic constants. It seems, that the long-wave dispersion gives essentially the isothermal elastic constants, as well in X-ray diffraction [61 H 1] as in neutron-diffraction [63 H 2]. This looks somewhat strange and is not quite well understood now. At any case, if one measures the dispersion curves as a function of temperature (for $T > \theta$), a linear extrapolation to $T = 0$ should give the true harmonic constants. But one can use also the complete dispersion curves to get knowledge of the c.p.'s of the crystal lattices. These dispersion curves are temperature dependent over the whole range of $\boldsymbol{q}$-values, due to the influence of anharmonicity. Therefore one has to use the linearly to $T = 0$ extrapolated dispersion-curves to obtain statements about the true c.p.'s of a crystal lattice. Sometimes dispersion-curves of room-temperature or other finite values have been used to determine c.p.'s. These c.p.'s. of course, contain the temperature and therefore anharmonic contributions; they cannot be used to get statements about the range of the interionic forces because one does not know what part of these "effective" c.p.'s is due to the temperature dependent anharmonic contributions and what part is due to the real harmonic c.p.'s. *Elastic constants and dispersion curves extrapolated linearly to* $T = 0$ *should be used to make reasonable statements on the harmonic (second order) c.p.'s and their ranges in the lattice.*

Another effect of anharmonicity on elastic properties is the appearance of the elastic constants of third order. These constants are zero in the harmonic approximation. Only in an anharmonic theory they occur. In our lowest order anharmonic approximation they are independent of temperature, as can be seen from the example in sect. 14. Therefore it is more convenient to calculate them by a pure mechanical procedure, as it is indicated in sect. 6. It can be done with the same procedure, only the expressions involved become rather complicated [60 L 1].

These third order constants determine the stress-dependence of the usual second order elastic constants. The experimental situation, underlying this concept, is the following one: Suppose the elastic medium is strained by an initial stress $S_{ij}$, the relation between this stress and the finite initial strain $\mathscr{V}_{ij}$ being (summation convention!)

$$S_{ij} = C_{ij,kl}\mathscr{V}_{kl} + \cdots = C_{ij,kl}u_{k|l} + \cdots, \qquad 2\mathscr{V}_{kl} = u_{k|l} + u_{l|k} + u_{i|k}u_{i|l} \tag{15.28}$$

where $u_{k|l}$ is the corresponding displacement gradient. Beginning with this deformed state, an additional stress $\sigma_{ij}$ is applied, producing an additional deformation $v_{ij}$, which can be assumed as infinitesimal, if the additional stress is sufficiently small, whereas the initial stress may be large. The coefficients between the additional stress and strain are the elastic constants $S_{ij,kl}$ (of second order, see sect. 6) for an initially stressed medium and depend, of course, on the initial stress. They have not the point group symmetries of the unstrained crystal, but those of the initially strained medium. We may write

$$\sigma_{ij} = S_{ij,kl}(S_{mn}) \cdot v_{kl}\,. \tag{15.29}$$

The $S_{ij,kl}(S_{mn})$ have to be calculated, using the theory of finite strain [37 M 1, 47 B 2]. For a homogeneous, isotropic pressure as the initial stress,

$$S_{ij} = -p\,\delta_{ij} \tag{15.30}$$

this has been done by *Birch* [47 B 1]. The evaluation is straightforward, but somewhat lengthy. A slight generalization to arbitrary initial stress gives

$$S_{ij,kl}(S_{mn}) = C_{ij,kl}(0) + D_{ij,kl,mn} \cdot u_{m|n}(S_{rs}) \tag{15.31}$$

where we have limited the expression to lowest order in $u_{m|n}$ or $S_{mn}$ resp., which is sufficient for nearly all the experiments having been done. $C_{ij,kl}(0) = C_{ij,kl}$ are the elastic constants measured in the usual way, beginning in a mechanically unstrained state. These constants occur in (15.28). The $D$-coefficients come out to be

$$\begin{aligned} D_{ij,kl,mn} = {} & C_{jl,mn} \cdot \delta_{ik} + C_{il,mn} \cdot \delta_{jk} - C_{ij,mn} \cdot \delta_{kl} + \\ & + C_{jn,kl} \cdot \delta_{im} + C_{in,kl}\,\delta_{jm} - C_{ij,kl} \cdot \delta_{mn} + \\ & + C_{ij,ln}\,\delta_{km} + C_{ij,kn}\,\delta_{lm} + C_{ij,kl,mn} \end{aligned} \tag{15.32}$$

which contains the elastic constants* of third order $C_{ij,kl,mn}$. All coefficients in (15.32) are related to the initially unstrained state. Thus

* The definition used here corresponds to that in sect. 6. The different constants $C_{ij,kl,mn}$ *are components of a six-rank tensor. Birch* and also *Seeger* and *Buck* [60 S 1] use another definition, in which the different constants *are not components of one unique tensor!* The relations between *Birch's* constants $C^B_{\alpha\beta\gamma}$ and ours $C_{\alpha\beta\gamma}$ are in *Voig't* notation for cubic crystals:

$$6C^B_{111} = C_{111};\quad 2C^B_{112} = C_{112};\quad C^B_{123} = C_{123};$$
$$C^B_{144} = 2C_{144};\quad C^B_{166} = 2C_{166};\quad C^B_{456} = 4C_{456}\,.$$

It should be mentioned that the $D_{ij,kl,mn}$ can be related to the $\bar{\bar{B}}_{ij,kl,mn}$ in (15.17) if one replaces strain and stress in (15.28) by thermal strain and stress and then uses (15.28) and (15.31).

from measurements of the stress-dependence of the elastic constants, the third order constants can be obtained, at least in principle. In (15.31) $u_{m|n}$ has to be expressed by $S_{mn}$. This cannot be done simply by inverting (15.28), because $u_{m|n}$ is not symmetric in general. Rather one has to use the procedure of sect. 6 to get knowledge on the stresses $S_{mn}$. Only in the case of isotropic pressure in cubic crystals this can be done simply, and this is the only case of practical interest; it is difficult to produce other stresses large enough to see the effect. In case of isotropic pressure (15.31) is simplified to

$$S_{ij,kl}(p) = C_{ij,kl} - \overline{D}_{ij,kl,mm} \cdot p \tag{15.31a}$$

where $\overline{D}_{ij,kl,mm}$ is related to (15.32) by means of (15.28) and (15.30) with $u_{k|l} = -\alpha \cdot \delta_{kl}$ (cubic crystal!). We obtain

$$\overline{D}_{ij,kl,mm} = \delta_{ik}\,\delta_{jl} + \delta_{il}\,\delta_{jk} - \delta_{ij}\,\delta_{kl} + \frac{\varkappa}{3}\left[C_{ij,kl} + C_{ij,kl,mm}\right] \tag{15.33}$$

or explicitly for cubic crystals* (in *Voigt's* notation)

$$\begin{aligned} -\frac{\mathrm{d}c_{11}}{\mathrm{d}p} &= \frac{\varkappa}{3}\left[2c_{11} + 2c_{12} + c_{111} + 2c_{112}\right], \\ -\frac{\mathrm{d}c_{12}}{\mathrm{d}p} &= \frac{\varkappa}{3}\left[-c_{11} - c_{12} + 2c_{112} + c_{123}\right], \\ -\frac{\mathrm{d}c_{44}}{\mathrm{d}p} &= \frac{\varkappa}{3}\left[c_{11} + 2c_{12} + c_{44} + 2c_{166} + c_{144}\right], \quad \frac{\varkappa}{3} = \frac{1}{c_{11} + 2c_{12}}. \end{aligned} \tag{15.34}$$

Measurement of the pressure dependence of the elastic constants therefore gives some combinations of the third order constants. Other methods of measuring the third order coefficients have been discussed by *Seeger* and *Buck* [60 S 1], by *Einspruch* and *Manning* [64 E 1] and *Melngailis*, *Maradudin* and *Seeger* [63 M 7]. If one studies the velocity of sound in an initially stressed medium one measures effective elastic constants (similar to those defined above), which depend on the initial stress and contain the third order constants. The calculation of these effective constants is simple, but rather lengthy and we refer to the papers [60 S 1, 64 E 1].

On the other hand, third order elastic constants in the energy of the crystal give rise to terms in the equation of motion for an elastic continuum, which are non-linear in the displacements gradients. The solution of this equation therefore contains higher harmonics. In transparent crystals the (longitudinal) sound waves can be used for producing a diffraction pattern of light, if the light propagates perpendicular to the sound wave. Now, if there are higher harmonics in this sound wave, the diffraction pattern deviates in a characteristic way from that of harmonic crystal, and thus gives certain statements about third order elastic constants. Detailed calculations have been done for the diffraction of light at longitudinal sound waves [63 M 7]. Also the damping of sound waves depends on the third order elastic constants and gives a possibility for measurements.

---

* See footnote p. 107.

## 16. Perturbation Expansion of the Free Energy and the Specific Heat

Though it is sufficient to use the quasiharmonic approximation in the calculation of anharmonic effects on mechanical properties of lattices, for the discussion of caloric properties, such as entropy or specific heat, a real anharmonic theory is needed, which takes into account the $\Phi_3$- and $\Phi_4$-term in (1.6) explicitly. A perturbation expansion of the free energy with respect to $\Phi_3$ and $\Phi_4$ has first been given in 1958 [58 L 1], using a quantum mechanical perturbation expansion proposed by *Leibfried* [I] and *Nakajima* [55 N 1]. Later this theory has been refined [61 L 1; 61 M 1, 2, 3]. The same result can be obtained by using the method of thermodynamic Greens functions [see Appendix; 63 C 2]. Before applying this method to the calculation of the lattice free energy, we will consider its general properties.

The lattice Hamiltonian consists of an "unperturbed" part $\mathcal{H}_0$ and the perturbation $W$, the anharmonicity, which we understand to consist of $\Phi_3$, $\Phi_4$ and, if necessary, $\Phi_1$. The zero Hamiltonian covers the kinetic energy of the lattice ions and the quasiharmonic part $\Phi_0 + \Phi_2$ of the potential energy. Thus we have the total Hamiltonian

$$\mathcal{H} = \mathcal{H}_0 + g\,W \tag{16.1}$$

where $g$ indicates the order of magnitude and is put equal to one afterwards. It should be emphasized, that *as well $\mathcal{H}_0$ as $W$ is proportional to $N$*, the number of unit cells in the periodicity volume (or in the crystal, if it is finite) and therefore to the volume itself. The partition function of a quantum mechanical system is given by*

$$\begin{aligned} Z(T,V) &= \operatorname{Trace} e^{-\beta\mathcal{H}(V)} = \sum_m \langle m\,|e^{-\beta\mathcal{H}(V)}|\,m\rangle \\ &= \sum_m \langle m\,|e^{-\beta(\mathcal{H}_0 + gW)}|\,m\rangle; \quad \beta = 1/kT \end{aligned} \tag{16.2}$$

if $|m\rangle$ denotes any complete set of eigenfunctions for the system under consideration. (16.2) defines the partition function with temperature $T$ and volume $V$ as the independent variables. Thus the Helmholtz free energy

$$F(T,V) = -\frac{1}{\beta}\ln Z(T,V) \tag{16.3}$$

is the corresponding thermodynamic function. We seek for an expansion of (16.2) or rather of (16.3) with respect to $g\,W$. Since $W$ is proportional to $N$ ($\approx 10^{20}$), $Z$ is proportional to $e^{-N}$, which would lead to no reasonable expansion of $Z$. But $F$ is the logarithm of $Z$ and therefore proportional to $N$. Thus every term in the expansion of $F$ or $\ln Z$ with respect to $g\,W$ is proportional to $N$; this gives a convenient and useful expansion of $F$. The first term has the meaning of a mean deviation of $F$ from its harmonic value according to the perturbation $W$, the second is the fluctuation in $F$ due to $W$ etc., *each term being exactly proportional to $N$*. This has not been realized sometimes. With $Z = Z(g)$ we have immediately

$$\ln Z(g) = \ln Z(0) + g\frac{Z'(0)}{Z(0)} + \frac{1}{2}g^2\left\{\frac{Z''(0)}{Z(0)} - \left[\frac{Z'(0)}{Z(0)}\right]^2\right\} + \cdots \tag{16.4}$$

* See any book on Quantum Mechanics and Quantum Mechanical Statistics.

with

$$Z'(0) = \frac{\partial Z(g)}{\partial g}\bigg|_{g=0}; \quad Z''(0) = \frac{\partial^2 Z(g)}{\partial g^2}\bigg|_{g=0}$$

and

$$F(T, V) = F_{\text{qh}}(T, V) - \frac{g}{\beta} \cdot \frac{Z'(0)}{Z(0)} - \frac{g^2}{2\beta}\left\{\frac{Z''(0)}{Z(0)} - \left[\frac{Z'(0)}{Z(0)}\right]^2\right\}. \tag{16.5}$$

$F_{\text{qh}}(T, V) = -\frac{1}{\beta}\ln Z(0)$ is the quasiharmonic value of the free energy and given by (15.3). What has to be calculated is $Z(0)$ and its derivatives with respect to $g$, i. e. the expansion of $Z$ with respect to $g$. Though in this series none of the terms need to be proportional to $N$, this *is* the case for $\ln Z$ and the terms of different order in (16.5). This is guaranteed by the form of the expansion and shall be emphasized again. Further, there is no trouble with the convergence of the expansion (16.5). Because most of the interaction between the ions is already contained in the harmonic (or quasiharmonic) term $\Phi_2$, $\Phi_3$, $\Phi_4$, etc. are "small" perturbations and the series (16.5) converges if the anharmonicities are sufficiently small. But this is a necessary assumption for any reasonable lattice theory in the presented form, and it has been proved to be sufficient for most purposes. However, it surely breaks down in the immediate neighborhood of the melting point and it seems also to be questionable whether phase transitions in the crystalline state can be explained properly in the framework of this theory.

The perturbation expansion of (16.2) is equivalent to that used in Diracs time-dependent perturbation theory where the time $t$ has been replaced by $-i\beta\hbar$*. Because $\mathscr{H}_0$ and $W$ do not commute, the expansion reads

$$Z(g) = \sum_m \left\langle m \left| e^{-\beta\mathscr{H}_0}\left\{1 - g\int_0^\beta \tilde{W}(\beta')\,d\beta' + \right.\right.\right. \\ \left.\left.\left. + g^2 \int_0^\beta\int_0^{\beta'} \tilde{W}(\beta')\,\tilde{W}(\beta'')\,d\beta'\,d\beta'' + \cdots\right\}\right| m\right\rangle \tag{16.6}$$

where

$$\tilde{W}(\beta) = e^{\beta\mathscr{H}_0}\,W\,e^{-\beta\mathscr{H}_0}$$

is the perturbation operator in the "interaction representation". Since $\exp(-\beta\mathscr{H}_0)$ appears as the unperturbed part, it is convenient to take the harmonic eigenstates for the complete but arbitrary system $|m\rangle$. We have thus

$$Z(0) = \sum_m e^{-\beta E_m}; \quad \mathscr{H}_0|m\rangle = E_m|m\rangle, \tag{16.7a}$$

$$\begin{aligned} Z'(0) &= -\sum_m \left\langle m \left| e^{-\beta\mathscr{H}_0}\int_0^\beta e^{\beta'\mathscr{H}_0}\,W\,e^{-\beta'\mathscr{H}_0}\right| m\right\rangle d\beta' \\ &= -\sum_m e^{-\beta E_m}\int_0^\beta W_{mm}\,d\beta' \\ &= -\beta\sum_m W_{mm}\,e^{-\beta E_m} \end{aligned} \tag{16.7b}$$

---

* See any of the standard books on Quantum Theory, e.g. *Becker-Sauter, Schiff, Merzbacher, Landau-Lifšič.*

and correspondingly

$$Z''(0) = 2 \sum_m \left\langle m \left| e^{-\beta \mathscr{H}_0} \int_0^\beta \tilde{W}(\beta')\, d\beta' \int_0^{\beta'} \tilde{W}(\beta'')\, d\beta'' \right| m \right\rangle$$

$$= 2 \sum_m \sum_n \left\langle m \left| e^{-\beta \mathscr{H}_0} \int_0^\beta \tilde{W}(\beta')\, d\beta' \right| n \right\rangle \left\langle n \left| \int_0^{\beta'} \tilde{W}(\beta'')\, d\beta'' \right| m \right\rangle$$

$$= 2 \sum_{m,n} W_{mn} W_{nm} \frac{e^{\beta(E_m - E_n)} - \beta(E_m - E_n) - 1}{(E_m - E_n)^2} e^{-\beta E_m}. \tag{16.7c}$$

This last expression is considerably simplified in the classical limit when all energy differences $(E_m - E_n) \ll 1/\beta = kT$ or, with other words, if $\beta \to 0$. Then we have

$$Z''(0) \to \beta^2 \sum_{mn} W_{mn} W_{nm} e^{-\beta E_m} = \beta^2 \sum_m (W^2)_{mm} e^{-\beta E_m}. \tag{16.7d}$$

In the classical limit the derivatives divided by $Z(0)$ are nothing else but the mean values of the powers of the perturbation:

$$\frac{Z'(0)}{Z(0)} = -\beta \langle W \rangle_{\mathrm{Av}}; \quad \frac{Z''(0)}{Z(0)} = \beta^2 \langle W^2 \rangle_{\mathrm{Av}} \tag{16.8}$$

whereas for lower temperatures there are some changes in the second order (and higher) terms, characteristic for quantum mechanics.

For the final calculation of the lattice free energy it is convenient to transform the lattice energy as given by (1.6) in "coordinate representation" to a representation in $q$-space (creation and annihilation operators for phonons). This unitary transformation is given by means of the eigenvectors of the equation of motion (15.4). It reads [I, II, 61 L 1]*

$$u_{i}^{\mu m} = \sqrt{\frac{\hbar}{2sNM_\mu}} \cdot \sum_{q\sigma}{}' \frac{1}{\sqrt{\omega(q\sigma)}} e_i^\mu(q\sigma) \cdot e^{iqAm} \left( b_\sigma^{q} - b_\sigma^{-q+} \right). \tag{16.9}$$

$b_\sigma^q$ is the annihilation, $b_\sigma^{q+}$ the creation-operator for the phonon $q, \sigma$ with the (minus-)commutation relations

$$\left[ b_\sigma^q, b_{\sigma'}^{q'+} \right]_- = \delta_{qq'} \delta_{\sigma\sigma'}; \quad \left[ b_\sigma^q, b_{\sigma'}^{q'} \right]_- = \left[ b_\sigma^{q+}, b_{\sigma'}^{q'+} \right]_- = 0. \tag{16.10}$$

The polarization vectors are normalized such that

$$\sum_{\mu i} e_i^{*\mu}(q\sigma)\, e_i^\mu(q\sigma') = s\, \delta_{\sigma\sigma'}; \quad \sum_\sigma e_i^{*\mu}(q\sigma)\, e_j^\nu(q\sigma) = s\, \delta_{\mu\nu} \delta_{ij}. \tag{16.11}$$

In this representation the zero Hamiltonian is that of a system of independent oscillators

$$\mathscr{H}_0 = \Phi_0\left( \ldots R_\mu^m \ldots \right) + \sum_{q\sigma} \hbar\, \omega(q\sigma) \left\{ b_\sigma^{q+} b_\sigma^q + 1/2 \right\} \tag{16.12}$$

* The center-of-mass motion (acoustical branches of dispersion curves with $q = 0$ and $\omega = 0$)! has to be dropped in the following, because it cannot be expressed by phonon-creation or annihilation operators. But this is uninteresting in the following.

where $\Phi_0$ and $\omega(\boldsymbol{q}\,\sigma)$ are the quasiharmonic quantities which are related to the temperature dependent mean equilibrium positions defined by (15.5). Transformation of the $\Phi_1$, $\Phi_3$, $\Phi_4$-terms in (1.6) gives

$$\Phi_1 = \sqrt{\frac{sN\hbar}{2}}\sum_{\sigma}{}' \frac{1}{\sqrt{\omega(0,\sigma)}}\,\Phi^{0}_{\sigma}\left(b^{0}_{\sigma} - b^{0+}_{\sigma}\right)$$

$$\Phi_3 = \frac{1}{3!}\sqrt{\frac{\hbar^3}{8sN}}\cdot\sum_{\substack{\boldsymbol{q}\,\boldsymbol{q}'\,\boldsymbol{q}''\\ \sigma\,\sigma'\,\sigma''}} \frac{1}{\sqrt{\omega(\boldsymbol{q}\sigma)\,\omega(\boldsymbol{q}'\sigma')\,\omega(\boldsymbol{q}''\sigma'')}}\,\Phi^{\boldsymbol{q}\,\boldsymbol{q}'\,\boldsymbol{q}''}_{\sigma\,\sigma'\,\sigma''}\times$$
$$\times\left(b^{\boldsymbol{q}}_{\sigma} - b^{-\boldsymbol{q}+}_{\sigma}\right)\left(b^{\boldsymbol{q}'}_{\sigma'} - b^{-\boldsymbol{q}'+}_{\sigma'}\right)\left(b^{\boldsymbol{q}''}_{\sigma''} - b^{-\boldsymbol{q}''+}_{\sigma''}\right)$$

$$\Phi_4 = \frac{\hbar^2}{4!\cdot 4sN}\sum_{\substack{\boldsymbol{q}\,\boldsymbol{q}'\,\boldsymbol{q}''\,\boldsymbol{q}'''\\ \sigma\,\sigma'\,\sigma''\,\sigma'''}} \frac{1}{\sqrt{\omega\,\omega'\,\omega''\,\omega'''}}\,\Phi^{\boldsymbol{q}\,\boldsymbol{q}'\,\boldsymbol{q}''\,\boldsymbol{q}'''}_{\sigma\,\sigma'\,\sigma''\,\sigma'''}\times \tag{16.13}$$
$$\times\left(b^{\boldsymbol{q}}_{\sigma} - b^{-\boldsymbol{q}+}_{\sigma}\right)\left(b^{\boldsymbol{q}'}_{\sigma'} - b^{-\boldsymbol{q}'+}_{\sigma'}\right)\left(b^{\boldsymbol{q}''}_{\sigma''} - b^{-\boldsymbol{q}''+}_{\sigma''}\right)\left(b^{\boldsymbol{q}'''}_{\sigma'''} - b^{-\boldsymbol{q}'''+}_{\sigma'''}\right)$$

with

$$\Phi^{0}_{\sigma} = \frac{1}{s}\sum_{\mu i}\frac{\Phi^{0}_{\substack{\mu\\ i}}}{\sqrt{M_\mu}}\,e_{\substack{\mu\\ i}}(0,\sigma) \tag{16.14}$$

$$\Phi^{\boldsymbol{q}\,\boldsymbol{q}'\,\boldsymbol{q}''}_{\sigma\,\sigma'\,\sigma''} = \frac{1}{sN}\sum_{\substack{\boldsymbol{m}\,\boldsymbol{n}\,\boldsymbol{p}\\ \mu\,\nu\,\varkappa\\ i\,j\,k}} \frac{\Phi^{\boldsymbol{m}\,\boldsymbol{n}\,\boldsymbol{p}}_{\substack{\mu\,\nu\,\varkappa\\ i\,j\,k}}}{\sqrt{M_\mu M_\nu M_\varkappa}}\,e_{\substack{\mu\\ i}}(\boldsymbol{q}\,\sigma)\,e_{\substack{\nu\\ j}}(\boldsymbol{q}'\,\sigma')\,e_{\substack{\varkappa\\ k}}(\boldsymbol{q}''\,\sigma'')\times$$
$$\times\, e^{i\boldsymbol{q}A\boldsymbol{m}+i\boldsymbol{q}'A\boldsymbol{n}+i\boldsymbol{q}''A\boldsymbol{p}}$$

$$\Phi^{\boldsymbol{q}\,\boldsymbol{q}'\,\boldsymbol{q}''\,\boldsymbol{q}'''}_{\sigma\,\sigma'\,\sigma''\,\sigma'''} = \frac{1}{sN}\sum_{\substack{\boldsymbol{m}\,\boldsymbol{n}\,\boldsymbol{p}\,\boldsymbol{r}\\ \mu\,\nu\,\varkappa\,\lambda\\ i\,j\,k\,l}} \frac{\Phi^{\boldsymbol{m}\,\boldsymbol{n}\,\boldsymbol{p}\,\boldsymbol{r}}_{\substack{\mu\,\nu\,\varkappa\,\lambda\\ i\,j\,k\,l}}}{\sqrt{M_\mu M_\nu M_\varkappa M_\lambda}}\,e_{\substack{\mu\\ i}}(\boldsymbol{q}\sigma)\,e_{\substack{\nu\\ j}}(\boldsymbol{q}'\sigma')\,e_{\substack{\varkappa\\ k}}(\boldsymbol{q}''\sigma'')\,e_{\substack{\lambda\\ l}}(\boldsymbol{q}'''\sigma''')\times$$
$$\times\, e^{i\boldsymbol{q}A\boldsymbol{m}+i\boldsymbol{q}'A\boldsymbol{n}+i\boldsymbol{q}''A\boldsymbol{p}+i\boldsymbol{q}'''A\boldsymbol{r}}\,.$$

The transformed c.p.'s have the following properties: Index pairs can be interchanged:

$$\Phi^{\boldsymbol{q}\,\boldsymbol{q}'\,\boldsymbol{q}''}_{\sigma\,\sigma'\,\sigma''} = \Phi^{\boldsymbol{q}'\,\boldsymbol{q}\,\boldsymbol{q}''}_{\sigma'\,\sigma\,\sigma''} = \Phi^{\boldsymbol{q}'\,\boldsymbol{q}''\,\boldsymbol{q}}_{\sigma'\,\sigma''\,\sigma}\,;\quad \Phi^{\boldsymbol{q}\,\boldsymbol{q}'\,\boldsymbol{q}''\,\boldsymbol{q}'''}_{\sigma\,\sigma'\,\sigma''\,\sigma'''} = \Phi^{\boldsymbol{q}'\,\boldsymbol{q}\,\boldsymbol{q}''\,\boldsymbol{q}'''}_{\sigma'\,\sigma\,\sigma''\,\sigma'''} = \cdots \tag{16.15}$$

which follows immediately with (2.1). Further it is ($\boldsymbol{r}$: order of c. p.)

$$\Phi^{-\boldsymbol{q}\,-\boldsymbol{q}'\,-\boldsymbol{q}''\ldots}_{\sigma\quad\sigma'\quad\sigma''\ldots} = (-1)^{\boldsymbol{r}}\,\Phi^{*\boldsymbol{q}\,\boldsymbol{q}'\,\boldsymbol{q}''\ldots}_{\ \sigma\,\sigma'\,\sigma''\ldots} \tag{16.16}$$

which is a consequence of (15.4b). The most important property follows from the translational invariance of the c.p.'s against translations by multiples of the basis-vectors (3.2). The sum over, $\boldsymbol{m}, \boldsymbol{n}, \boldsymbol{p}, \ldots$ can be replaced by a sum over $\boldsymbol{m}+\boldsymbol{h}, \boldsymbol{p}+\boldsymbol{h}, \ldots$, which gives a factor $\exp\{i(\boldsymbol{q} + \boldsymbol{q}' + \boldsymbol{q}'' + \cdots)\,A\,\boldsymbol{h}\}$. Since (16.14) must be independent of $\boldsymbol{h}$, this factor has to be equal to one, or, in other words, $\boldsymbol{q}+\boldsymbol{q}'+\boldsymbol{q}''+\cdots = 2\pi B\boldsymbol{m} = \boldsymbol{g}$, where $\boldsymbol{g}$ is a vector of the $2\pi$-reciprocal lattice. This can be expressed by saying that the transformed c.p.'s contain a Kronecker-factor (with the periodicity of the reciprocal lattice)

$$\Delta^{(\boldsymbol{r})}_{\boldsymbol{g}} = \Delta(\boldsymbol{q}+\boldsymbol{q}'+\boldsymbol{q}''+\cdots-\boldsymbol{g}) \tag{16.17}$$

for finite periodicity volumes*! Since the anharmonic terms represent a phonon-phonon-interaction, (16.14a) can be looked upon as the *conservation of pseudo- or quasi-momentum*, or *better as a selection rule in phonon-phonon-interaction processes*. Processes with $\boldsymbol{g} \neq 0$ are called Umklapp-processes. But which processes are normal ones and which are of Umklapp-type depends on the choice of the Brillouinzone (unit cell in reciprocal lattice). What is esssential is only the fact, that there are processes with $\boldsymbol{g} \neq 0$. So lattice thermal conductivity is strongly influenced by processes with $\boldsymbol{g} \neq 0$. But in the following discussions of this report this distinction is not needed and we refer for details to other papers. We indicate this property by writing the factor $\Delta_{\boldsymbol{g}}^{(r)}$ besides $\Phi^{\boldsymbol{q}\,\boldsymbol{q}'\,\boldsymbol{q}''\ldots}_{\sigma\,\sigma'\,\sigma''\ldots}$.

In (16.14a) this property of the c.p.'s has already been used; therefore (16.13a) contains only the relative motions of the different sublattices (without center-of-mass-motion), i. e. it contains only $3(s-1)$ terms. In sect. 15 we have seen that $\Phi_1$ is to be expected to be of order of $\Phi_3$, so that it can be handled just as the $\Phi_3$-term in the perturbation calculation. Both terms, $\Phi_1$ and $\Phi_3$ are odd in the phonon-operators and therefore give no contribution to the first order correction (16.7b). The only contribution to the first order term comes from the elements in $\Phi_4$, which have two creation and two annihilation operators (diagonal elements of $\Phi_4$). As we have assumed that the expansion (1.6) for the potential energy is such, that successive terms decrease in order of magnitude, we consider $\Phi_4$ only in linear approximation, i. e. in (16.7b) and therefore only the diagonal terms are of interest now. $\Phi_3$ and $\Phi_1$ enter only in the second order perturbation and all terms enter the expression. This leads to our statement in (15.1), that the free energy can be written in the form

$$F = F_{\text{qh}} + F_3(\Phi_3^2) + F_1(\Phi_1^2;\, \Phi_1\,\Phi_3) + F_4(\Phi_4)\,. \tag{16.18}$$

The detailed calculation of $F_4$, $F_3$ will only be scetched here. The diagonal elements of $\Phi_4$ stem from

$$\Phi_4 = \frac{6}{4!}\,\frac{\hbar^2}{4sN} \sum{}' \frac{\Delta_{\boldsymbol{g}}^{(4)}\,\Phi^{\boldsymbol{q}\,\boldsymbol{q}'\,\boldsymbol{q}''\,\boldsymbol{q}'''}_{\sigma\,\sigma'\,\sigma''\,\sigma'''}}{\sqrt{\omega\,\omega'\,\omega''\,\omega'''}} \cdot \Big\{ b^{-\boldsymbol{q}+}_{\sigma}\, b^{-\boldsymbol{q}'+}_{\sigma'}\, b^{\boldsymbol{q}''}_{\sigma''}\, b^{\boldsymbol{q}'''}_{\sigma'''} +$$

$$+\, b^{-\boldsymbol{q}+}_{\sigma}\, b^{\boldsymbol{q}''}_{\sigma''} \cdot \delta_{-\boldsymbol{q}'\boldsymbol{q}'''}\,\delta_{\sigma'\sigma'''} + b^{-\boldsymbol{q}'+}_{\sigma'}\, b^{\boldsymbol{q}'''}_{\sigma'''} \cdot \delta_{-\boldsymbol{q}\boldsymbol{q}''}\,\delta_{\sigma\sigma''} +$$

$$+\, \frac{1}{2}\,\delta_{-\boldsymbol{q},\boldsymbol{q}''}\,\delta_{\sigma\sigma''}\,\delta_{-\boldsymbol{q}'\boldsymbol{q}'''}\,\delta_{\sigma'\sigma'''}\Big\}$$

with either

$$\begin{matrix}\boldsymbol{q}\\ \sigma\end{matrix} = \begin{matrix}-\boldsymbol{q}''\\ \sigma''\end{matrix} \quad\text{and}\quad \begin{matrix}\boldsymbol{q}'\\ \sigma'\end{matrix} = \begin{matrix}-\boldsymbol{q}'''\\ \sigma'''\end{matrix} \quad\text{if}\quad \begin{matrix}\boldsymbol{q}\\ \sigma\end{matrix} \neq \begin{matrix}\boldsymbol{q}'\\ \sigma'\end{matrix}$$

or

$$\begin{matrix}\boldsymbol{q}\\ \sigma\end{matrix} = \begin{matrix}-\boldsymbol{q}'''\\ \sigma'''\end{matrix} \quad\text{and}\quad \begin{matrix}\boldsymbol{q}'\\ \sigma'\end{matrix} = \begin{matrix}-\boldsymbol{q}''\\ \sigma''\end{matrix} \quad\text{if}\quad \begin{matrix}\boldsymbol{q}\\ \sigma\end{matrix} \neq \begin{matrix}\boldsymbol{q}'\\ \sigma'\end{matrix}$$

* Otherwise it has to be replaced by a $\delta$-function.

or all indices are equal

$$\frac{\boldsymbol{q}}{\sigma}=\frac{\boldsymbol{q}'}{\sigma'}=\frac{-\boldsymbol{q}''}{\sigma''}=\frac{-\boldsymbol{q}'''}{\sigma'''}$$

therefore

$$\Phi_4=\frac{\hbar^2}{16sN}\left\{\sum_{\binom{q}{\sigma}\neq\binom{q'}{\sigma'}}{}' \frac{\Phi^{\boldsymbol{q}\,-\boldsymbol{q}\,\boldsymbol{q}'\,-\boldsymbol{q}'}_{\sigma\;\sigma\;\sigma'\;\sigma'}}{\omega\omega'}\left\langle 2b^{\boldsymbol{q}+}_{\sigma}b^{\boldsymbol{q}'+}_{\sigma'}b^{\boldsymbol{q}}_{\sigma}b^{\boldsymbol{q}'}_{\sigma'}+2b^{\boldsymbol{q}+}_{\sigma}b^{\boldsymbol{q}}_{\sigma}+1/2\right\rangle_{\mathrm{Av}}+\right.$$

$$\left.+\sum_{\boldsymbol{q}\sigma}\frac{\Phi^{\boldsymbol{q}\,-\boldsymbol{q}\,\boldsymbol{q}\,-\boldsymbol{q}}_{\sigma\;\sigma\;\sigma\;\sigma}}{\omega^2}\left\langle 2b^{\boldsymbol{q}+}_{\sigma}b^{\boldsymbol{q}+}_{\sigma}b^{\boldsymbol{q}}_{\sigma}b^{\boldsymbol{q}}_{\sigma}+2b^{\boldsymbol{q}+}_{\sigma}b^{\boldsymbol{q}}_{\sigma}+1/2\right\rangle_{\mathrm{Av}}\right\},$$

where the average means the thermal average over the phonon states, i. e. the sum over $m$ in (16.7a) divided by $Z(0)$. The frequencies, which occur in the phonon-distribution, are the quasiharmonic ones, i. e. they contain the mean equilibrium positions. But as the $\Phi_4$-term contains the anharmonicity already as a factor, the frequencies can be replaced by the true harmonic ones. With

$$\left\langle 2b^{\boldsymbol{q}+}_{\sigma}b^{\boldsymbol{q}'+}_{\sigma'}b^{\boldsymbol{q}}_{\sigma}b^{\boldsymbol{q}'}_{\sigma'}+b^{\boldsymbol{q}+}_{\sigma}b^{\boldsymbol{q}}_{\sigma}+b^{\boldsymbol{q}'+}_{\sigma'}b^{\boldsymbol{q}'}_{\sigma'}+1/2\right\rangle_{\mathrm{Av}}$$

$$=\left\langle 2n^{\boldsymbol{q}}_{\sigma}n^{\boldsymbol{q}'}_{\sigma'}+n^{\boldsymbol{q}}_{\sigma}+n^{\boldsymbol{q}'}_{\sigma'}+1/2\right\rangle_{\mathrm{Av}}=2\left(\bar{n}(\boldsymbol{q}\,\sigma)+1/2\right)\left(\bar{n}(\boldsymbol{q}'\,\sigma')+1/2\right)$$

and

$$\left\langle b^{\boldsymbol{q}+}_{\sigma}b^{\boldsymbol{q}+}_{\sigma}b^{\boldsymbol{q}}_{\sigma}b^{\boldsymbol{q}}_{\sigma}+2b^{\boldsymbol{q}+}_{\sigma}b^{\boldsymbol{q}}_{\sigma}+1/2\right\rangle_{\mathrm{Av}}$$

$$=\left\langle n^{\boldsymbol{q}}_{\sigma}\left(n^{\boldsymbol{q}}_{\sigma}-1\right)+2n^{\boldsymbol{q}}_{\sigma}+1/2\right\rangle_{\mathrm{Av}}=2\left(\bar{n}(\boldsymbol{q}\,\sigma)+1/2\right)^2$$

we finally obtain

$$F_4=\langle\Phi_4\rangle_{\mathrm{Av}}=\frac{1}{8sN}\sum_{\substack{\boldsymbol{q}\boldsymbol{q}'\\ \sigma\sigma'}}\frac{\Phi^{\boldsymbol{q}\,-\boldsymbol{q}\,\boldsymbol{q}'\,-\boldsymbol{q}'}_{\sigma\;\sigma\;\sigma'\;\sigma'}}{\omega^2(\boldsymbol{q}\sigma)\,\omega^2(\boldsymbol{q}'\sigma')}\varepsilon(\omega,T)\,\varepsilon(\omega',T) \tag{16.19}$$

$$\varepsilon(\omega,T)=\hbar\,\omega(\boldsymbol{q}\,\sigma)\left[\bar{n}(\boldsymbol{q}\,\sigma)+1/2\right]=\hbar\,\omega(\boldsymbol{q}\,\sigma)\left[\frac{1}{e^{\beta\hbar\omega}-1}+1/2\right]. \tag{16.20}$$

The calculation of $F_3$ is somewhat more involved [58 L 1, 61 L 1, 61 M 1]. With

$$\binom{\boldsymbol{q}}{\sigma}=\alpha;\quad\binom{\boldsymbol{q}'}{\sigma'}=\beta;\quad\binom{\boldsymbol{q}''}{\sigma''}=\gamma;$$

$$A_{\alpha\beta\gamma}=b_\alpha\,b_\beta\,b_\gamma-3b^{+}_{-\alpha}\,b_\beta\,b_\gamma;\quad G_{\alpha\beta\gamma}=\frac{1}{3!}\sqrt{\frac{\hbar^3}{8sN}}\,\frac{\Phi_{\alpha\beta\gamma}\Delta^{(3)}_{\boldsymbol{q}}}{\sqrt{\omega_\alpha\,\omega_\beta\,\omega_\gamma}}$$

we obtain

$$\Phi_3=\sum_{\alpha\beta\gamma}{}' G_{\alpha\beta\gamma}\left[A_{\alpha\beta\gamma}-A^{+}_{-\alpha-\beta-\gamma}+3\delta_{-\beta\gamma}(b_\alpha-b^{+}_{-\alpha})\right]. \tag{16.21}$$

Because of (16.17), the term $\sum_{\alpha\beta}G_{\alpha\beta-\beta}(b_\alpha-b^{+}_{-\alpha})$ in (16.21) contains only phonon-operators for the relative motion of the sublattices. This can be put together with $\Phi_1$ to give

$$\overline{\Phi_1}=\sqrt{\frac{sN\hbar}{2}}\cdot\sum_{\sigma}{}'\frac{1}{\sqrt{\omega(0,\sigma)}}\left[\Phi^{0}_{\sigma}+\frac{\hbar}{4sN}\sum_{\boldsymbol{q}'\sigma'}\frac{\Phi^{0\,\boldsymbol{q}'\,-\boldsymbol{q}'}_{\sigma\;\sigma'\;\sigma'}}{\omega(\boldsymbol{q}'\sigma')}\right]\left(b^{0}_{\sigma}-b^{0+}_{\sigma}\right) \tag{16.22}$$

and is different from zero only in non-primitive, non-parameterfree lattices. We consider the first terms of $\Phi_3$. $\Phi_1$ can be handled in the same manner. With (16.16) we have

$$\sum_{\alpha\beta\gamma} G_{\alpha\beta\gamma} A^{+}_{-\alpha-\beta-\gamma} = \sum_{\alpha\beta\gamma} G_{-\alpha-\beta-\gamma} A^{+}_{\alpha\beta\gamma} = -\sum_{\alpha\beta\gamma} G^{*}_{\alpha\beta\gamma} A^{+}_{\alpha\beta\gamma}$$

and

$$\Phi_3 = \sum_{\alpha\beta\gamma}{}' (G_{\alpha\beta\gamma} A_{\alpha\beta\gamma} + G^{*}_{\alpha\beta\gamma} A^{+}_{\alpha\beta\gamma}) . \tag{16.21a}$$

We have to calculate the matrix-elements $\langle m_\alpha, m_\beta, \ldots, |\Phi_3| m_\alpha, m_\beta, \ldots\rangle$ with the eigenstates of the zero-Hamiltonian for a system of independent oscillators. This gives*

$$\begin{aligned}
&\sum_{\alpha\beta\gamma}{}' G_{\alpha\beta\gamma}\langle m|A_{\alpha\beta\gamma}|n\rangle \\
&= \sum_{\alpha\neq\beta\neq\gamma\neq\alpha}{}' G_{\alpha\beta\gamma}\Big\{\sqrt{n_\alpha n_\beta n_\gamma}\,\langle m_\alpha|n_\alpha-1\rangle\langle m_\beta|n_\beta-1\rangle\langle m_\gamma|n_\gamma-1\rangle - \\
&\quad -3\sqrt{(n_{-\alpha}+1)\,n_\beta n_\gamma}\,\langle m_{-\alpha}|n_{-\alpha}+1\rangle\langle m_\beta|n_\beta-1\rangle\langle m_\gamma|n_\gamma-1\rangle\Big\}\times \\
&\quad\times \prod_{\delta\neq\alpha,\beta,\gamma}\langle m_\delta|n_\varepsilon\rangle + \\
&\quad+\sum_{\alpha\neq\gamma}{}' G_{\alpha\alpha\gamma}\Big\{3\sqrt{n_\alpha(n_\alpha-1)\,n_\gamma}\,\langle m_\alpha|n_\alpha-2\rangle\langle m_\gamma|n_\gamma-1\rangle - \\
&\quad-3\sqrt{(n_{-\alpha}+1)\,n_\gamma(n_\gamma-1)}\,\langle m_{-\alpha}|n_{-\alpha}+1\rangle\langle m_\gamma|n_\gamma-2\rangle - \\
&\quad-6\sqrt{(n_{-\alpha}+1)\,n_\alpha n_\gamma}\,\langle m_{-\alpha}|n_{-\alpha}+1\rangle\langle m_\alpha|n_\alpha-1\rangle\langle m_\gamma|n_\gamma-1\rangle\Big\}\times \\
&\quad\times\prod_{\delta\neq\alpha,\gamma}\langle m_\delta|n_\delta\rangle + \\
&\quad+\sum_\alpha G_{\alpha\alpha\alpha}\Big\{\sqrt{n_\alpha(n_\alpha-1)(n_\alpha-2)}\,\langle m_\alpha|n_\alpha-3\rangle - \\
&-3\sqrt{(n_{-\alpha}+1)(n_\alpha-1)}\,\langle m_{-\alpha}|n_{-\alpha}+1\rangle\langle m_\alpha|n_\alpha-2\rangle\Big\}\cdot\prod_{\delta\neq\alpha}\langle m_\delta|n_\delta\rangle
\end{aligned} \tag{16.23}$$

and since

$$\langle m_\alpha\ldots|A|n_\alpha\ldots\rangle = \langle n_\alpha\ldots|A^{+}|m_\alpha\ldots\rangle$$

we have

$$\begin{aligned}
|\langle m_\alpha\ldots|\Phi_3|n_\alpha\ldots\rangle|^2 &= \sum_{\alpha\beta\gamma;\alpha'\beta'\gamma'}\{G_{\alpha\beta\gamma}G^{*}_{\alpha'\beta'\gamma'}\times \\
&\times[\langle m|A_{\alpha\beta\gamma}|n\rangle\langle m|A_{\alpha'\beta'\gamma'}|n\rangle + \langle n|A_{\alpha\beta\gamma}|m\rangle\langle n|A_{\alpha'\beta'\gamma'}|m\rangle] + \\
&+[G_{\alpha\beta\gamma}G_{\alpha'\beta'\gamma'} + G^{*}_{\alpha\beta\gamma}G^{*}_{\alpha'\beta'\gamma'}]\langle m|A_{\alpha\beta\gamma}|n\rangle\langle n|A_{\alpha'\beta'\gamma'}|m\rangle\} .
\end{aligned} \tag{16.24}$$

Because of the orthogonality expressed by $\langle m_\alpha|n_\alpha-1\rangle$ etc. in (16.23) all mixed terms $(\alpha\,\beta\,\gamma)\neq(\alpha'\,\beta'\,\gamma')$ vanish, and by the same argument the whole last term. There are contributions only, if the set $(\alpha'\,\beta'\,\gamma')$ is equal to the set $(\alpha\,\beta\,\gamma)$. This gives a factor 3! if $\alpha\neq\beta\neq\gamma\neq\alpha$, a factor 2! if $\alpha=\alpha'$, $\beta\neq\gamma$ etc. There remain only relatively few terms, which can be

* Terms with $\gamma=-\beta$ can be handled in the same way and need no special investigation. They give contributions only if $\alpha=\binom{q}{\sigma}=\binom{0}{\sigma}$, which vanish in primitive lattices.

summed up after multiplication with the energy-factor in (16.7c), since the difference $n_\alpha - m_\alpha$ can only be 0, $\pm 1$, $\pm 2$, $\pm 3$ in view of the orthogonality factors in (16.23). Using the relations

$$\bar{n}_\alpha = \bar{n}_{-\alpha} = \frac{1}{e^{\beta\hbar\omega(\alpha)} - 1}\,; \quad \bar{n}_\alpha + 1 = \bar{n}_\alpha\, e^{\beta\hbar\omega(\alpha)}$$
$$\overline{n_\alpha^2} = 2\bar{n}_\alpha^2 + \bar{n}_\alpha\,; \quad \overline{n_\alpha^3} = 6\bar{n}_\alpha^3 + 6\bar{n}_\alpha^2 + \bar{n}_\alpha \tag{16.25}$$

we finally arrive at

$$F_3 = -\frac{1}{48 s N} \sum_{\substack{\mathbf{q}\mathbf{q}'\mathbf{q}'' \\ \sigma\sigma'\sigma''}} \frac{\left|\Phi^{\mathbf{q}\,\mathbf{q}'\,\mathbf{q}''}_{\sigma\,\sigma'\,\sigma''}\right|^2}{(\omega\,\omega'\,\omega'')^2}\, \Delta^{(3)}_{\mathbf{g}} \times$$
$$\times \left\{ \frac{\varepsilon_0\varepsilon_0'\varepsilon_0'' + 3\varepsilon_0\varepsilon'\varepsilon''}{\varepsilon_0 + \varepsilon_0' + \varepsilon_0''} + 3\,\frac{2\varepsilon_0\varepsilon'\varepsilon'' - \varepsilon\varepsilon'\varepsilon_0'' - \varepsilon_0\varepsilon_0'\varepsilon_0''}{\varepsilon_0 + \varepsilon_0' - \varepsilon_0''} \right\} \tag{16.26}$$

with

$$\varepsilon(\omega, T) = \hbar\,\omega\,(\bar{n}(\mathbf{q}\,\sigma) + 1/2)\,; \quad \varepsilon_0 = \hbar\,\omega\,(\mathbf{q}\,\sigma)/2\,;$$
$$\varepsilon' = \varepsilon(\omega', T)\,; \quad \omega' = \omega(\mathbf{q}'\,\sigma')\,.$$

A similar, but even simpler calculation leads to

$$F_1 = -\frac{1}{4}\, s\, N \cdot \sum_\sigma{}' \left| \frac{\bar{\Phi}^0_\sigma}{\omega(0, \sigma)} \right|^2 \tag{16.27}$$

which is *independent of temperature* and therefore does not give contributions to entropy, specific heats etc. The mixed terms, containing $\langle m\,|\bar{\Phi}_1|\, n\rangle\,\langle n\,|\bar{\Phi}_3|\, m\rangle$ and similar ones, vanish.

The quantity of most interest is the specific heat

$$C_V = -T\,\frac{\partial^2 F}{\partial T^2} \tag{16.28}$$

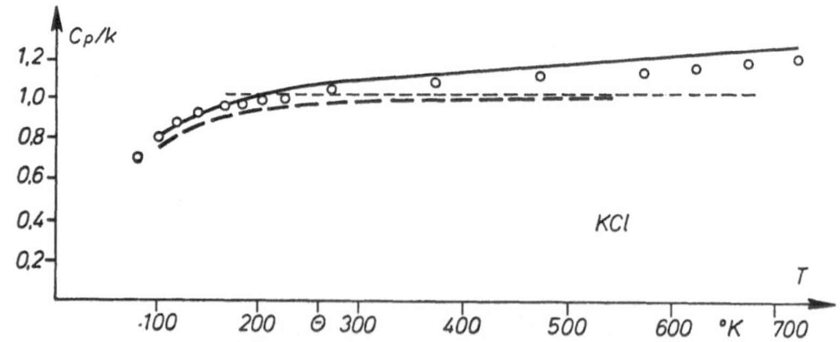


Fig. 5.8a. Specific heat at constant pressure as a function of temperature (high temperatures). Shown is the specific heat per ion for KCl. ∘∘∘ experimental values, ——— anharmonic theory with electrostatic and Born-Mayer-forces [61 L 1], — — — harmonic theory, - - - - - classical Dulong-Petit-value

for which we need to consider, apart from the quasiharmonic contribution, the terms $F_3$ and $F_4$. The discussion over the whole range of temperatures is rather involved. Only the limiting cases of high ($T > \theta$) and low ($T \to 0$) temperatures will be mentioned. At *high temperatures* we can expand with respect to $\hbar\omega/kT$ and reach at (with the harmonic contribution included)

$$C_V = 3\,s\,N\,k + \frac{k^2\,T}{6\,s\,N}\left[\sum_{\substack{\boldsymbol{q}\,\boldsymbol{q}'\,\boldsymbol{q}''\\ \sigma\,\sigma'\,\sigma''}} \left|\frac{\Phi^{\boldsymbol{q}\,\boldsymbol{q}'\,\boldsymbol{q}''}_{\sigma\,\sigma'\,\sigma''}}{\omega\,\omega'\,\omega''}\right|^2 \Delta^{(3)}_{\boldsymbol{g}} - \frac{3}{2}\sum_{\substack{\boldsymbol{q}\,\boldsymbol{q}'\\ \sigma\,\sigma'}} \frac{\Phi^{\boldsymbol{q}\,-\boldsymbol{q}\,\boldsymbol{q}'\,-\boldsymbol{q}'}_{\sigma\;\;\sigma\;\;\sigma'\;\;\sigma'}}{\omega^2\,\omega'^2}\right] \tag{16.29}$$

showing a linear increase (or decrease) with temperature above the classical Dulong-Petit-value $3\,s\,N\,k$. This linear dependence has been calculated already in simple cases by *Born* and *Brody* [22 B 1] and

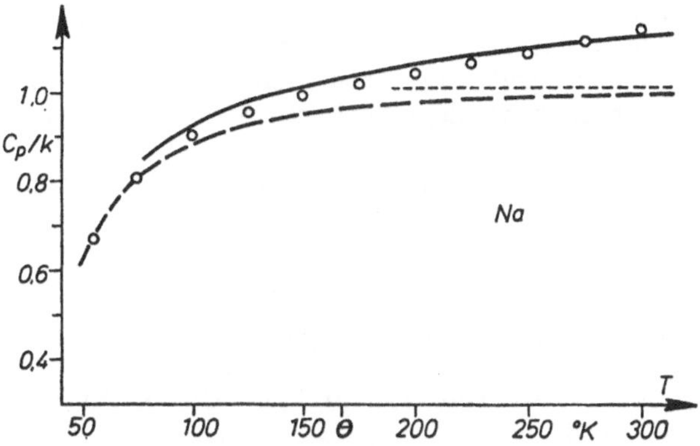


Fig. 5.8b. The same as fig. 5.8a, but for Na, assuming general forces between nearest neighbors [61 L 1]

*Schrödinger* [22 S 1], and can be obtained of course completely by classical calculation [61 L 1]. The evaluation of the sums in (16.29) is often difficult. In [61 L 1] we have given some approximation methods for this calculation and it has been shown later, [61 M 1, 2], that this approximations are good for the $\Phi_4$-term, but give numerical values with 10 or 20% in error for the $\Phi_3$-term. In [61 L 1] as well as in [61 M 1, 2] some models have been discussed in detail and we will not go into the details here.

In experiments one measures in general the specific heat at constant pressure, which is related to the specific heat at constant volume by means of the thermal expansion (15.10) and given by

$$C_p - C_V = 9 s^2 N\,T\,V_z \frac{\alpha^2(T)}{\varkappa^{\mathrm{is}}(T)}\,. \tag{16.30}$$

In our approximation $\alpha(T)$ is proportional to $\Phi_3$, therefore the isothermal compressibility can be replaced by its harmonic value, independent of temperature. $\alpha(T)$ we have seen to be (nearly) proportional to the specific heat in harmonic approximation (sect. 15). Therefore it is independent of temperature for $T > \theta$; the difference $C_p - C_V$ then and also $C_p$ is proportional to $T$ at high temperatures. The experiments show, that $C_p$ is always increasing with $T$ at $T > \theta$. As far as the interaction potential between the ions is known, the theory gives the same result. But $C_V$ is smaller, it may increase for some crystals and decrease for others.

At low temperatures the difference $C_p - C_V$ vanishes as $T^7$ and is uninteresting for most purposes.

At *low temperatures* the harmonic approximation gives the famous $T^3$-law of the specific heat. The question arises, whether this $T$-dependence is valid also for the anharmonic contributions $\Phi_3$ and $\Phi_4$. This has been investigated thoroughly in [61 L 1] and [61 M 1, 2, 3], and in fact all terms in the free energy give a $T^3$-dependence of the specific heat. In harmonic theory, one can define by $T^3$-dependence a Debye-Temperature $\tilde{\theta}$, which is related to the frequency-spectrum of the phonons and also to the elastic constants of the material [I]. Therefore one would try to define an anharmonic Debye-temperature by the coefficient in the anharmonic $T^3$-law of the specific heat and relate this to the anharmonic elastic constants. Before discussing this in detail, we will give a simple example, namely again the *monatomic linear lattice* with nearest neighbour interaction.

For the linear lattice we have the following anharmonic c.p.'s

$$\begin{aligned}
&\Phi^{01} = \Phi^{0\bar{1}} = -f\,;\\
&\Phi^{001} = -\Phi^{00\bar{1}} = -\Phi^{011} = \Phi^{0\bar{1}\bar{1}} = g\,; \qquad (16.31)\\
&\Phi^{0001} = \Phi^{000\bar{1}} = -\Phi^{0011} = -\Phi^{00\bar{1}\bar{1}} = -h\,.
\end{aligned}$$

Introducing this into the transformed c.p.'s, (16.14), we have (polarization vectors are unity now!)

$$\begin{aligned}
\Phi^{q\,q'\,q''} &= -i\frac{g}{f^{3/2}}\,\omega(q)\,\omega(q')\,\omega(q'')\sum_{\nu}(-1)^{\nu}\Delta\left(q+q'+q''-\frac{2\pi}{a}\nu\right)\\
\Phi^{q\,q'\,q''\,q'''} &= \frac{h}{f^{2}}\,\omega(q)\,\omega(q')\,\omega(q'')\,\omega(q''')\sum_{\nu}(-1)^{\nu}\Delta\left(q+q'+q''+q'''-\frac{2\pi}{a}\nu\right).
\end{aligned} \qquad (16.32)$$

If this is introduced into (16.19) we simply obtain

$$F_4 = \frac{h}{8f^2}\cdot\frac{U^2(T)}{N} \qquad (16.33)$$

[$U(T)$: internal energy in the harmonic approximation],

whereas the expression for $F_3$ becomes

$$\begin{aligned}
F_3 = -\frac{g^2}{48f^3}\cdot\frac{1}{N}\sum_{q\,q'\,q'',\nu}\Delta\left(q+q'+q''-\frac{2\pi}{a}\nu\right)\left\{\frac{\varepsilon_0\varepsilon_0'\varepsilon_0''+3\varepsilon_0\varepsilon'\varepsilon''}{\varepsilon_0+\varepsilon_0'+\varepsilon_0''}+\right.\\
\left.+3\,\frac{2\varepsilon_0\varepsilon'\varepsilon''-\varepsilon\varepsilon'\varepsilon_0''-\varepsilon_0\varepsilon_0'\varepsilon_0''}{\varepsilon_0+\varepsilon_0'-\varepsilon_0''}\right\}.
\end{aligned} \qquad (16.34)$$

Again the expression becomes simple only at high temperatures. Including the harmonic contribution we have for $T > \theta$ for the specific heat from (16.33, 34) or directly from (16.29)*

$$C_V = N k\left\{1+\frac{kT}{6f^3}\left(g^2-\frac{3}{2}hf\right)\right\}, \qquad (16.35)$$

* Note that $3s$ in the harmonic term and in (16.30) has to be replaced by one in a monatomic linear lattice; at the other places, $s = 1$.

[cf. 61 L 1, 61 M 1]. With (16.30) and (15.15) we have

$$C_p - C_V = N k \cdot \frac{g^2}{4 f^3} k T \tag{16.35a}$$

or

$$C_p = N k \left\{ 1 + \frac{k T}{12 f^3} (5 g^2 - 3 h f) \right\} . \tag{16.35b}$$

For low temperatures (16.34) has been evaluated by *Maradudin* et al. [61 M 1]. We will give only the result here:

$$F_3 = - \frac{0.4048}{6 \pi^3} \cdot \frac{g^2}{f^3} N \hbar^2 \omega_m^2 - \frac{1}{24} \cdot \frac{g^2}{f^3} N (k T)^2 . \tag{16.36b}$$

The $F_4$-contribution can be expressed by the (quasi-)harmonic value of the internal energy. For low temperatures we have

$$U(T) = \sum_q \hbar \omega(q \sigma) [\bar{n}(q \sigma) + 1/2] = \frac{N V_z}{2 \pi} \int_{-\pi/a}^{+\pi/a} dq \, \hbar \omega(q \sigma) \times$$
$$\times \left[ \frac{1}{2} + \sum_{\nu = 1}^{\infty} e^{-\nu \frac{\hbar \omega}{k T}} \right] \approx \frac{1}{\pi} N \hbar \omega_m + \frac{\pi}{3} N \frac{(k T)^2}{\hbar \omega_m} \tag{16.36a}$$

and therefore

$$F_4 = \frac{1}{8 \pi^2} \frac{h}{f^2} N \hbar^2 \omega_m^2 + \frac{1}{12} \frac{h}{f^2} N (k T)^2 . \tag{16.36c}$$

The Debye-temperature in the linear case is defined by the constant extrapolation of the long-wave-spectrum such that the number of the degrees of freedom is equal to $N$. This gives the Debye frequency $\omega_D$ and the Debye-temperature $\theta$:

$$\omega_D = \frac{\pi}{2} \omega_m ; \quad k \theta = \hbar \omega_D = \hbar \pi \sqrt{f / M} . \tag{16.37}$$

In our calculation we have started with the quasiharmonic approximation, i. e. all c.p.'s refer to the mean equilibrium position. Therefore $f(a)$ contains a term which depends on the thermal expansion (see p. 98) and it is

$$f(a) = \tilde{f} - \frac{\tilde{g}^2}{2 \tilde{f}^2} \cdot \frac{U(T)}{N} \Rightarrow \tilde{f} - \frac{\tilde{g}}{2 \pi \tilde{f}^2} \hbar \tilde{\omega}_m \quad \text{if} \quad T \to 0 . \tag{16.38}$$

Using (16.37) we derive from (16.36) the specific heat at low temperatures as

$$C_V = N k \left\{ \frac{\pi^2}{3} \cdot \frac{T}{\theta} + \frac{1}{12} \frac{k T}{f^3} (g^2 - 2 h f) \right\} . \tag{16.39}$$

By means of (16.38) $\theta$ is related to the Debye-temperature $\tilde{\theta}$ of the true harmonic approximation

$$\theta = \tilde{\theta} \left\{ 1 - \frac{1}{2 \pi^2} \frac{\tilde{g}^2}{\tilde{f}^3} k \tilde{\theta} \right\} . \tag{16.40}$$

The harmonic specific heat

$$C_V = N k \cdot \frac{\pi^2}{3} \cdot \frac{T}{\tilde{\theta}} \tag{16.41}$$

represents the Debye-formula for the low temperature limit in the linear case ($T$ instead of $T^3$). Now the anharmonic corrections can be described by introducing an effective Debye-temperature through

$$C_V = N k \cdot \frac{\pi^3}{3} \cdot \frac{T}{\theta_{\mathrm{eff}}} . \tag{16.41b}$$

This effective Debye-temperature can be gotten simply from (16.39) and (16.40):

$$\theta_{\mathrm{eff}} = \tilde{\theta} \left\{1 - \frac{1}{4\pi^2} \cdot \frac{k\tilde{\theta}}{\tilde{f}^3} (3\tilde{g}^2 - 2\tilde{h}\tilde{f})\right\} . \tag{16.42}$$

Also in threedimensional lattices the specific heat at low temperatures can be expressed by a Debye-law with an effective Debye-temperature, which in principle can be calculated from the c.p.'s. But the calculation in detail is rather involved for reasonable models. Instead of discussing different models, we will give some general remarks to the problem of *Debye-temperatures as influenced by anharmonicity.*

The general statement of the harmonic theory says, that the Debye-temperatures can be expressed by the velocity of sound or by the elastic constants [I, II]. This is, because at low temperatures only the small frequencies of lattice waves are excited, and these are proportional to the wave number $\boldsymbol{q}$. The constant of proportionality is the velocity of sound $c$. Therefore Debye [12 D 1] concluded, that one can use a spectrum which is derived under the assumption of $\omega \sim q$, which is just the result of the classical continuum theory of the elastic solid. And the velocity of sound is known to be related to the elastic constants. The question arises, whether these relations are true if anharmonic effects are taken into account, i. e. is it possible to express the Debye-temperature (in the limit $T \to 0$) by the elastic constants (in the limit $T \to 0$) in the presence of anharmonic terms in the potential energy (1.6)? The answer is positive and we will illustrate this first with the example of the linear lattice.

In the long-wave limit it holds

$$\omega = c \cdot q . \tag{16.43}$$

The sound velocity $c$ is related to the elastic constant [see 61 L 5, I, II]

$$c^2 = \frac{c^{\mathrm{ad}}(T)}{\varrho} = \frac{\tilde{f}\tilde{a}^2}{M} \left\{1 - \frac{U(T)}{4\tilde{f}^3 N} (3\tilde{g}^2 - 2\tilde{h}\tilde{f} + 4\tilde{g}\tilde{f}/\tilde{a})\right\} \tag{16.44}$$

if use is made of (15.21, 25, 15). $\varrho(T) = M/[b^2 \cdot a(T)]$ is the mass density in our example. The "Debye-temperature" $\theta_0$ is given by (16.37):

$$\begin{aligned} k\theta_0 &= \hbar\omega_{\mathrm{D}} = \hbar c q_{\mathrm{D}} = \hbar\pi \frac{c(T \to 0)}{a(T \to 0)} \\ &= \hbar\pi \sqrt{\tilde{f}/M} \left\{1 - \frac{U(T \to 0)}{2N\tilde{f}^3} (3\tilde{g}^2 - 2\tilde{h}\tilde{f})\right\} \end{aligned} \tag{16.45}$$

and with (16.37) and (16.36a)

$$k\tilde{\theta} = \hbar\pi\sqrt{\tilde{f}/M}; \quad U(T \to 0) = N k\tilde{\theta}/2\pi^2$$

and therefore

$$\theta_0 = \theta_{\mathrm{eff}} . \tag{16.46}$$

In fact, the Debye-temperature which determines the specific heat at low temperatures, $\theta_{\text{eff}}$ is equal to that calculated with the usual procedure from the true elastic constants (adiabatic or isothermal for $T \to 0$) at $T = 0$. These are, of course, not the elastic constants of the harmonic theory because of the interaction between anharmonicity and zero-point vibrations. This holds at least for the linear lattice*.

Later *Barron* and *Klein* [62 B 1] have given arguments which show that in fact in all lattices the Debye-temperature from the specific heat is equal to that from the true elastic constants at $T = 0$. These arguments are believed to be correct and we will repeat them here.

The basic concept is, that the lowest excited states at $T \to 0$ are the long elastic waves, so that the Helmholtz free energy in this limit must depend on the wave velocities just as in a harmonic crystal, but with effective velocities. Then anharmonic effects can enter in the elastic constants only through the interaction with the zero-point-energy, and hence they enter the frequencies of long waves by the same interaction.

The frequency dependence of the Helmholtz free energy for one oscillator is given by (see sect. 15)

$$\frac{\partial}{\partial \omega} f(\omega, T) = \frac{\varepsilon(\omega, T)}{\omega} \tag{16.47}$$

and therefore in the limit $T \to 0$ it must be

$$F(T) = \sum_{q\sigma} \frac{\varepsilon(\omega, T)}{\omega} \cdot \delta\omega \tag{16.48}$$

or

$$F(T) - F(0) = \sum_{\sigma q} \frac{\bar{\varepsilon}(\omega, T)}{\omega} \cdot \delta\omega \tag{16.48a}$$

where the thermal energy of an oscillator has been splitted in the zero-point term $\varepsilon_0(\omega)$ and the temperature dependent part $\bar{\varepsilon}(\omega, T)$. In the case of $F_4(T)$ we have from (16.19)

$$F_4(T) = \frac{1}{8sN} \sum_{qq'\sigma\sigma'} \frac{\Phi^{q\;-q\;q'\;-q'}_{\sigma\;\sigma\;\sigma'\;\sigma'}}{\omega^2\omega'^2} [\varepsilon_0(\omega) + \bar{\varepsilon}(\omega, T)]\,[\varepsilon_0(\omega') + \bar{\varepsilon}(\omega', T)]\,.$$

In the limit $T \to 0$ we have as the only temperature dependent part (lowest order terms in $T$)

$$F_4(T) - F_4(0) = \frac{1}{8sN} \sum_{\substack{qq'\\ \sigma\sigma'}} \frac{\Phi^{q\;-q\;q'\;-q'}_{\sigma\;\sigma\;\sigma'\;\sigma'}}{\omega^2\omega'^2} [\varepsilon_0(\omega)\,\bar{\varepsilon}(\omega', T) + \varepsilon_0(\omega')\,\bar{\varepsilon}(\omega, T)]$$

from which we can derive by comparison with (16.48a)

$$\delta\omega_4 = \frac{\hbar}{8sN} \sum_{q'\sigma'} \frac{\Phi^{q\;-q\;q'\;-q'}_{\sigma\;\sigma\;\sigma'\;\sigma'}}{\omega\omega'}\,. \tag{16.49a}$$

* It should be emphasized that this is contrary to what was stated in [61 L 1]. The wrong statements there arose from a trivial, but essential factor "two" which is missing in the first term of equ. (9.11). Thus all conclusions drawn from this formula are misleading and not correct.

Similarly we have from (16.26)

$$F_3(T) - F_3(0) = -\frac{1}{8sN} \sum_{q q' q'', \sigma\sigma'\sigma''} \left| \frac{\Phi^{q\,q'\,q''}_{\sigma\,\sigma'\,\sigma''}}{\omega\,\omega'\,\omega''} \right|^2 \Delta^{(3)}_g \times$$

$$\times \frac{\bar{\varepsilon}(\omega)\,\varepsilon_0(\omega')\,\varepsilon_0(\omega'') \cdot 2\,[\varepsilon_0(\omega') + \varepsilon_0(\omega'')]}{[\varepsilon_0(\omega) + \varepsilon_0(\omega'')]^2 - \varepsilon_0^2(\omega')} \,.$$

From a comparison with the limiting value for long waves ($\omega \to 0$) we derive

$$\delta\,\omega_3 = -\frac{\hbar}{8sN} \sum_{\substack{q' q'' \\ \sigma' \sigma''}} \frac{\left|\Phi^{q\,q'\,q''}_{\sigma\,\sigma'\,\sigma''}\right|^2}{\omega\,\omega'\,\omega''} \cdot \frac{1}{\omega' + \omega''} \,. \tag{16.49b}$$

This change of frequencies, as due to the change of the free energy by anharmonicity, has to be compared with the change of frequencies of long waves as calculated from the ground state energy. This can be done by simple perturbation theory, starting with the Hamiltonian as given by (16.12, 13), and gives just (16.49).

Therefore Debye's temperature taken from measurements of the specific heat for $T \to 0$ should be equal to that, which is calculated with the extrapolated harmonic elastic constants (see sect. 15, Fig. 5.7). If there is a disagreement in both values which should agree, it must be due to errors in measurement or approximations in calculation. However, in most cases which allow for a comparison, the measurements have been refined so much, that the elastic and specific heat values for $\theta$ do agree. There seems to be only one striking example of disagreement: the case of Be, as already mentioned in [61 L 1]. It might be, that this is due to an incorrect knowledge of the electronic contribution to the specific heat or also to errors in taking the limiting values of the elastic constants. New measurements would be desirable.

## 17. Further Discussions of Thermodynamic Properties

Now the question arises whether the approximations indicated in the preceding sections are sufficient or not, or in other words, do they describe the experimental behavior or not. The comparison can be done most simply with the temperature dependence of the lattice constant (thermal expansion) of the elastic constants and of the specific heat. All these quantities should vary linearly with temperature (above the Debye-temperature), provided the approximations are sufficient. We have given some examples, where the linear temperature dependence ($T > \theta$) is obvious (Fig. 5.4; 5.6; 5.8). But there can be found examples too, where there is a parabolic or even more complicated behavior (Fig. 5.9). The most sensitive quantities in this respect seem to be thermal expansion and specific heat. Deviations from the linear behavior occur in crystals with strong anharmonicities, in general those crystals where Debye-temperature and melting temperature differ by less than a factor of two.

The deviations become larger with increasing temperature, when temperature approaches the melting point. Apart from a few cases, in which $\theta \approx T_s$ (melting temperature), the higher order anharmonicities are essential only for $T > \theta$, that is in the classical limit. So it should be

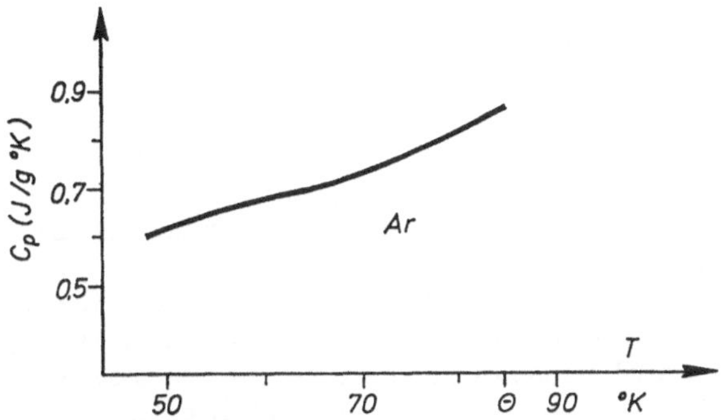


Fig. 5.9. Specific heat at constant pressure for argon, showing a nonlinear behavior at high temperatures. This might be due to higher order anharmonicity as well as to the formation of vacancies in the equilibrium state

sufficient to use classical statistical mechanics and thermodynamic functions in higher order calculations. We will not go into too much details here and rather discuss the general aspects as well as some recent papers on these problems.

We have to distinguish between two types of effects. Those which arise from higher order anharmonicities ($\Phi_5$, $\Phi_6$, ..., cf. Eq. 1.6) and those which arise from higher order approximations, compared with the approximations in sect. 13–16. These involve terms with $\Phi_3^4$, $\Phi_4^2$, ... as well as higher order terms in $\Phi_5$, $\Phi_6$, ... . In the next order of approximation we would have to take into account $\Phi_5^2$ and $\Phi_6$ as well as $\Phi_3^4$ and $\Phi_4^2$. It can be seen easily that these contributions lead to the same temperature dependence (in the classical limit), and that also the order of magnitude of the contributions of $\Phi_5^2$, $\Phi_6$, $\Phi_3^4$, $\Phi_4^2$ is the same one. The effect of these terms is a quadratic temperature dependence ($T > \theta$) for the lattice constant, the elastic constants, the specific heats and also for the piezoelectric constants and the dielectric constants.

In this connection we have to mention another effect, which leads to deviations from a linear temperature dependence. This is the contribution of lattice defects, especially of vacancies in the thermal equilibrium, to thermal expansion, elastic constants and specific heats. The equilibrium concentration of the vacancies depends exponentially on temperature, thus in a rough estimate the temperature dependence of all the quantities is an exponential one. It becomes comparable with anharmonic effects near the melting point. Thus in a quantitative discussion of the temperature dependence of elastic and caloric data of crystals higher order anharmonicities ($\Phi_5^2$, $\Phi_6$, $\Phi_3^4$, $\Phi_4^2$, ...) as well as effects of defects (vacancies) have to be taken into account.

We will discuss the anharmonic effects only. If we limit the procedure to the next order-effects, we just have to take into account the next expansion terms in (16.4–7). However, for the lattice constant and the elastic constants (derivatives of free energy with respect to strains, or coordinates) we need only one order less in the expansion of the logarithm of the partition function. We obtain the elastic constants from the second strain derivatives of $\Phi_0$, $F_2$, $F_3$ and $F_4$. The extension of the procedure is obvious. A classical calculation is sufficient.

For the specific heat, we need the next terms in the expansion (16.4–7). They give the contributions of $\Phi_3^4$; from the first two expansion terms in (16.4) we obtain, if properly inserted, the contributions of $\Phi_4^2$ as well as those of $\Phi_5^2$ and $\Phi_6$. Explicit calculations are lengthy, but straight-forward.

But instead of doing this procedure, it might be better, to use the general perturbation expansion (16.6) up to arbitrary high orders, to discuss the contributions of all the different processes $\Phi_3$, $\Phi_4$, $\Phi_5$, . . ., for example with a diagram technique, and, if possible, to sum up these terms in an appropriate way. Such a procedure is extremely valuable, if one is interested in general relations. But it has to be emphasized again, that if all orders in the perturbation expansion are considered, also all the orders in the anharmonicities have to be taken into account, or in other words, if the perturbation expansion is taken up to the $2\nu^{\text{th}}$ order, one has to take anharmonicities up to $\Phi_{2(\nu+1)}$. This holds, if one wants to compare experimental results with theoretical model calculations quantitatively. If we are interested only in general qualitative statements, say e. g. a statement on the temperature dependence, some of the terms may be sufficient. The situation is also somewhat different for dynamical quantities such as frequency-shifts, life-times etc. due to anharmonicity (see sect. 27).

A first step in working with diagram techniques has been done by *Cowley* [63 C 2], in connection with thermodynamic quantities. However, he only gave explicit expressions up to the same order as discussed in sections 13–16. Essentially the results agree with those given above. On the other hand, *Cowley* introduced "quasiharmonic" frequencies*, which describe the temperature dependent frequencies measured in thermal-neutron-spectroscopy (see sect. 27). The temperature dependent frequencies are then introduced in the harmonic free energy, instead of the harmonic frequencies. *Cowley* discusses the fact, that this procedure does not give the correct anharmonic free energy (16.19, 26). But this seems to be trivial. For using temperature dependent energy-eigenvalues in a harmonic theory (and this is what *Cowley* does) is not equivalent to calculating a partition function or a free energy by proper anharmonic

* The word "quasiharmonic" in *Cowleys* paper is very misleading, because the temperature dependent frequencies in his paper never can be obtained by a quasiharmonic approximation. Genuine anharmonic methods (cf. sect. 27) are necessary to obtain these frequencies. Our quasiharmonic approximation in sect. 15 starts with a true harmonic free energy, the frequencies of which depend on average positions, determined from the minimalization of the free energy.

methods. Only in the zero-temperature limit (mechanical limit) this might be justified. In any case one has to be cautious when dealing with temperature dependent frequencies in thermodynamics. In neutron spectroscopy, the frequencies (27.72) of course have a definite meaning.

The techniques of using Green-functions and diagrammatic summations is very similar to that which will be discussed in sect. 27. Details will not be given here. *Sham* [65 S 5] has given two very general formulas for the free energy in terms of the complete Green-function for the anharmonic crystal. But the difficulties in quantitative statements are not shifted by this procedure, because one would have to calculate the Green-function up to the order wanted and to solve a Dyson-equation similar to that given in (27.67).

Other, more quantitative calculations have been done by *Pathak* [65 P 1] and *Choquard* [67 C 1]. Both papers use the Green-function method to give some general relations and then sum up certain anharmonic diagrams to make quantitative statements. If only lowest order diagrams are considered, the results agree with those given in sect. 15, 16. Higher order diagrams describe the effects as mentioned in the beginning of this section.

Another interesting method has been given by *D. C. Wallace* [66 W 2]. In that paper the Green-functions are avoided; the method consists of a procedure which allows for a successive diagonalization of the anharmonic Hamiltonian. In every order, the diagonalized Hamiltonian defines renormalized frequencies, which can be used in statistical mechanics. The results agree again with those given in sect. 16 and in sect. 27 for frequency shifts and life-times.

*Högberg* [65 H 3] uses the method of functional derivatives to calculate the properties of anharmonic crystals. The results, of course, agree with those of other methods. There is one example which needs a very thorough investigation: solid Helium. This is for two reasons. The crystal exists only at very low temperatures, therefore quantum effects are extremely significiant; the anharmonicity is extremely large, and the zero point motion is also of more importance than in other crystals. The anharmonicity cannot be considered as a small perturbation. Many attempts for solving the problem have been made. We only mention two papers of *Horner* [67 H 1] and *Biem* [67 B 2], from which further references might be taken.

Another extension of the theory concerns the inclusion of external electrostatic fields. A simple discussion has been given in sect. 13 and 14. The constants related to the presence of an external field are the pyroelectric constants, the piezoelectric constants, the static susceptibility, the electrostrictive constants and other higher order constants. The piezoelectric constants and the susceptibility are different from zero already in the harmonic approximation [I, II]. Anharmonic effects give a temperature dependence to these quantities. With lowest order anharmonicity they are proportional to $T$ in the classical limit. This is confirmed by experiments [66 R 1].

Pyroelectric and electrostrictive constants are related to third order anharmonicity. In this approximation they are independent of temperature. Higher order anharmonicities make them vary with temperature. *Silverman* was the first to give lattice theoretical expressions for the electrostrictive constants [63 S 1]. The pyroelectric constant has been discussed in an old paper of *Born* [45 B 1]. In lattices without inversion symmetry, the behavior of the pyroelectric constant is similar to the thermal expansion coefficient. In a certain approximation it is proportional to the specific heat; similar phenomena as for the expansion coefficient may occur, the pyroelectric constant may change sign at low temperatures. These statements are mainly related to the lattice contribution to the electrical moment. The internal polarizability of the ions leads to another contribution to the pyroelectric moment. Recently *Thoma* [67 T 1] has reinvestigated the theory of the higher order constants, with special emphasis to the electrical phenomena. He also takes into account the surface charges in ionic crystals, which make the dipole-moment vanish in an equilibrium state. It turns out, that these surface charges play a dominant role.

We will conclude this chapter with these general remarks. There are lots of calculations which use these relations for special models. But the application of the general formalism to many cases is only a numerical task.

# VI. Point Defects in Crystal Lattices

## 18. Structure of Point Defects

In recent years the physics of lattices with defects has become interesting because the investigation of the behavior of defects turned out to be a powerful tool in understanding lattice properties. Here we will deal only with point defects, which can or must be handled with the methods of lattice theory. The importance of dislocations is also known; but the investigations are mainly done with the continuum theory of elasticity and they will not be discussed here. A large number of excellent books and reviews on dislocations exist [54 H 1, 55 S 2, 58 K 1, 64 F 1, 65 W 2]. Also, there is a large number of papers on point defects [55 S 2, 57 E 1, 63 D 2, 65 J 1, 66 S 2], but the essential features shall be reviewed shortly.

*Types of point defects:*

i) The simplest form of a point defect is the replacement of an ion of mass $M$ by an *isotope* of mass $M'$. As the chemical binding is determined by the electrons, there is no change in forces and therefore in the structure of the lattice by this replacement. However, the dynamical properties of the lattice are changed because the equation of motion now contains the mass $M'$ instead of $M$ in one position (or some positions, if there is a concentration of isotopes). These dynamical properties will be discussed in sect. 20ff.

ii) *Impurity ions.* If an ion of the ideal lattice is replaced by an ion of other chemical properties, i.e. with another number of electrons, this is called a (substitutional) impurity (e.g. the replacement of a Zn-atom in a Zn-lattice by a Cu-atom, or of a Na-ion in a NaCl-lattice by a K-ion). In this case there is not only a change in mass but also in the forces to the neighboring ions. The neighbors can be attracted or repelled and this leads to a lattice dilatation. The impurity can be looked upon as a deformation center for the whole lattice. Impurities might be located also at interstitial positions. We consider these impurities as interstitials.

iii) *Vacancies.* One ion of the lattice has left its position and is placed on the surface of the crystal. There is one empty lattice position. If the ionic forces between neighboring ions are mainly repulsive, then the neighbors to the "vacancy" now are missing the repulsion of the removed ion and they are drawn toward the empty place. This situation happens to be in metals [57 K 1, 60 S 2, 62 S 2]. On the other hand, if the forces between neighbors are mainly attractive the ions surrounding the vacancy are repelled. This is the case in alkali halides [52 B 1, 57 F 1]. In general one has to know the forces very accurately for quantitative calculations.

iv) *Interstitials.* Instead of eliminating one ion from its lattice position, an additional ion may be placed on some position between the ideal lattice points. There are stable and unstable positions for such an interstitial and it is often difficult to know, which positions are stable. It may depend very sensitively on the potential energy of the ions.

v) *Defect pairs.* Besides these single point defects there may be complexes; and because there is always a finite concentration of point defects, there is a certain probability for the defects of forming larger complexes. The details depend on the interaction energy of the defects and on temperature, of course. There is an activation energy for the migration of the defects, at higher temperatures they can move more easily than at lower temperatures. Such complexes may be two neighboring vacancies, or a vacancy-interstitial pair at a certain distance, which is called *Frenkel-defect.* Sometimes the attraction between defects may be large enough that clusters of defects with many ions in it are formed. In case of impurity-clusters this is called precipitation.

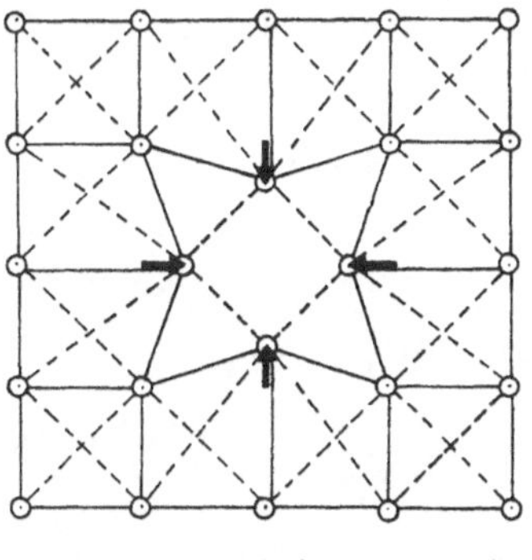

Fig. 6.1. A simple model of a vacancy. Indicated are the displacements from the ideal lattice positions due to the vacancy

vi) Especially in alkali-halides there is the possibility of removing, say, a Cl-ion from a NaCl-lattice and to replace it by an electron. The electron can be excited from its ground state by electromagnetic radiation (light); this leads to absorption of radiation, the crystals take a certain colour. Such *colour center* and related centers are discussed in a preceding article by *H. Pick* [66 P 1] and we will not go into details. In the following we will only shortly review the mechanical properties of single point defects and the methods which can be used for their investigation.

As a first example for the calculation of defect structures we consider the simple case of an impurity or vacancy in a primitive lattice (Fig. 6.1). The extension to non-primitive lattices can be done immediately. A pure lattice theoretical method has been given by *Kanzaki* [57 K 1], and somewhat simplified and generalized by *Flinn* and *Maradudin* [62 F 1]. The defect is brought into the lattice by two steps:

1) The displacement field is produced so that it is equal in magnitude and direction to the displacement field of the defect, but the ion in the center is still that of the ideal lattice. The change of energy is then

$$\Delta\Phi_1 = \sum_{mi} \Phi_i^m u_i^m + \frac{1}{2}\sum_{\substack{mn\\ij}} \Phi_{i\,j}^{mn} u_i^m u_j^n + \cdots \tag{18.1}$$

$\Phi_i^m$, $\Phi_{i\,j}^{mn}$ are the coupling-parameters of the ideal lattice, referring to the equilibrium positions of the ideal lattice. Thus $\Phi_i^m = 0$. $u_i^m$ are the displacements.

2) Now, after having reached the final displacement state, the central ion is removed and replaced by the impurity, or, in case of a vacancy, the position remains empty. One has to break all the bonds between the central ion and the host lattice ions and to introduce new bonds between defect ion and host ions. The change in energy can be written as*

$$\Delta\Phi_2 = \Delta\Phi_0 + \sum_{\substack{m\neq 0\\ i}} F_i^m u_i^m + \frac{1}{2}\sum_{\substack{m\neq 0\\ ij}} H_{i\,j}^{mm} u_i^m u_j^m + \cdots. \tag{18.2}$$

The coefficients $\Delta\Phi_0$, $F_i^m$, $H_{i\,j}^{mm}$ have to be calculated from a knowledge of the potential energy between the ions. Simple examples will be given below. The total change in potential energy is then

$$\Delta\Phi = \Delta\Phi_1 + \Delta\Phi_2$$
$$= \frac{1}{2}\sum_{\substack{mn\\ij}} \Phi_{i\,j}^{mn} u_i^m u_j^n + \Delta\Phi_0 + \sum_{\substack{m\neq 0\\ i}} F_i^m u_i^m + \frac{1}{2}\sum_{\substack{m\neq 0\\ ij}} H_{i\,j}^{mm} u_i^m u_j^m + \cdots. \tag{18.3}$$

The new equilibrium positions are determined by the minimum of the potential energy, i.e.

$$\frac{\partial\Delta\Phi}{\partial u_i^m} = 0 = \sum_{nj} \Phi_{i\,j}^{mn} u_j^n + F_i^m + \sum_j H_{i\,j}^{mm} u_j^m + \cdots \quad \text{for} \quad m \neq 0\,. \tag{18.4}$$

* It is assumed, that the position of the defect ion is that of the ideal lattice ion, i. e. it remains a center of inversion in primitive lattices. The generalization to other cases is not difficult. Further we assume already central-forces. Otherwise we would have to use different superscripts on $H$ and $u$.

This equation has to be solved for the $u_i^m$. We neglect higher than linear terms now. We define a *static Greens function by*

$$\sum_{nj} \Phi_{i\,j}^{m\,n} G_{j\,k}^{n\,p} = \Delta_{i\,k}^{m\,p}; \quad G_{j\,k}^{n\,p} = G_{j\;\;k}^{n-p\,0} \tag{18.5}$$

$\Delta_{i\,k}^{m\,p}$ is a Kronecker-symbol for $\boldsymbol{m}\,\boldsymbol{p}$ and $i\,k$. The Greens function is the reciprocal of the coupling-matrix $\Phi_{i\,j}^{m\,n}$ and can be calculated from the properties of the ideal lattice and thus is known at least in principle. We consider the last two terms of (18.4) as inhomogeneities to the homogeneous equation

$$\sum_{nj} \Phi_{i\,j}^{m\,n} u_j^n = 0\,. \tag{18.6}$$

A solution of the homogeneous equation is $u_j^n \equiv u_j$, independent of $\boldsymbol{n}$, because of (2.6). This means a motion of the center of mass of the crystal and is uninteresting in the following. A solution of the inhomogeneous equation is then

$$u_i^m = -\sum_{0 \neq nj} G_{i\,j}^{m\,n} \left\{ F_j^n + \sum_k H_{j\,k}^{n\,n} u_k^n \right\} \tag{18.7}$$

This is an "integral" equation for the $u_i^m$, where the boundary conditions have to be included. The solution can be simplified by reasonable assumptions about the range of the forces. Let us suppose, that the interaction takes place only between nearest neighbors. Then $F_j^n$ and $H_{i\,k}^{n\,n}$ are different from zero only for $\boldsymbol{n}$ neighbor to zero. The solution of (18.7) reduces to the solution of $z$ algebraic equations* for $\boldsymbol{m}$, also being neighbor to zero. $z$ is the number of neighbors. The other displacements of the lattice can than be calculated from the neighbor displacements according to (18.7).

As an example we consider first the linear lattice with nearest neighbor interaction. We have to calculate $F_i^m$ and $H_{i\,j}^{m\,m}$ from the change $\Delta\Phi_2$ in the potential. Let $\varphi(R)$ be the potential between ideal lattice ions and $\psi(R)$ between impurity and neighbor. Then

$$\begin{aligned}\Delta\Phi_2 &= \sum_{m \neq 0} \{\psi(R^m + u^m) - \varphi(R^m + u^m)\} \\ &= \sum_{m \neq 0} \{\psi(R^m) - \varphi(R^m)\} + \sum_{m \neq 0} \{\psi'(R^m) - \varphi'(R^m)\} u^m + \\ &\quad + \frac{1}{2} \sum_{m \neq 0} \{\psi''(R^m) - \varphi''(R^m)\} u^m u^m.\end{aligned} \tag{18.8}$$

The displacement of the defect ion has been put equal to zero already. Because of our limitation to nearest neighbors we have ($a$: lattice constant)

$$\begin{aligned} F^1 &= -F^{-1} = \psi'(a) - \varphi'(a) \\ H^{11} &= H^{-1-1} = \psi''(a) - \varphi''(a)\,, \quad \text{all others zero;} \end{aligned} \tag{18.9}$$

and from symmetry $u^1 = -u^{-1}$.

---

* Because of the symmetry of the defect, some of the equations are not independent, e. g. inversion symmetry reduces the number of independent equations to $z/2$, etc.

From (18.7)

$$\begin{aligned} u^m &= -G^{m1}\{F^1 + H^{11}u^1\} - G^{m-1}\{F^{-1} + H^{-1-1}u^{-1}\} \\ &= (G^{m-1} - G^{m1})(F^1 + H^{11}u^1)\,. \end{aligned} \tag{18.10}$$

With $m = 1$ we have

$$u^1 = -u^{-1} = \frac{(G^{20} - G^{00})\,F^1}{1 - (G^{20} - G^{00})\,H^{11}} \tag{18.11a}$$

and therefore

$$u^m = -u^{-m} = \frac{(G^{m+10} - G^{m-10})\,F^1}{1 - (G^{20} - G^{00})\,H^{11}}\,. \tag{18.11b}$$

We are left with the calculation of the static Greens function, which is the limiting case of the dynamic Greens function

$$\sum_{nj} \Phi^{mn}_{ij}\, G^{np}_{jk} - \lambda G^{mp}_{ik} = \Delta^{mp}_{ik}\,; \quad \lambda = M\omega^2\,. \tag{18.12}$$

It can be gotten by expanding $G^{np}_{jk}$ with respect to the solutions of the ideal lattice equation (right hand side of (18.12) equal to zero), inserting this into (18.12) and solving for the expansion coefficients. This is discussed in detail in section 21, and we will use the result already here:

$$G^{mn}_{ij}(\omega) = \frac{1}{NM}\sum_{q\sigma} \frac{e_i(q\,\sigma)\,e_j^*(q\,\sigma)}{\omega^2(q\,\sigma) - \omega^2 - i\epsilon}\,\mathrm{e}^{i q (R^m - R^n)}\,. \tag{18.13}$$

In case of a linear lattice we have for $\omega^2 = 0$ the static Greens function

$$G^{mn} = \frac{a}{2\pi}\int_{-\pi/a}^{+\pi/a} \mathrm{d}q\,\frac{\mathrm{e}^{i q a (m-n)}}{2f(1 - \cos q\,a)} = -\frac{|m-n|}{2f}\,; \quad f = \varphi''(a)\,. \tag{18.13a}$$

Introducing (18.9) and (18.13a) into (18.11) we have

$$u^m = -u^{-m} = -\frac{\psi'(a)}{\psi''(a)}\,; \quad m > 0\,. \tag{18.14}$$

$\varphi'(a) = 0$ is the equilibrium condition of the ideal lattice. The displacement is constant for all the atoms. In the limits of our model, this can be understood easily. The defect ion displaces the neighbors according to the changed energy, but as the potentials between the other ions remain unchanged, their distance is also unchanged and the *displacement is only that of the neighbors*. The constant displacement of the ions is related to the boundary conditions, tacitly assumed in the solution of (18.7) with (18.13), namely free ends in an infinite lattice. It can be positive or negative, depending only on the sign of $\psi'(a)$, for $\psi''(a)$ has to be positive in order to have a stable lattice. In a model for a vacancy $\psi(a)$ has to be zero, and (18.14) becomes unreasonable. This is clear, because a linear model does not allow in a consistent way for a vacancy. Apart from this, the linear lattice is a poor model for relaxation of ions by defects.

But a three-dimensional model can be discussed similarly. It is simply

$$\begin{aligned}
\Delta\Phi_0 &= \sum_{m\neq 0}\{\psi(\boldsymbol{R}^m) - \varphi(\boldsymbol{R}^m)\};\\
F_i^m &= (B^m/R^m)\, X_i^m;\\
H_{i\,j}^{m\,m} &= (A^m - B^m/R^m)\, X_i^m X_j^m \Big/ (R^m)^2 + (B^m/R^m)\,\delta_{ij};\\
A^m &= \psi''(\boldsymbol{R}^m) - \varphi''(\boldsymbol{R}^m);\quad B^m = \psi'(\boldsymbol{R}^m) - \varphi'(\boldsymbol{R}^m);
\end{aligned} \tag{18.15}$$

if the defect ion is a center of inversion and the interaction is limited to central forces. Further we restrict to nearest and next nearest neighbor interaction. In cubic crystals the displacements of the first and second neighbors satisfy

$$u_i^{m_1} = \alpha_1 X_i^{m_1} \Big/ l_1;\quad u_i^{m_2} = \alpha_2 X_i^{m_2} \Big/ l_2\,. \tag{18.16}$$

$\alpha_1, \alpha_2$ are the absolute displacements, $l_1, l_2$ the distances and $z_1, z_2$ the numbers of first and second neighbors. We use further

$$A_k = \psi''(l_k) - \varphi''(l_k);\quad B_k = \psi'(l_k) - \varphi'(l_k)\,. \tag{18.15a}$$

Inserting (18.15) and (18.16) into (18.7) gives after some calculation for the determination of the $\alpha_k$

$$\begin{aligned}
\{1 + S_{11} A_1\}\,\alpha_1 + S_{12} A_2 \alpha_2 &= -S_{11} B_1 - S_{12} B_2\\
S_{21} A_1 \alpha_1 + \{1 + S_{22} A_2\}\,\alpha_2 &= -S_{21} B_1 - S_{22} B_2
\end{aligned} \tag{18.17}$$

and for the displacements $u_i^m$ with $|\boldsymbol{m}| > |\boldsymbol{m}_1|, |\boldsymbol{m}_2|$

$$\begin{aligned}
u_i^m = &-(B_1 + A_1\alpha_1)\frac{1}{l_1}\sum_{n_1 j} G_{i\,j}^{m\,n_1} X_j^{n_1} -\\
&-(B_2 + A_2\alpha_2)\frac{1}{l_2}\sum_{n_2 j} G_{i\,j}^{m\,n_2} X_j^{n_2}\,.
\end{aligned} \tag{18.18}$$

It is

$$S_{11}\cdot z_1 l_1^2 = \sum_{m_1 n_1 ij} X_i^{m_1} G_{i\,j}^{m_1\,n_1} X_j^{n_1};\quad S_{22}\cdot z_2\cdot l_2^2 = \sum_{m_2 n_2 ij} X_i^{m_2} G_{i\,j}^{m_2\,n_2} X_j^{n_2}$$

$$S_{12}\cdot z_1 l_1 l_2 = S_{21}\cdot z_2 l_1 l_2 = \sum_{m_1 n_2 ij} X_i^{m_1} G_{i\,j}^{m_1\,n_2} X_j^{n_2} \tag{18.17a}$$

where the sums in (18.18) and (18.17a) extend over first ($\boldsymbol{m}_1$) and second ($\boldsymbol{m}_2$) neighbors of the impurity ion. In deriving (18.17) it has been used that none of the neighbors of a certain order is distinguished from the others. The remaining problem is the calculation of the Greens function. This has to be done for every (cubic) crystal model separately. The only case in which it has been done is a face-centered cubic lattice with central-force interaction between nearest neighbors [62 F 1]. In that case (18.17, 18) can be simplified further: the remaining equations are

$$\alpha_1 = -\frac{S_{11} B_1}{1 + S_{11} A_1} \tag{18.19a}$$

$$u_i^m = -\frac{B_1}{1 + S_{11} A_1}\frac{1}{l_1}\sum_{n_1 j} G_{i\,j}^{m\,n_1} X_j^{n_1}\qquad |\boldsymbol{m}| > |\boldsymbol{n}_1|\,. \tag{18.19b}$$

In the case of a vacancy we have $\psi(\boldsymbol{R}^m) = 0$, and because for nearest neighbor interaction the equilibrium condition of the ideal lattice is $\varphi'(l_1) = 0$, $B_1$ vanishes and all the displacements are zero. Thus in a nearest neighbor model there is no displacement of the host lattice ions due to a vacancy. This is clear, because cutting a bond with $\varphi'(l) = 0$ does not exert a force on the ion. But if second neighbors are taken into account, the equilibrium condition is $z_1 l_1 \varphi'(l_1) + z_2 l_2 \varphi'(l_2) = 0$ and the $B_i$ are different from zero ($z_1 l_1 B_1 + z_2 l_2 B_2 = 0$ in case of a vacancy).

To get an insight into the displacements at large distances from the defect one can use the asymptotic expansion of $G^{m\,n_1}_{i\;j}$. If $|\boldsymbol{m}|$ is large compared to $|\boldsymbol{n}_1|$ we can write

$$G^{m\,n_1}_{i\;j} = G^{m-n_1\,0}_{i\qquad j} = G^{m\,0}_{i\;j} - \sum_k \frac{\partial G^{m\,0}_{i\;j}}{\partial X^m_k} X^{n_1}_k .$$

Now, because of the inversion symmetry $\sum_{n_1 j} G^{m\,0}_{i\;j} X^{n_1}_j = 0$ and in cubic crystals we have further

$$\sum_{n_1} X^{n_1}_j X^{n_1}_k = \delta_{jk} \cdot z_1 l_1^2/3; \quad \sum_{n_2} X^{n_2}_j X^{n_2}_k = \delta_{jk} \cdot z_2 l_2^2/3$$

and therefore with $|\boldsymbol{m}| \to \infty$

$$u^m_i = \underbrace{\frac{1}{3}\{z_1 l_1 (B_1 + A_1 \alpha_1) + z_2 l_2 (B_2 + A_2 \alpha_2)\}}_{M} \sum_j \frac{\partial G^{m\,0}_{i\;j}}{\partial X^m_k} . \tag{18.20}$$

The asymptotic form of the Greens function is not easy to calculate and it has been done only in the above mentioned case [62 F 1]. It is rather easy to see that it is proportional to $1/R^m$, therefore $u^m_i$ is proportional to $(R^m)^{-2}$, but the details depend on the crystal structure. However, far away from the defect one can use the continuum theory of elasticity and replace the Greens function by that of an elastic medium. But even the elastic Greens function can be calculated exactly only in a few cases (isotropic or hexagonal crystals).

A few remarks to the above described method shall be added. The range of interaction often is not limited to very near neighbors. Then it is not easy to solve (18.7) by the preceding method. One can use the following procedure. (18.7) can be written formally as

$$u^m_i = -\sum_{nj} G^{m\,n}_{i\;j} F^n_j + \sum_{\substack{np\\ jkl}} G^{m\,n}_{i\;j} H^{n\,n}_{j\;k} G^{n\,p}_{k\;l} F^p_l - + \cdots \tag{18.22}$$

which is an iteration formula for the $u^m_i$. But the use of this formula is very tedious. Other methods have been developed for the investigation of defects. This is also necessary because of the limitations of the above method. In (18.4) we have limited the equation to linear terms in the displacements. This is a reasonable assumption only if the displacements due to the defects are small, which is the case for some impurities or perhaps vacancies, but not for interstitials. In principle the method can be extended to higher terms in the $u_i^m$, but this would enlarge the numerical work enormously*; other methods have been developed.

Most of the calculations have been done by dividing the crystal into three regions: the first contains the neighbors of the defect ion up to a certain order (third or fourth neighbors). In this region the ionic displacements due to the defect are assumed to be large. The second region is a transition to the third one; this last part is treated with the elastic theory using a displacement field for the description. The displacements are small for larger distances of the defect ion. The transition region is used to fit the lattice theoretical description in the interior to the elastic behavior at larger distances. First we discuss the elastic region.

As an example we start again with (18.7) and central forces. Inserting (18.15) we have

$$u_i^m = -\sum_{nj} G_{i\,j}^{mn} \left\{ \frac{B^n}{R^n} \left( X_j^n + u_j^n \right) + \sum_k \left( A^n - \frac{B^n}{R^n} \right) \frac{X_j^n X_k^n}{(R^n)^2} u_k^n \right\}. \quad (18.23)$$

The sum over $n$ extends over the range of the forces. Thus we can find an ion $m$ which is outside of this range or even "far" away. This is the elastic region. We then can use the expansion of $G_{i\,j}^{mn}$ with respect to $n$, i.e. (18.20). The term with $G_{i\,j}^{m0}$ again vanishes, if the defect has inversion symmetry. We are then left with

$$u_i^m = -\sum_{jk} M_{kj} \frac{\partial G_{i\,j}^{m0}}{\partial X_k^m} \quad (18.24)$$

and

$$M_{kj} = -\sum_n \left\{ \frac{B^n}{R^n} X_k^n \left( X_j^n + u_j^n \right) + \left( A^n - \frac{B^n}{R^n} \right) \frac{X_k^n X_j^n}{(R^n)^2} \sum_l X_l^n u_l^n \right\} \quad (18.25)$$

which in cubic crystals reduces to the form given by (18.21): $M_{kj} = M \cdot \delta_{kj}$. The $u_j^n$ have to be calculated from the equation for small $n$, as indicated above. They are the displacements in the first region. If these have been calculated, the elastic displacement field is known, provided we know the static Greens function. From the derivation of (18.24) the meaning of $M_{jk}$ becomes clear. In (18.7) we have forces, $F_j^n$ and $\sum_k H_{j\,k}^{nn} u_k^n$ which produce the displacements $u_i^m$, their magnitude being given by a convolution with the Greens function. However, as we have seen, the term containing the Greens functions vanishes in the elastic region; what enters, is the derivative of the Greens function multiplied with moments of forces or pairs of forces (force dipoles). Thus the elastic displacement field of point defects can be described by the field of force dipoles, the net force being zero. As can be seen from (18.25), they need not necessarily

* It should be mentioned, that the above formulation can be used as a variational method. In (18.3) one makes a reasonable Ansatz for the $u$ with free parameters, which will be determined by minimalization of $\Delta\,\Phi$. In this way a knowledge of Greens function is not necessary.

be symmetric. The antisymmetric part defines the torque of the force dipole. If all the $u_j^n \sim X_j^n$ then $M_{jk}$ would be symmetric. Of course, as we deal with the elastic displacement field, the Greens function $G_{ij}^{m0}$ can be replaced by the elastic Greens function $G_{ij}(\boldsymbol{R})$, and (18.24) reads

$$u_i(\boldsymbol{R}) = -\sum_{jk} M_{kj} \cdot \frac{\partial G_{ij}(\boldsymbol{R})}{\partial X_k} . \tag{18.24a}$$

The elastic Greens function is defined by the elastic equation of motion (in the stationary case):

$$-C_{ij,kl}\, G_{mk'jl}(\boldsymbol{R}) - \varrho\,\omega^2\, G_{im}(\boldsymbol{R}) = \delta_{im}\,\delta(\boldsymbol{R}) \tag{18.26}$$

where $C_{ij,kl}$ are the elastic constants, $\varrho$ is the mass density and $|j$ means the derivative with respect to $X_j$.

The calculation of $G_{mi}$ is difficult in general*. In a closed form it can be done only for isotropic or hexagonal media [53 K 1]. For anisotropic cubic materials an expansion of the static Greens function has been given [55 K 1, 60 S 2, 62 S 2]. The main features can be seen already in the isotropic case; expanding the Greens function with respect to the solutions of the homogeneous equation (18.26), inserting this into (18.26) and determining the expansion coefficients from satisfying (18.26) we immediately obtain the representation

$$G_{ij}(\boldsymbol{R}) = \frac{1}{(2\pi)^3\,\varrho} \int \mathrm{d}\boldsymbol{q}\, e^{i\boldsymbol{q}\boldsymbol{R}} \sum_\sigma \frac{e_i(\boldsymbol{q}\,\sigma)\, e_j(\boldsymbol{q}\,\sigma)}{\omega^2(\boldsymbol{q}\,\sigma) - \omega^2 - i\,\epsilon} \tag{18.27}$$

$\boldsymbol{q}$: wave number vector, $\sigma$: polarisation type of elastic wave, $\boldsymbol{e}(\boldsymbol{q}\sigma)$: polarisation vector. The eigenfrequencies $\omega^2(\boldsymbol{q}\sigma)$ are the solutions of

$$\|C_{ij,kl}\, q_j\, q_l - \varrho\,\omega^2\,\delta_{ik}\| = 0 . \tag{18.28}$$

In (18.27) we have already chosen the path of integration as to have outgoing waves in scattering problems $(-i\epsilon)$. In an isotropic medium we have

$$C_{ij,kl} = \lambda\,\delta_{ij}\,\delta_{kl} + \mu(\delta_{ik}\,\delta_{jl} + \delta_{il}\,\delta_{jk}) \tag{18.29}$$

$\lambda = c_{12}$; $\mu = c_{44}$ are the Lamé constants. The solutions of (18.28) are then

$$\begin{aligned} &\varrho\,\omega^2(\sigma = 1) = (\lambda + 2\mu)\, q^2 = \varrho\, c_l^2\, q^2; \quad e_i(\sigma = 1) = q_i/q \\ &\varrho\,\omega^2(\sigma = 2, 3) = \mu q^2 = \varrho\, c_t^2\, q^2; \sum_{\sigma=2}^{3} e_i(\sigma)\, e_j(\sigma) = \delta_{ij} - q_i\, q_j/q^2 . \end{aligned} \tag{18.30}$$

Introducing this into (18.27), the static Greens function can be easily calculated to be

$$G_{ij}(\boldsymbol{R}) = \frac{1}{8\pi\,\mu(\lambda + 2\mu)} \left\{(\lambda + 3\mu)\,\frac{\delta_{ij}}{R} + (\lambda + \mu)\,\frac{X_i X_j}{R^3}\right\}. \tag{18.31}$$

* It should be mentioned, that the static Greens function in the theory of elasticity is often called the fundamental integral of the elastic equation [53 K 1, 55 K 1, 60 S 2].

From (18.31), together with (18.24a), it follows that the displacement field of point defects varies as $R^{-2}$ with increasing distance from the defect. This is valid also for other crystal symmetries, but the angular dependence is much more complicated.

All these statements hold for a single defect. If we have a concentration of defects, we have to superpose the displacement fields of the different defects. At large distances from the region with defects we can use again (18.24), where $M_{kj}$ now represents the effective force dipoles of the defect concentration. In cubic crystals with isotropic defects or with anisotropic defects with statistically distributed orientations we have always

$$M_{kj} = M \cdot \delta_{kj} \,.$$

With this form we have, if the material is in addition isotropic,

$$u_i(\boldsymbol{R}) = \frac{M}{4\pi(\lambda + 2\mu)} \cdot \frac{X_i}{R^3}. \tag{18.32}$$

For anisotropic cubic material, *Seeger* et. al. have derived the corresponding expression:

$$u_i(\boldsymbol{R}) = M \left\{ d_0 \frac{X_i}{R^3} + d_1 \frac{X_i^3}{R^5} + d_2 \frac{X_i^5}{R^7} + d_3 \frac{X_i}{R^7} (X_1^2 X_2^2 + X_2^2 X_3^2 + X_3^2 X_1^2) \right\}. \tag{18.33}$$

The $d$-coefficients are rather complicated functions of the three elastic constants and must be evaluated numerically [60 S 2, 62 S 2]. For the details we refer to the different papers on this subject [57 E 1, 57 K 1, 62 F 1].

In (18.20) we have limited the expansion of $G_{i\ j}^{m\ n}$ to the first term in $X_j^n$. In special cases, however, it might turn out, that this is not sufficient for a quantitative discussion of the displacement field. Then higher terms have to be included. The next term, e.g., would give rise to a contribution

$$u_i^m = - \sum_{klj} M_{lkj} \frac{\partial^2 G_{i\ j}^{m\ 0}}{\partial X_k^m \, \partial X_l^m} \tag{18.34}$$

which is the displacement field of a force quadrupole. Its contribution can be calculated as above, provided the static Greens function (fundamental integral) is known.

The main difficulty lies in the investigation of the displacements in the neighborhood of the defect ions, if these displacements are so large, that they cannot be handled as described by the method discussed above. This is the case especially if the defect ions have positions not being lattice positions, i.e. interstitials. In many cases it is not even known which is the true position of the interstitial. Very careful calculations of the displacement energy would indicate the stable positions of interstitials; but then it is necessary to know the interacting potential between the ions very accurately. It has turned out, that very small changes in the parameters of the potential sometimes lead to quite different stable positions of the interstitials.

To find the equilibrium positions of the neighboring atoms, the potential energy has to be calculated with the displaced positions as parameters. The energy contains the energy of that region, which is treated lattice theoretically, the energy of the elastic region and the interaction energy of both parts. It is convenient to treat this energy also with lattice theory. The question arises which choice for the boundary of the elastic continuum has to be made. But if the displacement energy of the defect is strongly localized in the vicinity of the defect (lattice region), the position of the boundary does not influence the energy appreciably. In the case of a symmetric point defect (substitutional impurity, vacancy, interstitial with cubic symmetry) the boundary can be chosen as a spherical surface. If further the range of the interacting potential is small, the assumption is justified, that only the ions on the surface of the continuum region participate on the forces of the lattice region. The elastic displacement field is that of a free continuum with given displacements at the surface, as discussed above.

The lattice part of the energy has to be calculated thoroughly, without expansion with respect to the displacements, if these are not small. Thus nearly all accurate values have been gotten by numerical methods. We refer to the papers for the details. The main points are the following.

In *van der Waals-crystals* the potential is described by a power-potential, mainly a Lennard-Jones-6-12-potential. The calculations have been done mainly for vacancies, but also for impurities and interstitials. For vacancies the displacements of the nearest neighbors are directed inward. The order of magnitude is some percents of the lattice constant. The second neighbors are displaced outwards in general. At interstitials the displacements of neighbors are mainly outward; it is assumed that the interstitial position in van der Waals-crystals is a cubic symmetry position (cube center in face-centered crystals). This need not be true (see for metals).

The displacements in the neighborhood of impurities may be inward or outward, depending on the size and the potential of the impurity ion.

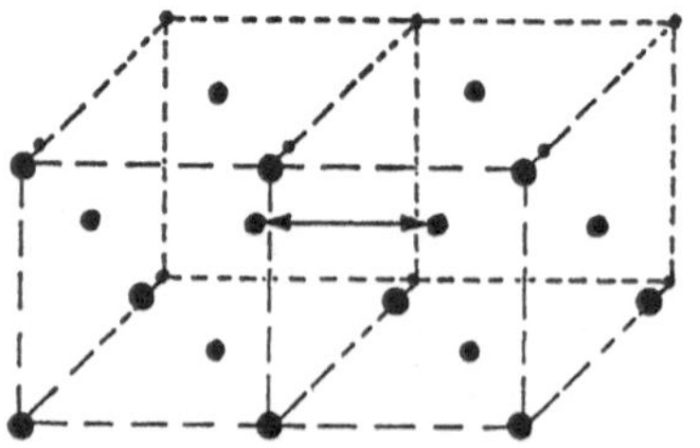

Fig. 6.2. Interstial in a face-centered-cubic lattice; shown is the dumbbell-configuration

The most refined calculations have been done for vacancies and interstitials in *metals* [60 S 2, 61 B 1, 62 S 2, 64 J 2, 65 J 1]. The main difficulty is the poor knowledge of the potential energy. It is generally considered as a sum of a repulsive Born-Mayer-potential (exponential) or Bohr potential (exponentially screened Coulomb-potential) and the attractive contribution of the „free" electrons. The parameters of the potential are determined from elastic constants, scattering experiments and other experimental data. Calculations for vacancies are not so difficult and give an inward displacement for the nearest neighbors, up to maximal 10% of the lattice constant. The calculation of the interstitial displacements has been more difficult, as it was not clear, which position is the most stable one. Investigations [61 B 1, 62 S 2, 66 J 2] for different possible interstitial-configurations seem to indicate, that in face-centered metals, at least in Cu and similar ones, the so called dumbbell-configuration (Fig. 6.2) is the stable one. This is confirmed by experiments in Ni. The displacements are outward for the neighbors of the dumbbell. But the stable configuration depends very sensitively on the potential and another position might be stable in other metals.

The potential in *ionic crystals*, especially in alkali halides, is better known. The main contribution is from the coulomb-interaction and the Born-Mayer-repulsive forces. But also the polarization of the ions plays a great role, especially when the ions are displaced from their ideal lattice positions. A large number of calculations have been done, mainly for alkali halides. At vacancies, the nearest neighbor displacement is outward. This is a consequence of the lack of the coulomb attraction between vacant ion and its neighbors. The coulomb interaction in this case dominates over the other contributions to the potential. This is different for the situation at $F$-centers, e. g.; there is no charge missing at the center, but the repulsion between the replaced ion and its neighbors vanishes. In this case there is an inward displacement of the neighbors. The displacements at impurities depend on the special kind of the impurity ion. Investigations have been done with differently charged impurities (e. g. NaCl with $K^+$, $Ca^{++}$, $N^+$, . . . centers). In recent years molecular impurity centers have become of great interest. This is e. g. a $OH^-$ or $CN^-$ or $NO_2^-$ center in KCl. No calculations of the structure of these centers have been done; the potential is not known well enough for reasonable calculations. What is known on the structure of these centers originates from experimental investigations. The center of mass of the molecule in general will not be identical with the ideal lattice point. Also the orientation of the molecule is an essential feature for the discussion of the structure. It may be different for different molecules in the same alkali halide, and even different for the same molecule in different alkali halides of the same type. It is likely that it depends rather strongly on the ionic or molecular radii and their induced or permanent electric moment. A full understanding needs a complete knowledge of the interionic and intermolecular forces. Interstitials in alkali halides are generally supposed to be situated in the midth of the cube edges. With this assumption calculations of the structure have been done [52 B 1, 57 F 1, 63 F 2]. But it is not sure, that these positions are the stablest ones in any case.

Few calculations have been done on the structure of point defects in homopolar crystals (Si, Ge) [61 S 1]. The difficulties lie in the poor knowledge of the interaction potential and in the fact that the ideal lattice positions are no centers of inversion symmetry. Thus there are only few symmetry requirements for the direction of the displaced ions or, in other words, one has a large number of parameters to be determined by the minimalization of the potential energy.

The effect of temperature on the ionic displacement at point defects has been investigated first by *Fischer* [59 F 1]. At finite temperatures the free energy or free enthalpy has to be minimalized rather than the potential energy. As we have seen with the discussion of anharmonic effects (chapter V), temperature effects on the lattice structure (thermal expansion) enter only through anharmonic terms in the potential energy. Thus it is clear that we cannot describe temperature effects with the procedure at the beginning of this section. The free energy includes a knowledge of the vibration spectrum of the crystal. But it is sufficient again for structural questions to use the quasiharmonic approximation. This has been shown by general considerations [59 F 1, 61 L 1]. Quantitative calculations have been done for a central force model. The displacements of the neighboring ions to a vacancy, relative to the temperature dependent lattice constant, increase with increasing temperature.

## 19. Change of Volume and Energy of Defects

In all the preceding calculations we have tacitly assumed that the crystal containing the point defect is infinitely extended. In fact it has a finite boundary and we have to look whether this has an influence on the quantities of interest.

The change of volume of a crystal due to the displacements of the ions can be gotten by integrating the displacement field over the surface of a sphere. But this volume change $\Delta V_\infty$ is that which corresponds to an infinite crystal. In the case of an isotropic medium we have from (18.24 a) and (18.31)

$$\Delta V_\infty = \frac{1}{3(\lambda + 2\mu)} \sum_j M_{jj} . \tag{19.1}$$

This is not correct for a finite medium. According to the displacement field in the crystal there is a stress field. If we limit the crystal to a finite region and use the displacement field as described in sect. 18, the corresponding stress field would not vanish at the surface of the finite crystal. However, in a finite free crystal the stresses have to vanish at the surface. To establish this surface condition, appropriate stresses (and therefore displacements) have to be added which just cancel the induced surface stresses. This image term has been first investigated by *Eshelby* [57 E 1]. The surface stress is given by

$$\sigma_{ij}(\boldsymbol{R}_0) = \sum_{kl} C_{ij,kl}\, u_{k|l}(\boldsymbol{R}_0) \tag{19.2}$$

where $C_{ij,kl}$ are the elastic constants, as given by (18.29) for isotropic material and $u_{k|l}(\boldsymbol{R}_0)$ is the derivative of the displacement field as given by (18.24 a) and (18.31). We consider the isotropic case with

$M_{jk} = M\,\delta_{jk}$; this is the model of sect. 18. A simple calculation gives [$\boldsymbol{R}_0$ is the coordinate of the surface of the finite crystal in (19.3, 4)!]

$$u_{k|l}(\boldsymbol{R}_0) = \frac{M}{4\pi(\lambda + 2\mu)} \cdot \frac{1}{R_0^3}\{\delta_{kl} - 3\,X_{0k}\,X_{0l}/R_0^2\} \tag{19.3}$$

$$-\sigma^I_{ij}(\boldsymbol{R}_0) = 2\,\mu \cdot u_{i|j}(\boldsymbol{R}_0) = \frac{2\,\mu\,M}{4\pi(\lambda + 2\mu)} \cdot \frac{1}{R_0^3}\{\delta_{ij} - 3\,X_{0i}\,X_{0j}/R_0^2\}\,. \tag{19.4}$$

The minus sign indicates, that negative stresses have to be applied to cancel the induced ones. The change in volume due to these stresses in an isotropic or cubic material then is*

$$\Delta\,V_I = \frac{\varkappa}{3}\sum_{ij}\int X_j\,\sigma^I_{ij}\,\mathrm{d}f_i = -\frac{2}{3}\varkappa\,\mu\sum_{ij}\int u_{i|j}\,X_j\,\mathrm{d}f_i = \frac{4}{3}\,\frac{\varkappa\,\mu\,M}{\lambda + 2\mu} \tag{19.5}$$

where $\varkappa = 3/(3\lambda + 2\mu)$ is the compressibility of the crystal (reciprocal bulk modulus). With (19.1) and (19.5) we have

$$\Delta\,V = \Delta\,V_\infty + \Delta\,V_I = \frac{3M}{3\lambda + 2\mu} = \varkappa\,M \quad \text{or} \quad \Delta V/\Delta\,V_\infty = \varkappa(\lambda + 2\mu)\,. \tag{19.6}$$

A slight generalization of (19.6) to non-isotropic defects in cubic materials gives

$$\Delta\,V = \frac{\varkappa}{3}\sum_j M_{jj}\,. \tag{19.7}$$

These additional stresses cause a homogeneous strain of the crystal, given by (19.3); from this it follows that the contribution of these additional strains to the displacements of the ions due to the defect and to the energy of the defect is of the order of $R_0^{-3}$, which is negligible for the crystals which we have assumed to be large. We have to take into account the image term only for the calculation of the volume change, where it contributes nearly a factor of two.

So far we have dealt with a single defect only. However, in fact there is always a concentration of defects. *Fischer* and *Hahn* [63 F 1] have shown, that in case of many defects, we can use (19.7) multiplied by the number of defects, provided the defects are distributed homogeneously and the interaction of the defects can be neglected. According to the preceding section the influence of defects to the surrounding continuum can be described by force dipoles $M_{ij}$. If the dipole is situated at $\boldsymbol{R}'$, the exerted force density in $\boldsymbol{R}$ is given by

$$f_i(\boldsymbol{R}, \boldsymbol{R}') = -\,M_{ij}\frac{\partial}{\partial X_j}\,\delta(\boldsymbol{R} - \boldsymbol{R}')\,. \tag{19.8}$$

If there is a defect density $n(\boldsymbol{R}')$, the total force density at $\boldsymbol{R}$ is given by

$$f_i(\boldsymbol{R}) = \int n(\boldsymbol{R}')\,f_i(\boldsymbol{R}, \boldsymbol{R}')\,\mathrm{d}\boldsymbol{R}' = -\,M_{ij}\frac{\partial}{\partial X_j}\int n(\boldsymbol{R}')\,\delta(\boldsymbol{R} - \boldsymbol{R}')\,\mathrm{d}\boldsymbol{R}'$$
$$= -\,M_{ij}\frac{\partial}{\partial X_j}\,n(\boldsymbol{R}) \tag{19.9}$$

* This holds as far as the linear theory of elasticity is valid. In the non-linear theory additional terms arise.

where the *interaction of defects* has been *neglected* already. For a *homogeneous distribution* of defects we have

$$n(\boldsymbol{R}) = \begin{cases} n_D = N_D/V & \text{inside } V \\ 0 & \text{outside } V \end{cases} \tag{19.10}$$

where $V$ is the volume of the crystal and $N_D$ the number of defects. $n(\boldsymbol{R})$ is a step-function with the step at the surface of the crystal. Inserting this into the second integral of (19.9), replacing $\frac{\partial}{\partial X_j}\delta(\boldsymbol{R}-\boldsymbol{R}')$ by $-\frac{\partial}{\partial X'_j}\delta(\boldsymbol{R}-\boldsymbol{R}')$ and using Gauss' theorem, we have immediately

$$f_i(\boldsymbol{R}) = M_{ij}\cdot\frac{N_D}{V}\int_{(S)}\delta(\boldsymbol{R}-\boldsymbol{R}')\,\mathrm{d}S'_j \tag{19.11}$$

where the integration is over the surface of the crystal. The total force on the surface can be gotten by integrating (19.11) over a volume element at the surface, for (19.11) contains only the force density at the surface. Thus we have for an element $\mathrm{d}S_j$ in the surface

$$\mathrm{d}F_i = \frac{N_D}{V}M_{ij}\,\mathrm{d}S_j = \sigma_{ij}\,\mathrm{d}S_j \tag{19.12}$$

$\sigma_{ij}$ is the tensor of the induced surface stresses. It is constant over the surface and therefore causes a homogeneously strained state in the crystal. (19.12) replaces the above formulas, valid for one defect, in the case of a homogeneously distributed concentration of defects. In the case of inhomogeneously distributed defects or in the case of large interaction between defects, additional terms to (19.12) may arise. This must be investigated separately.

If the defects have cubic symmetry or if they are orientated statistically in a cubic material, we have *

$$M_{ij} = M\,\delta_{ij}\,.$$

Then

$$\sigma_{ij} = \frac{N_D}{V}\cdot M\cdot\delta_{ij} = -p\,\delta_{ij}$$

is equivalent with an isotropic pressure. The change in volume is (in linear theory)

$$\Delta V = -\varkappa p V = \varkappa M\cdot N_D \tag{19.13}$$

which is (19.6, 7), multiplied with the number of defects.

The validity of (19.13) depends on the validity of the linear theory of elasticity. In the immediate neighborhood of a defect the displacements may be so large, that the linear theory is not valid. But the region of this nonlinearity is only a few lattice constants, so that it can be neglected, if the crystal is large and the concentration of defects is small. (19.13) is valid in most cases.

* In case of statistically orientated defects, $M$ is the average about all possible directions of orientation.

(19.13) is the change of volume due to the pure defect. Now for example, if one wants to consider the change in mass-density of the crystal due to the defects, another contribution arises. Let us consider the case of vacancies. (19.13) represents the change in volume due to displacement field of the defects. But the removed ion has to be considered as gone to the surface (Schottky-defect) and thus enlarges the volume by one atomic volume ($V_z$ in case of Bravais-lattices). Thus the effective volume change for Schottky defects (vacancies) is

$$\Delta V_{\text{eff}} = (\varkappa M + V_z)\, N_{\text{D}} \,. \tag{19.14}$$

Similar considerations have to be done for other defects.

*Fischer* and *Hahn* have shown further, that the mean change of the lattice constant, due to defects, and determined with X-ray scattering, is identical with the mean change of the lattice constant, calculated from (19.13). The general proof is not difficult [63 F 1]. Here we will not go into the details.

The calculation of *defect energies* (Formation energy etc.) can be done best with the formalism of sect. 18 using Greens function. But this is limited, as mentioned, to small displacements of the neighboring atoms. The energy of the defect is simply given by (18.3) with the displacements $u_i^m$ determined from (18.4) or (18.7). Multiplying (18.4) by $u_i^m$, summing over $m$ and $i$ and inserting this into (18.3) we have

$$\Delta \Phi = \Delta \Phi_0 + \frac{1}{2} \sum_{mi} F_i^m u_i^m \tag{19.15}$$

$u_i^m$ are the displacements as determined by (18.7), $\Delta \Phi_0$ has to be calculated from the given or assumed potential between the ions. In the case of central forces and the defect ion being in a lattice position we have (18.15a)

$$\Delta \Phi_0 = \sum_{m \neq 0} \{\psi(\boldsymbol{R}^m) - \varphi(\boldsymbol{R}^m)\} \tag{19.16}$$

$\varphi(\boldsymbol{R}^m)$ is the potential between host lattice ions.

(19.15) represents the energy of the defect. It is not what is called the formation energy. E.g. in the case of a vacancy the ion is removed from its lattice position (to infinity, energy $\Delta \Phi$) and then brought to the surface of the crystal (from infinity). This last process is connected with a gain of energy, namely just the binding energy of one lattice ion. In the case of a vacancy ($\psi(\boldsymbol{R}^m) \equiv 0$) it is just $-\frac{1}{2}\Delta \Phi_0$. Thus the formation energy of a vacancy is given by

$$E_{\text{v,f}} = \frac{1}{2}\left\{\Delta \Phi_0 + \sum_{m,i} F_i^m u_i^m\right\}. \tag{19.17}$$

Similar considerations give the formation energies of defects in other cases. A substitutional impurity is the removal of a host lattice ion (to infinity) and the introduction of an impurity ion (from infinity). Thus $\Delta\Phi$ represents just the formation energy in the case of impurities.

Another quantity of interest is the migration energy of defects. The defects migrate from one equilibrium position to another one by passing through a saddle-point of the lattice potential for the defect. Sometimes the saddle-point configuration can be guessed from symmetry considerations, but in many cases different possible configurations have to be investigated in order to find the saddle-point with the lowest energy. The difference between the energy in the saddle-point configuration $\Delta\Phi_S$ and the energy in the equilibrium configuration $\Delta\Phi_E$ is the migration energy of the defect

$$E_{\mathrm{m}} = \Delta\Phi_S - \Delta\Phi_E\,. \tag{19.18}$$

Because in the saddle-point-configuration the displacements of the neighboring atoms are often larger than in the equilibrium configuration, the above method using Greens function is not adequate, at least not in the form of the linear theory. Rather one has to make more accurate calculations using numerical methods as mentioned on page 133 for the calculation of displacements and energies in saddle-point configurations. For details we refer again to a number of papers [e.g. those mentioned in sect. 18, 19]. In the case of central force interaction between nearest and second neighbors only we have with the formulae and abbreviations of sect. 18

$$\Delta\Phi_0 = z_1[\psi(l_1) - \varphi(l_1)] + z_2[\psi(l_2) - \varphi(l_2)] \tag{19.19}$$

and

$$\sum_{m,i} F_i^m u_i^m = B_1 z_1 \alpha_1 + B_2 z_2 \alpha_2 \tag{19.20}$$

for defects with cubic symmetry in cubic Bravais-crystals. As an illustration we will discuss the simplest model numerically.

We consider a face-centered cubic Bravais-crystal with Lennard-Jones-6-12-potential between first and second neighbors:

$$\varphi(R) = 4\varepsilon\left\{\left(\frac{\sigma}{R}\right)^{12} - \left(\frac{\sigma}{R}\right)^{6}\right\}. \tag{19.21}$$

The equilibrium condition for the ideal lattice demands $\sigma = 0,9\,l$; $l = l_1$ is the nearest neighbor distance, $l_2 = \sqrt{2}\,l = a$ is the lattice constant and second neighbor distance. The unit volume is $V_z = a^3/4 = l^3/\sqrt{2}$. The energy constant $\varepsilon$ is equal to $1{,}085 \cdot 10^{-2}$ eV in Argon e.g.

The defect is assumed to be a vacancy. Then $\psi(R) \equiv 0$ and $\varphi(R)$ is given by (19.21). A straight forward calculation gives

$$\begin{aligned} \varphi(l_1) &= -0.997\,\varepsilon; & \varphi(l_2) &= -0.246\,\varepsilon \\ B_1 &= 0.687\,\varepsilon/l; & B_2 &= -0.971\,\varepsilon/l \\ A_1 &= -84.83\,\varepsilon/l^2; & A_2 &= 2.825\,\varepsilon/l^2. \end{aligned} \tag{19.22}$$

The main difficulty arises from the calculation of the Greens function. This has to be done numerically, but it has been done only for a nearest neighbor model (*Maradudin, Flinn* [62 F 1]), which is not appropriate for the description of a vacancy. Now Greens function is an integral over the spectral distribution of the lattice frequencies, which is supposed not to be changed essentially by the influence of second neighbors. Therefore we take Maradudins and Flinns Greens function also for our second neighbor calculation. Probably the error will be very small. With these functions we have

$$S_{11} = 0.003\,64\,l^2/\varepsilon;\quad S_{22} = 0.004\,07\,l^2/\varepsilon;$$
$$S_{21} = 2S_{12} = 0.000\,147\,l^2/\varepsilon\,. \tag{19.23}$$

Inserting (19.22, 23) into (18.17) we get

$$\alpha_1 = -0.0035\,l,\quad \alpha_2 = +0.0038\,l \tag{19.24}$$

which agrees fairly well with *Kanzaki*'s [57 K 1] result:

$$\alpha_1 = -0.00315\,l,\quad \alpha_2 = +0.00405\,l\,. \tag{19.24 a}$$

*Kanzaki* has discussed the same model, not using the Greens functions method, but rather solving (18.4) by iteration, which, indeed, is completely equivalent. The nearest neighbors of the vacancy are thus displaced inward, the second neighbors are displaced outward (cf. page 136). The force-dipole-moment is given by

$$M = -1.214\,\varepsilon\,. \tag{19.25}$$

For the energies involved we have

$$\Delta\Phi_0 = +13.44\,\varepsilon;\quad \sum_{mi} F_i^m u_i^m = -0.051\,\varepsilon \tag{19.26a}$$

and therefore

$$\Delta\Phi = 13.41\,\varepsilon = 0.147\ \mathrm{eV}\,, \tag{19.26b}$$

and the formation energy of a vacancy

$$E_{v,f} = \frac{1}{2}\left\{\Delta\Phi_0 + \sum_{mi} F_i^m u_i^m\right\} = 6.69\,\varepsilon = 0.0725\ \mathrm{eV}\,. \tag{19.26c}$$

The numerical values are related to Argon.

The formula (19.1) for the change of the volume $\Delta V_\infty$ is valid only for isotropic media, whereas the above model does not lead to isotropy. However, the effect of anisotropy is not very essential and we can use (19.1) approximately. The Lamé-constants $\lambda$, $\mu$ can be related to certain averages of the cubic elastic constants using an averaging method of *Voigt* [10 V 1].*

$$\lambda = \bar{c}_{12} = \frac{1}{5}(c_{11} + 4c_{12} - 2c_{44})$$
$$\mu = \bar{c}_{44} = \frac{1}{5}(c_{11} - c_{12} + 3c_{44})\,. \tag{19.27}$$

* *Kröner* [58 K 2] has given a somewhat improved method of averaging the elastic constants. This gives in the example given $\lambda = 34{,}18\,\varepsilon/V_z$; $\mu = 29{,}46\,\varepsilon/V_z$.

With (19.21) we can calculate the elastic constants from (6. 23, 34) and obtain (central-forces ⇒ Cauchy relation!)

$$c_{11} = 75.86\,\varepsilon/V_z;\quad c_{12} = c_{44} = 42.8\,\varepsilon/V_z;\quad 1/\varkappa = 53.82\,\varepsilon/V_z$$
$$\lambda = \mu = 32.92\,\varepsilon/V_z. \tag{19.28}$$

Using (19.25) and (19.28) it is found

$$\Delta V_\infty = -0.012\,55\,V_z;\quad \Delta V = -\,0.0225\,V_z;$$
$$\Delta V/\Delta V_\infty = 1.78 \tag{19.29}$$

whereas Kanzaki finds

$$\Delta V = -0.020\,V_z. \tag{19.29a}$$

The effective change for a Schottky defect is

$$\Delta V_{\text{eff}} = 0.9775\,V_z. \tag{19.30}$$

We have given here only a very short review on the structure of defects; in the following discussion of the dynamical properties we need only these few facts. Detailed knowledge involves large numerical calculations, which cannot be discussed here.*

## 20. Dynamical Properties of Defects. General Remarks. Localized Modes

If there are defects in the lattice, the equation of motion for the lattice ions can be written as

$$M_\mu^{m}\,\ddot{u}_{\mu\,i}^{m} = -\sum_{n\nu j} \Phi'^{\,m\,n}_{\mu\,\nu}{}_{i\,j}\, u_{\nu\,j}^{n} - \cdots. \tag{20.1}$$

Here we have assumed the adiabatic approximation again to be valid [27 B 1, 51 B 1]. This can be done with the same arguments as for the ideal lattice. Further we will restrict our remarks to the harmonic approximation for the moment. $M_\mu^{m}$ is the mass of the ion at $\boldsymbol{m}$, $\mu$, which now may be different for different $\boldsymbol{m}$. $\Phi'^{\,m\,n}_{\mu\,\nu}{}_{i\,j}$ are the coupling parameters of the lattice with defects. These defects cause the invariance of the displacement pattern against translations by multiples of basis vectors $\boldsymbol{a}^{(k)}$ not to be valid now. That means, the relations of sect. 2 remain valid in defect lattices, whereas those of sect. 3 do not. As the relation (3.2) guarantees the solutions of the ideal lattice equation of motion to be plane waves, the solutions of (20.1) cannot be exact planes waves.

Let us assume a single point defect in the lattice. If now a plane wave coming from the left e. g., meets the defect, this wave becomes scattered, leading to outgoing waves in all the directions. Thus the concept of *scattering states* has to be introduced for the description of the dynamics

* In defect crystals, the elastic constants change too. A general procedure of calculating the elastic constants of defect crystals is given in [67 L 2].

of defect lattices. On the other hand, a point defect in an otherwise infinite medium may allow for eigenstates with (exponentially) decreasing amplitudes when the distance from the defect increases, or in other words: as the solutions of (20.1) need not be plane waves with real wave vector $\boldsymbol{q}$, the vector $\boldsymbol{q}$ might become complex, giving an exponential decrease for the amplitudes. Such states are *localized states*.

If there are more defects in the lattice, there is a certain interaction between the defects. This gives a contribution to scattering and localized states and has to be taken into account, at least if the concentration of the defects becomes large, if the mean distance between the defects is not large compared to the lattice constant. This leads to the investigation of the *dynamics of statistically distributed defects* in the lattice.

Also *surface problems* belong to this category of lattice phenomena, for a lattice surface can be looked upon as a plane defect in an otherwise infinite lattice. Here we will give a review on the methods and results connected with the dynamics of defects in lattices. It is convenient to rewrite (20.1) in order to introduce the deviations from the ideal equation of motion for the lattice ions. A simple calculation gives

$$\sum_{n\nu j} \Phi^{\boldsymbol{m}\,\boldsymbol{n}}_{\substack{\mu\ \nu\\ i\ j}} u^{\boldsymbol{n}}_{\substack{\nu\\ j}} + M_\mu \ddot{u}^{\boldsymbol{m}}_{\substack{\mu\\ i}} = -\left(M^{\boldsymbol{m}}_\mu - M_\mu\right) \ddot{u}^{\boldsymbol{m}}_{\substack{\mu\\ i}} - \sum_{n\nu j}\left(\Phi'^{\boldsymbol{m}\,\boldsymbol{n}}_{\substack{\mu\ \nu\\ i\ j}} - \Phi^{\boldsymbol{m}\,\boldsymbol{n}}_{\substack{\mu\ \nu\\ i\ j}}\right) u^{\boldsymbol{n}}_{\substack{\nu\\ j}}\,. \tag{20.2a}$$

$M^{\boldsymbol{m}}_\mu - M_\mu$ is the deviation from the ideal lattice mass $M_\mu$ (independent of $\boldsymbol{m}$), $\Phi'^{\boldsymbol{m}\,\boldsymbol{n}}_{\substack{\mu\ \nu\\ i\ j}} - \Phi^{\boldsymbol{m}\,\boldsymbol{n}}_{\substack{\mu\ \nu\\ i\ j}}$ the corresponding deviation for the coupling constants. The right hand side of (20.2a) can be looked upon as a perturbation of the ideal lattice equation (left side) or as an inhomogeneity to the ideal lattice equation. Introduction of a perturbation operator

$$J^{\boldsymbol{m}\,\boldsymbol{n}}_{\substack{\mu\ \nu\\ i\ j}} = -\left(M^{\boldsymbol{m}}_\mu - M_\mu\right) \delta^{\boldsymbol{m}\,\boldsymbol{n}}_{\substack{\mu\ \nu\\ i\ j}} \frac{\partial^2}{\partial t^2} + \Phi^{\boldsymbol{m}\,\boldsymbol{n}}_{\substack{\mu\ \nu\\ i\ j}} - \Phi'^{\boldsymbol{m}\,\boldsymbol{n}}_{\substack{\mu\ \nu\\ i\ j}} \tag{20.3a}$$

gives

$$\sum_{n\nu j} \Phi^{\boldsymbol{m}\boldsymbol{n}}_{\substack{\mu\ \nu\\ i\ j}} u^{\boldsymbol{n}}_{\substack{\nu\\ j}} + M_\mu \ddot{u}^{\boldsymbol{m}}_{\substack{\mu\\ i}} = \sum_{n\nu j} J^{\boldsymbol{m}\boldsymbol{n}}_{\substack{\mu\ \nu\\ i\ j}}(t)\, u^{\boldsymbol{n}}_{\substack{\nu\\ j}}\,. \tag{20.2b}$$

This formulation is especially convenient in stationary problems where the time-dependence can be eliminated by the assumption*

$$u^{\boldsymbol{m}}_{\substack{\mu\\ i}} \sim \mathrm{e}^{-i\omega t}\,. \tag{20.4}$$

Then

$$J^{\boldsymbol{m}\boldsymbol{n}}_{\substack{\mu\ \nu\\ i\ j}}(\omega) = \left(M^{\boldsymbol{m}}_\mu - M_\mu\right) \omega^2 \delta^{\boldsymbol{m}\boldsymbol{n}}_{\substack{\mu\ \nu\\ i\ j}} + \Phi^{\boldsymbol{m}\boldsymbol{n}}_{\substack{\mu\ \nu\\ i\ j}} - \Phi'^{\boldsymbol{m}\boldsymbol{n}}_{\substack{\mu\ \nu\\ i\ j}} \tag{20.3b}$$

and

$$\sum_{n\nu j} \Phi^{\boldsymbol{m}\boldsymbol{n}}_{\substack{\mu\ \nu\\ i\ j}} u^{\boldsymbol{n}}_{\substack{\nu\\ j}} - M_\mu \omega^2 u^{\boldsymbol{m}}_{\substack{\mu\\ i}} = \sum_{n\nu j} J^{\boldsymbol{m}\boldsymbol{n}}_{\substack{\mu\ \nu\\ i\ j}}(\omega) \cdot u^{\boldsymbol{n}}_{\substack{\nu\\ j}}\,. \tag{20.2c}$$

Because of the time-dependence in (20.3a), this formulation is sometimes inconvenient in time-dependent problems. Another possibility is to

* This corresponds to a Fourier-transformation of the equation with respect to time.

multiply Eq. (20.1) by $M_\mu / M^{m}_{\mu}$ and then to introduce mass- and coupling-*deviations*. We get again an equation of type (20.2b)

$$\sum_{n\nu j} \Phi^{mn}_{\substack{\mu\nu\\ ij}}\, u^{n}_{\substack{\nu\\ j}} + M_\mu\, \ddot{u}^{m}_{\substack{\mu\\ i}} = \sum_{n\nu j} \hat{J}^{mn}_{\substack{\mu\nu\\ ij}}\, u^{n}_{\substack{\nu\\ j}} \qquad (20.2\text{d})$$

but with time-independent perturbation

$$\hat{J}^{mn}_{\substack{\mu\nu\\ ij}} = \frac{M^{m}_{\mu} - M_\mu}{M^{m}_{\mu}}\, \Phi^{mn}_{\substack{\mu\nu\\ ij}} + \frac{M_\mu}{M^{m}_{\mu}} \left(\Phi^{mn}_{\substack{\mu\nu\\ ij}} - \Phi'^{mn}_{\substack{\mu\nu\\ ij}}\right). \qquad (20.3\text{c})$$

The disadvantage now is the non-symmetry of $\hat{J}^{mn}_{\substack{\mu\nu\\ ij}}$ in the pairs $m\,\mu\,i$ and $n\,\nu\,j$ whereas (20.3a, b) is symmetric in these pairs. Both formulations are equivalent and it is a question of convenience only whether (20.3 a, b) or (20.3c) is used. If one is interested in stationary problems of single point defects (localized states, scattering states), (20.3b) seems to be most adequate. In time-dependent problems or in problems with a statistical distribution of defects, (20.3c) is probably more convenient.

The general solution can be given in terms of Greens function, similar to the procedure in the static case (Sect. 18). We define the Greens function by

$$\sum_{n\nu j} \Phi^{mn}_{\substack{\mu\nu\\ ij}}\, G^{np}_{\substack{\nu\varkappa\\ jk}}(t) + M_\mu\, \ddot{G}^{mp}_{\substack{\mu\varkappa\\ ik}}(t) = \delta^{mp}_{\substack{\mu\varkappa\\ ik}}\, \delta(t) \qquad (20.5\text{a})$$

in the time-dependent case or by the corresponding equation

$$\sum_{n\nu j} \Phi^{mn}_{\substack{\mu\nu\\ ij}}\, G^{np}_{\substack{\nu\varkappa\\ jk}}(\omega) - M_\mu\, \omega^2\, G^{mp}_{\substack{\mu\varkappa\\ ik}}(\omega) = \delta^{mp}_{\substack{\mu\varkappa\\ ik}}, \qquad (20.5\text{b})$$

for stationary problems. A special solution $\overset{s}{u}{}^{m}_{\substack{\mu\\ i}}$ of the inhomogeneous equation (20.2) can then be gotten by a convolution of the Greens function with the inhomogeneity:

$$\overset{s}{u}{}^{m}_{\substack{\mu\\ i}}(t) = \sum_{\substack{n\nu j\\ p\varkappa k}} \int \mathrm{d}t'\, G^{mn}_{\substack{\mu\nu\\ ij}}(t-t')\, J^{np}_{\substack{\nu\varkappa\\ jk}}(t')\, u^{p}_{\substack{\varkappa\\ k}}(t') \qquad (20.6\text{a})$$

in the time-dependent case or

$$\overset{s}{u}{}^{m}_{\substack{\mu\\ i}} = \sum_{\substack{n\nu j\\ p\varkappa k}} G^{mn}_{\substack{\mu\nu\\ ij}}(\omega)\, J^{np}_{\substack{\nu\varkappa\\ jk}}(\omega)\, u^{p}_{\substack{\varkappa\\ k}} \qquad (20.6\text{b})$$

in the stationary case. The general solution is a sum of a general solution of the corresponding homogeneous equation $\overset{\circ}{u}{}^{m}_{\substack{\mu\\ i}}$ and the solution $\overset{s}{u}{}^{m}_{\substack{\mu\\ i}}$ of the inhomogeneous equation. However, because the inhomogeneity contains the displacements, the $u^{p}_{\substack{\varkappa\\ k}}$ at the right hand side of (20.6) are again the sum of $\overset{\circ}{u}{}^{p}_{\substack{\varkappa\\ k}}$ and $\overset{s}{u}{}^{p}_{\substack{\varkappa\\ k}}$; so the general solution is (e.g. in the stationary case)

$$\overset{\circ}{u}{}^{m}_{\substack{\mu\\ i}} + \overset{s}{u}{}^{m}_{\substack{\mu\\ i}} = \overset{\circ}{u}{}^{m}_{\substack{\mu\\ i}} + \sum_{\substack{n\nu j\\ p\varkappa k}} G^{mn}_{\substack{\mu\nu\\ ij}}\, J^{np}_{\substack{\nu\varkappa\\ jk}} \left(\overset{\circ}{u}{}^{p}_{\substack{\varkappa\\ k}} + \overset{s}{u}{}^{p}_{\substack{\varkappa\\ k}}\right) \qquad (20.7)$$

or

$$\mathring{\boldsymbol{u}} + \dot{\boldsymbol{u}} = \mathring{\boldsymbol{u}} + \mathbf{G}\,\mathbf{J}\,(\mathring{\boldsymbol{u}} + \dot{\boldsymbol{u}}) \tag{20.7a}$$

which is an integral or sum equation for the determination of $\dot{\boldsymbol{u}}$. The solution of (20.2), of course, is connected with certain boundary conditions. The boundary conditions can be satisfied by choosing the appropriate Greens function. In the stationary scattering problem, e.g. the boundary condition is the following: $\mathring{\boldsymbol{u}}$ represents an incoming plane wave; then $\dot{\boldsymbol{u}}$ must be outgoing scattered waves. One has to choose what is called the outgoing Greens function. In the discussion of localized states one has to use a Greens function which leads to decreasing amplitudes (with increasing distance from the defect).

The formal solution of (20.7) is

$$\dot{\boldsymbol{u}} = \frac{\mathbf{G}\,\mathbf{J}}{\mathbf{1} - \mathbf{G}\,\mathbf{J}}\,\mathring{\boldsymbol{u}} \tag{20.8a}$$

or

$$\boldsymbol{u} = \mathring{\boldsymbol{u}} + \dot{\boldsymbol{u}} = \mathring{\boldsymbol{u}} + \frac{\mathbf{G}\,\mathbf{J}}{\mathbf{1} - \mathbf{G}\,\mathbf{J}}\,\mathring{\boldsymbol{u}} = \frac{1}{\mathbf{1} - \mathbf{G}\,\mathbf{J}}\,\mathring{\boldsymbol{u}}\,. \tag{20.8b}$$

If the Greens function is known, (20.8) represents the whole solution of the problem; thus all difficulties lie in finding the appropriate Greens function, which is a numerical problem for every lattice or model of a lattice. Some general properties of the Greens function will be discussed in the next section. For the moment we assume it as known.

The scattering problem results in the calculation of the scattering amplitude. $\mathring{\boldsymbol{u}}$ represents an incoming plane wave with given wave vector $\boldsymbol{q}$, polarization $\sigma$ and frequency $\omega(\boldsymbol{q}\sigma)$ in the stationary case. We have to use the asymptotic expansion of the Greens function. Then the coefficients of $e^{iqR}/R$ define the scattering amplitude; $R$ is the distance in which the scattered wave is observed, assuming the defect in the origin. Difficulties arise from the denominator $\mathbf{1} - \mathbf{G}\,\mathbf{J}$, which in most cases is not easy to handle. Sometimes it is useful to approximate (20.8) by the expansion (Born series)

$$\boldsymbol{u} = \mathring{\boldsymbol{u}} + \sum_{\nu=1}^{\infty} (\mathbf{G}\,\mathbf{J})^{\nu}\,\mathring{\boldsymbol{u}} \tag{20.9}$$

valid for small perturbations $\mathbf{J}$ or better, if all the elements of $\mathbf{G}\,\mathbf{J}$ are small compared to unity. The neglect of all higher order terms gives the first Born approximation

$$\boldsymbol{u} = \mathring{\boldsymbol{u}} + \mathbf{G}\,\mathbf{J}\,\mathring{\boldsymbol{u}}\,. \tag{20.10}$$

But it might happen, that $\mathbf{1} - \mathbf{G}\,\mathbf{J}$ becomes very small, so that every expansion is a bad approximation. This happens if the real part of $\mathbf{1} - \mathbf{G}\,\mathbf{J}$ vanishes and the imaginary part is also small. Such scattering states have large scattering amplitudes and are called *resonance states*. They are similar to localized states in a certain respect. The details are discussed in sect. 22.

*Localized states* have decreasing amplitudes with increasing distance from the defect. Therefore they cannot contain wave like contributions, which arise from the solutions $\mathring{\boldsymbol{u}}$ of the homogeneous equation. In order to have nonvanishing amplitudes $\boldsymbol{u}$ or $\dot{\boldsymbol{u}}$ for vanishing $\mathring{\boldsymbol{u}}$ $(\mathbf{1} - \mathbf{GJ})^{-1}$ must contain singularities, or in other words the determinant

$$\|\mathbf{1} - \mathbf{GJ}\| = 0 . \tag{20.11}$$

This is the first condition for localized states, being defined by the real poles of the scattering amplitude. The second condition arises from the fact, that the amplitudes have to be truly decreasing. This gives certain conditions for the Greens function or its parameters, the frequencies. Let us take the Greens function in a special representation, using an expansion with respect to the eigenvectors of the homogeneous equation

$$G^{m\,p}_{\substack{\mu\,\varkappa\\ i\,k}}(\omega) = \frac{1}{Ns\sqrt{M_\mu M_\varkappa}} \sum_{\boldsymbol{q}\sigma} e_{\substack{\mu\\ i}}(\boldsymbol{q}\,\sigma)\, e^*_{\substack{\varkappa\\ k}}(\boldsymbol{q}\,\sigma)\, \mathrm{e}^{i\boldsymbol{q}(\boldsymbol{R}^m - \boldsymbol{R}^p)} \cdot a(\boldsymbol{q}\,\sigma) \tag{20.12}$$

because from (20.5) it follows that $G$ must be a function of $\boldsymbol{R}^m - \boldsymbol{R}^p$. Inserting (20.12) into (20.5b), using the secular equation of the ideal lattice

$$\omega^2(\boldsymbol{q}\,\sigma)\, e_{\substack{\mu\\ i}}(\boldsymbol{q}\,\sigma) = \sum_{mn\nu j} \frac{\Phi^{m\,n}_{\substack{\mu\,\nu\\ i\,j}}}{\sqrt{M_\mu M_\nu}}\, \mathrm{e}^{i\boldsymbol{q}(\boldsymbol{R}^n - \boldsymbol{R}^m)}\, e_{\substack{\nu\\ j}}(\boldsymbol{q}\,\sigma) \tag{20.13}$$

multiplying by $\{M_\varkappa/M_\mu\}^{1/2}\, e^*_{\substack{\mu\\ i}}(\boldsymbol{q}'\,\sigma')\, \mathrm{e}^{-i\boldsymbol{q}'\boldsymbol{R}^m}$ and summing over $\boldsymbol{m}$, $\mu$, $\boldsymbol{i}$ we have

$$a(\boldsymbol{q}\,\sigma) = \frac{1}{\omega^2(\boldsymbol{q}\,\sigma) - \omega^2}$$

or

$$G^{m\,p}_{\substack{\mu\,\varkappa\\ i\,k}}(\omega) = \frac{1}{sN\sqrt{M_\mu M_\varkappa}} \sum_{\boldsymbol{q}\sigma} \frac{e^{\mu}_{i}(\boldsymbol{q}\,\sigma)\, e^{\varkappa *}_{k}(\boldsymbol{q}\,\sigma)}{\omega^2(\boldsymbol{q}\,\sigma) - \omega^2}\, \mathrm{e}^{i\boldsymbol{q}(\boldsymbol{R}^m - \boldsymbol{R}^p)}. \tag{20.14}$$

If frequencies and polarization vectors of the ideal lattice are known, $\mathbf{G}$ can be calculated. In the limit of an infinite crystal, the sum over $\boldsymbol{q}$ can be replaced by an integral over the first Brillouinzone. If $\omega$ has a certain value in the region of ideal lattice frequencies, there is a pole in the region of integration, to which there corresponds a special real value of $\boldsymbol{q}$. This gives a contribution to the Greens function and from the form (20.14) it can be seen, that the Greens function has a wave-like, not a localized character. This holds for every branch $\sigma$. Therefore the frequencies $\omega$ necessarily must lie outside the frequency branches of the ideal lattice frequencies in order to have localized states. These states can be characterized by complex $\boldsymbol{q}$-values.

In very special cases it might be, that there are localized states with frequencies in the region of ideal lattice frequencies. It might happen, if the maximum frequency of a certain branch $\sigma'$ is higher than the maximum frequencies of all other branches $\sigma \neq \sigma'$, and the branch $\sigma'$ is separated from branches $\sigma \neq \sigma'$ for all values of $\boldsymbol{q}$. If then the product of that part of $\mathbf{G}$ which is related to $\sigma'$ with $\mathbf{J}$ is zero (orthogonality of a certain part of **G** with **J**), then there might be localized states with frequencies between the maximum frequencies of branches $\sigma \neq \sigma'$ and that of branch $\sigma'$. But such a case is not known to us for point defects in ionic lattices*.

Thus in Bravais-lattices localized states have frequencies above the band of ideal lattice frequencies; in nonprimitive lattices there might be localized states with frequencies in the gap between acoustical and optical frequency bands or in the gap between different optical bands or above the highest optical frequency, provided the bands in question do not overlap and the gaps vanish. A very simple argument, why this is so, is the following. Suppose the defect lattice ion is excited to vibrate with a frequency belonging to the ideal lattice frequencies, then the lattice is able to pick up this frequency and to carry over the displacements through the whole lattice without damping. If the frequencies are outside the lattice frequencies, the lattice cannot pick them up.

Even if Greens function is known, the determination of the localized states is difficult because of the high dimensionality of the problem. It can be seen immediately, that the dimension of the secular equation (20.11) is given by the dimension of **J** which in turn is determined by the range of the defect, or of the forces of the defect ion. We have seen in sect. 18, that the defect ion causes a displacement field which decreases as $R^{-2}$ with distance $R$ from the point defect. Thus the exact range of the defect is infinite. But the change in force-constants at larger distances from the defect is very small and influences the localized states only negligably. In most cases it will be sufficient to limit the range of the defect to the range of interacting forces, generally assumed to act between nearest and next-nearest neighbors. In fact only such models have been discussed. The secular problem is thus of dimension $3\,(1 + z_1 + z_2 + \cdots)$, if there is one central defect ion and $z_1$ next neighbors, $z_2$ second neighbors etc.

Another possibility of simplifying the solution of (20.11) is related to the symmetry of the defect. Though the translation symmetry of the ideal lattice is cancelled by the defect, the point group symmetry of the lattice site is conserved, at least in general. Thus a vacancy or substitutional impurity has the point group symmetry of the host lattice**. An interstitial might have also the point group symmetry of the host lattice or a lower symmetry (dumbbell-configuration of an interstitial in cubic-face-centered lattices). Group theory provides a powerful tool in using the symmetry of the defect for the determination of the symmetries of the localized eigenstates. The procedure has been explained largely in another review on this subject and we refer to this paper for details [64 L 1]. The dimension of the problem to be solved is thus reduced to a small number, say 1 to 6 in general.

---

* This statement is related only to ionic lattices. In molecular lattices with librational and translational degrees of freedom [sec. 10—12], localized states of librational motions can have frequencies in translational bands and vice versa [65 D 2]. Also surface states have frequencies in the band of ideal lattice frequencies.

** There are exceptions, if the electronic ground state of the impurity in the lattice is degenerated (Jahn-Teller-effect).

The only problem we are left with is the calculation of Greens function, which has been done only for the following three models: i) a simple cubic lattice with nearest-neighbor central and non-central interaction [60 M 1] and ii) a cubic-face-centered lattice with nearest-neighbor-central interaction [64 K 1], and iii) a cubic-body-centered lattice model [66 K 1].

A simple approximation method for the calculation of frequencies of localized states is by means of a variational method. The stationary equation corresponding to (20.1) can be formulated as a variational equation, we introduce

$$\boldsymbol{\Phi}' = \left\{\Phi'^{m\,n}_{\mu\,\nu}{}_{i\,j}\right\}; \quad \mathbf{M}' = \left\{M^{m}_{\mu}\,\delta^{m\,n}_{\mu\,\nu}{}_{i\,j}\right\}$$

as the operators describing the defect and

$$\boldsymbol{u} = \left\{u^{m}_{\mu}{}_{i}\right\}$$

as the state vector. Then

$$\mathbf{M}'\,\omega^2\,\boldsymbol{u} = \boldsymbol{\Phi}'\,\boldsymbol{u} \tag{20.15}$$

is the stationary equation to (20.1). The corresponding variational expression to be extremalized is

$$\Omega^2[\boldsymbol{u}] = \frac{(\boldsymbol{u},\,\boldsymbol{\Phi}'\,\boldsymbol{u})}{(\boldsymbol{u},\,\mathbf{M}'\,\boldsymbol{u})}, \tag{20.16}$$

called Rayleigh's quotient. Choosing a trial vector $\boldsymbol{u}$ with free parameters, inserting this into (20.16) and varying with respect to the parameters determines the best fit for these parameters. Using these fitted parameters, the quotient (20.16) gives the best approximation of frequencies with the chosen trial vector. The details of this method are again discussed in [64 L 1]. In the calculation of localized states we expect decreasing amplitudes with increasing distance from the defect. If the states are rather strongly localized, we can assume that only displacement amplitudes of near neighbors are different from zero. Thus a trial state vector contains amplitudes being different from zero only for near neighbors, e. g. in the range of interacting forces. This limits the number of free parameters.

We have seen, that the continuously distributed frequencies of the lattice cover the region from $\omega = 0$ to a maximum frequency $\omega_m$. The localized states have frequencies above $\omega_m$ (in Bravais-lattices, the extrapolation to other lattices is obvious). Thus a variational procedure would be an approximation to the highest localized frequency, giving a smaller value than the exact one. But using the symmetry of the defect, we can choose the trial vectors already as to have the exact symmetry of the possible eigenstates. With this choice we cannot only approximate the localized state with the highest frequency but also all others. It is possible to approximate all eigenstates by the variational procedure, the frequencies thus calculated being smaller as the exact ones. This procedure has been used in most cases for the determination of localized states [63 L 1, 64 D 1]. Other methods are essentially equivalent to a variational procedure [61 L 3, 63 L 2]. Exact calculations have been done only for linear lattices (by different methods)

[56 M 1, 57 H 2, 64 L 1] and the three above mentioned three-dimensional cases (with Greens-functions) [63 L 1, 66 K 1]. There exist a number of review papers on the different models which have been investigated [56 L 2, 64 L 1, 64 M 1, 2; 66 M 2, 66 L 1]. Here we will discuss only some general features of the localized states, using the simplest possible models.

The simplest form of a defect is an isotopic defect, which we assume to be in the origin of the coordinate system. The mass of the isotope may be $M'_\mu$, that of the host lattice ions $M_\mu$ and the relative change is

$$\varepsilon = (M'_\mu - M_\mu)/M_\mu \,. \tag{20.17}$$

Then

$$J^{m\,n}_{\substack{\mu\,\nu\\ i\,j}} = \varepsilon\, M_\mu\, \omega^2\, \delta^{m\,n}_{\substack{\mu\,\nu\\ i\,j}}\, \delta^{m\,0}_{\mu\,1} \tag{20.18}$$

if the isotope belongs to the sublattice $\mu = 1$. From (20.8) we have for localized states ($\mathring{u} \equiv 0$)

$$\mathring{u}^{m}_{\substack{\mu\\ i}} = \varepsilon\, M_1\, \omega^2\, G^{m\,0}_{\substack{\mu\,1\\ i\,j}}\, \mathring{u}^{0}_{\substack{1\\ j}} \,. \tag{20.19}$$

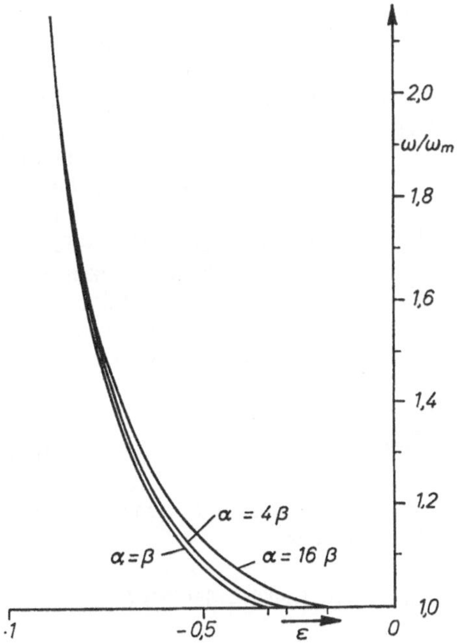


Fig. 6.3. Frequencies of localized modes in a simple cubic lattice with nearest neighbor interaction, due to an isotopic defect with relative change of mass $\varepsilon = (M' - M)/M$; $\alpha$ and $\beta$ are central and non-central force constants

This gives the amplitudes of all the ions except the isotopic one. For $m = 0$, $\mu = 1$ we have the secular equation determining the frequency as function of $\varepsilon$:

$$\left\| \delta_{ij} - \varepsilon\, M_1 \omega^2\, G^{0\,0}_{\substack{1\,1\\ i\,j}} \right\| = 0 \,. \tag{20.20}$$

In cubic crystals each of the coordinate directions is equivalent, therefore $G^{0\,0}_{\substack{1\,1\\i\,j}}$ is diagonal with respect to the lower indices and all the diagonal elements are equal. The localized state of an isotopic defect is triply degenerated with frequency from

$$1 = \varepsilon M_1 \omega^2 G^{00}_{\substack{11\\11}} . \qquad (20.20\,a)$$

Consider first a Bravais-lattice. The indices corresponding to $\mu$, $\nu$ can be dropped. The frequencies of localized states are larger than all the ideal frequencies. Therefore from (20.14) it can be seen that $G^{00}_{11} < 0$ for localized states. Thus $\varepsilon < 0$ or $M' < M$ in order to have localized states at isotopes in Bravais-lattices. For $\varepsilon \to -1$ or $M' \to 0$ the isotopic degrees of freedom vanish and the frequency of the localized state must reach infinity. Fig. 6.3 shows the frequency versus $\varepsilon$ for a simple cubic crystal with next-neighbor interaction, where the Greens function is known [60 M 1].

In non-primitive lattices the connection between frequency and relative change in mass is qualitatively the same for localized states with frequencies above the highest optical band. However, in the band gaps there might be localized states in case of $\varepsilon < 0$ as well as $\varepsilon > 0$; in lattices with two ions in the unit cell the gap states occur for $\varepsilon < 0$, if the isotope belongs to the sublattice with the heavier mass; they occur for $\varepsilon > 0$, if the isotope belongs to the sublattice with the lighter mass. The details can be found in some other review papers [64 L 1, 66 L 1, 66 M 2].

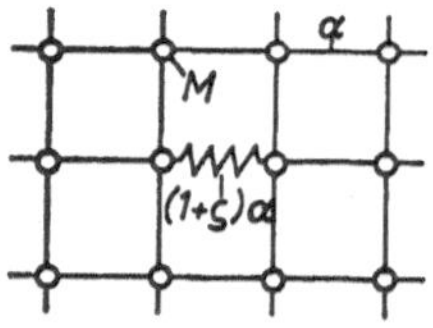

Fig. 6.4. Simple cubic lattice with a single central force constant different from the ideal lattice force constants

To discuss the effect of force-constant changes, we use again the simplest model: a simple cubic lattice with nearest neighbor interaction. The coupling parameters of the ideal lattice are [I]

$$\Phi^{0\;100}_{i\;j} = -\begin{Bmatrix} \alpha & 0 & 0 \\ 0 & \beta & 0 \\ 0 & 0 & \beta \end{Bmatrix} . \qquad (20.21)$$

Now suppose only one force constant between ion $\boldsymbol{m} = 0$ and $\boldsymbol{m} = 100$ is changed (Fig. 6.4). Then it is

$$J^{\boldsymbol{m}\,\boldsymbol{n}}_{i\;j} = \alpha \zeta \delta_{ij} \delta_{i1} \{\delta^{\boldsymbol{m}\,0} - \delta^{\boldsymbol{m}\,100}\} \{\delta^{\boldsymbol{n}\,100} - \delta^{\boldsymbol{n}\,0}\} , \qquad (20.22)$$

where

$$\zeta = (\alpha' - \alpha)/\alpha \qquad (20.23)$$

is the relative change of the force constant. Using*

$$G^{\boldsymbol{m}\,\boldsymbol{n}}_{i\;j} = G^{\boldsymbol{m}-\boldsymbol{n}\;0}_{i\qquad j} \quad \text{and} \quad G^{\boldsymbol{m}\,\boldsymbol{n}}_{i\;j} = +\,G^{-\boldsymbol{m}\,-\boldsymbol{n}}_{i\qquad j} \qquad (20.24)$$

* The second relation only holds in lattices where each ion is a center of inversion (Bravais-lattices e. g.).

and the abbreviation

$$A = \alpha\zeta\left(G^{00}_{11} - G^{0\,100}_{1\,1}\right) \tag{20.25}$$

we have the secular equation

$$\begin{Bmatrix} A+1 & -A \\ -A & A+1 \end{Bmatrix} \begin{Bmatrix} u^0_1 \\ u^{100}_1 \end{Bmatrix} = 0\,. \tag{20.26}$$

Symmetry requires, or direct inspection shows the eigenstates to be*

$$\begin{Bmatrix} u^0_1 \\ u^{100}_1 \end{Bmatrix} = \overset{A_u\text{-state}}{\frac{1}{\sqrt{2}}\begin{Bmatrix} 1 \\ 1 \end{Bmatrix}} \text{ or } \overset{A_g\text{-state}}{\frac{1}{\sqrt{2}}\begin{Bmatrix} 1 \\ -1 \end{Bmatrix}}\,. \tag{20.27}$$

Thus (20.26) can be diagonalized by means of a unitary matrix

$$U = \frac{1}{\sqrt{2}}\begin{Bmatrix} 1 & 1 \\ 1 & -1 \end{Bmatrix}\,. \tag{20.28}$$

The frequencies then follow from

$$\begin{aligned} A_u&: \quad 1 = 0 \text{ (impossible!)} \\ A_g&: \quad 1 = -2A = -2\alpha\zeta\left(G^{00}_{11} - G^{0\,100}_{1\quad 1}\right)\,. \end{aligned} \tag{20.29}$$

There is no localized $A_u$-state, which is clear, because the force-constant is not stressed in this motion. From (20.14) $G^{00}_{11} < 0$ and $\left|G^{00}_{11}\right| > \left|G^{0\,100}_{1\quad 1}\right|$. Further $\alpha > 0$ from the stability of the lattice. Therefore the $A_g$-state

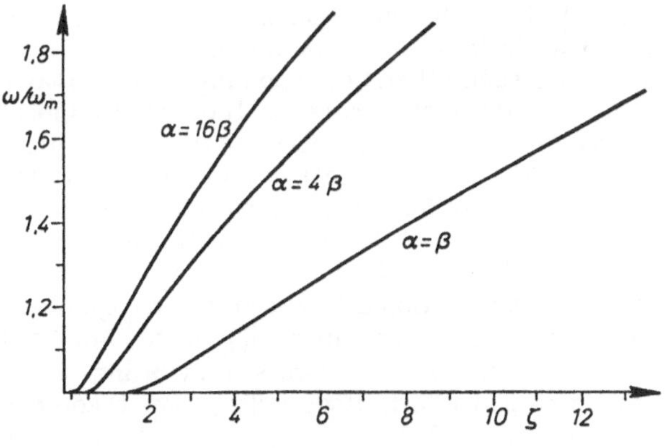


Fig. 6.5. Frequencies of localized states in a simple cubic lattice with a single force constant $\alpha'$ different from ideal lattice values. $\varphi = (\alpha' - \alpha)/\alpha$

occurs, if $\zeta > 0$ and sufficiently large. The force-constants at the defect must be larger than those of the host lattice to have localized states. This is valid also for more complicated defects. Fig. 6.5 shows the calculated frequencies as function of $\zeta$.

* $A_g$, $A_u$ are group-theoretical notations for vibration states (see [64 L 1]). g denotes states which are symmetric with respect to inversion, u denotes states, which are antisymmetric. A means, that there is no degeneracy.

A simple consideration already shows the general features. In an Einstein model [I], the frequency of lattice atoms is given by

$$\omega^2 \sim \alpha/M \tag{20.30}$$

where $\alpha$ represents the force-constants and $M$ the masses. Localized states have frequencies above the ideal frequency band (in Bravais-lattices). To have such frequencies from (20.30)* the force-constants near the defect must become sufficiently large and the masses of the defect ions sufficiently small. This holds qualitatively for every model. Only in the band-gaps of nonprimitive lattices more complicated situations arise.

Many models have been discussed in detail and we refer to the papers for the details. Calculations have been done for impurities, vacancies and interstitials in Bravais-lattices as well as in ionic and homopolar lattices. The general features are always the same, quantitative results depend sensitively on the defects or the models used. In Bravais-lattices it turns out, that a vacancy should not have localized states with reasonable assumptions on the force-constants [64 L 1, 66 M 2, 66 L 1]. On the other hand, localized states should occur near interstitials and in many cases also at impurities. In non-primitive lattices there might be localized states at vacancies with frequencies in the band-gaps; the same holds for interstitials and impurities [64 L 1, 66 M 2, 66 L 1]. Most quantitative calculations have been done using the linear model of a lattice. These were the first investigations on the dynamics of defect lattices at all [56 M 1, 57 H 2].

In molecular lattices the situation is slightly different. There are librational and translational types of motion (sect. 10—12), which in general interact with each other. Correspondingly there may be localized modes of librational and translational type. It occurs, that localized modes of librational type can have frequencies in the region of translational frequencies of the ideal molecular lattice and vice versa. This has been investigated quantitatively for a linear molecular lattice [65 D 2].

In recent investigations, molecular impurities in ionic lattices have become of considerable interest. Such molecular defects again might have librational and translational localized modes. Quantitative calculations are difficult because the interacting forces are hardly known. But the general features can be discussed by using simple pictures, without much numerical effort. Especially group theoretical methods provide a powerful tool [64 L 1].

Experiments on localized states have been done in many cases. They use the interaction of phonons, especially the localized phonons, with other kinds of radiation, mainly with photons (infrared-absorption) and neutrons (scattering of thermal neutrons at lattice ions). The theoretical background of these interactions is reviewed in chapter VII. Infrared-absorption-measurements have been done mainly for infrared-active states in ionic lattices [60 S 3, 62 F 2, 65 F 1, 2], using $H^-$- and $D^-$-ions

* The Einstein-model is a good approximation for localized states. It can be looked upon as the simplest variational Ansatz.

as impurities.* Other sorts of defects have been investigated also (molecular impurities etc.) [66 S 4, 66 P 2]. Localized modes should give a contribution to Raman-absorption, if the modes are Raman-active. Measurements have been done in a few cases only [64 S 1].

Neutron scattering experiments could give quite a lot of information on localized states, if certain conditions are satisfied (sect. 27). Up till now only two experiments have been performed [62 M 1, 66 M 3]. In these experiments all the localized states can be investigated, because all the vibrational states contribute to neutron-scattering, contrary to optical absorption measurements, which only give information on active states.

Before dealing with the scattering of phonons at defects and the interaction of defects, we will discuss the Greens function and its properties in some detail, because the discussion can be done best in terms of the Greens function.

## 21. The Phonon Greens Function

For the investigations of the dynamics of lattices we have seen the Greens function to be a powerful tool. Especially in scattering problems the Greens function formulation is the best one. We will discuss certain aspects of the one-phonon-Greens function here**. The starting point is the equation of motion for the lattice displacements in the harmonic approximation:

$$\sum_{\boldsymbol{n}\nu j} \Phi_{\substack{\mu\ \nu \\ i\ j}}^{\boldsymbol{m}\,\boldsymbol{n}} u_{\substack{\nu \\ j}}^{\boldsymbol{n}} + M_\mu \ddot{u}_{\substack{\mu \\ i}}^{\boldsymbol{m}} = 0 . \tag{21.1}$$

The displacements $u_{\substack{\mu \\ i}}^{\boldsymbol{m}}$ can be expanded with respect to the eigensolutions of this equation [I, II, (16.9)]

$$u_{\substack{\mu \\ i}}^{\boldsymbol{m}} = \sqrt{\frac{\hbar}{2\,s\,N\,M_\mu}} \sum_{\boldsymbol{q}\sigma}{}' \frac{1}{\sqrt{\omega(\boldsymbol{q}\,\sigma)}} e_{\substack{\mu \\ i}}(\boldsymbol{q}\sigma)\, e^{i\,\boldsymbol{q}\boldsymbol{R}^{\boldsymbol{m}}} \left[ b_\sigma^{\boldsymbol{q}}(t) - b^{+\,-\boldsymbol{q}}_{\ \ \sigma}(t) \right], \tag{21.2}$$

where the equation of motion for the amplitudes $b_\sigma^{\boldsymbol{q}}$ is given by

$$\dot{b}_\sigma^{\boldsymbol{q}} + i\,\omega(\boldsymbol{q}\sigma)\, b_\sigma^{\boldsymbol{q}} = 0; \quad \dot{b}_\sigma^{\boldsymbol{q}+} - i\,\omega(\boldsymbol{q}\sigma)\, b_\sigma^{\boldsymbol{q}+} = 0 . \tag{21.3}$$

The expansion in (21.2) is with respect to running waves; the center of mass-motion is not contained in (21.2), but has to be treated separately. It is irrelevant in the following.

In introducing quantum-mechanics, the displacements become operators for the displacement field. Then also the $b_\sigma^{\boldsymbol{q}}$, $b^{+\,\boldsymbol{q}}_{\ \sigma}$ become operators;

* Not all the localized states are infrared-active. The active states must have a dipole-moment, i. e. the corresponding displacement field must transform as a polar vector in coordinate transformations (see [64 L 1]).

** for zero temperature, which is the pure mechanical Greens function. For the thermodynamic Greens functions see Appendix.

they have the meaning of annihilation and creation operators with the commutation relations [II]

$$\left[b_\sigma^{\boldsymbol{q}}, b_{\sigma'}^{\boldsymbol{q}'+}\right]_- = \delta_{\boldsymbol{q}\boldsymbol{q}'}\,\delta_{\sigma\sigma'};\quad \left[b_\sigma^{\boldsymbol{q}}, b_{\sigma'}^{\boldsymbol{q}'}\right]_- = \left[b_\sigma^{\boldsymbol{q}+}, b_{\sigma'}^{\boldsymbol{q}'+}\right]_- = 0\,. \tag{21.4}$$

The Hamiltonian of the free phonon system is

$$\mathscr{H} = \sum_{\boldsymbol{m} i \mu} \frac{1}{2 M_\mu} \left(\overset{\boldsymbol{m}}{p}{}_i^\mu\right)^2 + \frac{1}{2} \sum_{\substack{\boldsymbol{m}\boldsymbol{n}\\ \mu\nu i j}} \Phi_{i\,j}^{\overset{\boldsymbol{m}\boldsymbol{n}}{\mu\,\nu}}\, \overset{\boldsymbol{m}}{u}{}_i^\mu\, \overset{\boldsymbol{n}}{u}{}_j^\nu \tag{21.5a}$$

expressed with the field operators $\overset{\boldsymbol{m}}{u}{}_i^\mu$, $\overset{\boldsymbol{m}}{p}{}_i^\mu = M_\mu \overset{\boldsymbol{m}}{\dot{u}}{}_i^\mu$ or

$$\mathscr{H} = \sum_{\boldsymbol{q}\sigma} \hbar\,\omega(\boldsymbol{q}\sigma)\left[b_\sigma^{+\boldsymbol{q}}\, b_\sigma^{\boldsymbol{q}} + 1/2\right] \tag{21.5b}$$

if expressed by the annihilation and creation operators.* The eigenstates of (21.5b) can be characterized by the occupation numbers of the phonons $n(\boldsymbol{q}\sigma) = 0, 1, 2, \ldots$ and can be written

$$|\ldots n_{\boldsymbol{q}\sigma}\ldots\rangle = \prod_{\boldsymbol{q}\sigma} \frac{1}{\sqrt{n(\boldsymbol{q}\,\sigma)!}} \left\{b_\sigma^{+\boldsymbol{q}}\right\}^{n(\boldsymbol{q}\,\sigma)} |0\rangle \tag{21.6}$$

where $|0\rangle$ is the zero-phonon (ground) state. The time-dependence of the operators in the Heisenberg-picture, which we will use here, then is

$$b_\sigma^{\boldsymbol{q}}(t) = e^{\frac{i}{\hbar}\mathscr{H} t}\, b_\sigma^{\boldsymbol{q}}(0)\, e^{-\frac{i}{\hbar}\mathscr{H} t};\quad \overset{\boldsymbol{m}}{u}{}_i^\mu(t) = e^{\frac{i}{\hbar}\mathscr{H} t}\, \overset{\boldsymbol{m}}{u}{}_i^\mu(0)\, e^{-\frac{i}{\hbar}\mathscr{H} t}\,. \tag{21.7}$$

The commutation relations for the field operators follow from the above relations:

$$\begin{aligned}
&\left[\overset{\boldsymbol{m}}{u}{}_i^\mu(t), \overset{\boldsymbol{n}}{u}{}_j^\nu(t')\right]_- \\
&\quad = -\frac{i\hbar}{sN\sqrt{M_\mu M_\nu}} \sum_{\boldsymbol{q}\sigma} \frac{\sin[\omega(\boldsymbol{q}\,\sigma)\,(t-t')]}{\omega(\boldsymbol{q}\,\sigma)}\, e_i^\mu\, e_j^{\nu *}(\boldsymbol{q}\sigma)\, e^{i\boldsymbol{q}(\boldsymbol{R}^m - \boldsymbol{R}^n)}\,. \\
&\left[\overset{\boldsymbol{m}}{\dot{u}}{}_i^\mu(t), \overset{\boldsymbol{n}}{u}{}_j^\nu(t')\right]_- \\
&\quad = -\frac{i\hbar}{sN\sqrt{M_\mu M_\nu}} \sum_{\boldsymbol{q}\sigma} \cos[\omega(\boldsymbol{q}\sigma)\,(t-t')]\, e_i^\mu\, e_j^{\nu *}(\boldsymbol{q}\sigma)\, e^{i\boldsymbol{q}(\boldsymbol{R}^m - \boldsymbol{R}^n)}\,.
\end{aligned} \tag{21.8}$$

Further we need the orthogonality of the eigenvectors

$$\begin{aligned}
\frac{1}{sN} \sum_{\boldsymbol{m}\mu i} e_i^\mu(\boldsymbol{q}\sigma)\, e^{i\boldsymbol{q}\boldsymbol{R}^m}\, e_i^{\mu *}(\boldsymbol{q}'\sigma')\, e^{-i\boldsymbol{q}'\boldsymbol{R}^m} &= \delta_{\boldsymbol{q}\boldsymbol{q}'}\,\delta_{\sigma\sigma'} \\
\sum_{\mu i} e_i^\mu(\boldsymbol{q}\sigma)\, e_i^{\mu *}(\boldsymbol{q}\sigma') &= s\,\delta_{\sigma\sigma'}
\end{aligned} \tag{21.9a}$$

* The zero-point energy $\sum_{\boldsymbol{q}\sigma} \hbar\,\omega(\boldsymbol{q}\,\sigma)/2$ is not essential here and can be dropped. Further, the center-of-mass-motion is ignored in the following.

and the completeness

$$\frac{1}{sN}\sum_{\boldsymbol{q}\sigma} e_i^{\mu}(\boldsymbol{q}\sigma)\, e_j^{\nu *}(\boldsymbol{q}\sigma)\, e^{i\boldsymbol{q}(\boldsymbol{R}^m-\boldsymbol{R}^n)} = \delta^{m\,n}_{\mu\,\nu\atop i\,j}\sum_{\sigma} e_i^{\mu}(q\sigma)\, e_j^{\nu *}(q\sigma) = s\,\delta_{\mu\nu}\,\delta_{ij}. \tag{21.9b}$$

The polarization vectors are chosen to obey [I]:

$$e_i^{\mu}(-\boldsymbol{q}\sigma) = -e_i^{\mu *}(\boldsymbol{q}\sigma). \tag{21.10}$$

Eliminating the time-dependence from the $b_\sigma^{\boldsymbol{q}}$ in (21.2) and relabeling in the second term $-\boldsymbol{q}$ into $\boldsymbol{q}$ we have

$$u_i^{\mu\,m}(t) = \sqrt{\frac{\hbar}{2sNM_\mu}}\sum_{\boldsymbol{q}\sigma}\frac{1}{\sqrt{\omega(\boldsymbol{q}\sigma)}}\Big\{b_\sigma^{\boldsymbol{q}}\, e_i^{\mu}(\boldsymbol{q}\sigma)\, e^{i[\boldsymbol{q}\boldsymbol{R}^m-\omega(\boldsymbol{q}\sigma)t]} + b^{+}{}_\sigma^{\boldsymbol{q}}\, e_i^{\mu *}(\boldsymbol{q}\sigma)\, e^{-i[\boldsymbol{q}\boldsymbol{R}^m-\omega(\boldsymbol{q}\sigma)t]}\Big\}. \tag{21.11}$$

The Greens function can be defined by* (20.5a):

$$\sum_{n\nu j}\Phi^{m\,n}_{\mu\,\nu\atop i\,j}\, G^{n\,p}_{\nu\,\varkappa\atop j\,k} + M_\mu\, \ddot G^{m\,p}_{\mu\,\varkappa\atop i\,k} = \delta^{m\,p}_{\mu\,\varkappa\atop i\,k}\,\delta(t-t'). \tag{21.12}$$

We will show that a solution of (21.12) for the Greens function is given by

$$\begin{aligned} G^{m\,n}_{\mu\,\nu\atop i\,j}(t-t') &= \frac{i}{\hbar}\Big\langle 0\Big| T\, u_i^{\mu\,m}(t)\, u_j^{\nu\,n}(t')\Big|0\Big\rangle \\ &= \frac{i}{\hbar}\Big\langle 0\Big| u_i^{\mu\,m}(t)\, u_j^{\nu\,n}(t')\Big|0\Big\rangle\cdot\Theta(t-t') + \\ &\quad + \frac{i}{\hbar}\Big\langle 0\Big| u_j^{\nu\,n}(t')\, u_i^{\mu\,m}(t)\Big|0\Big\rangle\cdot\Theta(t'-t). \end{aligned} \tag{21.13}$$

$T$ is the time-ordering operator, which means that the displacement field operator with the larger time argument has to be put in front of that with the smaller argument; this fact can be expressed also with the step-function

$$\Theta(t) = \begin{cases} 1 & t > 0 \\ 0 & t < 0. \end{cases} \tag{21.14}$$

---

* The dots mean derivation with respect to $t$ (not $t'$). The Greens function is defined sometimes in a slightly different way (e. g. *Abrikosov* [63 A 1]):

$$\Phi^{mn}\, G^{np} + M_\mu\, \ddot G^{mp} = -\Phi^{mp}\,\delta(t-t')$$

or

$$G^{np} + (\Phi^{-1})^{nm}\, M_\mu\, \ddot G^{mp} = -\delta^{np}\,\delta(t-t')$$

with

$$G^{mp} = -\frac{i}{\hbar}\langle 0\,|T\,\varphi^m(t)\,\varphi^p(t')|\,0\rangle.$$

The corresponding field operator is then

$$\varphi_i^{\mu\,m}(t) = \sqrt{\frac{\hbar}{2sNM_\mu}}\sum_{\boldsymbol{q}\sigma}\sqrt{\omega(\boldsymbol{q}\sigma)}\,\Big\{b_\sigma^{\boldsymbol{q}}\, e_i^{\mu}(\boldsymbol{q}\sigma)\, e^{i\boldsymbol{q}[\boldsymbol{R}^m-\omega(\boldsymbol{q}\sigma)t]} + \text{c. c.}\Big\}$$

which is a kind of a velocity field. The choice of the definition is a matter of convenience without physical meaning. One can always transform from one definition to another one.

Derivation of (21.13) with respect to $t$ gives

$$\begin{aligned}\dot{G}^{mn}_{\mu\nu\atop ij} &= \frac{i}{\hbar}\left\langle 0\left|\dot{u}^{m}_{\mu\atop i}(t)\, u^{n}_{\mu\atop j}(t')\right|0\right\rangle \Theta(t-t') + \frac{i}{\hbar}\left\langle 0\left|u^{n}_{\nu\atop j}(t')\,\dot{u}^{m}_{\mu\atop i}(t)\right|0\right\rangle\cdot\Theta(t'-t)\\ &\quad+\frac{i}{\hbar}\left\langle 0\left|\left[u^{m}_{\mu\atop i}(t), u^{n}_{\nu\atop j}(t)\right]_{-}\right|0\right\rangle \delta(t-t') \qquad (21.15\text{a})\\ &= \frac{i}{\hbar}\left\langle 0\left|T\,\dot{u}^{m}_{\mu\atop i}(t)\, u^{n}_{\nu\atop j}(t')\right|0\right\rangle,\end{aligned}$$

because $u^{m}_{\mu\atop i}(t)$ and $u^{n}_{\nu\atop j}(t)$ commute (eq. 21.8). Because further

$$\left[\dot{u}^{m}_{\mu\atop i}(t), u^{n}_{\nu\atop j}(t)\right]_{-} = \frac{\hbar}{i}\frac{1}{M_\mu}\delta^{mn}_{\mu\nu\atop ij}$$

we have with a similar calculation

$$\ddot{G}^{mn}_{\mu\nu\atop ij}(t-t') = \frac{i}{\hbar}\left\langle 0\left|T\,\ddot{u}^{m}_{\mu\atop i}(t)\, u^{n}_{\mu\atop j}(t')\right|0\right\rangle + \frac{1}{M_\mu}\delta^{mn}_{\mu\nu\atop ij}\,\delta(t-t')\,. \qquad (21.15\text{b})$$

Inserting this into (21.12) we see immediately that (21.13) satisfies (21.12). The Greens function (21.13) is called the *outgoing Greens function.* The reason can be seen later. The Greens function depends only on $t-t'$ which is a consequence of the homogeneity of the system in time. Therefore we put $t'=0$ in the following. Further, because of the homogeneity of the lattice, $G^{mn}_{\mu\nu\atop ij}$ depends only on $\boldsymbol{m}-\boldsymbol{n}$.

Introducing (21.2) into (21.13) and using

$$\begin{aligned}&\langle 0\mid b^+b\mid 0\rangle = \langle 0\mid bb\mid 0\rangle = \langle 0\mid b^+b^+\mid 0\rangle = 0 \quad \text{for every}\quad \boldsymbol{q}\sigma\,,\\ &\left\langle 0\left|b^{\boldsymbol{q}}_{\sigma}\, b^{\boldsymbol{q}'+}_{\sigma}\right|0\right\rangle = \delta_{\boldsymbol{q}\boldsymbol{q}'}\,\delta_{\sigma\sigma'}\end{aligned} \qquad (21.16)$$

we have

$$G^{mn}_{\mu\nu\atop ij}(t) = \frac{i}{2sN\sqrt{M_\mu M_\nu}}\sum_{\boldsymbol{q}\sigma}\frac{1}{\omega(\boldsymbol{q}\sigma)}\, e^{\mu}_{i}\, e^{\nu *}_{j}(\boldsymbol{q}\sigma)\, e^{i\boldsymbol{q}(\boldsymbol{R}^m-\boldsymbol{R}^n)-i\omega(\boldsymbol{q}\sigma)|t|} \qquad (21.17)$$

This is the Greens function as a function of coordinate and time. Sometimes it is more convenient to use the Greens function in the momentum and frequency representation (Fourier-transform). We have seen in sect. 20, that in stationary problems the frequency dependent function is of special interest.

We define the corresponding transformations by*

$$\begin{aligned}\overline{M}\,G(\boldsymbol{q}\sigma;\boldsymbol{q}'\sigma';t) &= \frac{1}{sN}\sum_{\substack{mn\\ \mu\nu ij}} e^{\mu *}_{i}(\boldsymbol{q}\sigma)\, e^{-i\boldsymbol{q}\boldsymbol{R}^m}\times\\ &\quad\times\sqrt{M_\mu M_\nu}\, G^{mn}_{\mu\nu\atop ij}(t)\, e^{\nu}_{j}(\boldsymbol{q}'\sigma')\, e^{i\boldsymbol{q}'\boldsymbol{R}^n}\end{aligned} \qquad (21.18)$$

The time-transformation is

$$\begin{aligned}G^{mn}_{\mu\nu\atop ij}(\omega) &= \lim_{\varepsilon\to 0+}\int_{-\infty}^{+\infty} G^{mn}_{\mu\nu\atop ij}(t)\, e^{i\omega t-\varepsilon|t|}\,dt\\ G^{mn}_{\mu\nu\atop ij}(t) &= \frac{1}{2\pi}\int_{\mathfrak{C}} G^{mn}_{\mu\nu\atop ij}(\omega)\, e^{-i\omega t}\,d\omega\,.\end{aligned} \qquad (21.19)$$

---

* $\overline{M}$ is a conveniently chosen mass average.

As a consequence of $G^{mn}_{\mu\,\nu\atop i\,j} = G^{m-n\,0}_{\ \mu\ \nu\atop \ i\ j}$ we have with (21.17)

$$G(\boldsymbol{q}\sigma;\boldsymbol{q}'\sigma';t) = G(\boldsymbol{q}\sigma,t)\,\delta_{\boldsymbol{q}\boldsymbol{q}'}\,\delta_{\sigma\sigma'}\,. \tag{21.20}$$

The transformed functions are

$$G(\boldsymbol{q}\sigma;t) = \frac{i}{2M\,\omega(\boldsymbol{q}\,\sigma)}\,\mathrm{e}^{-i\omega(\boldsymbol{q}\sigma)|t|} \tag{21.21a}$$

$$G^{mn}_{\mu\,\nu\atop i\,j}(\omega) = \frac{1}{sN\sqrt{M_\mu M_\nu}}\sum_{\boldsymbol{q}\,\sigma}\frac{e_i^\mu(\boldsymbol{q}\sigma)\,e_j^{\nu*}(\boldsymbol{q}\sigma)}{\omega^2(\boldsymbol{q}\sigma)-\omega^2-i\varepsilon}\,\mathrm{e}^{i\,\boldsymbol{q}(\boldsymbol{R}^m-\boldsymbol{R}^n)} \tag{21.21b}$$

which, of course, is identical with (20.14), and

$$G(\boldsymbol{q}\sigma;\omega) = \frac{1}{\overline{M}\{\omega^2(\boldsymbol{q}\,\sigma)-\omega^2-i\,\varepsilon\}}\,. \tag{21.21c}$$

From (21.21b, c) it can be seen, that $G(\omega)$ contains poles in the complex $\omega$-plane. Therefore in (21.19) we have to specify the path $\mathfrak{C}$. The path

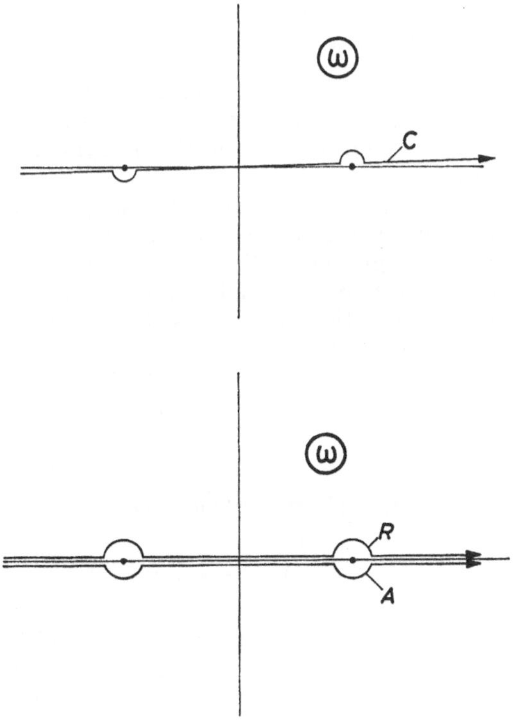


Fig. 6.6. Paths of integration for different Green functions

which leads from (21.21c) to (21.21a) is shown in Fig. 6.6. This makes it clear why this function is the outgoing Greens function. In scattering problems just this function satisfies the condition of outgoing scattered waves (sect. 22).

We can choose other paths, shown in Fig. 6.6. The corresponding functions are the *advanced and retarded Greens functions.* We just give their different forms, the *upper sign* belonging to the *retarded*, the *lower sign* to the *advanced* function

$$G_{R,A}{}^{m\,n}_{\mu\,\nu}_{i\,j}(t) = \pm \frac{i}{\hbar}\left\langle 0\left|\left[u^{m}_{\mu\,i}(t), u^{n}_{\nu\,j}(0)\right]_-\right|0\right\rangle \cdot \Theta(\pm t)$$

$$= \pm \frac{1}{s N \sqrt{M_\mu M_\nu}} \sum_{q\sigma} \frac{1}{\omega(q\sigma)} e^{\mu}_{i}(q\sigma)\, e^{\nu *}_{j}(q\sigma)\, e^{i q (R^m - R^n)} \times$$

$$\times \sin[\omega(q\sigma)\, t] \cdot \Theta(\pm t)\,.$$

$$G_{R,A}(q\sigma, t) = \pm \frac{1}{\overline{M}\, \omega(q\sigma)} \cdot \sin[\omega(q\sigma)\, t] \cdot \Theta(\pm t) \tag{21.22}$$

$$G_{R,A}{}^{m\,n}_{\mu\,\nu}_{i\,j}(\omega) = \frac{1}{s N \sqrt{M_\mu M_\nu}} \sum_{q\sigma} \frac{e^{\mu}_{i}\, e^{\nu *}_{j}(q\sigma)}{\omega^2(q\sigma) - (\omega \pm i\varepsilon)^2}\, e^{i q (R^m - R^n)}$$

$$G_{R,A}(q\sigma, \omega) = \frac{1}{\overline{M}\{\omega^2(q\sigma) - (\omega \pm i\varepsilon)^2\}}\,.$$

The half sum of the advanced and retarded function corresponds to an integration which is equal to taking Cauchys principle value (Fig. 6.6). These functions are useful mainly in thermodynamic considerations (finite temperature).

The poles in (21.21b, c) have a simple meaning. If $\omega = \omega(q\sigma)$ there is just one phonon with wave-vector $q$, polarization $\sigma$ and frequency $\omega$, which propagates through the lattice. Therefore the Greens function corresponding to such a phonon is called a *propagator.*

The static Greens function, used in sect. 18, 19, is just the limit $\omega \to 0$ of (21.21b, c), or, according to (21.19), the "time average" of the time-dependent Greens function. Thus, if we have the general function, we also have the static Greens function. The complex conjugate of (21.21b) is

$$G^{m\,n\,*}_{\mu\,\nu}{}_{i\,j}(\omega) = \frac{1}{s N \sqrt{M_\mu M_\nu}} \sum_{q\sigma} \frac{e^{\mu}_{i}(q\sigma)\, e^{\nu *}_{j}(q\sigma)}{\omega^2(q\sigma) - \omega^2 + i\varepsilon}\, e^{i q (R^m - R^n)}$$

or

$$G^{m\,n *}_{\mu\,\nu}{}_{i\,j}(\omega; i\varepsilon) = G^{m\,n}_{\mu\,\nu}{}_{i\,j}(\omega; -i\varepsilon)\,. \tag{21.23}$$

This can be seen by taking the complex conjugate of (21.21b) using (21.10) and relabeling the sum over $q$ by $-q$. If the number of ions in the crystal (periodicity volume) is large, we can replace the sum in (21.21b) by an integral in $q$-space and rewrite

$$G^{m\,n}_{\mu\,\nu}{}_{i\,j}(\omega) = \frac{V_z}{s(2\pi)^3 \sqrt{M_\mu M_\nu}} \sum_{\sigma} \int d\,\omega^2(q\sigma) \int_{\omega^2(q\sigma) = \text{const.}} \frac{dO}{|\text{grad}_q\, \omega^2(q\sigma)|} \times$$

$$\times \frac{e^{\mu}_{i}\, e^{\nu *}_{j}\, e^{i q (R^m - R^n)}}{\omega^2(q\sigma) - \omega^2 - i\varepsilon}$$

where the integral over $\mathrm{d}O$ means integration over the surface $\omega^2(\boldsymbol{q}\sigma) = \text{const.}$ in $\boldsymbol{q}$-space. Now

$$\frac{1}{\omega^2(\boldsymbol{q}\sigma)-\omega^2 \mp i\varepsilon} = \mathrm{P}\frac{1}{\omega^2(\boldsymbol{q}\sigma)-\omega^2} \pm i\pi\,\delta\left(\omega^2(\boldsymbol{q}\sigma)-\omega^2\right)$$

and therefore

$$G^{m\,n}_{\mu\,\nu}{}_{i\,j}(\omega) = \frac{V_z}{s(2\pi)^3\sqrt{M_\mu M_\nu}}\,\mathrm{P}\int\limits_{\omega^2=\text{const.}}\sum_\sigma \mathrm{d}\omega^2(\boldsymbol{q}\sigma)\frac{\mathrm{d}O}{|\mathrm{grad}_q\,\omega^2(\boldsymbol{q}\sigma)|}\times$$
$$\times\frac{e^\mu_i e^{\nu*}_j\,\mathrm{e}^{i\boldsymbol{q}(\boldsymbol{R}m-\boldsymbol{R}n)}}{\omega^2(\boldsymbol{q}\sigma)-\omega^2}+$$
$$+\,i\pi\frac{V_z}{s(2\pi)^3\sqrt{M_\mu M_\nu}}\sum_\sigma\int\limits_{\omega^2(q\sigma)=\omega^2}\frac{\mathrm{d}O\,e^\mu_i e^{\nu*}_j\,\mathrm{e}^{i\boldsymbol{q}(\boldsymbol{R}m-\boldsymbol{R}n)}}{|\mathrm{grad}_q\,\omega^2(\boldsymbol{q}\sigma)|}.$$

Using (21.23) we have

$$\mathrm{Im}\,G^{m\,n}_{\mu\,\nu}{}_{i\,j}(\omega) = \pi\cdot\frac{V_z}{s(2\pi)^3\sqrt{M_\mu M_\nu}}\sum_\sigma\int\limits_{\omega^2(q\sigma)=\omega^2}\frac{\mathrm{d}O\,e^\mu_i e^{\nu*}_i\,\mathrm{e}^{i\boldsymbol{q}(\boldsymbol{R}m-\boldsymbol{R}n)}}{|\mathrm{grad}_q\,\omega^2(\boldsymbol{q}\sigma)|}. \tag{21.24}$$

Inserting (21.24) into the real part of $G$, we obtain the *dispersion relation*

$$\mathrm{Re}\,G^{m\,n}_{\mu\,\nu}{}_{i\,j}(\omega) = \frac{1}{\pi}\mathrm{P}\int\frac{\mathrm{d}\omega'^2\cdot\mathrm{Im}\,G^{m\,n}_{\mu\,\nu}{}_{i\,j}(\omega')}{\omega'^2-\omega^2}. \tag{21.25}$$

Putting ${}^{m}_{\mu}{}_{i} = {}^{n}_{\nu}{}_{i}$ in (21.24) and summing over $\mu$, $i$ we obtain

$$\begin{aligned}\mathrm{Im}\sum_{\mu i} M_\mu\,G^{0\,0}_{\mu\,\mu}{}_{i\,j}(\omega) &= \pi\frac{V_z}{(2\pi)^3}\sum_\sigma\int\limits_{\omega^2(q\sigma)=\omega^2}\frac{\mathrm{d}O}{|\mathrm{grad}_q\,\omega^2(\boldsymbol{q}\sigma)|}\\ &= 3s\pi\,g(\omega^2) = \frac{3s\pi}{2\omega}z(\omega)\,.\end{aligned} \tag{21.26}$$

Here $g(\omega^2)$ is the squared and $z(\omega)$ the linear frequency distribution of the ideal lattice normalized to one. This relation is especially simple for cubic Bravais-crystals. Then $G^{00}_{11} = G^{00}_{22} = G^{00}_{33}$ and

$$z(\omega) = 2\omega\,g(\omega^2) = \frac{2\omega M}{\pi}\,\mathrm{Im}\,G^{00}_{11}(\omega) \tag{21.26a}$$

(21.26) can be derived directly from (21.21b) also.

We will mention some other properties of the causal Greens function, which are of practical use. As already mentioned (page 159), it is

$$G^{m\,n}_{\mu\,\nu}{}_{i\,j} = G^{m-n\,0}_{\mu\quad\nu}{}_{i\quad j}. \tag{21.27}$$

From (21.1) or better from the stationary equation (20.5b) it can be seen immediately that the

$$\text{point group symmetries of } G^{m\,n}_{\mu\,\nu}{}_{i\,j}(\omega) \text{ are those of } \Phi^{m\,n}_{\mu\,\nu}{}_{i\,j}. \tag{21.28}$$

This simplifies calculations in some cases.

The Greens function $G^{m\,n}_{\mu\,\nu\atop i\,j}(\omega)$ in our definition is (negative) definite, for frequencies $\omega^2$ larger than all the ideal lattice frequencies $\omega^2(\boldsymbol{q}\sigma)$, as can be seen simply from (21.21b, c). Also with $\omega^2$ in the band-gaps of non-primitive lattices the Greens function has a certain definiteness. This contains just the region of localized states. For localized states $[\omega^2 \neq \omega^2(\boldsymbol{q}\sigma)$ for all $\boldsymbol{q}\sigma]$, $\varepsilon$ (21.21b, c) can be put equal to zero; then $G^{m\,n}_{\mu\,\nu\atop i\,j}$ is hermitean. Because the perturbation $J^{m\,n}_{\mu\,\nu\atop i\,j}$ is always hermitean, this together with the definiteness of $G^{m\,n}_{\mu\,\nu\atop i\,j}$ for localized states guarantees the existence of eigenvectors for the equation (20.11) determining the localized states. The product of two hermitean matrices is not necessarily similar to a diagonal matrix, and thus eigenstates may not exist if one of the factors is not negative or positive definite.

*Asymptotic Expansion*

In scattering problems it is useful to know the asymptotic expansion of the Greens function, specially for the stationary case (21.21b). In the limit of infinite periodicity volume $N \to \infty$ we have

$$G^{m\,n}_{\mu\,\nu\atop i\,j}(\omega) = \frac{V_z}{s\,(2\pi)^3 \sqrt{M_\mu M_\nu}} \sum_{\sigma'} \int_{\text{B. Z.}} \mathrm{d}\boldsymbol{q}' \frac{e^\mu_i\, e^{\nu *}_j\,(\boldsymbol{q}'\,\sigma')}{[\omega^2(\boldsymbol{q}'\,\sigma') - \omega^2 - i\,\varepsilon]}\, \mathrm{e}^{i\,\boldsymbol{q}'\,(\boldsymbol{R}^m - \boldsymbol{R}^n)} \tag{21.29}$$

where the integral extends over the first Brillouin-zone.

Now it is

$$\frac{1}{\omega^2(\boldsymbol{q}'\,\sigma') - \omega^2 - i\,\varepsilon} = i \int_0^\infty \exp\{-i\omega^2(\boldsymbol{q}'\sigma')\,t + i\,\omega^2 t - \varepsilon\,t\}\,\mathrm{d}t$$

and therefore

$$G^{m\,n}_{\mu\,\nu\atop i\,j}(\omega) = \frac{i\,V_z}{s\,(2\pi)^3 \sqrt{M_\mu M_\nu}} \sum_{\sigma'} \int_{\text{B. Z.}} \mathrm{d}\boldsymbol{q}' \int_0^\infty \mathrm{d}t\, e_{\mu\atop i}\, e_{\nu *\atop j}\,(\boldsymbol{q}'\sigma')\, e^{i\varphi - \varepsilon t}$$

$$\varphi = [\omega^2(\boldsymbol{q}\sigma) - \omega^2(\boldsymbol{q}'\sigma')]\,t + \boldsymbol{q}\,(\boldsymbol{R}^m - \boldsymbol{R}^n)\,.$$

Here we have written $\omega^2(\boldsymbol{q}\sigma)$ for $\omega^2$, which represents the frequency as function of $\boldsymbol{q}, \sigma$ for the incoming waves in scattering problems. If $\boldsymbol{R}^m - \boldsymbol{R}^n \to \infty$ only that part of the integral conbributes essentially, where the phase $\varphi$ is constant (method of stationary phases). The stationary value is therefore given by

$$\omega^2(\boldsymbol{q}'\,\sigma') = \omega^2(\boldsymbol{q}\,\sigma)\,, \tag{21.30a}$$

$$t\{\mathrm{grad}_q\, \omega^2(\boldsymbol{q}'\,\sigma')\}_{q'} = \boldsymbol{R}^m - \boldsymbol{R}^n\,; \quad t > 0\,. \tag{21.30b}$$

If there are many values $\boldsymbol{q}'\,\sigma'$ which satisfy (21.30), all the different contributions have to be summed up. Using $\boldsymbol{\varkappa}$ and $\tau$ for the differences in $\boldsymbol{q}$ and $t$ from the values defined by (21.30), we have the expansion of $\varphi$:

$$\varphi = \boldsymbol{q}'\,(\boldsymbol{R}^m - \boldsymbol{R}^n) - \tau\,\boldsymbol{\varkappa}\,\mathrm{grad}_{q'}\,\omega^2(\boldsymbol{q}'\sigma') - \frac{1}{2}\,t\,\omega^2_{|ij}(\boldsymbol{q}'\,\sigma')\,\varkappa_i\,\varkappa_j + \cdots$$

with

$$\omega^2_{|ij} = \frac{\partial^2 \omega^2}{\partial q_i \, \partial q_j} \, .$$

We assume further, that the polarization vectors $e^{\mu}_i(\boldsymbol{q}\,\sigma)$ change slowly with $\boldsymbol{q}'$, so that we can replace them by their values at (21.30), neglecting all higher contributions. The limits of the integral can be replaced by $-\infty$ or $+\infty$, respectively, because contributions originate from the vicinity of (22.30) only.

We have*

$$G^{mn}_{\substack{\mu\\i}\,\substack{\nu\\j}} = \frac{i\,V_z}{s\,(2\pi)^3 \sqrt{M_\mu M_\nu}} \sum_{\sigma'(\boldsymbol{q}')} e^{\mu}_{i}(\boldsymbol{q}'\,\sigma')\, e^{\nu\,*}_{j}(\boldsymbol{q}'\,\sigma')\, \mathrm{e}^{i\boldsymbol{q}'(\boldsymbol{R}^m - \boldsymbol{R}^n)} \times$$
$$\times \iint\limits_{-\infty}^{+\infty} \mathrm{d}\boldsymbol{\varkappa}\,\mathrm{d}\tau \cdot \exp\left\{-\, i\,\tau\,\boldsymbol{\varkappa}\,\mathrm{grad}_{q'}\omega^2(\boldsymbol{q}'\,\sigma') - \frac{1}{2}\, i\, t\, \omega^2_{|ij}\varkappa_i\,\varkappa_j\right\} .$$

The coordinate-system $\varkappa_i$ can be replaced by another one $\varrho_l$ using an orthogonal transformation:

$$\varkappa_i = \sum_l u_{il}\,\varrho_l\,; \quad \sum_l u_{jl}\,u_{il} = \delta_{ij}\,; \quad |\det u| = 1\, .$$

The new system is supposed to have the following properties: i) the $\varrho_1$-axis is parallel to the direction of $\mathrm{grad}_{q'}\,\omega^2(\boldsymbol{q}'\,\sigma')$, therefore

$$|\boldsymbol{\varkappa}\,\mathrm{grad}\,\omega^2(\boldsymbol{q}'\,\sigma')| = \varrho_1\,|\mathrm{grad}\,\omega^2(\boldsymbol{q}'\,\sigma')|\, .$$

ii) mixed derivatives vanish:

$$\frac{\partial^2\omega^2}{\partial\varrho_2\,\partial\varrho_3} \sum_{l,m} u_{2l}\,u_{3m}\,\frac{\partial^2\omega^2}{\partial\varkappa_l\,\partial\varkappa_m}$$

This can be assumed without loss of generality. Then

$$\int\limits_{-\infty}^{+\infty} \exp\{-\,i\,\tau\,\varrho_1\,|\mathrm{grad}\,\omega^2|\}\,\mathrm{d}\tau = \frac{2\pi}{|\mathrm{grad}\,\omega^2|}\,\delta(\varrho_1)\, ,$$

$$\int\limits_{-\infty}^{+\infty} \exp\{\pm\, i\,\chi^2\,\varrho^2\}\,\mathrm{d}\varrho = \frac{\sqrt{\pi}}{\chi}\exp\{\pm\, i\,\pi/4\} \quad \text{(Fresnel's integral)}\, .$$

With

$$t = \frac{|R^m - R^n|}{|\mathrm{grad}_{q'}\,\omega^2\,(\boldsymbol{q}'\,\sigma')|}\ ; \quad \omega^2_{|ii} = \frac{\partial^2\omega^2}{\partial\varrho_i^2}\ ; \quad \varepsilon_i = \mathrm{sgn}\,(\omega^2_{|ii})\, ,$$

we have finally

$$G^{mn}_{\substack{\mu\\i}\,\substack{\nu\\j}}(\omega) = \frac{i\,V_z}{2\pi s\sqrt{M_\mu M_\nu}} \sum_{\sigma'(\boldsymbol{q}')} \frac{e^{\mu}_{i}(\boldsymbol{q}'\,\sigma')\,e^{\nu\,*}_{j}(\boldsymbol{q}'\,\sigma')}{|\omega^2_{|22}\,\omega^2_{|33}|^{1/2}}\,\mathrm{e}^{-i\frac{\pi}{4}(\varepsilon_2+\varepsilon_3)} \times \times \frac{\mathrm{e}^{i\boldsymbol{q}'(\boldsymbol{R}^m - \boldsymbol{R}^n)}}{|\boldsymbol{R}^m - \boldsymbol{R}^n|} \tag{21.31}$$

in the limit $|\boldsymbol{R}^m - \boldsymbol{R}^n| \to \infty$.

* $\sum_{\sigma'(q')}$ indicates that we have to sum over all the $\boldsymbol{q}'$, $\sigma'$-values satisfying (21.30)

Even in this limiting case the explicit calculation is rather difficult, for it depends on the dispersion curves of the frequencies and on the polarisation vectors at $\boldsymbol{q}'\,\sigma'$-positions defined by (21.30).

The explicit calculation can be done in one simple case by means of straightforward integration: this is in the *elastic limit of lattice theory for an isotropic medium.*

In investigations where only slowly varying displacement fields occur this function can be used with great success. Such cases are the static displacements of ions due to defects, far off from these defects (sect. 18, 19) and dynamical displacements in the case of long waves (small frequencies). This gives appropriate approximations even in non-isotropic media; at least one gets a qualitative insight in what happens really.

The elastic equation of motion in homogeneous media is ($\varrho$: mass density)

$$-\sum_{jkl}\frac{\partial}{\partial X_j}C_{ij,kl}\,u_{k|l}+\varrho\,\ddot{u}_i=-\sum_{jkl}C_{ij,kl}\,u_{k|jl}+\varrho\,\ddot{u}_i=0\,. \qquad (21.32)$$

The Greens function is defined by

$$-\sum_{jkl}C_{ij,kl}\,G_{km|jl}(\boldsymbol{R},t)+\varrho\,\ddot{G}_{im}(\boldsymbol{R},t)=\delta_{im}\,\delta(\boldsymbol{R})\,\delta(t)\,, \qquad (21.33\text{a})$$

or in stationary problems:

$$-\sum_{jkl}C_{ij,kl}\,G_{km|jl}(\boldsymbol{R},t)-\varrho\,\omega^2\,G_{im}(\boldsymbol{R},\omega)=\delta_{im}\,\delta(\boldsymbol{R})\,. \qquad (21.33\text{b})$$

In isotropic media it is

$$C_{ij,kl}=\lambda\,\delta_{ij}\,\delta_{kl}+\mu\,(\delta_{ik}\,\delta_{jl}+\delta_{il}\,\delta_{jk}) \qquad (21.34)$$

with the Lamé-constants

$$\begin{aligned}\lambda&=c_{12}=c_{11}-2c_{44}\,;\quad \mu=c_{44}=\varrho\,c_t^2\\ \lambda+2\,\mu&=c_{11}=\varrho\,c_l^2\,.\end{aligned} \qquad (21.35)$$

$c_l$ and $c_t$ are longitudinal and transverse velocities of sound. Frequencies and polarizations are given by

$$\begin{aligned}&\omega^2(\boldsymbol{q},\sigma=1)=c_l^2\,q^2\,; && e_i(\sigma=1)=q_i/q\\ &\omega^2(\boldsymbol{q},\sigma=2,3)=c_t^2\,q^2\,; && \sum_j q_i\,e_j(\sigma=2,3)=0\\ & && \sum_{\sigma=2,3}e_i(\sigma)\,e_j^*(\sigma)=\delta_{ij}-q_i\,q_j/q^2\,.\end{aligned} \qquad (21.36)$$

Inserting this into the representation

$$G_{ij}(\boldsymbol{R},\omega)=\frac{1}{(2\pi)^3\,\varrho}\sum_\sigma\int_{-\infty}^{+\infty}\mathrm{d}\boldsymbol{q}\,\frac{e_i(\sigma)\,e_j^*(\sigma)}{\omega^2(\boldsymbol{q}\,\sigma)-\omega^2-i\,\varepsilon}\,e^{i\boldsymbol{q}\boldsymbol{R}}, \qquad (21.37)$$

we receive by straight forward integration

$$
\begin{aligned}
G_{ij}(\boldsymbol{R}, \omega) &= \frac{\delta_{ij}}{4\pi\varrho c_t^2} \cdot \frac{1}{R} e^{i|\omega| R/c_t} + \frac{1}{4\pi\varrho\omega^2} \frac{\partial^2}{\partial X_i \partial X_j} \cdot \frac{1}{R} \left(e^{i|\omega| R/c_t} - e^{i|\omega| R/c_l}\right) \\
&= \frac{\delta_{ij}}{4\pi\varrho\omega^2 R^2} \{-[1 - i|\omega| R/c_t - \omega^2 R^2/c_t^2]\, e^{i|\omega| R/c_t} + \\
&\quad + [1 - i|\omega| R/c_l]\, e^{i|\omega| R/c_l}\} + \\
&\quad + \frac{X_i X_j}{4\pi\varrho\omega^2 R^5} \{[3 - 3i|\omega| R/c_t - \omega^2 R^2/c_t^2]\, e^{i|\omega| R/c_t} - \\
&\quad - [3 - 3i|\omega| R/c_l - \omega^2 R^2/c_l^2]\, e^{i|\omega| R/c_l}\}\,.
\end{aligned}
\tag{21.38}
$$

The asymptotic expansion of this elastic Greens-function can be seen easily to be

$$
\begin{aligned}
G_{ij}(R \to \infty; \omega) &= \frac{\delta_{ij}}{4\pi\varrho c_t^2} \cdot \frac{1}{R} e^{i|\omega| R/c_t} + \\
&\quad + \frac{X_i X_j}{4\pi\varrho R^3} \left\{\frac{1}{c_l^2} e^{i|\omega| R/c_l} - \frac{1}{c_t^2} e^{i|\omega| R/c_t}\right\}.
\end{aligned}
\tag{21.39}
$$

The static limit follows from $\omega R \to 0$

$$
\begin{aligned}
G_{ij}(R; \omega \to 0) &= \frac{\delta_{ij}}{8\pi\varrho R}\left(\frac{1}{c_t^2} + \frac{1}{c_l^2}\right) + \frac{X_i X_j}{8\pi\varrho R^3}\left(\frac{1}{c_t^2} - \frac{1}{c_l^2}\right) + \\
&\quad + i\frac{\delta_{ij}}{12\pi\varrho} |\omega| \left(\frac{2}{c_t^3} + \frac{1}{c_l^3}\right).
\end{aligned}
\tag{21.40}
$$

From this equation the value for $\omega = 0$ can be seen directly, which has been used in sect. 18. Further, as the other expansion coefficients are proportional to powers of $\omega R$, it follows [Eq. (21.26a)]

$$
z(\omega) = \frac{2\omega}{\pi} M \operatorname{Im} G_{11}(0, \omega) = \frac{V_z}{6\pi^2} \omega^2 \left(\frac{2}{c_t^3} + \frac{1}{c_l^3}\right) \tag{21.41}
$$

which represents just the frequency spectrum of an isotropic medium [I].

This Greens function can be used in static and small frequency calculations. Of course, in the high-frequency region the lattice structure becomes essential and (21.38) is a bad or even wrong approximation. For example, localized modes cannot be calculated in the case of a continuous medium because the concept of localized modes is strongly connected to a maximum frequency of the spectrum, i.e. to a discrete structure of the crystal.

It might be, that one wants to use the elastic Greens-function as an approximation to the lattice Greens-function, imposing a cut-off-frequency $\omega_D$ (Debye-frequency) on the spectrum in the calculation of the Greens-function. But such a procedure contains some inconsistence, which is connected with the fact, that a real continuum has no cut-off of the spectrum (no smallest length) and that the introduction of a cut-off is an artificial one, which has even a wrong analytic behavior at $\omega = \omega_D$. This holds in one- and twodimensional cases as well. The following happens:

1) Whereas the correct lattice Greens-function $G_{ij}^{0\,m}(\omega)$ for $\omega > \omega_m$ (maximum lattice frequency) decreases exponentially with increasing distance from the defect $|\boldsymbol{m}|$, the "elastic Greens-function" with a cut-off $\omega_D$ decreases only with some inverse power of the distance for $\omega > \omega_D$. The power depends on the dimension of the lattice. In calculating $G_{ij}^{0\,0}$ however, one can use the Debye-spectrum as a certain approximation, which is reasonable for $\omega \gg \omega_D$, but unreasonable for $\omega \approx \omega_D$. For higher superscripts $\boldsymbol{m} \neq 0$, the "approximation" is even worse. The frequencies of localized states can be approximated in a better and even simpler way by using the variational procedure mentioned in sect. 21 [64 L 1].

2) In a one-dimensional monatomic crystal the situation is extremely worse for frequencies in the band of lattice frequencies. In that case for $\omega < \omega_m$ the lattice Greens-function $G^{00}(\omega)$ is purely imaginary, whereas the "elastic Greens-function with cut-off-frequency" has a real and an imaginary part for $\omega < \omega_D$, whereas the real part vanishes for $\omega_D \to \infty$, i.e. for a true continuum. This has the consequence that in a monatomic linear lattice with an isotopic defect of a heavy mass, there is no resonance state, whereas in a pseudo-continuum with cut-off-frequency there "is" a resonance state, and in a true continuum there is again no resonance. This shows, that the concept of a cut-off-frequency contradicts the properties of a continuum (see sect. 22).

3) In the threedimensional case one can use the „elastic Greens-function" as an approximation for the lattice Greens-function especially for $G_{ij}^{0\,0}(\omega)$ as long as $\omega$ is sufficiently small compared to $\omega_m$ or $\omega_D$. That means, as long as the result for $G_{ij}^{0\,0}(\omega)$ does not depend very much on $\omega_D$. For $\omega \approx \omega_D$ the use of the elastic spectrum gives a wrong behavior for $G_{ij}^{0\,0}$. In any case $G_{ij}^{0\,0}(\omega)$ has a real and an imaginary part for $\omega < \omega_D$.

4) These statements do not affect the famous use of the Debye-spectrum in calculating specific heats, for example. It should be emphasized only, that one has to be cautious when imposing a cut-off-frequency to a continuum, or when replacing a lattice spectrum by an elastic one.

The time-dependent isotropic Greens function can be obtained from the Fourier-transformation of (21.38). It is

$$\begin{aligned} -i4\pi\varrho^2 G_{ij}(\boldsymbol{R},t) &= (\delta_{ij} - X_i X_j/R^2)\,\frac{1}{c_t}\cdot\frac{1}{R^2 - c_t^2 t^2 + i\varepsilon} + \\ &+ \frac{X_i X_j}{R^2 c_l}\,\frac{1}{R^2 - c_l^2 t^2 + i\varepsilon} + \\ &+ (\delta_{ij} - 3X_i X_j/R^2)\,\frac{1}{R^2} \times \left\{\frac{1}{c_l} - \frac{1}{c_t} - \right. \\ &\left. - \frac{|t|}{2R}\left[\operatorname{artgh}\left(\frac{|c_l t|}{R} - i\varepsilon\right) - \operatorname{artgh}\left(\frac{|c_t t|}{R} - i\varepsilon\right)\right]\right\}. \end{aligned} \tag{21.42}$$

## 22. Scattering of Phonons at Defects

The quantities we will take for describing the scattering process are, as usual, the scattering amplitude and the differential or the total cross section, respectively. They are all defined by the coefficients of the asymptotic spherical wave. Using

$$\tau = (1 - \mathbf{GJ})^{-1} \tag{22.1}$$

we have from the general solution of the defect equation of motion (20.8b) and the asymptotic expansion (21.31)

$$\begin{aligned} u_{\mu i}^{\boldsymbol{m}} &= \mathring{u}_{\mu i}^{\boldsymbol{m}} + G_{\mu i \, \nu j}^{\boldsymbol{m}\,\boldsymbol{n}} (J\tau)_{\mu j \, \varkappa k}^{\boldsymbol{n}\,\boldsymbol{p}} \mathring{u}_{\varkappa k}^{\boldsymbol{p}} \\ &= \mathring{u}_{\mu i}^{\boldsymbol{m}} + \frac{i V_z}{2\pi s \sqrt{M_\mu M_\nu}} \sum_{\sigma'(\boldsymbol{q}')} \frac{e_i^\mu e_j^{\nu *}(\boldsymbol{q}'\sigma')}{|\omega^2_{|22}\, \omega^2_{|33}|^{1/2}} \mathrm{e}^{-i\frac{\pi}{4}(\varepsilon_2+\varepsilon_3)} \times \\ &\quad \times \frac{\mathrm{e}^{i\boldsymbol{q}'(\boldsymbol{R}^{\boldsymbol{m}}-\boldsymbol{R}^{\boldsymbol{n}})}}{|R^{\boldsymbol{m}} - R^{\boldsymbol{n}}|} (J\tau)_{\mu i\, \varkappa k}^{\boldsymbol{n}\,\boldsymbol{p}} \mathring{u}_{\varkappa k}^{\boldsymbol{p}} \end{aligned} \tag{22.2}$$

which represents the solution of the stationary problem. $\mathring{u}_{\mu i}^{\boldsymbol{m}}$ is the incoming plane wave, with given frequency $\omega(\boldsymbol{q}\sigma)$, polarization $\sigma$ and direction of wave-vector $\boldsymbol{q}$. It is one of the solutions of the ideal lattice equation of motion (21.1)

$$\mathring{u}_{\mu i}^{\boldsymbol{m}} \sim \frac{1}{\sqrt{M_\mu}} e_i^\mu(\boldsymbol{q}\sigma)\, \mathrm{e}^{i\boldsymbol{q}\boldsymbol{R}^{\boldsymbol{m}}} . \tag{22.3}$$

As can be seen from (22.2), the sum over $\boldsymbol{n}$ extends over the range of the defect, which is a small "vector" compared with the large "vector" $\boldsymbol{m}$ for the position of asymptotic observation. Therefore we can neglect $\boldsymbol{R}^{\boldsymbol{n}}$ in the denominator of the spherical wave in (22.2). Inserting (22.3) into (22.2) we have

$$\begin{aligned} u_{\mu i}^{\boldsymbol{m}}(\boldsymbol{q}\sigma) &= \mathring{u}_{\mu i}^{\boldsymbol{m}}(\boldsymbol{q}\sigma) + \sum_{\sigma'(\boldsymbol{q}')} F(\boldsymbol{q}\sigma; \boldsymbol{q}'\sigma') \frac{1}{\sqrt{M_\mu}} e_i^\mu(\boldsymbol{q}'\sigma') \frac{\mathrm{e}^{i\boldsymbol{q}'\boldsymbol{R}^{\boldsymbol{m}}}}{R^{\boldsymbol{m}}} \\ &= \mathring{u}_{\mu i}^{\boldsymbol{m}}(\boldsymbol{q}\sigma) + \sum_{\sigma'(\boldsymbol{q}')} F(\boldsymbol{q}\sigma; \boldsymbol{q}'\sigma') \frac{\mathring{u}_{\mu i}^{\boldsymbol{m}}(\boldsymbol{q}'\sigma')}{R^{\boldsymbol{m}}} \end{aligned} \tag{22.4}$$

with

$$F(\boldsymbol{q}\sigma, \boldsymbol{q}'\sigma') = \frac{i V_z}{2\pi s} \cdot \frac{\mathrm{e}^{-i\pi/4(\varepsilon_2+\varepsilon_3)}}{|\omega^2{}_{|22}\, \omega^2{}_{|33}|^{1/2}} \langle \boldsymbol{q}'\sigma' | J\tau | \boldsymbol{q}\sigma \rangle , \tag{22.4a}$$

$$\langle \boldsymbol{q}'\sigma' | J\tau | \boldsymbol{q}\sigma \rangle = \sum_{\substack{\boldsymbol{m}\boldsymbol{p} \\ \nu\varkappa jk}} e_j^{\nu *}(\boldsymbol{q}'\sigma')\, \mathrm{e}^{-i\boldsymbol{q}'\boldsymbol{R}^{\boldsymbol{n}}} \frac{1}{\sqrt{M_\mu M_\nu}} (J\tau)_{\nu j\, \varkappa k}^{\boldsymbol{n}\,\boldsymbol{p}} e_k^\varkappa(\boldsymbol{q}\sigma)\, \mathrm{e}^{i\boldsymbol{q}\boldsymbol{R}^{\boldsymbol{p}}} . \tag{22.4b}$$

The sum in (22.4) extends, as already mentioned, over all the values $\sigma', \boldsymbol{q}'$ which satisfy (21.30). The phase factor in the spherical wave in (22.4) does not only depend on the absolute value of $\boldsymbol{R}^{\boldsymbol{m}}$, because of the anisotropy of crystal lattices. The phase of the outgoing waves depends on the direction of the wave relative to the crystal axes and can be determined by symmetry considerations. However, in simple cases the phase is nearly or completely independent of the direction (e.g. in isotropic media). The scattering amplitude describes the scattering of an incoming plane wave of frequency $\omega$, direction $\boldsymbol{q}$ and polarization $\sigma$ into an outgoing wave of the same frequency $\omega$, but with different direction $\boldsymbol{q}'$ and polarization $\sigma'$.

The differential cross section $d\sigma$ is defined by

$$S_0 \, d\sigma = \boldsymbol{S}_s \, |R^m|^2 \, d\Omega \, \boldsymbol{e}_F \, . \tag{22.5}$$

$S_0$ is the incoming flux, $\boldsymbol{S}_s$ is the outgoing flux through the angle $d\Omega$ in the direction $\boldsymbol{e}_F$. The flux is given by the energy of the wave times its group velocity $\boldsymbol{v}(\boldsymbol{q}\sigma) = \text{grad}\,\omega(\boldsymbol{q}\sigma)$ in the direction of the wave. As the energy is proportional to a function of the frequencies, being constant, times the squared wave amplitudes, we arrive at

$$d\sigma = \frac{1}{|\text{grad}\,\omega^2(\boldsymbol{q}\,\varrho)|} \sum_{\sigma'(\boldsymbol{q}')} |\text{grad}\,\omega^2(\boldsymbol{q}'\sigma')| \cdot |F(\boldsymbol{q}\sigma;\boldsymbol{q}'\sigma')|^2 \cdot d\Omega \tag{22.6}$$

(22.6) can be read in another way. Instead of considering the cross-section for the scattering into a solid angle $d\Omega$, we ask for the energy which is lost by the scattering of the wave due to a certain surface element $dO_\omega$ of the surface $\omega^2(\boldsymbol{q}'\sigma') = \omega^2(\boldsymbol{q}\sigma) = \text{const}$. It is

$$\frac{d\Omega}{|\omega_{f22}^2\,\omega_{f33}^2|} = \frac{dO_\omega}{|\text{grad}\,\omega^2(\boldsymbol{q}'\,\sigma')|^2} \tag{22.7}$$

and thus the energy transfer from the wave $\boldsymbol{q}\sigma$ into the wave $\boldsymbol{q}'\sigma'$ is given by

$$d\sigma(\boldsymbol{q}\sigma \to \boldsymbol{q}'\sigma') = \frac{V_s^2}{(2\pi\,s)^2} \cdot \frac{|\langle \boldsymbol{q}'\,\sigma'\,|J\,\tau|\,\boldsymbol{q}\,\sigma\rangle|^2\,dO_\omega}{|\text{grad}\,\omega^2(\boldsymbol{q}\,\sigma)|\,|\text{grad}\,\omega^2(\boldsymbol{q}'\,\sigma')|} \, . \tag{22.8}$$

Using (20.3a, b) and (21.27) we have

$$(J\tau)^{m\,n}_{\substack{\mu\,\nu\\ i\,j}} = (J\tau)^{n\,m}_{\substack{\nu\,\mu\\ j\,i}} \quad \text{and} \quad \langle \boldsymbol{q}'\sigma' \,|\, J\tau \,|\, \boldsymbol{q}\sigma\rangle^* = \langle \boldsymbol{q}\sigma \,|\, J\tau^* \,|\, \boldsymbol{q}'\sigma'\rangle \tag{22.9}$$

because $J$ is real. Therefore

$$d\sigma(\boldsymbol{q}\sigma \to \boldsymbol{q}'\sigma') = d\sigma(\boldsymbol{q}'\sigma' \to \boldsymbol{q}\sigma) \, , \tag{22.10}$$

if the surface elements $dO_\omega$ and $dO_{\omega'}$ are chosen equal. With (page 160/1)

$$\frac{dO_\omega}{|\text{grad}\,\omega^2(\boldsymbol{q}'\,\sigma')|} = \delta\,[\omega^2 - \omega^2(\boldsymbol{q}'\sigma')]\,dV_{(\boldsymbol{q}'\sigma')} = \frac{1}{\pi}\,\text{Im}\,\frac{dV_{(\boldsymbol{q}'\sigma')}}{\omega^2(\boldsymbol{q}'\,\sigma') - \omega^2 - i\,\varepsilon}$$

we have from (22.8) for the total cross-section

$$\sigma_{\text{tot}}(\boldsymbol{q}\sigma) = \frac{V_z}{(2\pi\,s)^2} \cdot \frac{(2\pi)^3}{|\text{grad}\,\omega^2(\boldsymbol{q}\,\sigma)|} \cdot \frac{1}{N\pi}\,\text{Im} \sum_{(\boldsymbol{q}'\sigma')} \frac{|\langle \boldsymbol{q}'\,\sigma'|\,J\,\tau|\,\boldsymbol{q}\,\sigma\rangle|^2}{\omega^2(\boldsymbol{q}'\,\sigma') - \omega^2 - i\,\varepsilon}$$

where the integral over the Brillouin-zone has been replaced by the sum over all the $\boldsymbol{q}'$-values. Using the Greens function (21.21b) with the definition of $|\boldsymbol{q}\sigma\rangle$ from (22.4b)

$$G^{m\,n}_{\substack{\mu\,\nu\\ i\,j}} = \frac{1}{s\,N} \sum_{\boldsymbol{q}'\sigma'} \frac{|\boldsymbol{q}'\,\sigma'\rangle\,\langle \boldsymbol{q}'\,\sigma'|}{\omega^2(\boldsymbol{q}'\,\sigma') - \omega^2 - i\,\varepsilon}$$

we have

$$\sigma_{tot}(\boldsymbol{q}\sigma) = \frac{2V_z}{s\,|\mathrm{grad}\,\omega(\boldsymbol{q}\,\sigma)|}\,\mathrm{Im}\langle \boldsymbol{q}\sigma\,|\,J\tau G(J\tau)^*\,|\,\boldsymbol{q}\sigma\rangle$$

and with

$$\mathrm{Im}\langle \boldsymbol{q}\sigma\,|\,J\tau\,G\,J\tau^*\,|\,\boldsymbol{q}\sigma\rangle = -\,\mathrm{Im}\langle \boldsymbol{q}\sigma\,|\,J\tau^*\,|\,\boldsymbol{q}\sigma\rangle$$

and (22.4a)

$$= \mathrm{Im}\langle \boldsymbol{q}\sigma\,|\,J\tau\,|\,\boldsymbol{q}\sigma\rangle$$

$$\sigma_{tot} = -\,4\pi\,\frac{|\omega^2_{|22}\,\omega^2_{|33}|^{1/2}}{|\mathrm{grad}\,\omega^2(\boldsymbol{q}\,\sigma)|}\,\mathrm{Im}\,i\,\mathrm{e}^{i\pi/4(\varepsilon_2+\varepsilon_3)}\,F(\boldsymbol{q}\sigma;\,\boldsymbol{q}\sigma)\,. \tag{22.11}$$

This is the generalization of the *optical theorem* to cases where $\omega^2$ is not proportional to $q^2$. $F(\boldsymbol{q}\sigma;\,\boldsymbol{q}\sigma)$ is the scattering amplitude in the direction of the incoming wave, with conservation of polarization.

If $\omega^2 \sim q^2$ for all $\boldsymbol{q}$ and $\sigma$, then

$$\sigma_{tot} = \frac{4\pi}{q}\,\mathrm{Im}\,F(\boldsymbol{q}\sigma;\,\boldsymbol{q}\sigma) \tag{22.11a}$$

which is the more familiar form.

The main problem is the calculation of the reciprocal of $\mathbf{1} - \mathbf{G}\,\mathbf{J} = \tau^{-1}$ and the frequency derivatives in (22.4). In the most cases this can be done only numerically. But the essential feature of the scattering problem can be seen already with the model of an isotropic continuous medium; the scattering phenomena are not characteristic for lattice phenomena.

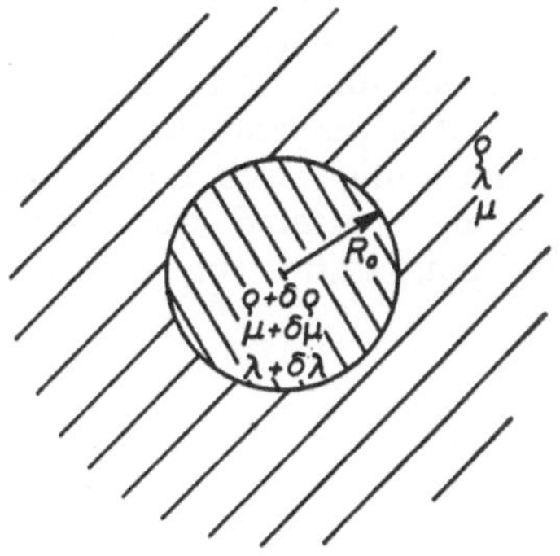


Fig. 6.7. A "defect" in an isotropic elastic continuum. In a certain region, mass density and elastic constants are different from the ideal values

Therefore we discuss first the following model (Fig. 6.7): In a continuous medium there is a spherical region of radius $R_0$ with a change in mass-density $\delta\varrho$ and a change in the isotropic elastic constants $\delta\lambda$, $\delta\mu$. The change is supposed to have the form of a step-function $\Theta(R_0 - R)$*. The equation of motion (21.32) becomes in the stationary case

$$-\sum_{jkl} C_{ij,kl}\,u_{k|jl} - \varrho\,\omega^2\,u_i = J_i(\boldsymbol{R})$$

$$= \{\omega^2\cdot\delta\varrho\,u_i + \sum_k\,[(\delta\lambda + \delta\mu)\,u_{k|ik} + \delta\mu\cdot u_{i|kk}]\}\,\Theta(R_0 - R) -$$

$$-\sum_k\,[\delta\lambda(X_i/R)\,u_{k|k} + \delta\mu(X_k/R)\,(u_{i|k} + u_{k|i})]\,\delta(R_0 - R)\,. \tag{22.12}$$

The second term on the right hand side arises from the derivation of the elastic constants in (21.32), if they are not constant. The solution of (22.12) can be given in terms of Greens function as

$$u_i(\boldsymbol{R}) = \mathring{u}_i(\boldsymbol{R}) + \dot{u}_i(\boldsymbol{R}) \tag{22.13a}$$

$$\dot{u}_i(\boldsymbol{R}) = \int_{-\infty}^{+\infty} G_{ij}(\boldsymbol{R} - \boldsymbol{R}')\,J_j(\boldsymbol{R}')\,\mathrm{d}\boldsymbol{R}'\,, \tag{22.13b}$$

---

* Presumably this assumption is somewhat unrealistic. But it is sufficient in the following discussion.

where $\overset{i}{u}_i(\boldsymbol{R})$ is the incoming wave. The perturbation $J_j(\boldsymbol{R}')$ contains, according to (22.12), the sum of $\overset{i}{\boldsymbol{u}}$ and $\overset{s}{\boldsymbol{u}}$.

*Borns Approximation*

The solution of (22.13) is simple, if we can neglect $\overset{s}{\boldsymbol{u}}$ at the right hand side of (22.12); this gives the first Born approximation. Since we are interested in the scattering amplitude only, we replace Greens function by (21.39). The incoming wave is supposed to be

$$u_j(\boldsymbol{R}') = e_j\, \mathrm{e}^{i\boldsymbol{q}\boldsymbol{R}' - i\omega t}; \quad \omega > 0 . \tag{22.14}$$

$\boldsymbol{e}$ is the polarization of the incoming wave, $\boldsymbol{q}$ its wave vector, which satisfies $q = \omega/c_a$, where $c_a$ is the wave velocity of the incoming wave. For longitudinal waves we have $\boldsymbol{e} = \boldsymbol{q}/q$ and $c_a = c_l$, for transverse waves $\boldsymbol{e}$ is perpendicular to $\boldsymbol{q}$ and $c_a = c_t$. We define the direction of the incoming wave by $\boldsymbol{n} = \boldsymbol{q}/q$ and the asymptotic direction of the outgoing scattered wave by $\boldsymbol{n}' = \boldsymbol{R}/R$ (Fig. 6.8). A wave which is longitudinal after the scattering process has polarization parallel to $\boldsymbol{n}'$,

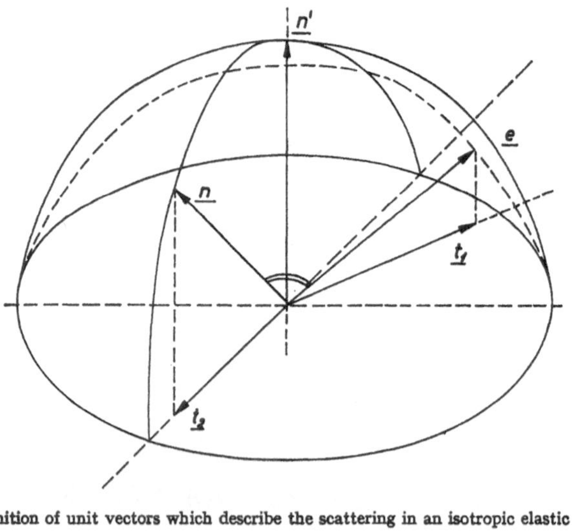


Fig. 6.8. Definition of unit vectors which describe the scattering in an isotropic elastic continuum

a transverse wave has polarization perpendicular to $\boldsymbol{n}'$, which can be described by the vectors

$$\boldsymbol{t}_1 = \boldsymbol{e} - \boldsymbol{n}'(\boldsymbol{e}\boldsymbol{n}') \quad \text{or} \quad \boldsymbol{t}_2 = \boldsymbol{n} - \boldsymbol{n}'(\boldsymbol{n}\boldsymbol{n}') \tag{22.15}$$

or by both. A straightforward calculation, using (22.12) and (22.13) then gives for long waves,

$$qR_0 \ll 1; \quad V_0 = \frac{4\pi}{3} R_0^3 , \tag{22.16}$$

$$\dot{u}_i(\boldsymbol{R}) = \frac{V_0}{4\pi\varrho c_l^2} n_i'\{\delta\varrho\cdot\omega^2(\boldsymbol{e}\boldsymbol{n}') - q\,\omega/c_l\,[\delta\lambda(\boldsymbol{e}\boldsymbol{n}) + 2\delta\mu(\boldsymbol{e}\boldsymbol{n}')(\boldsymbol{n}\boldsymbol{n}')]\}\times$$
$$\times\frac{e^{i\omega R/c_l}}{R} +$$
$$+\frac{V_0}{4\pi\varrho c_t^2}\{\delta\varrho\,\omega^2 t_{1i} - q\,\omega/c_t\cdot\delta\mu[(\boldsymbol{n}\boldsymbol{n}')t_{1i} + (\boldsymbol{e}\boldsymbol{n}')t_{2i}]\}\,,\times$$
$$\times\frac{e^{i\omega R/c_t}}{R}\,. \tag{22.17}$$

The first term represents that part of the incoming wave, which is scattered into a longitudinal outgoing wave, the second term is the part, which is transverse after scattering. The factors of the spherical wave are essentially the scattering amplitudes. We obtain the cross-section by taking the squared amplitudes, multiplying by the quotient of outgoing and incoming group velocities and integrating over the solid angle, i.e. all directions of $\boldsymbol{n}'$. We have for the cross-section from the initial wave into the longitudinal part

$$\sigma(a\to l) = \frac{V_0^2 q}{4\pi\varrho^2 c_l^3\omega}\left\{\frac{1}{3}(\delta\varrho)^2\omega^4 + \frac{q^2\omega^2}{c_l^2}\left[(\boldsymbol{e}\boldsymbol{n})^2\left((\delta\lambda)^2 + \frac{4}{3}\delta\lambda\delta\mu + \frac{4}{5}(\delta\mu)^2\right) + \right.\right.$$
$$\left.\left. + (\boldsymbol{e}\times\boldsymbol{n})^2\frac{4}{15}(\delta\mu)^2\right]\right\} \tag{22.18}$$

and into the transverse part

$$\sigma(a\to t) = \frac{V_0^2 q}{4\pi\varrho^2 c_t^3\omega}\left\{\frac{2}{3}(\delta\varrho)^2\omega^4 + \frac{q^2\omega^2}{c_t^2}\delta\mu^2\left[\frac{2}{3} - \frac{2}{15}(\boldsymbol{e}\boldsymbol{n})^2 - \frac{4}{15}(\boldsymbol{e}\times\boldsymbol{n})^2\right]\right\}.$$

The total cross-section is the sum of both:

$$\sigma_{\text{tot}} = \sigma(a\to l) + \sigma(a\to t)\,. \tag{22.19}$$

For incoming longitudinal waves we have

$$q = \omega/c_l;\quad (\boldsymbol{e}\boldsymbol{n}) = 1;\quad (\boldsymbol{e}\times\boldsymbol{n}) = 0 \tag{22.20a}$$

for incoming transverse waves

$$q = \omega/c_t;\quad (\boldsymbol{e}\boldsymbol{n}) = 0;\quad (\boldsymbol{e}\times\boldsymbol{n})^2 = 1\,. \tag{22.20b}$$

The cross-section is proportional to the fourth power of the frequency. This is called Rayleigh-scattering and has been discussed several times, mainly in connection with the contribution of defects to thermal conductivity [58 K 3]. The fourth power law always occurs if the dispersion of the waves involved is $\omega\sim q$. Since the elastic limit is in agreement with lattice theory for long waves, the scattering law $\sim\omega^4$ is also valid in the lattice theoretical description if $q\to 0$. This is the most important result, but it is not sufficient in all the cases (see below). If we compare $\sigma(t\to l)$ and $\sigma(l\to t)$ according to (22.18) we find*

$$2c_l^2\sigma(t\to l) = c_t^2\sigma(l\to t)$$

* Contrary to (22.10) in this relation the squares of the sound velocities occur. This is, because in (22.10) the differential cross-section refers to equal surface elements on the $\omega$-surfaces, whereas in this formula the cross-sections refer to equal solid angles $\mathrm{d}\Omega$.

in contradiction to (22.10). But this is only apparent, because we have not distinguished between both the transverse states after scattering, whereas before scattering we have assumed that there is a definite transverse polarization. This gives just a factor two.

In the preceding discussion we have used the first Born approximation. It is apparent that the result is reasonable only for small frequencies ($qR_0 \ll 1$) and for small perturbations ($\delta\varrho/\varrho$, $\delta\mu/\mu$, $\delta\lambda/\lambda \ll 1$). Otherwise Borns approximation fails. We will discuss what might happen with large perturbations with an even simpler example.

*Resonance Scattering*

In order to avoid too lengthy formulae, we limit the defect to a change in mass-density only:

$$J_i(\boldsymbol{R}) = \delta\varrho \cdot \omega^2 \cdot u_i(\boldsymbol{R}) \cdot \Theta(R_0 - R) \,. \tag{22.21}$$

Inserting this into (22.13) we obtain the integral equation

$$\begin{aligned} \overset{s}{u}_i(\boldsymbol{R}) = \delta\varrho\,\omega^2 \sum_j \int_{-\infty}^{+\infty} G_{ij}(\boldsymbol{R}-\boldsymbol{R}')\,\overset{s}{u}_j(\boldsymbol{R}')\,\Theta(R_0-R')\,\mathrm{d}\boldsymbol{R}' + \\ + \delta\varrho\,\omega^2 \sum_j \int_{-\infty}^{+\infty} G_{ij}(\boldsymbol{R}-\boldsymbol{R}')\,\overset{\circ}{u}_j(\boldsymbol{R}')\,\Theta(R_0-R')\,\mathrm{d}\boldsymbol{R}' \,. \end{aligned} \tag{22.22}$$

The inhomogeneous part is known with a given incoming wave $\overset{\circ}{u}_i(R')$. In the first part we need a knowledge of $\overset{s}{u}_i(\boldsymbol{R}')$ in the region $0 < R' < R_0$. We make the assumption that the wavelengths of incoming and scattered wave are large compared to $R_0$: $qR_0 \ll 1$ as above, so that $\overset{s}{u}_i(\boldsymbol{R}')$ can be looked upon as nearly constant in the region $R_0$. We obtain the equation for the determination of $\overset{s}{u}_i(0)$:

$$\begin{aligned} \overset{s}{u}_i(0) - \delta\varrho \cdot \omega^2 \int_{-\infty}^{+\infty} G_{ij}(\boldsymbol{R}')\,\Theta(R_0-R')\,\mathrm{d}\boldsymbol{R}' \cdot \overset{s}{u}_j(0) \\ = \delta\varrho\,\omega^2 \int_{-\infty}^{+\infty} G_{ij}(\boldsymbol{R}')\,\overset{\circ}{u}_j(R')\,\Theta(R_0-R')\,\mathrm{d}\boldsymbol{R}' \,. \end{aligned}$$

Using the expansion (21.40) for $G_{ij}(\boldsymbol{R}')$ in the long wave-length-limit we have

$$\begin{aligned} \overset{s}{u}_i(0)\left\{1 - \delta\varrho\,\omega^2 \frac{1}{\sigma\varrho}\left(\frac{1}{c_l^2} + \frac{2}{c_t^2}\right) R_0^2 - i\,\delta\varrho\,\omega^3 \frac{R_0^3}{9\varrho}\left(\frac{1}{c_l^3} + \frac{2}{c_t^3}\right)\right\} \\ = e_i \frac{\delta\varrho\,\omega^2 R_0^2}{6\varrho}\left(\frac{2}{c_l^2} + \frac{1}{c_t^2}\right) + \cdots \end{aligned}$$

where $e_i$ is the polarization vector of the incoming wave. Defining

$$\begin{aligned} \omega_R^2 &= \frac{6}{\varepsilon R_0^2}\left(\frac{2}{c_l^2} + \frac{1}{c_t^2}\right)^{-1}; \quad \varepsilon = \delta\varrho/\varrho \\ \gamma &= \frac{2}{\varepsilon R_0}\left(\frac{2}{c_l^3} + \frac{1}{c_t^3}\right)\left(\frac{2}{c_l^2} + \frac{1}{c_t^2}\right)^{-2} \end{aligned} \tag{22.23}$$

we have

$$\dot{u}_i(0) = \frac{\omega^2}{\omega_R^2 - \omega^2 - i \cdot 2\gamma\,\omega^3/\omega_R^2}\, e_i(\boldsymbol{q}\,\sigma) = f_R\, e_i(\boldsymbol{q}\,\sigma)\,, \tag{22.24}$$

which shows a resonance behavior at $\omega = \omega_R$. This value can be used for the determination of the asymptotic behavior of (22.22). Using the asymptotic form (21.39) and replacing $\dot{u}_i(R')$ in (22.22) by $\dot{u}_i(0)$ we finally obtain

$$\dot{u}_i(\boldsymbol{R}) = \frac{V_0\,\varepsilon\,\omega^2}{4\pi}(1 + f_R)\left\{t_{1i}\frac{e^{i\,\omega R/c_t}}{R\,c_t^2} + n_i'(\boldsymbol{e}\,\boldsymbol{n}')\frac{e^{i\,\omega R/c_l}}{R\,c_l^2}\right\}. \tag{22.25}$$

$V_0$, $\boldsymbol{t}_1$ and $\boldsymbol{n}'$ are defined as before. With (22.23) we obtain for the total cross-section

$$\sigma_{\text{tot}} = 2\,V_0\,\varepsilon\,\gamma\left(\frac{\omega}{\omega_R}\right)^4 \cdot \frac{q}{\omega}\,|1 + f_R|^2\,. \tag{22.26}$$

$\omega/q$ defines the group velocity of the incoming wave ($c_l$ or $c_t$). Since the imaginary part in $f_R$ is essential only near $\omega \approx \omega_R$, and we have limited to lowest order in frequency, we can rewrite

$$1 + f_R = \frac{\omega_R^2 + \cdots}{\omega_R^2 - \omega^2 - i \cdot 2\gamma\,\omega_R}\,. \tag{22.27}$$

In this derivation no assumption on the magnitude of $\delta\,\varrho = \varepsilon\,\varrho$ has been made. However, we restricted the discussion to small values of $\omega$ or more precisely, to

$$q\,R_0 \ll 1$$

and this implies also

$$\omega_R\,R_0/c \ll 1\,, \quad \text{or} \quad \varepsilon = \delta\,\varrho/\varrho \quad \text{sufficiently large}\,. \tag{22.28}$$

If $\omega_R \gg \omega$ (22.26) agrees with the result of Borns approximation, of course. Whereas the Born approximation does not satisfy the optical theorem, the solution (22.25, 26) does, provided we use $f_R$ in the form as defined in (22.24).

The form (22.26) of the cross-section exhibits a strong resonance phenomenon, which is the stronger, the larger the relative change in mass-density is. It occurs at small frequencies, the resonance frequency and the width decrease with increasing $\varepsilon$. Such a resonance scattering has been mentioned first by *Brout* and *Vissher* [62 B 4], for a lattice with isotopic defects. But, as is shown by our result, it is not characteristic for lattices; it also occurs in continuous media under appropriate conditions, which resemble those of lattice theory.

$\omega_R$ and $\gamma$ can be related to the Debye-frequency $\omega_D$, i. e. the cut-off-frequency of the spectrum, if we impose the lattice structure on the continuum. With [I]

$$\overline{\left(\frac{1}{c^3}\right)} = \frac{1}{3}\left(\frac{2}{c_t^3} + \frac{1}{c_l^3}\right) = \frac{6\pi^2}{V_z\,\omega_D^3}\,; \qquad V_z = \frac{4\pi}{3}R_0^3$$

and*

$$\overline{\left(\frac{1}{c^2}\right)} = \frac{1}{3}\left(\frac{2}{c_t^2} + \frac{1}{c_l^2}\right) = 0.93\,\overline{\left(\frac{1}{c^3}\right)}^{2/3}$$

* This relation is not exact. The value 0.93 is a mean value for the possible proportions of $c_t/c_l$

we have

$$\omega_R^2 = \omega_D^2/(2.72\,\varepsilon)\,; \quad \gamma = \omega_D/(3.15\,\varepsilon)\,. \tag{22.29}$$

This shows that the resonance is rather strong (width small compared to $\omega_D$), if $\varepsilon \gtrsim 2$. The resonance occurs at small frequencies. The mass-density has to be larger in the defect region in order to have a resonance scattering. In sect. 20 we have seen, that localized modes occur in the contrary case, if the masses are smaller than those of the host lattice. This is what is to be expected.

The same procedure can be used to seek for resonances due to a change in the elastic constants $\delta\lambda$ and $\delta\mu$. The calculation is somewhat more involved, but can be done. Resonances will occur, if $\delta\lambda$ and $\delta\mu$ are smaller than the constants of the ideal lattice, or are even negative. The latter would mean, that the defect region is a region of instability, which can exist only if it is clamped by the surrounding medium.

The situation is not very different if we use lattice theory for a description of scattering. Of course, the scattering amplitude is different for large frequencies; but the resonances occur at small frequencies and should not differ much in their behavior for a continuum and a lattice. We give two simple examples, first that of an isotopic defect in a cubic Bravais-lattice. The relative change in mass is $\varepsilon = (M' - M)/M$. According to (20.18) we have

$$J_{i\ j}^{m\ n} = \varepsilon\, M\, \omega^2\, \delta_{i\ j}^{m\ n}\, \delta^{m0} \tag{22.30a}$$

and

$$\tau_{i\ j}^{m\ n} = \left(\frac{1}{1 - G J}\right)_{i\ j}^{m\ n} = \delta_{i\ j}^{m\ n} + A_{i\ j}^{m\ 0}\, \delta^{n0} \tag{22.30b}$$

with

$$A_{i\ j}^{m0} = \frac{\varepsilon\, M\, \omega^2}{1 - \varepsilon\, M\, \omega^2\, G_{11}^{00}}\, G_{i\ j}^{m0} \tag{22.30c}$$

because $G_{ij}^{00} = G_{11}^{00} \cdot \delta_{ij}$ in cubic lattices. Further we have $(J\,\tau)_{i\ j}^{m\ n} = \varepsilon\, M\, \omega^2 \left(\delta_{ij} + A_{ij}^{00}\right) \delta^{m0}\, \delta^{n0}$. Inserting this into (22.4) we obtain

$$F(\boldsymbol{q}\,\sigma; \boldsymbol{q}'\,\sigma') = \frac{i\,V_z}{2\pi}\, \frac{e^{-i\pi/4\,(\epsilon_2 + \epsilon_3)}}{|\omega^2|_{22}\, \omega^2|_{33}|^{1/2}} \cdot \frac{\varepsilon\, \omega^2}{1 - \varepsilon\, M\, \omega^2\, G_{11}^{00}} \sum_i e_i^*(\boldsymbol{q}'\,\sigma')\, e_i(\boldsymbol{q}\,\sigma)\,. \tag{22.31}$$

From the dispersion behavior and the Greens function $G_{11}^{00}$ the scattering amplitude can be calculated. Because $G_{11}^{00}$ contains a real and an imaginary part, it is clear that the denominator of (22.31) shows a resonance behavior. If we put

$$D(\omega) = 1 - \varepsilon\, M\, \omega^2\, G_{11}^{00} \tag{22.32}$$

the resonance frequency is given by

$$\mathrm{Re}\, D(\omega = \omega_R) = 0 \tag{22.33a}$$

whereas the width of the resonance is

$$\gamma = \frac{\operatorname{Im} D(\omega_R)}{\operatorname{Re} D'(\omega_R)}; \quad D'(\omega_R) = \left(\frac{\partial D}{\partial \omega}\right)_{\omega=\omega_R}. \tag{22.33b}$$

The equation for the resonance frequency is just the same one as for the determination of the localized modes, if one remembers, that the Greens function for localized states [$\omega \neq \omega(\boldsymbol{q}\,\sigma)$ for all the $\boldsymbol{q}$, $\sigma$-values] is real. The difference between localized states and resonance states is twofold: localized states have frequencies outside the bands of ideal lattice frequencies (in ionic crystals) and zero width (in harmonic theory), whereas resonance states have frequencies in the region of lattice frequencies and finite width, already in the harmonic theory. The resonance is the more pronounced, the smaller the resonance frequency is. This follows from the fact, that the imaginary part contains the spectral density of states. From (21.25) and (21.26) we have

$$\operatorname{Im} G_{11}^{00} = \frac{\pi}{M} g(\omega^2); \quad \operatorname{Re} G_{11}^{00} = \frac{1}{M} \mathrm{P} \int_0^{\omega_m^2} \frac{g(\omega'^2)\,\mathrm{d}\omega'^2}{\omega'^2 - \omega^2}. \tag{22.34}$$

In the Debye-approximation it is

$$g(\omega^2) = \frac{3}{2}\frac{\omega}{\omega_D^3} \qquad 0 < \omega < \omega_D$$

$$= 0 \qquad \text{otherwise}$$

and therefore*

$$\operatorname{Im} G_{11}^{00} = \frac{3\pi}{2M} \cdot \frac{\omega}{\omega_D^3}; \quad \operatorname{Re} G_{11}^{00} = \frac{3}{M\omega_D^2}\left\{1 - \frac{\omega}{\omega_D} \operatorname{artgh} \frac{\omega}{\omega_D}\right\}.$$

The resonance frequency follows from

$$\frac{1}{3\varepsilon} = \left(\frac{\omega_R}{\omega_D}\right)^2 \left\{1 - \frac{\omega_R}{\omega_D} \operatorname{artgh} \frac{\omega_R}{\omega_D}\right\}, \tag{22.35a}$$

the width from

$$\frac{\pi}{4}\left(\frac{\omega_R}{\omega_D}\right)^2 = \frac{\gamma}{\omega_D}\left\{1 - \frac{\omega_R^2}{2(\omega_D^2 - \omega_R^2)} - \frac{3\,\omega_R}{2\,\omega_D} \operatorname{artgh} \frac{\omega_R}{\omega_D}\right\}. \tag{22.35b}$$

We expect the resonances to exist at small frequencies. This leads to

$$\omega_R^2 = \omega_D^2/3\varepsilon; \quad \gamma = \omega_D \cdot \pi/12\varepsilon, \tag{22.36}$$

in accordance with (22.29). The differences in the numerical values are related to putting $V_s = 4\pi R_0^3/3$. A slightly different choice would give even better agreement.

Equation (22.32) allows for a somewhat different reading. Instead of asking for the zeros of the real part of $D$ for real $\omega$, we can ask for the zeros of $D(\omega)$ with complex $\omega = \omega_R - i\gamma$:

$$D(\omega_R - i\gamma) = 0. \tag{22.37}$$

* The term with "artgh" results from the cut-off of the elastic spectrum at $\omega_D$. Where this term is essential, the use of the elastic spectrum is no longer allowed (see page 166). $\omega_D$ in front of the curly brackets and in $\operatorname{Im} G_{11}^{00}$ is only an abbreviation for the parameters which enter the spectrum for small $\omega$ ($\to 0$).

For small $\omega_R$, $\gamma$ this gives the same result as above. Complex frequencies give states with a time-dependence $\exp[-i\,\omega_R\,t - \gamma\,t]$ i. e. exponentially decaying states. The life-time of these states is just the reciprocal width of the resonance states.

A complete lattice theoretical calculation of (22.31) is possible only for a simple cubic lattice with nearest neighbor interaction. In other cases the Greens function has not been calculated [65 M 2]*. For this model we have

$$e_i(\boldsymbol{q}\,\sigma) = \delta_{i\sigma} \tag{22.38}$$
$$M\,\omega^2(\boldsymbol{q}, \sigma = 1) = 4\,\alpha \sin^2 q_1\,a/2 + 4\,\beta(\sin^2 q_2\,a/2 + \sin^2 q_3\,a/2)$$
$$\text{and cyclic}.$$

The total cross-section for an incoming wave $\boldsymbol{q}\,\sigma$ is from (22.11) and (22.31)

$$\sigma_{\text{tot}} = \frac{2\,V_z}{|\operatorname{grad}\omega^2(\boldsymbol{q}\,\sigma)|}\operatorname{Im}\frac{\varepsilon\,\omega^2}{1 - \varepsilon\,M\,\omega^2\,G_{11}^{00}}\,. \tag{22.39}$$

Fig. 6.9 shows the resonance frequency as a function of mass-change $\varepsilon$, compared with what follows from Debyes approximation. One has to keep

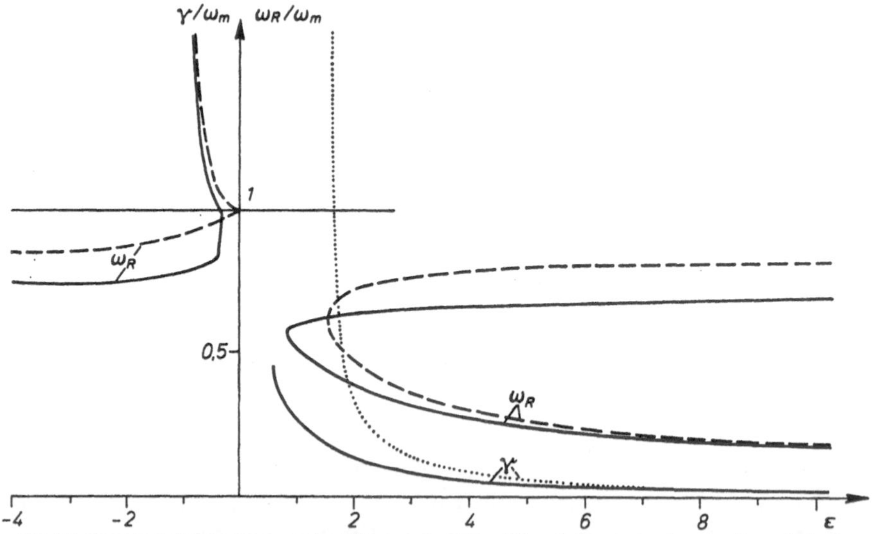


Fig. 6.9. Resonance frequencies $\omega_R$ and widths $\gamma$ as function of the relative change in mass, for an isotopic defect in a simple cubic lattice. Full lines with the tabulated Greens function [65 M 2], broken lines using a Debye approximation in calculating the Greens function

in mind, that the Debye-approximation is reasonable only for small $\omega$, say $\omega \leqq \omega_D/2$ (see remark on page 166). The curves have been drawn

* It should be emphasized, that the lattice is not stable for the values $\alpha = \beta$, for which the Greens function has been calculated only ($0 < \omega < \omega_m$). It would be stable for $\alpha > 2\,\beta > 0$. But as the Greens function has not been calculated for these values, we use the numerical values of the unstable model. The qualitative behavior is not changed when going to stable models, only quantitative differences occur.

for larger values, also using (22.35) but this is a poor approximation and definitely wrong near $\omega/\omega_m \approx 1$. Fig. 6.10 shows the poles of $D^{-1}(\omega)$, eq. 22.32 in the complex $\omega$-plane, with $\varepsilon$ as parameter. For $\omega \geqq \omega_D/2$ this has again only a qualitative meaning, because the Debye-approximation has been used. That part of the curve in Fig. 6.9, which has

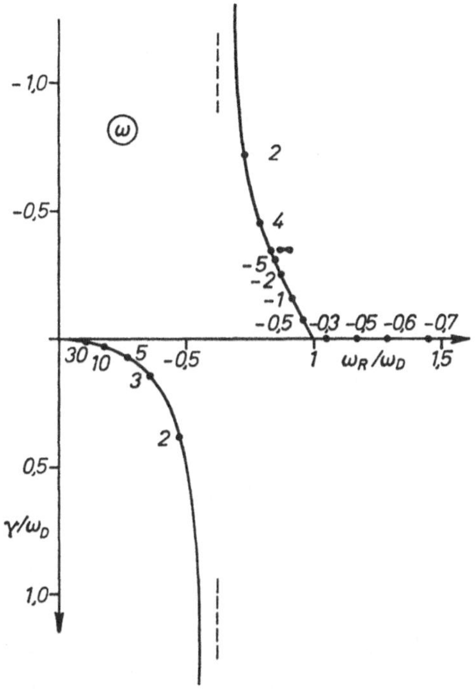


Fig. 6.10. Poles of the scattering amplitude in the complex $\omega$-plane. The numbers at the curve give the values of $\varepsilon = (M' - M)/M$ as a parameter

$\partial\omega_R/\partial\varepsilon > 0$ belongs the negative widths $\gamma$ of the resonance. Resonances are possible for $\infty > \varepsilon \gtrsim 0.8$ with frequencies $0 < \omega_R \lesssim 0.57\,\omega_m$ and with $\varepsilon < -0.33$, or frequencies $\omega_R \gtrsim 0.99\,\omega_m$. For $\varepsilon < -0.34$, $\omega_R > \omega_m$ the widths of the resonances are zero and we have localized states [compare sect. 21, Fig. 6.3]. Fig. 6.11 shows the total scattering cross section for an incoming longitudinal wave with $q = \{q_1, 0, 0\}$. The pronounced resonance for $\varepsilon = 5$ can be seen, whereas $\varepsilon = 1$ shows no pronounced behavior. The resonance has mainly disappeared. At the maximum frequency for waves in the (100)-direction, $\omega^2 = \omega_m^2/3$ the group velocity vanishes, thus $\sigma_{tot}$ being infinite. This is no resonance. The Debye-approximation fails at large frequencies. If the incoming waves have frequencies between $0.99\,\omega_m$ and $\omega_m$, there occurs a very weak resonance if $-0.34 \lesssim \varepsilon \lesssim -0.33$. This need not be the case in other lattices. In this special model it arises from the behavior of the spectral density at $\omega \approx \omega_m$ where it becomes small again.

As already mentioned on page 166, this resonance which is connected with a single mass change in the threedimensional lattice, does not occur on a one-dimensional lattice; and it does not occur in a corresponding model of a one-dimensional continuum (without cut-off frequency). If one imposes a cut-off-frequency to the spectrum of the continuum, this leads to a "resonance", but in any case, this is not a reasonable physical model. The nature of this single-mass-resonance in a three-dimensional lattice is somewhat different from other resonances. E. g. if there are two heavy masses at different positions in a linear lattice, there is a resonance. But this has a different character compared with that described above. The situation for changes in force-constants is similar to the two-mass resonance.

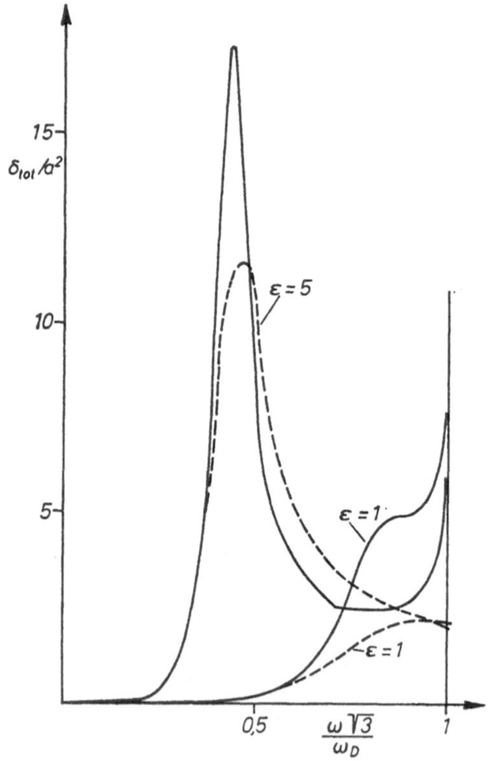


Fig. 6.11. Total scattering cross section for the scattering of plane waves at an isotopic defect in a simple cubic lattice. The incoming wave propagates in the (100)-direction. It is assumed $\alpha = \beta$. Full line exact, broken line with a Debye approximation used in the Green function

The difference between the threedimensional and the one-dimensional lattice, containing a single heavy mass, can be seen more or less evidently by the following consideration: To have a resonance, one needs a restoring force and a damping force. The damping force occurs when the heavy mass moves in the lattice, as well in an one-dimensional as in a three-dimensional lattice. In an one-dimensional lattice there is no restoring force; if one spring is strained in a static displacement, the restoring force is proportional $1/N$, which vanishes for $N \to \infty$ ($N$: number of

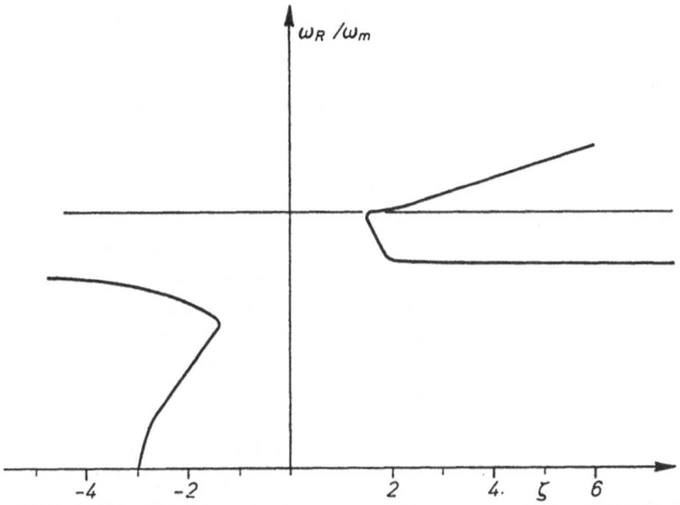


Fig. 6.12. Resonance frequency as function of a change in force constant, due to a single force-constant change in a simple cubic lattice

atoms in the linear lattice). In a three-dimensional lattice the restoring force is not of the order $1/N$, because the lattice-ions are connected by a number of springs in all directions, not only by one spring in one direction.

To see the influence of a change in force-constants on the scattering of phonons, we use the simple example of sect. 20. The occurrence of resonances needs a large scattering amplitude for certain frequencies. The only quantity which can lead to a pronounced behavior is the $\tau = (\mathbf{1} - \mathbf{G}\,\mathbf{J})^{-1}$ term in (22.4), giving large terms if the determinant $\mathbf{1} - \mathbf{G}\,\mathbf{J}$ becomes very small. Because the point group symmetries of $\mathbf{G}\,\mathbf{J}$ are always the same, for frequencies inside the region of band frequencies as well as for those outside, we can use the group-theoretical procedure of section 20 (for details see [64 L 1, 65 M 2]) to reduce the secular equation to a simpler form. The only difference now is the imaginary part of $\mathbf{G}$, which vanishes in the case of localized states. Thus the resonance condition is that the

real parts of the reduced secular equation,
$\|\mathbf{1} - \mathbf{G}\,\mathbf{J}\|$ have to vanish for frequencies in the region (22.40)
of the ideal lattice spectrum.

For the simple example of sect. 20, where we have changed only one force-constant of the lattice, we have the resonance condition [see (20.29)]

$$1 = -2\,\alpha\,\zeta\,\mathrm{Re}\left\{G^{00}_{11} - G^{0\,100}_{1\,1}\right\}. \tag{22.41}$$

Fig. 6.12 shows the resonance frequency as function of $\zeta$. Now $\partial \omega_R/\partial \zeta$ has to be positive to allow for positive widths. Resonances occur if for small frequencies $0 < \omega_R \lesssim 0.58\,\omega_m$, if $-\infty < \zeta \lesssim -1.35$, and for $0.99\,\omega_m \lesssim \omega_R$ if $\zeta \gtrsim 1.50$. For $\zeta \gtrsim 1.54$, $\omega_R > \omega_m$ the states are localized. $\zeta < -1$ means however, that the force-constant involved becomes negative; such a situation would be unstable, if the rest of the lattice would not clamp this spring. This is not a feature of this very special model; calculations for a more complicated model give similar results [65 M 2], and one can easily see by general considerations that resonances are most likely to occur if force-constants become negative, though this is not a necessary condition in any case.

Though it is rather simple to calculate resonance frequencies, the calculation of the cross-section implies much numerical work. Whereas the determination of the resonances only needs the zeros of the real part of the determinant $\|\mathbf{1} - \mathbf{G}\,\mathbf{J}\|$, the cross-section needs the inversion of the whole matrix $(\mathbf{1} - \mathbf{G}\,\mathbf{J})$. The procedure of simplifying this is the following: One seeks for the point group symmetries of $\mathbf{1} - \mathbf{G}\,\mathbf{J}$, determines the unitary transformation $\mathbf{U}$, which transforms $\mathbf{1} - \mathbf{G}\,\mathbf{J}$ into its irreducible components $\mathbf{U}(\mathbf{1} - \mathbf{G}\,\mathbf{J})\,\mathbf{U}^{-1}$ and calculates the reciprocal of these irreducible submatrices. This involves the inversion of lower rank matrices than the inversion of the original one; in general the rank is lower than maximal six, most irreducible submatrices have rank one to three. If this has been done, the inverse matrix can be transformed back to the original representation. But one can describe the scattering also in the representation, in which the $(\mathbf{1} - \mathbf{G}\,\mathbf{J})^{-1}$-matrix is reduced to its irreducible sets. These irreducible sets all have a certain transformation behavior, corresponding to the decomposition of scattered waves into partial waves. In cubic crystals with defects of the complete cubic symmetry e. g., the partial amplitudes which belong to $A_{1g}$-irreducible representations describe $S$-wave scattering, those of $F_{1u}$-representations $P$-wave scattering, those of $E_g$- and $F_{2g}$-representations $D$-wave scattering and so on. The characteristic data for the most important defects have been given in [64 L 1]. Calculations have been done only in one simple example [65 M 2]. The most discussions of scattering at defects have used an isotopic defect as a model [62 B 4, 63 T 2].

The scattering amplitudes of phonons at defects are essential for the determination of the defect contribution to thermal resistance. Investigations have been done theoretically and experimentally in a number of papers.

The explicit calculation of thermal conductivity $\varkappa$ is tedious. A first approximation is given by using a formula of *Debye* [14 D 1], according to which

$$\varkappa \sim \int \frac{v(\omega)}{\sigma(\omega)}\, c(\omega)\, z(\omega)\, \mathrm{d}\omega\,. \tag{22.42}$$

$\sigma(\omega)$ is the scattering cross section for the phonons, $v(\omega)$ their group velocity, $c(\omega)$ the specific heat and $z(\omega)$ the spectral density. For $\omega \to 0$ it is $v(\omega) = \text{const.}$, $c(\omega) = \partial\varepsilon(\omega)/\partial T \sim \text{const.}$ and $z(\omega) \sim \omega^2$. For the scattering at point defects we have seen in (22.18, 19), that there is a very small scattering cross section for $\omega \to 0$, being proportional to $\omega^4$. Thus the integrand in (22.42) varies as $\omega^{-2}$; this results in a diverging value for the thermal conductivity as far as it is due to the scattering at point defects. To get a reasonable result, other scattering processes have to be discussed at the same time. Such processes are the phonon-phonon-scattering and the scattering of phonons at the boundaries of the crystals. If one adds the cross-sections of the different processes in (22.42), one indeed gets a finite value for $\varkappa$. A somewhat better procedure for the discussion of thermal conductivity is the use of Boltzmann's equation.

The total Hamiltonian of the system can be written as

$$\mathscr{H} = \mathscr{H}_0 + \Phi_3 + J_0 + J_2 \,. \tag{22.43}$$

$\mathscr{H}_0 = \sum_{\boldsymbol{q}\sigma} \hbar\,\omega(\boldsymbol{q}\sigma) \left\{b^{+\boldsymbol{q}}_{\sigma} b^{\boldsymbol{q}}_{\sigma} + 1/2\right\}$ is the Hamiltonian of the ideal harmonic crystal; $\Phi_3$ is the perturbation due to the anharmonicity of the crystal (phonon-phonon-interaction). It is given by (16.13), containing products of three creation or annihilation operators. $J_0$ describes the point defects $\left(\text{it is essentially } J^{mn}_{i\,j}\, u^m_i\, u^n_j\right)$ and is quadratic in $b^{\boldsymbol{q}}_{\sigma}, b^{+\boldsymbol{q}}_{\sigma}$, whereas $J_2$ describes the interaction of phonons with the boundary, being again quadratic in $b^{+\boldsymbol{q}}_{\sigma}, b^{\boldsymbol{q}}_{\sigma}$, (in a first approximation). The transition probability due to the interactions then is, in Born's approximation

$$\omega_{i\to f} = \frac{2\pi}{\hbar}\,\delta(\varepsilon_i - \varepsilon_f)\,|\langle \mathrm{f}\,|\,\Phi_3 + J_0 + J_2\,|\,\mathrm{i}\rangle|^2 \,, \tag{22.44}$$

where i, f represent initial and final states of the phonons, specified by the set of occupation numbers: $|\mathrm{i}\rangle = \left|\ldots n^{\boldsymbol{q}}_{\sigma} \ldots\right\rangle$. $\varepsilon_{\mathrm{i}}, \varepsilon_{\mathrm{f}}$ are the corresponding energies. The main features can be seen already by neglecting all the interference terms between the different scattering processes. It can be shown, that this approximation is not quite bad [65 T 2]. All the different contributions then enter the Boltzmann equation additively. The Boltzmann equation states, that the changes in time of the mean occupation numbers due to interaction and convection satisfy [I]:

$$\dot{\bar{n}}_{\mathrm{C}}(\boldsymbol{q}\sigma) + \dot{\bar{n}}_{\mathrm{I}}(\boldsymbol{q}\sigma) = 0 \,. \tag{22.45}$$

The convection term is given by* [I]:

$$\dot{\bar{n}}_{\mathrm{C}}(\alpha) = -\,(\boldsymbol{v}_\alpha \cdot \operatorname{grad} T) \cdot \frac{\hbar\,\omega_\alpha}{4\,k\,T^2 \sinh^2 \hbar\,\omega_\alpha/2\,k\,T} \,. \tag{22.46}$$

The anharmonic interaction term can be taken from [I, equ. (93.6)] or [64 L 2, equ. (5.10)]; using the deviations

$$\nu_\alpha = n_\alpha - \bar{n}^T_\alpha \tag{22.47}$$

* In the following we often abbreviate $\boldsymbol{q}\,\sigma$ by $\alpha$, $\boldsymbol{q}'\,\sigma'$ by $\beta$, etc.

from the thermal equilibrium values for the description of the change in $n_\alpha$ we have

$$\left\{\frac{d}{dt}\nu_\alpha\right\}_{anh} = -\frac{\pi\hbar}{8N}\times$$
$$\times\sum_{\beta\gamma}\frac{|\Phi_{\alpha\beta\gamma}|^2\{\delta(\omega_\alpha-\omega_\beta-\omega_\gamma)+\delta(\omega_\beta-\omega_\alpha-\omega_\gamma)+\delta(\omega_\gamma-\omega_\alpha-\omega_\beta)\}}{\omega_\alpha\,\omega_\beta\,\omega_\gamma\sinh(\hbar\,\omega_\alpha/2kT)\sinh(\hbar\,\omega_\beta/2kT)\sinh(\hbar\,\omega_\gamma/2kT)}\times$$
$$\times\{\nu_\alpha\sinh^2(\hbar\omega_\alpha/2kT)+\nu_\beta\sinh^2(\hbar\omega_\beta/2kT)+\nu_\gamma\sinh^2(\hbar\omega_\gamma/2kT)\}\,. \tag{22.48}$$

The interaction of the phonons with defects is described by (22.8)

$$d\sigma(\alpha\to\beta)=\frac{V_z}{(2\pi s)^2}\frac{|\langle\beta\,|J\,\tau|\,\alpha\rangle|^2}{|\mathrm{grad}\,\omega_\alpha^2|}\,\delta(\omega_\alpha^2-\omega_\beta^2)\,dV_{(\beta)}\,. \tag{22.49}$$

This gives the energy transfer from the wave $\alpha$ into the wave $\beta$, therefore a decrease in the occupation number $n_\alpha$ and an increase in $n_\beta$. With the incoming energy flux

$$S_0=\frac{1}{V}v_\alpha\,\hbar\,\omega_\alpha\,n_\alpha$$

we have for the transition from $n_\alpha$ into $n_\beta$

$$-\frac{d}{dt}[\hbar\,\omega_\alpha n_\alpha]=S_0\,d\sigma(\alpha\to\beta)$$

or

$$-\frac{d}{dt}n_\alpha=\frac{1}{V}v_\alpha n_\alpha\,d\sigma(\alpha\to\beta) \tag{22.50a}$$

and for the whole decrease of $n_\alpha$

$$-\frac{d}{dt}n_\alpha=\frac{1}{V}v_\alpha\,n_\alpha\,\sigma_{tot}(\alpha)\,. \tag{22.50b}$$

The change in $n_\beta$ is given by a similar equation:

$$\frac{d}{dt}n_\beta=\frac{1}{V}v_\alpha\,n_\alpha\,d\sigma(\alpha\to\beta)\,. \tag{22.50c}$$

This is the energy going from $\alpha$ into $\beta$. The total change in $n_\alpha$ is composed of the decrease due to (22.50b) and the increase due to the scattering from all the $\beta$ into $\alpha$, described by (22.50c) if $\alpha$ and $\beta$ are interchanged and if it is summed over all processes; thus totally

$$\frac{d}{dt}n_\alpha=-\frac{1}{V}v_\alpha\,n_\alpha\,\sigma_{tot}(\alpha)+\frac{1}{V}\sum_{\sigma'}\int dq'\,v(q'\sigma')\,n(q'\sigma')\,d\sigma(q'\sigma'\to\alpha)$$

and with (22.48) and (22.10ff.) and $v_\alpha=|\mathrm{grad}\,\omega_\alpha|=\frac{1}{2\omega_\alpha}|\mathrm{grad}\,\omega_\alpha^2|$

$$\frac{d}{dt}n_\alpha=\frac{1}{2V\omega_\alpha}\left(\frac{V_z}{2\pi s}\right)^2\int d\beta\,|\langle\beta\,|J\tau\,|\,\alpha\rangle|^2\,\delta(\omega_\alpha^2-\omega_\beta^2)\cdot\{n_\beta-n_\alpha\}\,. \tag{22.51}$$

This holds for a single defect. In case of a small concentration of defects, we have to multiply by the number of defects $n_D$, if interactions between different defects can be neglected. $p=n_D/N=n_D\,V_z/V$ is the ionic concentration of defects. For small deviations from the thermal equilibrium we have (22.47) and finally, because there is no change of $n_\alpha$ in thermal equilibrium

$$\left\{\frac{\mathrm{d}}{\mathrm{d}t}\nu_\alpha\right\}_{\mathrm{def}} = \frac{p\,V_z}{8\pi^2 s^2 \omega_\alpha} \int \mathrm{d}\beta |\langle\beta\,|J\tau\,|\,\alpha\rangle|^2\,\delta(\omega_\alpha^2 - \omega_\beta^2)\cdot\{n_\beta - n_\alpha\}\,. \tag{22.52}$$

The boundary scattering can be described approximately by the use of a relaxation time, assuming

$$\left\{\frac{\mathrm{d}}{\mathrm{d}t}\nu_\alpha\right\}_{\mathrm{bound}} = -\frac{1}{\tau_{\mathrm{B}}}\cdot\nu_\alpha;\quad \tau_{\mathrm{B}} = L/\bar{v}\,. \tag{22.53}$$

$L$ is a characteristic length of the crystal, $\bar{v}$ a mean velocity of sound.

The sum of (22.48), (22.52) and (22.53) gives the total change in $n_\alpha$ due to interaction processes.

The energy current is given by

$$S_i = -\sum_j \varkappa_{ij}(\mathrm{grad}\,T)_j = \frac{1}{V}\sum_\alpha \hbar\omega_\alpha(\bar{n}_\alpha^T + \nu_\alpha)\,v_{\alpha,i}\,. \tag{22.54 a}$$

$\varkappa_{ij}$ is the tensor of thermal conductivity. Because of

$$\omega(\boldsymbol{q}\sigma) = \omega(-\boldsymbol{q}\sigma)\,;\quad \boldsymbol{v}(\boldsymbol{q}\sigma) = -\boldsymbol{v}(-\boldsymbol{q}\sigma)$$

(22.54 a) simplifies to

$$-\sum_j \varkappa_{ij}(\mathrm{grad}\,T)_j = \frac{1}{V}\sum_\alpha \hbar\omega_\alpha\nu_\alpha\cdot v_{\alpha,i}\,. \tag{22.54 b}$$

In thermal equilibrium the energy current vanishes, as it must be. If we have solved (22.45) for the $\nu_\alpha$, we could calculate $\varkappa_{ij}$. The equation for the $\nu_\alpha$ is a linear inhomogeneous system, which can be solved easily only if all the non-diagonal elements can be neglected (relaxation-time approximation). But in general the non-diagonal elements are not small compared to the diagonal ones. A better solution would be by means of a variational procedure. Every inhomogeneous equation allows for a variational procedure. The equation for the determination of the $\nu_\alpha$ can be rewritten as

$$\frac{\hbar\omega_\alpha}{T^2}(\boldsymbol{v}_\alpha, \mathrm{grad}\,T) + \sum_\beta P_{\alpha\beta}\,\nu_\beta = 0\,. \tag{22.55}$$

$P_{\alpha\beta}$ is an abbreviation for all the scattering processes. It can be shown easily, that

$$P_{\alpha\beta} = P_{\beta\alpha}\,. \tag{22.56}$$

Therefore eq. (22.55) can be looked upon as the condition that the functional

$$G(\ldots\nu_\alpha\ldots) = -\frac{1}{2}\sum_{\alpha,\beta} P_{\alpha\beta}\,\nu_\alpha\,\nu_\beta - \frac{1}{T^2}\sum_\alpha \hbar\,\omega_\alpha(\boldsymbol{v}_\alpha, \mathrm{grad}\,T)\cdot\nu_\alpha \tag{22.57}$$

is a maximum. The maximal value of $G$ is

$$\begin{aligned} G_{\max} &= \frac{1}{2}\sum_{\alpha\beta} P_{\alpha\beta}\,\nu_\alpha\,\nu_\beta = -\frac{1}{2T^2}\sum_\alpha \hbar\omega_\alpha(\boldsymbol{v}_\alpha,\ \mathrm{grad}\,T)\ \nu_\alpha \\ &= \frac{V}{2T^2}\sum_{ij} \varkappa_{ij}(\mathrm{grad}\,T)_i\,(\mathrm{grad}\,T)_j \end{aligned} \tag{22.58}$$

according to (22.54a). Thus the thermal conductivity is directly related to the maximal value of $G$. In cubic crystals it is simply

$$G_{\max} = \varkappa \cdot V (\operatorname{grad} T)^2 / 2 T^2 . \tag{22.58a}$$

Using a trial function with free parameters for the $\nu_\alpha$, from the maximalization of this trial function we get an approximate value of $\varkappa$, which is lower than the correct value.

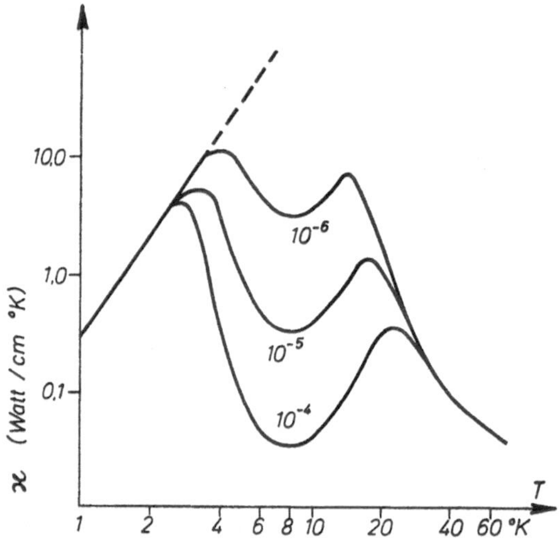


Fig. 6.13. Thermal conductivity as function of temperature. The dip is due to resonance scattering at an isotopic defect with $\varepsilon = 5$, calculated approximately according to [65 T 2]

This procedure with the simplest trial function for the $\nu_\alpha$ has been used to calculate the thermal conductivity of a crystal with isotopic (mass) defects [63 T 2, 65 T 2]. Some of the results are shown in Fig. 6.13. The dip is strongly related to the resonance phenomenon in the phonon scattering at defects. Similar results have been obtained by *Elliott* and *Taylor* [64 E 2]. Apart from the dip, occuring only if strong resonance scattering is present, the defect contribution consists in a general lowering of the thermal conductivity over the whole temperature region [see 66 F 1].

Resonances have been observed experimentally by *Pohl* and coworkers [63 P 1, 66 S 3, 66 P 3]. They have investigated the thermal conductivity in alkali halides containing different kinds of defects. In the system KCl with $J$-impurities a slight indentation is found, but not a real dip (Fig. 6.14). In the case of KCl with Li-impurities there is a dip (Fig. 6.15); but this is surely not due to a mass-resonance. It might be that it corresponds to a resonance which originates from a change in force-constants. We have seen (page 180) that such resonances mainly occur, if the defect force-constants are negative. In the case of Li-impuritis in KCl such force-constants can be explained only by rather special assumptions on the potential for Li in KCl. There must be at

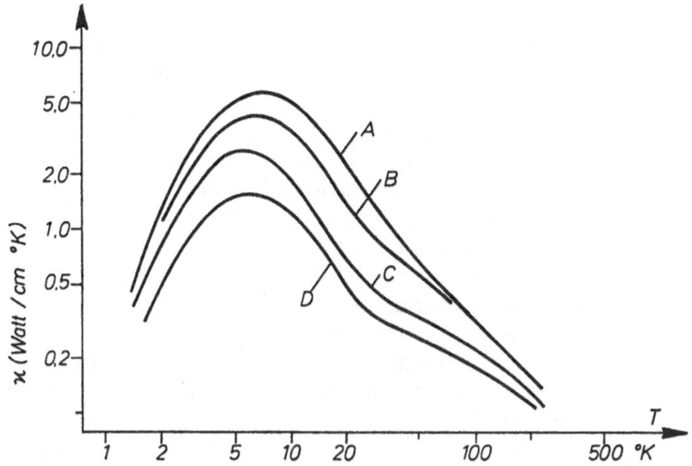


Fig. 6.14. Experimental results for the thermal conductivity as function of temperature, according to [63 P 1], for a KCl-crystal with J-impurities. The curves are: *A:* undoped KCl; *B:* KCl with 6,2 · $10^{-3}$ mol-% KJ; *C:* KCl with 7,75 · $10^{-3}$ mol-% KJ; *D:* KCl with 0,31 mol-% KJ

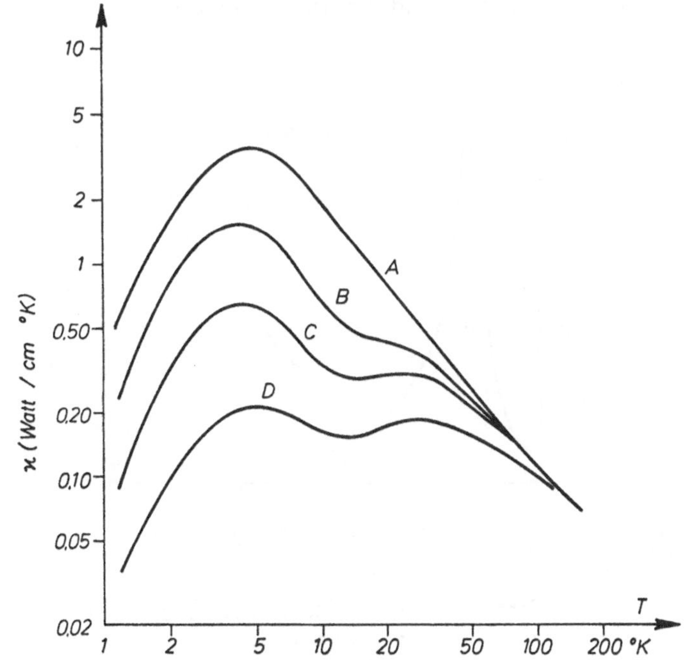


Fig. 6.15. Experimental results for the thermal conductivity as function of temperature, according to [66 P 2] for a KBr-crystal with Li-impurities. The curves are: *A:* undoped KBr; *B:* KBr with 2,2 · $10^{-3}$ mol-% LiBr; *C:* KBr with 3,7 · $10^{-3}$ mol-% LiBr; *D:* KBr with 0,13 mol-% LiBr

least a maximum in the potential for Li between neighboring Cl-ions. Then there would be a tunnel-splitting of the lowest vibration levels. The transition between such levels leads to a resonance. A more detailed knowledge is necessary for definite statements.

Other resonances, *Pohl* and coworkers have found, are due to librational motions of molecular impurities in alkali-halides. There are certain equilibrium positions for the orientation of such molecules. If the temperature is very low, they only librate about their equilibrium positions. These librational states for molecular impurities in ionic crystals are quasi-localized resonance states, having small resonance frequencies. Phonon scattering at these impurities can be described by the theory outlined above and give corresponding results.

Another possibility for the investigation of resonance states is related to infrared-and Raman-measurements. As stated on page 179, the resonance frequencies can be calculated from the zeros of the real part of $\|\mathbf{1} - \mathbf{GJ}\|$. This equation has the same point group symmetries as that for localized states; this is $\|\mathbf{1} - \mathbf{GJ}\| = 0$ with real $\mathbf{G}$. The reduction of resonance states gives the same irreducible parts, having the same infrared- and Raman-activities as the corresponding localized states. Thus a resonance state which is related to a change of mass of one ion in an ionic lattice should be infrared-active. These resonance-states have been observed in alkali-halides, e.g. in NaCl with Ag-impurities. The resonance frequency and the width of the resonance agree fairly well with

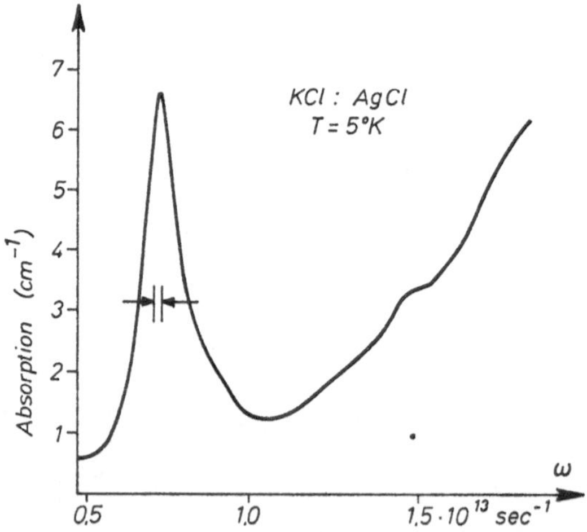


Fig. 6.16. Absorption coefficient for KCl with Ag-impurities, according to *Sievers* [66 S 4]. It shows strong resonance behavior at low frequencies. For comparison the Debye-frequency $\omega_D = 3{,}14 \cdot 10^{13}\ \text{sec}^{-1}$ (see also fig. 7.11)

the values resulting from the simple calculation on page 175 (see Fig. 6.16). Of course there is an additional width from the interaction of various defects, and from anharmonicity. Raman-active states, however, are related to changes in force-constants and occur only if such changes are present in the lattice. Also such states have been found experimentally.

## 23. Interaction of Defects. Spectral Density of Lattices with Defects and Related Properties

For a first insight in what happens if there are more than one defect in a lattice we discuss again the simplest case of this kind. Let us assume two isotopes (change in mass only) with relative mass $\varepsilon = (M' - M)/M$ at lattice sites **0** and $\boldsymbol{h}$ in a Bravais lattice. We consider the stationary problem. The perturbation is given by

$$J_{i\ j}^{m\,n} = \varepsilon M \omega^2 \delta_{ij} \{\delta^{0m} \delta^{0n} + \delta^{hm} \delta^{nh}\}; \quad \boldsymbol{h} \neq 0\,. \tag{23.1}$$

According to (20.11), (22.40) the frequencies of the localized states and of the resonances are determined by

$$\|\mathbf{G}\mathbf{J} - 1\| = 0 \tag{23.2}$$

or explicitely with (23.1)

$$\begin{Vmatrix} \varepsilon M \omega^2 G_{i\ i}^{0\,0} - 1 & \varepsilon M \omega^2 G_{i\ i}^{0\,h} \\ \varepsilon M \omega^2 G_{i\ i}^{0\,h} & \varepsilon M \omega^2 G_{i\ i}^{0\,0} - 1 \end{Vmatrix} = 0 \quad \text{for every } i; \tag{23.2a}$$

the properties of $\mathbf{G}$ have been used in (23.2a). If the two isotopes are separated by a large distance ($\boldsymbol{h} \to \infty$), $G_{i\ i}^{0\,h} \to 0$ and we have just twice the equation for a single isotope. There is no interaction between both and all the states (localized as well as resonance) are doubly degenerated for every $i$. This is clear from the physical principles. If the distance between both isotopes is finite (or rather small) the degenerated states split into two states; the magnitude of the splitting depends on the distance and is the largest, if the isotopes are in nearest neighbor positions. This is clear also. Equation (23.2a) can be reduced to two one-dimensional problems immediately by symmetry consideration*:

$$\left\{\varepsilon M \omega^2 \left(G_{i\ i}^{0\,0} + G_{i\ i}^{0\,h}\right) - 1\right\} \cdot \left\{\varepsilon M \omega^2 \left(G_{i\ i}^{0\,0} - G_{i\ i}^{0\,h}\right) - 1\right\} = 0 \tag{23.2b}$$

$$\text{for every } i = 1, 2, 3\,.$$

In cubic Bravais-crystals, $G_{1\,1}^{0\,0} = G_{2\,2}^{0\,0} = G_{3\,3}^{0\,0}$, that means every localized (or resonance) state of a single isotope is triply degenerated, thus the total degeneracy of two infinitely separated isotopes is six. Because $G_{i\ i}^{0\,h}$ for $\boldsymbol{h} \neq 0$ is not independent of $i$, the states do not split only into two triply degenerated states, but into more states with less degeneracy. In the three cubic Bravais crystals the splitting is as follows:

| Position of isotopes neighbor | number of states and degeneracy for simple cubic | b.c.c. | f.c.c. |
|---|---|---|---|
| nearest | 2 simple<br>2 double | 2 triple | 2 simple<br>2 double |
| second | 2 simple<br>2 double | 2 simple<br>2 double | 2 simple<br>2 double |
| third | 2 triple | 2 simple<br>2 double | 2 simple<br>2 double |
| ∞ | 1 sixfold | 1 sixfold | 1 sixfold |

* This holds only if the two isotopes are equal.

If one makes special assumptions on the force constants there might be an additional accidental degeneracy. This holds for localized states as well as for resonances. Cases may occur, where one or two of the splitted states are resonance states, the others are localized; this happens if the spectral density of the ideal lattice frequencies is small enough as to allow for resonance states at large frequencies.

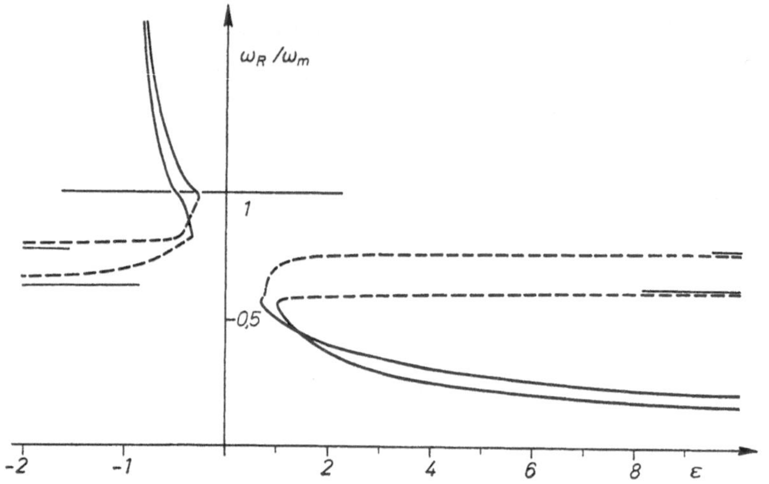


Fig. 6.17. Resonance frequencies $\omega_R$ for two isotopes $\varepsilon = (M' - M)/M$ in nearest neighbor position in a simple cubic lattice. There are two resonances now.

To get a little quantitative insight we consider again a simple cubic lattice with nearest neighbor interaction. The Greens function is tabulated only for $\alpha = \beta$ (see page 176), which we use for quantitative calculations. Then $G^{0\boldsymbol{h}}_{ii}$ is independent of $i$ and the two states are triply degenerated. $G^{0\boldsymbol{h}}_{ii}$ is real for localized states and the frequencies can be calculated immediately from (23.2b) as functions of $\varepsilon$. For the resonance states we can use (23.2b) for the calculation of imaginary frequencies or the real part to get the resonance frequencies. Fig. 6.17 shows the two $\omega_R$ as function of $\varepsilon$ for $\boldsymbol{h} = (100)$, in which the splitting is largest.

| For | there are (is) | with |
|---|---|---|
| $\infty > \varepsilon \gtrsim 1$ | two resonance states | $0 < \omega_R \lesssim 0.56\,\omega_m$ |
| $1 \gtrsim \varepsilon \gtrsim 0.675$ | one resonance state | $0.51\,\omega_m \lesssim \omega_R \lesssim 0.58\,\omega_m$ |
| $0.675 \gtrsim \varepsilon \gtrsim -0.255$ | nothing | |
| $-0.255 \gtrsim \varepsilon \gtrsim -0.257$ | one resonance state | $0.99\,\omega_m \lesssim \omega_R \lesssim \omega_m$ |
| $-0.257 \gtrsim \varepsilon \gtrsim -0.332$ | one localized state | $\omega_m \lesssim \omega_R$ |
| $-0.332 \gtrsim \varepsilon \gtrsim -0.495$ | one resonance and<br>one localized state | $0.83\,\omega_m \lesssim \omega_R \lesssim \omega_m$<br>$\omega_m \lesssim \omega_R$ |
| $-0.495 \gtrsim \varepsilon > -1$ | two localized states | $\omega_m \lesssim \omega_R$ . |

The region for resonance frequencies near the upper limit of the ideal frequency band is larger than in the case of one single isotope. However, the resonances at high frequencies are again less pronounced than those at low frequencies. On the other hand, because of the finite width of the resonance states, the low frequency resonances do overlap, even in the case of the largest splitting. This can be seen by using the widths of Fig. 6.9, which is an approximate value even in the case of two isotopes. The high frequency resonance generally occurs together with a localized state (zero width) and therefore has a smaller width than the width of the two low-frequency- resonances.

If the isotopes have a larger separation, things are similar. Only the splitting is smaller and the low-frequency-resonances overlap even more. Raising the restriction $\alpha = \beta$, even more states occur, but for the resonance states this means not true splitting because the widths are too large (in most cases at least). Only true localized states become more separated [64 L 1].

The high-frequency-resonance is strongly related to the properties of the ideal lattice spectrum at these frequencies. Therefore in other than simple cubic lattices these resonances might not occur or might be very weak, whereas the low-frequency-resonances should be present in every lattice if $\varepsilon$ has the right order of magnitude.

High-frequency-resonances as well as low-frequency ones should give contributions to thermal resistance. Because the widths of the low-frequency resonances are relatively large, in thermal resistance the splitting should not be resolvable. The high-frequency-resonances should be essential in thermal resistance at higher temperatures (above 100° K). But they are relatively weak and it is questionable whether they lead to a pronounced behavior in thermal resistance.

If there are changes of force-constants in the lattice, similar phenomena should arise. But the quantitative calculation is very tedious. What happens qualitatively can be seen without much numerical work.

Because the splitting due to interactions is relatively small, for larger concentrations ($\gtrsim 1\%$) of defects one should use statistical calculations for the determination of the spectral density of defect lattices. Once the spectral density is known, most of the properties of the defect lattice can be evaluated. Equ. (21.24) relates the Greens function to the spectral distribution of lattice frequencies. For if we form

$$\operatorname{Im} \sum_{m\mu i} M_\mu G^{m\,m}_{\substack{\mu\,\mu\\ i\,i}}(\omega^2 + i\epsilon) = \pi \cdot \frac{N V_z}{(2\pi)^3} \sum_\sigma \int_{\omega^2 = \text{const.}} \frac{dO}{|\text{grad}_q\, \omega^2(\boldsymbol{q}\,\sigma)|}$$

$$= 3sN \cdot \pi\, g(\omega^2)\,. \tag{23.3}$$

$g(\omega^2)$ is just the squared frequency spectrum, normalized to one. In (23.3) the Greens function is related to the ideal lattice equation of motion (21.12). But in the derivation none of the properties of $\mathbf{G}$, which are related to the ideal lattice $\left(\text{e.g. } G^{m\,m}_{\substack{\mu\,\mu\\ i\,i}} = G^{0\,0}_{\substack{\mu\,\mu\\ i\,i}}\right)$ have been used. Thus (23.3) holds for the Greens function of a non-ideal lattice as well, provided it is defined appropriately. In tensor notation, the stationary Greens function of the ideal lattice was defined by (20.5)

$$\boldsymbol{\Phi}\mathbf{G} - \mathbf{M}\omega^2\mathbf{G} = \boldsymbol{\Delta}; \quad \mathbf{M} = \left\{M_\mu\, \delta_{\mu\ \nu}^{m\ n}{}_{i\ j}\right\}; \quad \boldsymbol{\Delta} = \left\{\delta_{\mu\ \nu}^{m\ n}{}_{i\ j}\right\}. \tag{23.4}$$

Using the perturbation as given by (20.3c) we define the Greens function **R** of the defect lattice by*

$$(\boldsymbol{\Phi} - \mathbf{J})\,\mathbf{R} - \mathbf{M}\omega^2\,\mathbf{R} = \boldsymbol{\Delta}\,. \tag{23.5}$$

**R** satisfies all the general relations which does **G**. Thus the spectral density of the defect lattice is given by

$$3sN\pi\, g_{\mathrm{D}}(\omega^2) = \mathrm{Im} \sum_{m\mu i} M_\mu\, R_{\mu\ \mu}^{m\ m}{}_{i\ i}\,(\omega^2 + i\epsilon)\,. \tag{23.6a}$$

The sum in (23.6a) represents nothing else as the trace of **R**. Thus we can write

$$g_{\mathrm{D}}(\omega^2) = \frac{1}{3sN} \cdot \frac{1}{\pi}\,\mathrm{Trace\ Im}\,\mathbf{M}\,\mathbf{R}(\omega^2 + i\epsilon)\,. \tag{23.6b}$$

The trace is independent of the representation, therefore (23.6b) holds also for the Fourier-transform (wave-vector representation) of (23.6b). Because of the definition of **J** in (20.3c), **M** is the mass-tensor of the ideal lattice.

**R** can be related to the Greens-function of the ideal lattice. An alternative way of writing (23.4) is

$$\mathbf{G} = (\boldsymbol{\Phi} - \mathbf{M}\omega^2)^{-1}; \quad \boldsymbol{\Phi} - \mathbf{M}\omega^2 = \mathbf{G}^{-1}\,. \tag{23.7}$$

Inserting this into (23.5) we receive

$$(\mathbf{1} - \mathbf{G}\mathbf{J})\,\mathbf{R} = \mathbf{G} \quad \text{or} \quad \mathbf{R} = (\mathbf{1} - \mathbf{G}\mathbf{J})^{-1}\,\mathbf{G}\,. \tag{23.8}$$

This, of course, is again valid in every representation. The response **R** can be calculated from the ideal lattice function **G** and the perturbation expansion

$$\mathbf{R} = \mathbf{G} + \mathbf{G}\mathbf{J}\mathbf{G} + \mathbf{G}\mathbf{J}\,\mathbf{G}\mathbf{J}\,\mathbf{G} + \cdots. \tag{23.8a}$$

The spectral distribution of the defect lattice is

$$g_{\mathrm{D}}(\omega^2) = \frac{1}{3sN\pi}\,\mathrm{Im\ Trace}\,\mathbf{M}(\mathbf{1} - \mathbf{G}\mathbf{J})^{-1} \cdot \mathbf{G}(\omega^2 + i\epsilon) \tag{23.9}$$

where the frequencies occur in **G** only. (23.9) can be expressed in an alternative way. (23.8) can be written

$$\mathbf{R} = \mathbf{G} + \frac{\mathbf{G}\mathbf{J}}{1 - \mathbf{G}\mathbf{J}} \cdot \mathbf{G}$$

or

$$\begin{aligned}\mathrm{Trace}\,\mathbf{M}\mathbf{R} &= \mathrm{Trace}\,\mathbf{M}\mathbf{G} + \mathrm{Trace}\,\mathbf{M}\frac{\mathbf{G}\mathbf{J}}{1 - \mathbf{G}\mathbf{J}}\,\mathbf{G} \\ &= \mathrm{Trace}\,\mathbf{M}\mathbf{G} + \mathrm{Trace}\,\mathbf{G}\mathbf{M}\mathbf{G}\mathbf{J}\frac{1}{1 - \mathbf{G}\mathbf{J}}\,.\end{aligned}$$

---

* **R** is sometimes called response-function, because it describes the response of the lattice to the introduction of defects.

From definition (21.21b, c) it follows

$$\mathbf{GMG} = \frac{\mathrm{d}\mathbf{G}}{\mathrm{d}\omega^2}\,.$$

Further it holds for any matrix $\mathbf{D}(\omega^2)$

$$\frac{\mathrm{d}}{\mathrm{d}\omega^2}\ln\|\mathbf{D}\| = \mathrm{Trace}\,\mathbf{D}^{-1}\frac{\mathrm{d}\mathbf{D}}{\mathrm{d}\omega^2}$$

and therefore

$$\mathrm{Trace}\,\mathbf{GMGJ}\,\frac{1}{1-\mathbf{GJ}} = -\frac{\mathrm{d}}{\mathrm{d}\omega^2}\ln\|\mathbf{1}-\mathbf{GJ}\|\,.$$

Trace $\mathbf{MG}$ represents the spectral distribution of the ideal lattice. We have finally

$$g_{\mathrm{D}}(\omega^2) - g(\omega^2) = -\frac{1}{3sN}\cdot\frac{1}{\pi}\,\mathrm{Im}\,\frac{\mathrm{d}}{\mathrm{d}\omega^2}\ln\|\mathbf{1}-\mathbf{GJ}\|\,. \qquad (23.10)$$

Both forms (23.9) and (23.10) can be used for the calculation of the spectrum. Though we have used $\mathbf{J}$ in the form (20.3c) in the derivation of (23.10), it can be shown easily that (23.10) holds also, if $\mathbf{J}$ is used in the form (20.3a). Of course, the result must be independent from the definition of the perturbation $\mathbf{J}$. (23.10) involves the knowledge of the determinant $\|\mathbf{1}-\mathbf{GJ}\|$ which is related to the determination of localized states and resonance states. The form (23.10) will be useful only if there is a small number of defect ions in the lattice. For larger (statistical) concentrations of defects (23.6b) will be the appropriate form.

In deriving (23.10) it has been assumed tacitly that both $g(\omega^2)$ and $g_{\mathrm{D}}(\omega^2)$ are normalized to one.

Thus the integral

$$\mathrm{Im}\int_0^\infty \frac{\mathrm{d}}{\mathrm{d}\omega^2}\ln\|\mathbf{1}-\mathbf{GJ}\|\,\mathrm{d}\omega^2 = \mathrm{Im}\ln\|\mathbf{1}-\mathbf{GJ}\|\,\Big|_0^\infty$$

has to be zero. For $\omega^2\to\infty$ the Greens function is real, and $G(\omega^2\to\infty)\to 0$, thus the imaginary part vanishes. For $\omega^2\to 0$, according to (23.8) $\ln\|\mathbf{1}-\mathbf{GJ}\|\to\ln\|\boldsymbol{\Phi}-\mathbf{J}\| - \ln\|\boldsymbol{\Phi}\|$; now $\boldsymbol{\Phi}-\mathbf{J}$ and $\boldsymbol{\Phi}$ are positive-definite because of the stability of the lattice (apart from the three center-of-mass degrees of freedom, which are irrelevant) and therefore $\ln\|\mathbf{1}-\mathbf{GJ}\|$ is real, and the integral in question vanishes.

For localized states $\mathbf{G}$ and $\|\mathbf{1}-\mathbf{GJ}\|$ is real, the imaginary part vanishes; *but* if $\omega$ is just equal to one of the localized frequencies $\omega_{\mathrm{loc}}$ $\|\mathbf{1}-\mathbf{GJ}\|$ is just zero and changes its sign when $\omega$ passes $\omega_{\mathrm{loc}}$. Thus at one side of $\omega_{\mathrm{loc}}$ the determinant is positive, at the other side negative; the imaginary part of the logarithm is a step-function, its derivative a $\delta$-function at the position of the localized state. Each localized state contributes a $\delta$-function to the spectral density. In the region of the ideal lattice frequencies, $\mathbf{G}$ has a continuous imaginary part and so has the contribution of the defects to the spectral density. In some cases further contributions arise from discontinuities in the ideal lattice spectrum, where the determinant $\|\mathbf{1}-\mathbf{GJ}\|$ might change its sign*.

* This happens in the case of a monatomic linear lattice, where the spectral density has a discontinuity at the maximum frequency. See [66 L 1].

If there is a larger concentration of defects in a lattice, it is best to use (23.6b) for the determination of the spectral density. In most cases it will be sufficient to assume a statistical distribution of defects in a lattice. Then one can define an average defect Greens function $\overline{\mathbf{R}}$

$$\overline{\mathbf{R}} = \mathbf{G} + \mathbf{G}\overline{\mathbf{J}}\mathbf{G} + \mathbf{G}\,\overline{\mathbf{J}\mathbf{G}\mathbf{J}}\,\mathbf{G} + \cdots \tag{23.11}$$

where the average process is over the statistical distribution [*Langer*, 61 L 4]. The procedure is simplest in the $(\boldsymbol{q}\sigma)$-representation of the $\mathbf{R}, \mathbf{G}, \mathbf{J}$-matrices. The corresponding transformations from the $\binom{m}{\mu \atop i}$ representation are defined by (21.18) *.

$$\begin{aligned} R^{\boldsymbol{q}\boldsymbol{q}'}_{\sigma\sigma'} &= \frac{1}{s N \overline{M}} \sum_{\substack{mn\\ \mu\nu ij}} e^{\mu *}_{i}(\boldsymbol{q}\sigma) \mathrm{e}^{-i\boldsymbol{q}\boldsymbol{R}m} \sqrt{M_\mu M_\nu}\, R^{\,m\,n}_{\mu\,\nu \atop i\,j}\, e^{\nu}_{j}(\boldsymbol{q}'\sigma')\, \mathrm{e}^{i\boldsymbol{q}'\boldsymbol{R}n} \\ J^{\boldsymbol{q}\boldsymbol{q}'}_{\sigma\sigma'} &= \frac{\overline{M}}{sN} \sum_{\substack{mn\\ \mu\nu ij}} e^{\mu *}_{i}(\boldsymbol{q}\sigma)\, \mathrm{e}^{-i\boldsymbol{q}\boldsymbol{R}m} \frac{1}{\sqrt{M_\mu M_\nu}} J^{\,m\,n}_{\mu\,\nu \atop i\,j}\, e^{\nu}_{j}(\boldsymbol{q}'\sigma')\, \mathrm{e}^{i\boldsymbol{q}'\boldsymbol{R}n}. \end{aligned} \tag{23.12}$$

Using this and (21.20, 21c) we obtain

$$\begin{aligned} \overline{R^{\boldsymbol{q}\boldsymbol{q}'}_{\sigma\sigma'}} = G(\boldsymbol{q}\sigma) \Big\{ & \delta_{\boldsymbol{q}\boldsymbol{q}'}\,\delta_{\sigma\sigma'} + \overline{J^{\boldsymbol{q}\boldsymbol{q}'}_{\sigma\sigma'}} + \sum_{\boldsymbol{q}_1\sigma_1} \overline{J^{\boldsymbol{q}\boldsymbol{q}_1}_{\sigma\sigma_1} J^{\boldsymbol{q}_1\boldsymbol{q}'}_{\sigma_1\sigma'}}\, G(\boldsymbol{q}_1\sigma_1) + \\ & + \sum_{\substack{\boldsymbol{q}_1\boldsymbol{q}_2\\ \sigma_1\sigma_2}} \overline{J^{\boldsymbol{q}\boldsymbol{q}_1}_{\sigma\sigma_1} J^{\boldsymbol{q}_1\boldsymbol{q}_2}_{\sigma_1\sigma_2} J^{\boldsymbol{q}_2\boldsymbol{q}'}_{\sigma_2\sigma'}}\, G(\boldsymbol{q}_1\sigma_1)\, G(\boldsymbol{q}_2\sigma_2) + \cdots \Big\} G(\boldsymbol{q}'\sigma'). \end{aligned} \tag{23.13}$$

The main difficulty arises from the statistical average over the defects. This can be done in simple cases only, one of which is again the isotopic mixture of two different masses. It is

$$J^{\boldsymbol{q}\boldsymbol{q}'}_{\sigma\sigma'} = \frac{\overline{M}\,\omega^2(\boldsymbol{q}'\sigma')}{sN} \sum_{m\mu i} \eta^{m}_{\mu}\, e^{*}_{\mu \atop i}(\boldsymbol{q}\sigma)\, e_{\mu \atop i}(\boldsymbol{q}'\sigma')\, \mathrm{e}^{i(\boldsymbol{q}'-\boldsymbol{q})\boldsymbol{R}m} \tag{23.14}$$

where

$$\varepsilon^{m}_{\mu} = \frac{M_\mu - M^m_\mu}{M_\mu}; \qquad \eta^{m}_{\mu} = \frac{\varepsilon^m_\mu}{1+\varepsilon^m_\mu}$$

and $\overline{M}$ is an arbitrary mass, which can be chosen appropriately. The averaging process can be performed to give**

$$\overline{J^{\boldsymbol{q}\boldsymbol{q}'}_{\sigma\sigma'}} = \bar{\eta}\,\overline{M}\,\omega^2(\boldsymbol{q}'\sigma')\,\delta_{\boldsymbol{q}\boldsymbol{q}'}\,\delta_{\sigma\sigma'}; \qquad E^{\boldsymbol{q}\boldsymbol{q}_1}_{\sigma\sigma_1} = \sum_i e^{*}_i(\boldsymbol{q}\sigma)\, e_i(\boldsymbol{q}_1\sigma_1)$$

$$\begin{aligned} \overline{J^{\boldsymbol{q}\boldsymbol{q}_1}_{\sigma\sigma_1} J^{\boldsymbol{q}_1\boldsymbol{q}'}_{\sigma_1\sigma'}} = & \frac{1}{N}(\overline{\eta^2} - \bar{\eta}^2)\,\overline{M}^2\,\omega^2(\boldsymbol{q}'\sigma')\,\omega^2(\boldsymbol{q}_1\sigma_1)\cdot E^{\boldsymbol{q}\boldsymbol{q}_1}_{\sigma\sigma_1} E^{\boldsymbol{q}_1\boldsymbol{q}}_{\sigma_1\sigma'}\cdot\delta_{\boldsymbol{q}\boldsymbol{q}'} + \\ & + \bar{\eta}^2\,\overline{M}^2\,\omega^4(\boldsymbol{q}\sigma)\,\delta_{\boldsymbol{q}\boldsymbol{q}_1}\,\delta_{\boldsymbol{q}_1\boldsymbol{q}'}\,\delta_{\sigma\sigma_1}\,\delta_{\sigma_1\sigma'} \end{aligned}$$

---

* The different mass-factors are related to the fact, that $\mathbf{G}$ and $\mathbf{R}$ are reciprocals of $\boldsymbol{\Phi} - \mathbf{M}\,\omega^2$ or $\boldsymbol{\Phi} - \mathbf{J} - \mathbf{M}\,\omega^2$, resp.

** This holds for statistical distributions of isotopes in Bravais-lattices. Otherwise one has to use averages for every sublattice separately: $\bar{\varepsilon} \to \bar{\varepsilon}_\mu$. The procedure is somewhat more tedious, but can be done similarly. In quadratic and higher averages one term of order $1/N$ has been neglected. It arises from the fact, that if one position has been occupied by an isotope, the second isotope has only $(N-1)$-positions available for occupation. In the limit $N \to \infty$ this term is irrelevant.

$$\overline{J^{\boldsymbol{q}\boldsymbol{q}_1}_{\sigma\sigma_1} J^{\boldsymbol{q}_1\boldsymbol{q}_2}_{\sigma_1\sigma_2} J^{\boldsymbol{q}_2\boldsymbol{q}'}_{\sigma_2\sigma'}} = \frac{1}{N^2}\left(\overline{\eta^3} - 3\bar{\eta}\,\overline{\eta^2} + 2\bar{\eta}^3\right) \overline{M}^3\, \omega^2(\boldsymbol{q}'\sigma')\, \omega^2(\boldsymbol{q}_1\sigma_1)\, \omega^2(\boldsymbol{q}_2\sigma_2) \times$$
$$\times\, E^{\boldsymbol{q}\boldsymbol{q}_1}_{\sigma\sigma_1} E^{\boldsymbol{q}_1\boldsymbol{q}_2}_{\sigma_1\sigma_2} E^{\boldsymbol{q}_2\boldsymbol{q}}_{\sigma_2\sigma'} \cdot \delta_{\boldsymbol{q}\boldsymbol{q}'} +$$
$$+ \frac{1}{N}\bar{\eta}\left(\overline{\eta^2} - \bar{\eta}^2\right) \overline{M}^3\, \omega^2(\boldsymbol{q}'\sigma')\, \omega^4(\boldsymbol{q}_1\sigma_1) \times$$
$$\times\, E^{\boldsymbol{q}\boldsymbol{q}_1}_{\sigma\sigma_1} E^{\boldsymbol{q}_1\boldsymbol{q}}_{\sigma_1\sigma'}\, \delta_{\boldsymbol{q}\boldsymbol{q}'}\, \delta_{\boldsymbol{q}_1\boldsymbol{q}_2}\, \delta_{\sigma_1\sigma_2} +$$
$$+ \frac{2}{N}\bar{\eta}\left(\overline{\eta^2} - \bar{\eta}^2\right) \overline{M}^3\, \omega^4(\boldsymbol{q}'\sigma')\, \omega^2(\boldsymbol{q}_2\sigma_2) \times$$
$$\times\, E^{\boldsymbol{q}\boldsymbol{q}_2}_{\sigma\sigma_2} E^{\boldsymbol{q}_2\boldsymbol{q}}_{\sigma_2\sigma'}\, \delta_{\boldsymbol{q}\boldsymbol{q}_1}\, \delta_{\boldsymbol{q}_1\boldsymbol{q}'}\, \delta_{\sigma_1\sigma'} +$$
$$+ \bar{\eta}^3\, \overline{M}^3\, \omega^6(\boldsymbol{q}\sigma)\, \delta_{\boldsymbol{q}\boldsymbol{q}'}\, \delta_{\boldsymbol{q}_1\boldsymbol{q}_2}\, \delta_{\boldsymbol{q}_2\boldsymbol{q}'}\, \delta_{\sigma\sigma_1}\, \delta_{\sigma_1\sigma_2}\, \delta_{\sigma_2\sigma'}\,. \tag{23.15}$$

Equs. (23.13, 15) allow for a graphical description [61 L 4]. As mentioned each free phonon Greens function represents a free phonon propagator, which may be denoted by a solid line. Each $J^{\boldsymbol{q}\boldsymbol{q}'}_{\sigma\sigma'}$ means an interaction with a defect (isotope) and may be denoted by a dashed line. The final $\boldsymbol{q}'$

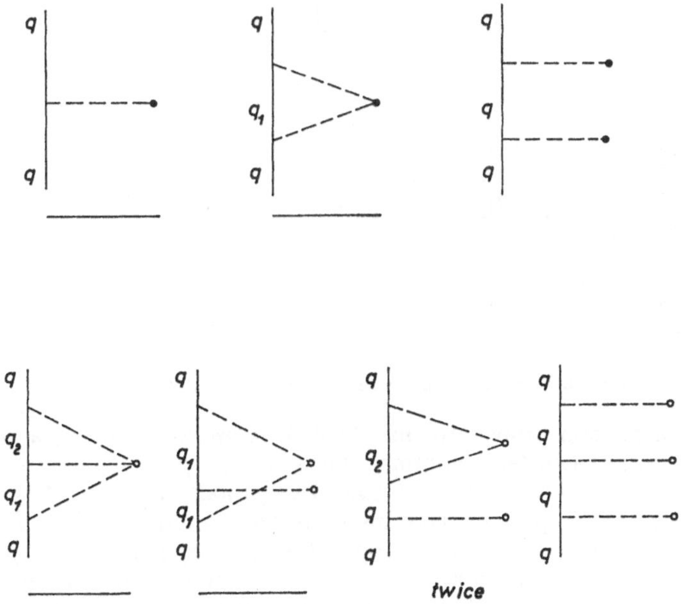


Fig. 6.18a. Diagrams which contribute to a statistical spectrum due to a concentration of isotopes in simple lattices. The underlined diagrams are proper diagrams. Diagrams of first to third order.

is always equal to the initial $\boldsymbol{q}$, which can be seen from (23.15). Contributions with higher order averages ($\overline{\eta^2}$; $\overline{\eta^3}$; . . .) arise from terms $\boldsymbol{m} = \boldsymbol{n}$, etc. and have the meaning of a twofold interaction of one phonon with the same isotope, etc. The phonon line crosses the corresponding interaction line twice. Fig. 6.18 shows the graphical representation in the order of the contributions in (23.15). There are two types of diagrams: those which can be separated into two unconnected diagrams when

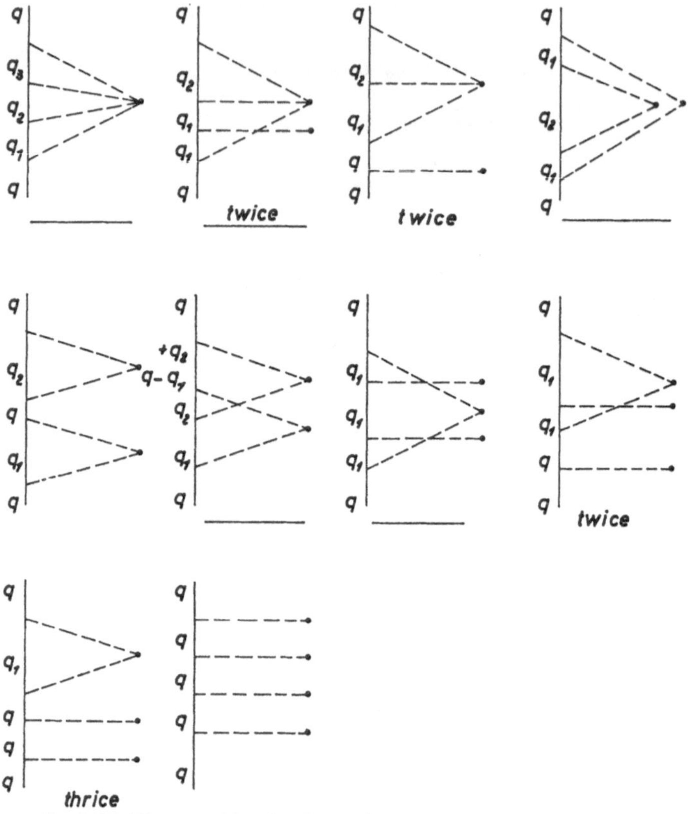


Fig. 6.18 b. Diagrams of fourth order for the situation explained in fig. 6.18 a

cutting a phonon propagator line (called *improper* diagrams) and those which can not (*proper* diagrams). There is one proper diagram of the first and second order, and there are two proper diagrams of the third order, e.g. The improper diagrams represent a multiplication of two or more proper diagrams. For example, the sixth diagram is the product of the first and second diagram. This can be seen immediately from (28.15). It is a mixed product, therefore the product occurs twice. Because all the improper diagrams can be obtained from products of the proper ones, all the improper diagrams can be represented by the proper ones, if all powers and mixed powers of the proper diagrams are taken into account. If

the sum over all the proper diagrams is denoted by $-\overline{M}\,\Omega(\boldsymbol{q}\sigma)$ (23.16) the diagonal elements of (23.13) can be written as

$$\overline{R^{qq}_{\sigma\sigma}} = G(\boldsymbol{q}\sigma) \cdot \sum_{\nu=0}^{\infty} \{-\overline{M}\,\Omega(\boldsymbol{q}\sigma)\,G(\boldsymbol{q}\sigma)\}^{\nu} = \frac{G(\boldsymbol{q}\,\sigma)}{1+\overline{M}\,\Omega(\boldsymbol{q}\,\sigma)\,G(\boldsymbol{q}\,\sigma)}\,. \tag{23.17}$$

The factors of this geometric series in the proper diagrams are just the right ones as to give the correct numbers of the improper diagrams. This can be seen simply by counting and comparison. Using (21.21c) we have

$$\overline{R^{qq}_{\sigma\sigma}} = \frac{1}{\overline{M}\,\{\omega^2(\boldsymbol{q}\,\sigma) + \Omega\,(\boldsymbol{q}\,\sigma) - \omega^2 - i\,\epsilon\}}\,, \tag{23.17a}$$

from which the spectral density can be calculated. The meaning of $\Omega(\boldsymbol{q}\sigma)$ can be seen. It represents the change in the frequencies due to the introduction of defects. (23.17a) is exact, provided in the determination of $\Omega(\boldsymbol{q}\sigma)$ all the proper diagrams have been taken into account. This is a difficulty which has not been overcome up till now.

We use a simple example for illustration. Suppose there are two statistically distributed masses $M_1$ and $M_2$ with concentration $p_1$ and $p_2$, $p_1 + p_2 = 1$; we assume $p_1 > p_2 = p$. $\overline{M}$ can be chosen arbitrarily and we will use $\overline{M} = M_1$. Then

$$\varepsilon^1 = 0;\quad \varepsilon^2 = \frac{M_2 - M_1}{M_1} = \varepsilon;\quad \eta = \frac{\varepsilon}{1+\varepsilon};\quad \overline{\eta^{\varrho}} = \sum_{\nu=1}^{2} p_\nu (\eta^\nu)^\varrho = p \cdot \eta^\varrho\,. \tag{23.18}$$

In the summation of the proper diagrams we keep only those which are linear in the concentration, thus limiting ourselves to small concentra-

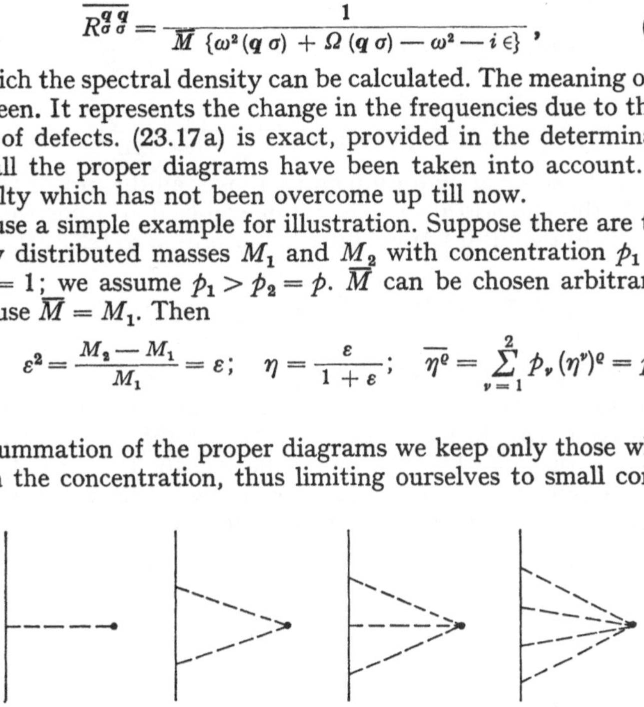

Fig. 6.19. Diagrams giving contributions which are linear in the concentration of isotopes

tions of mass $M_2$. Then only the first diagram of every order contributes (Fig. 6.19) and we obtain

$$\begin{aligned} -M_1\,\Omega(\boldsymbol{q}\sigma) &= p\,\eta\,M_1\,\omega^2(\boldsymbol{q}\sigma) + \\ &+ p\,\eta^2\,M_1\,\omega^2(\boldsymbol{q}\sigma) \sum_{ij} e_i^*(\boldsymbol{q}\sigma) \left\{\delta_{ij} + M_1\,\omega^2\,G^{00}_{ij}\right\} e_j(\boldsymbol{q}\sigma) + \\ &+ p\,\eta^3 M_1\,\omega^2(\boldsymbol{q}\sigma) \sum_{ijk} e_i^*(\boldsymbol{q}\sigma) \left\{\delta_{ik} + M_1\,\omega^2 G^{00}_{ik}\right\}\left\{\delta_{kj} + M_1\,\omega^2 G^{00}_{kj}\right\} e_j(\boldsymbol{q}\sigma) + \cdots \end{aligned} \tag{23.19}$$

where we have used (21.21b)

$$\begin{aligned} \delta_{ij} + M_1\,\omega^2\,G^{00}_{ij}(\omega) &= \frac{1}{N}\sum_{\boldsymbol{q}\sigma} \frac{e_i(\boldsymbol{q}\,\sigma)\,e_j^*(\boldsymbol{q}\,\sigma)\,M_1\,\omega^2(\boldsymbol{q}\,\sigma)}{M_1\,\{\omega^2(\boldsymbol{q}\,\sigma) - \omega^2 - i\,\epsilon\}} \\ &= \frac{1}{N}\sum_{\boldsymbol{q}\sigma} e_i(\boldsymbol{q}\sigma)\,M_1\,\omega^2(\boldsymbol{q}\sigma)\,G(\boldsymbol{q}\sigma) \cdot e_j^*(\boldsymbol{q}\sigma)\,. \end{aligned}$$

The sum in (23.19) is again a geometric series, giving

$$-M_1\,\Omega(\boldsymbol{q}\sigma) = p\,\eta\,M_1\,\omega^2(\boldsymbol{q}\sigma)\left\{1+\eta\left[\sum_{ij} e_i^*(\boldsymbol{q}\sigma)\left(\frac{1+M_1\,\omega^2\,G^{00}}{1-\eta(1+M_1\,\omega^2\,G^{00})}\right)_{ij} e_j(\boldsymbol{q}\sigma)\right]\right\}.$$

A further simplification arises in cubic lattices. Then $G^{00}_{ij} = G^{00}_{11}\cdot\delta_{ij}$ and $\sum_i e_i^*(\boldsymbol{q}\sigma)\,e_j(\boldsymbol{q}\sigma) = 1$ thus

$$-M_1\,\Omega(\boldsymbol{q}\sigma) = p\,\varepsilon\,M_1\,\omega^2(\boldsymbol{q}\sigma)\cdot\frac{1}{1-\varepsilon\,M_1\,\omega^2\,G^{00}_{11}} \tag{23.20}$$

and

$$\overline{R^{\boldsymbol{q}\boldsymbol{q}}_{\sigma\sigma}} = \frac{1}{M_1\{\omega^2(\boldsymbol{q}\,\sigma) - \omega^2 - p\,\varepsilon\,\omega^2(\boldsymbol{q}\,\sigma)\,[1-\varepsilon\,M_1\,\omega^2\,G^{00}_{11}]^{-1} - i\,\epsilon\}}\,. \tag{23.21}$$

The rest is the summation (or integration) over all the $\boldsymbol{q}\sigma$. As $\overline{\mathbf{R}}$ depends on $\boldsymbol{q}\sigma$ only through $\omega(\boldsymbol{q}\sigma)$ the integration can be done using the spectral distribution of the ideal lattice. Finally [according to (23.6b)]

$$g_D(\omega^2) = \frac{1}{\pi}\,\mathrm{Im}\int_0^\infty \overline{R}(\omega^2;\,\omega'^2)\cdot g(\omega'^2)\,\mathrm{d}\omega'^2 \tag{23.22}$$

where $\omega'$ replaces $\omega(\boldsymbol{q}\sigma)$ and $g(\omega'^2)$ is the spectrum of the ideal lattice. The integration can be performed analytically only in simple cases (linear lattice). This has been done by *Langer* [61 L 4]. Other investigations have been done by machine calculations [60 D 1, 61 D 1, 62 B 5]. Further the influence of other diagrams as those which are linear in $p$ has been discussed* [63 D 3]. It turns out that it is not negligible if

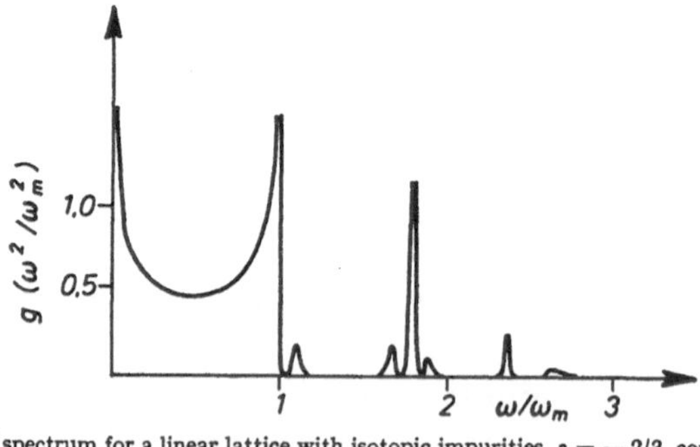


Fig. 6.20. Statistical spectrum for a linear lattice with isotopic impurities, $\varepsilon = -2/3$, concentration $p = 0{,}1$; the peaks for $\omega > \omega_m$ are related to agglomerates of isotopes [60 D 1]

concentrations are larger (percent or more). Some features can be seen from (23.21) already; for $1 - \varepsilon M_1 \omega^2 G^{00}_{11} = 0$ is just the condition for the

* Another approximation would be to replace the $G$-function in the denominator of (23.21) by $R$, having thus an implicit equation for $R$. This has to be solved then. As the $G$ in (23.21) is related to the phonon progation between two interactions, this means, that one takes into account all those diagrams, which have independent interactions between connected ones (e. g. the second third order diagram and the second and fourth order diagram in Fig. 6.18).

localized modes or for resonance modes, if the real part is used. In this case we have a strong resonance in **R** and we can suppose that $g_D(\omega^2)$ shows again this resonance, having an accumulation of frequencies at the corresponding frequency. That is what the numerical calculations just show (Fig. 6.20).

A change in the force constants can be handled in the same way. To make the writing easier and calculations more obvious, we only discuss the model of a linear lattice, where there are two kinds of springs $f_1$, $f_2$ distributed statistically over the lattice. We use again relative changes

$$\zeta_1 = 0; \quad \zeta_2 = (f_2 - f_1)/f_1 = \zeta; \quad \overline{\zeta^\varrho} = p\zeta^\varrho . \tag{23.23}$$

The spring between ions $m$ and $m+1$ is denoted by $\zeta^m$, that between $m-1$ and $m$ by $\zeta^{m-1}$. Then (there are no $\sigma$-indices now)

$$J^{qq'} = \frac{f_1}{N} \sum_m \mathrm{e}^{-i(q-q')am} \{\zeta^m (\mathrm{e}^{iq'a} - 1) + \zeta^{m-1} (\mathrm{e}^{-iq'a} - 1)\} . \tag{23.24}$$

With the frequencies

$$M\omega^2 = 4f_1 \sin^2 qa/2 \tag{23.25}$$

of the ideal lattice we obtain easily

$$\begin{aligned} \overline{J^{qq'}} &= -\zeta M \omega^2(q)\, \delta_{qq'} \\ \overline{J^{qq_1} J^{q_1q'}} &= \frac{1}{N} (\overline{\zeta^2} - \zeta^2) M^2 \omega^2(q')\, \omega^2(q_1)\, \delta_{qq'} + \\ &\quad + \zeta^2 M^2 \omega^2(q)\, \delta_{qq_1} \delta_{q_1q'} . \end{aligned} \tag{23.26}$$

These expressions are completely identical with (23.15). Thus the same graphical procedure can be used. In the approximation linear in $p$ we receive (23.19) with $-\zeta$ instead of $\eta$, and, of course, without polarization vectors. Now it is

$$\frac{1}{N} \sum_{q\sigma} \frac{\omega^2(q\sigma)}{\omega^2(q\sigma) - \omega^2 - i\epsilon} = \frac{2f_1}{NM} \sum_{q\sigma} \frac{1 - \cos qa}{\omega^2(q\sigma) - \omega^2 - i\epsilon} = 2f_1 (G^{00} - G^{01})$$

and therefore

$$-\Omega(q\sigma) = -p\zeta\omega^2(q\sigma) \frac{1}{1 + 2f_1\zeta(G^{00} - G^{01})} \tag{23.27}$$

or

$$\overline{R^{qq}} = \frac{1}{M\{\omega^2(q\sigma) - \omega^2 + p\zeta\omega^2(q\sigma)\,[1 + 2f_1\zeta(G^{00} - G^{01})]^{-1} - i\epsilon\}} . \tag{23.28}$$

Again the condition for localized modes or resonance modes is contained in (23.27, 28) which can be seen by a comparison with (20.29) and (22.41). Because

$$2f_1(G^{00} - G^{01}) = 1 + M\omega^2 G^{00}$$

(23.28) can be brought into the form (23.21), if the force-constants are described by

$$\zeta' = -\frac{\zeta}{1+\zeta} = (f_1 - f_2)/f_2 . \tag{23.23a}$$

The calculations of *Langer* can be used when $\varepsilon$ is replaced by $\zeta'$ (or $\eta$ by $-\zeta$). In threedimensional lattices the calculations are a little more involved, but similar results can be obtained. Qualitatively the same phenomena arise. The integration of (23.21) according to (23.22) can be done simply for the linear lattice. In that case $G^{00}$ is either real ($\omega > \omega_m$) or purely imaginary ($\omega < \omega_m$). It is given by

$$G^{00} = \frac{-1}{M_1 \sqrt{\omega^2 (\omega^2 - \omega_m^2)}} = \frac{1}{M_1} \bar{g}(\omega^2) + \frac{i\pi}{M_1} g(\omega^2), \qquad (23.29)$$

which can be seen immediately using (21.21b) and (23.25). If $\omega > \omega_m$, $G^{00}$ real, we obtain immediately

$$g_D(\omega^2) = \frac{1}{M_1} \int_0^{\omega_m^2} g(\omega'^2) \cdot \delta\left[\omega'^2\left(1 - \frac{p\varepsilon}{1 - \varepsilon\omega^2 \bar{g}(\omega^2)}\right) - \omega^2\right] d\omega'^2$$

from which it can be seen that this integral is different from zero only if $\omega^2$ is larger than the localized frequency $\omega_{loc}^2$, defined by $1 - \varepsilon\omega^2\bar{g}(\omega^2) = 0$ (page 151). This implies negative $\varepsilon$. Otherwise $g_D(\omega^2) = 0$ for $\omega^2 > \omega_m^2$. Further there exists an upper limit for the possible values of $\omega^2$ having $g_D(\omega^2) = 0$. This value depends on the concentration $p$ and is the larger the larger $p$ is. Integration of the last equation gives

$$g_D(\omega^2) = \frac{1}{M_1} \cdot \frac{1}{1 - \frac{p\varepsilon}{1 - \varepsilon\omega^2\bar{g}}}\, g\left(\frac{\omega^2}{1 - \frac{p\varepsilon}{1 - \varepsilon\omega^2\bar{g}}}\right)$$

which can be written

$$g_D(\omega^2) = \frac{1}{\pi M_1} \operatorname{Re}\left\{-\omega^2\left[\omega^2 - \omega_m^2 + p\varepsilon \frac{\omega_m^2\sqrt{\omega^2 - \omega_m^2}}{\sqrt{\omega^2 - \omega_m^2} + \varepsilon\omega}\right]\right\}^{-1/2} \quad \omega > \omega_m. \qquad (23.30\text{a})$$

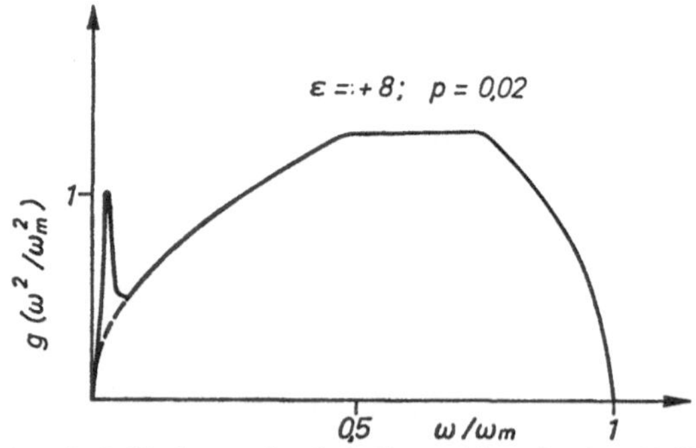


Fig. 6.21. Model spectrum for lattice frequencies; the lattice contains a fraction of 0,02 isotopes with $\varepsilon = 8$ Simple cubic lattice [64 M 1]

For $\omega < \omega_m$ the denominator in (23.21) contains an imaginary part (apart from $i\epsilon$) and $i\epsilon$ can be dropped.

The integration then gives

$$g_D(\omega^2) = \frac{1}{\pi M_1} \operatorname{Re}\left\{\omega^2\left[\omega_m^2 - \omega^2 - p\varepsilon \frac{\omega_m^2\sqrt{\omega_m^2 - \omega^2}}{\sqrt{\omega_m^2 - \omega^2} - i\varepsilon\omega}\right]\right\}^{-1} \quad 0 < \omega < \omega_m. \qquad (23.30\text{b})$$

The lattice spectrum in the region between $0 < \omega < \omega_m$ is only changed slightly, at least in this approximation for small concentrations. The most essential feature is the change of the singularity in the spectrum at $\omega = \omega_m$. Whereas in the ideal linear lattice this is proportional to $(\omega_m^2 - \omega^2)^{-1/2}$, it is now proportional to $(\omega_m^2 - \omega^2)^{-1/4}$. This can be seen by expanding (23.30b) about the singular position.

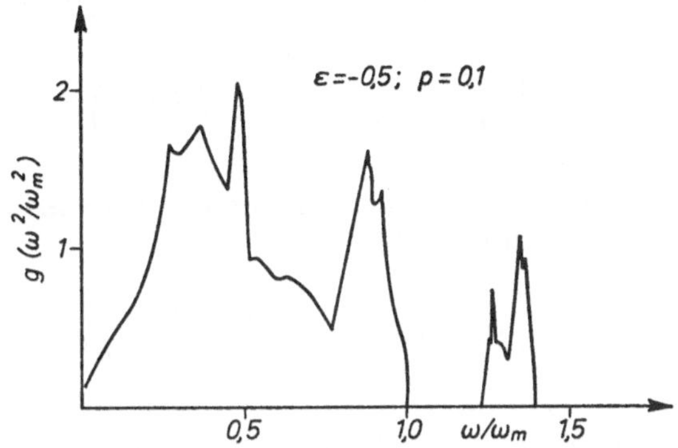


Fig. 6.22. The same as fig. 6.21, for a face-centered cubic lattice with isotopes. Concentration is $p = 0{,}1$. The occurence of a band of localized modes is obvious [64 M 1]

Equation (23.28) leads to the same spectral behavior as was pointed out already. In threedimensional crystals the corresponding integrations have been done by machines. The essential features are analogous to those of the linear lattice. As well as localized modes give a strong accumulation of frequencies, the resonance modes do. Fig. 6.21, 22 show some of the calculated spectra [64 M 1].

At larger concentrations this simple method fails. Higher order contributions in the diagrams have to be taken into account. This has been done, at least approximately, by *Davies* [63 D 3]. The correction leads to a less pronounced behavior of the localized states, the spectral density being smeared out over a larger region in the vicinity of localized frequencies. The changes in the region $0 < \omega < \omega_m$ are again less significant.

Apart from these more or less analytical methods a number of lattice models, mainly one- and twodimensional ones, have been investigated by machine calculations [*Dean* and *Martin*, 60 D 1, 61 D 1, 62 B 5]. Without going into more details, we give a few figures, showing the characterists of these spectra (Fig. 6.22).

Once the spectral density of the defect lattice is known, thermodynamic and related properties, as far as they depend on the lattice frequencies, can be calculated. Suppose such a quantity is given by

$$F = \sum_{\boldsymbol{q}\sigma} f[\omega^2(\boldsymbol{q}\sigma)]\,. \tag{23.31}$$

For example, $F$ may be the vibrational part of the free energy, or of the entropy or of the specific heat and so on. Using the spectral distribution, we can rewrite (23.31) into

$$F = 3sN \int f(\omega^2)\, g(\omega^2)\, \mathrm{d}\omega^2 \tag{23.32}$$

or for the change in $F$ due to the introduction of defects into the lattice

$$\Delta F = 3sN\int_0^\infty f(\omega^2)\,[g_D(\omega^2) - g(\omega^2)]\,\mathrm{d}\omega^2 \tag{23.33}$$

or*, using (23.10)

$$\begin{aligned}\Delta F &= -\frac{1}{\pi}\int_0^\infty f(\omega^2)\,\mathrm{Im}\,\frac{\mathrm{d}}{\mathrm{d}\omega^2}\ln\|1 - GJ\|\,\mathrm{d}\omega^2 \\ &= +\frac{1}{\pi}\int_0^\infty \mathrm{Im}\ln\|1 - GJ\|\cdot\left(\frac{\mathrm{d}f(\omega^2)}{\mathrm{d}\omega^2}\right)\mathrm{d}\omega^2\end{aligned} \tag{23.33a}$$

if the integrated part vanishes at the limits (see page 191). Alternative forms can be gotten with (23.7, 8):

$$\begin{aligned}\Delta F &= -\frac{1}{\pi}\int_0^\infty f(\omega^2)\,\mathrm{Im}\,\frac{\mathrm{d}}{\mathrm{d}\omega^2}\ln\frac{\|\Phi - J - M\,\omega^2\|}{\|\Phi - M\,\omega^2\|}\,\mathrm{d}\omega^2 \\ &= \frac{1}{\pi}\int_0^\infty f(\omega^2)\,\mathrm{Im}\,\frac{\mathrm{d}}{\mathrm{d}\omega^2}\{\ln\|R\| - \ln\|G\|\}\,\mathrm{d}\omega^2 \\ &= -\frac{1}{2\pi i}\int_{\mathfrak{C}} f(\omega^2)\,\{\ln\|R\| - \ln\|G\|\}\,\mathrm{d}\omega^2\end{aligned} \tag{23.33b}$$

where $\mathfrak{C}$ denotes a path in the complex $\omega^2$-plane, which includes all the zeros of the arguments of the logarithm, but none of their poles. If we denote the Greens function of a single isotope with relative mass $\varepsilon$ by $R(\varepsilon)$ we can define a self-energy of the isotope in an otherwise ideal lattice by

$$E_S = -\frac{1}{2\pi i}\int_{\mathfrak{C}} f(\omega^2)\,\frac{\mathrm{d}}{\mathrm{d}\omega^2}\{\ln\|R(\varepsilon)\| - \|\ln G\|\}\,\mathrm{d}\omega^2 \tag{23.34}$$

Here we have to take [I]

$$f(\omega^2) = \frac{1}{2}\hbar\omega + kT\ln\{1 - \mathrm{e}^{-\hbar\omega/kT}\} \tag{23.35}$$

which leads to the free energy when integrated. The first term corresponds to the zero-point energy, the second term gives the temperature-dependent contribution. Correspondingly the interaction energy between two isotopes $\varepsilon_1$ and $\varepsilon_2$ can be defined as the difference between the energies with $\varepsilon_1$ and $\varepsilon_2$ at finite distance and that for $\varepsilon_1$ and $\varepsilon_2$ at infinite distance:

$$\begin{aligned}E_I = -\frac{1}{2\pi i}\int_{\mathfrak{C}} f(\omega^2)\,\frac{\mathrm{d}}{\mathrm{d}\omega^2}\{&\ln[\|R(\varepsilon_1,\varepsilon_2)\|\cdot\|G\|] \\ &- \ln[\|R(\varepsilon_1)\|\,\|R(\varepsilon_2)\|]\}\,\mathrm{d}\omega^2.\end{aligned} \tag{23.36}$$

* Here again, a decomposition into irreducible parts is possible; the procedure is obvious.

The generalization of these equations to other types of defects is obvious, as well as to other thermodynamic functions. For calculations of the entropy for example instead of $f(\omega^2)$ we have to use

$$S(\omega^2) = k\{(1+n)\ln(1+n) - n\ln n\}; \qquad n(\omega) = \frac{1}{e^{\hbar\omega/kT} - 1}. \tag{23.37}$$

Detailed calculations again involve extended calculations. They have been done for special models by different authors [56 M 1, 60 M 2]. Here we will merely state the essential points.

Defects with the same sign of $\varepsilon$ ($\varepsilon_1 > 0$, $\varepsilon_2 > 0$ or $\varepsilon_1 < 0$ and $\varepsilon_2 < 0$) attract each other, those with different signs have a repulsive interaction. This holds for low as well as high temperatures, as long as only harmonic vibrations are considered [56 M 1]. For $T \to 0$ the interaction energy is proportional to $R^{-7}$ (in three-dimensional) or $R^{-3}$ (in one-dimensional) lattices; here $R$ is a measure for the distance between the isotopes [60 M 2]. At high temperatures (above Debye-temperature) the dependence on the distance is much more involved, but the interaction decreases with increasing distance. Thus the vibrational part of the interaction energy favors a precipitation of similar defects (equal $\varepsilon$). This is confirmed by calculations of *Prigogine, Bingen* and *Jeener*, who have calculated the energy of lattices with defects as a function of the degree of order of these defects [54 P 1]. For the details of these calculations we will refer to the original papers mentioned.

Other effects being strongly related to the lattice vibration spectrum are correlations. Those of interest are the displacement-correlations between the displacements of different ions at different times

$$\left\langle u^{m}_{\mu i}(t)\, u^{n}_{\nu j}(t')\right\rangle_T = \frac{1}{2\pi}\int_{-\infty}^{+\infty} U^{mn}_{\mu i\,\nu j}(\omega)\, \mathrm{e}^{-i\omega(t-t')}\,\mathrm{d}\omega \tag{23.38}$$

and the velocity- (or momentum-) correlations

$$\left\langle \dot{u}^{m}_{\mu i}(t)\, \dot{u}^{n}_{\nu j}(t')\right\rangle_T = \frac{1}{2\pi}\int_{-\infty}^{+\infty} \omega^2\, U^{mn}_{\mu i\,\nu j}\, \mathrm{e}^{-i\omega(t-t')}\,\mathrm{d}\omega\,. \tag{23.39}$$

The average is over the thermal distribution of the lattice vibrations, defects included, if they are present in the lattice. The correlations depend on $t - t'$ only because the system is homogeneous in time. If the lattice is an ideal one, in addition they depend only on $\boldsymbol{m} - \boldsymbol{n}$; but in defect lattices, there is a true dependence on $\boldsymbol{m}$ and $\boldsymbol{n}$, because the homogeneity of the lattice is destroyed. From the Fourier-transformed correlation $U^{mn}_{\mu i\,\nu j}(\omega)$ both correlations can be calculated.

In defect lattices the equation of motion is given by (20.2). To get the stationary eigenvectors of (20.2), we can eliminate the time-dependence always by using (20.4). Suppose we had found the exact eigen-solutions of the stationary equation of motion with defects present. These solutions are denoted by $\overset{0}{u}{}^{m}_{\mu i}(\alpha)$, where $\alpha$ describes the eigenstate.

It contains the localized states as well as resonance states and "wave-like" states, but the $\overset{0}{u}{}^{m}_{\mu i}(\alpha)$ are no longer real plane waves because the translation invariance of the lattice is destroyed by the defects. We can decompose the displacements with respect to these eigenstates:

$$u^{m}_{\mu i}(t) = \frac{1}{\sqrt{M_\mu}} \sum_\alpha a_\alpha(t)\, \overset{0}{u}{}^{m}_{\mu i}(\alpha)$$

$$\sum_{m\mu i} \overset{0}{u}{}^{m}_{\mu i}(\alpha)\, \overset{0}{u}{}^{*m}_{\mu i}(\beta) = \delta_{\alpha\beta}\,; \quad \sum_\alpha \overset{0}{u}{}^{m}_{\mu i}(\alpha)\, \overset{0}{u}{}^{*n}_{\nu j}(\alpha) = \delta^{mn}_{\mu\nu\, ij}\,. \tag{23.40}$$

Because of the equation of motion the $a_\alpha$ satisfy

$$\ddot{a}_\alpha + \omega_\alpha^2 a_\alpha = 0\,, \tag{23.41}$$

which represents a system of independent oscillators, some of which may be "localized" ones. Therefore the decomposition in creation and annihilation operators can be done with the same arguments as for ideal lattices, though the $\overset{0}{u}$ are no longer plane waves. All the statements which are related to oscillator properties hold in defect lattices as well. We rewrite (23.40) correspondingly into* [see (21.11)]

$$u^{m}_{\mu i}(t) = \sqrt{\frac{\hbar}{2M_\mu}} \sum_\alpha{}' \frac{1}{\sqrt{\omega_\alpha}} \left[b_\alpha\, \overset{0}{u}{}^{m}_{\mu i}(\alpha)\, e^{-i\omega_\alpha t} + b_\alpha^+\, \overset{0}{u}{}^{*m}_{\mu i}(\alpha)\, e^{i\omega_\alpha t}\right]. \tag{23.42}$$

Inserting this into (23.38) and using

$$\langle b_\alpha b_\beta^+\rangle_T = (\bar{n}_\alpha + 1)\,\delta_{\alpha\beta}; \quad \langle b_\alpha^+ b_\beta\rangle_T = \bar{n}_\alpha \cdot \delta_{\alpha\beta}$$

$$\langle b_\alpha b_\beta\rangle_T = \langle b_\alpha^+ b_\beta^+\rangle_T = 0; \quad \bar{n}_\alpha = \frac{1}{e^{\hbar\omega_\alpha/kT} - 1} \tag{23.43}$$

we obtain

$$\left\langle u^{m}_{\mu i}(t)\, u^{n}_{\nu j}(0)\right\rangle = \frac{\hbar}{2\sqrt{M_\mu M_\nu}} \sum_\alpha{}' \frac{1}{\omega_\alpha}\, \overset{0}{u}{}^{m}_{\mu i}(\alpha)\, \overset{0}{u}{}^{*n}_{\nu i}(\alpha) \times$$
$$\times \{(\bar{n}_\alpha + 1)\, e^{-i\omega_\alpha t} + \bar{n}_\alpha\, e^{i\omega_\alpha t}\}$$

and the Fourier-transform

$$U^{mn}_{\mu\nu\, ij}(\omega) = \frac{2\pi\hbar}{\sqrt{M_\mu M_\nu}} \sum_\alpha{}' \frac{1}{2\omega_\alpha}\, \overset{0}{u}{}^{m}_{\mu i}(\alpha)\, \overset{0}{u}{}^{*n}_{\nu i}(\alpha) \times$$
$$\times \{(\bar{n}_\alpha + 1)\,\delta(\omega_\alpha - \omega) + \bar{n}_\alpha\,\delta(\omega_\alpha + \omega)\}$$
$$= \frac{2\pi\hbar}{\sqrt{M_\mu M_\nu}} [\bar{n}(\omega) + 1]\, \mathrm{sgn}\,\omega \sum_\alpha{}' \overset{0}{u}{}^{m}_{\mu i}(\alpha) \cdot \delta(\omega_\alpha^2 - \omega^2)\, \overset{0}{u}{}^{*n}_{\nu j}(\alpha)$$
$$= 2\hbar[\bar{n}(\omega) + 1]\, \mathrm{sgn}\,\omega \cdot \mathrm{Im} \sum_\alpha{}' \frac{\overset{0}{u}{}^{m}_{\mu i}(\alpha)\, \overset{0}{u}{}^{*n}_{\nu j}(\alpha)}{\sqrt{M_\mu M_\nu}\, (\omega_\alpha^2 - \omega^2 - i\epsilon)}\,.$$

The expression in the imaginary part is just the definition of the Greens-function for the lattice with defects included, i.e. $(\Phi - J - M\omega^2 - i\epsilon)^{-1}$ which has been denoted by $R^{mn}_{\mu\nu\, ij}$ in equ. (23.5). Thus

$$U^{mn}_{\mu\nu\, ij}(\omega) = 2\hbar\, \mathrm{sgn}\,\omega\, [\bar{n}(\omega) + 1]\, \mathrm{Im}\, R^{mn}_{\mu\nu\, ij}(\omega^2 + i\epsilon)\,. \tag{23.44}$$

* Again center-of-mass-motions have to be excluded ($\omega_\alpha = 0$).

$\mathbf{R}(\omega)$ can be calculated with the procedure described above*. In ideal lattices $\mathbf{R}$ can be replaced by $\mathbf{G}$, which depends only on $\boldsymbol{m} - \boldsymbol{n}$, and so does $\mathbf{U}$.

With (23.44) we obtain the correlation (23.38) in the form

$$\left\langle u^{m}_{\mu i}(t)\, u^{n}_{\nu j}(0)\right\rangle_T = \frac{\hbar}{\pi}\int_0^\infty \mathrm{d}\omega\,\{\cos\omega t\cdot \mathrm{ctgh}\,\hbar\omega/2kT - i\sin\omega t\}\times \times \mathrm{Im}\, R^{mn}_{\mu i\, \nu j}(\omega^2 + i\epsilon) \tag{23.45}$$

whereas for the velocity correlation the integrand has to be multiplied by $\omega^2$.

A simple alternative form can be gotten for the correlations at equal time ($t = 0$). We introduce the socalled dynamical matrix

$$\mathbf{D} = \mathbf{M}^{-1/2}(\boldsymbol{\Phi} - \mathbf{J})\,\mathbf{M}^{-1/2}\,.$$

Then

$$\mathbf{R}(\omega^2 + i\epsilon) = \mathbf{M}^{-1/2}(\mathbf{D} - \omega^2 - i\epsilon)^{-1}\,\mathbf{M}^{-1/2}\,,$$

$$\mathrm{Im}\,\mathbf{R}(\omega^2 + i\epsilon) = \pi\,\mathbf{M}^{-1/2}\,\delta(\mathbf{D} - \omega^2)\,\mathbf{M}^{-1/2}\,.$$

Introducing this into (23.45) for $t = 0$ we obtain

$$\left\langle u^{m}_{\mu i}(0)\, u^{n}_{\nu i}(0)\right\rangle_T = \frac{\hbar}{2}\left\{\mathbf{M}^{-1/2}\,\mathbf{D}^{-1/2}\,\mathrm{ctgh}\,\frac{\hbar\,\mathbf{D}^{1/2}}{2kT}\cdot\mathbf{M}^{-1/2}\right\}^{mn}_{\mu i\,\nu j} \tag{23.45a}$$

or in the classical limit ($T >$ Debye-temperature)

$$\left\langle u^{m}_{\mu i}(0)\, u^{n}_{\nu i}(0)\right\rangle_T = kT\cdot\{\mathbf{M}^{-1/2}\,\mathbf{D}^{-1}\,\mathbf{M}^{-1/2}\}^{mn}_{\mu i\,\nu j} = kT\cdot\{(\boldsymbol{\Phi} - \mathbf{J})^{-1}\}^{mn}_{\mu i\,\nu j}\,. \tag{23.45b}$$

This form has been given first by *Born* in the theory of $X$-ray diffraction by vibrating crystals [43 B 1]. A corresponding form can be given for the velocity correlation:

$$\left\langle \dot u^{m}_{\mu i}(0)\, \dot u^{n}_{\nu j}(0)\right\rangle_T = \frac{\hbar}{2}\left\{\mathbf{M}^{-1/2}\,\mathbf{D}^{1/2}\,\mathrm{ctgh}\,\frac{\hbar\,\mathbf{D}^{1/2}}{2kT}\,\mathbf{M}^{-1/2}\right\}^{mn}_{\mu i\,\nu j} \tag{23.45c}$$

or in the classical limit

$$\left\langle \dot u^{m}_{\mu i}(0)\, \dot u^{n}_{\nu j}(0)\right\rangle_T = kT\,(\mathbf{M}^{-1})^{mn}_{\mu i\,\nu j}\,. \tag{23.45d}$$

In the classical limit the velocity-correlation only depends on the masses, the displacement-correlations on the force-constants (see page 205/6).

For a single isotopic defect at the origin of the lattice (cubic Bravais-lattice) we can use (22.30) together with (23.8), having

$$R^{mn}_{i\,j}(\omega^2 + i\epsilon) = G^{mn}_{i\,j}(\omega^2 + i\epsilon) + \frac{\varepsilon M(\omega^2 + i\epsilon)}{1 - \varepsilon M(\omega^2 + i\epsilon)\,G^{00}_{11}}\sum_k G^{m0}_{i\,k}\,G^{n0}_{k j}\,. \tag{23.46}$$

The first part represents the contribution of the ideal lattice, the second part is the change due to the isotopic defect.

---

* Another derivation can be given, using thermodynamic Greens functions (see Appendix).

Because the Greens functions decrease with increasing argument $\boldsymbol{m}$ or $\boldsymbol{m}-\boldsymbol{n}$, resp., the influence of the defect on the correlation becomes smaller with increasing distance of the correlating atoms from the defect. The *auto-correlation* ($\boldsymbol{m}=\boldsymbol{n}$) is of special interest (chapter VII). We have

$$R_{i\,j}^{\boldsymbol{m}\,\boldsymbol{m}}(\omega^2) = G_{ij}^{00} + \frac{\varepsilon M \omega^2}{1-\varepsilon M \omega^2 G_{11}^{00}} \sum_k G_{i\,k}^{\boldsymbol{m}0} G_{k\,j}^{\boldsymbol{m}0} \qquad (23.47)$$

showing the largest deviation from the ideal lattice value again for the isotope itself.

For a little more quantitative insight let us discuss the auto-correlation for the isotope itself in a cubic lattice. We have simply

$$R_{ij}^{00}(\omega^2) = \frac{1}{1-\varepsilon M \omega^2 G_{11}^{00}} G_{11}^{00} \cdot \delta_{ij} \qquad (23.47\,\mathrm{a})$$

with, according to (21.25, 26)

$$G_{11}^{00}(\omega^2 + i\epsilon) = \frac{1}{M} \mathrm{P} \int \frac{d\omega'^2 g(\omega'^2)}{\omega'^2 - \omega^2} + i\frac{\pi}{M} g(\omega^2) = \frac{1}{M}\bar{g} + \frac{i\pi}{M} g\,. \quad (23.48)$$

In the region of ideal lattice frequencies $[g(\omega^2) \neq 0]$ we now can take the limit $\epsilon \to 0$ and obtain:

$$\mathrm{Im}\, R_{11}^{00} = \frac{\pi}{M} g(\omega^2) \frac{1}{(1-\varepsilon\omega^2\bar{g})^2 + (\pi\varepsilon\omega^2 g)^2}\,; \quad \omega^2 < \omega_m^2\,. \quad (23.49\,\mathrm{a})$$

If $\omega^2 > \omega_m$, an imaginary part arises in the following way. We put $\omega^2 + i\epsilon = z$ and have

$$\mathrm{Im}\, R_{11}^{00}(z) = \mathrm{Im} \frac{G_{11}^{00}(\omega^2)}{1-\varepsilon M \omega^2 G_{11}^{00}(\omega^2) - \varepsilon M \frac{d}{dz}[z\, G_{11}^{00}(z)]_{z=\omega^2} \cdot i\epsilon + \cdots}$$

In the limit $\epsilon \to 0$ there is an imaginary part only if

$$1 - \varepsilon M \omega^2 G_{11}^{00}(\omega^2) = 0; \quad \omega^2 > \omega_m^2\,,$$

which is just the condition for the localized state. Therefore

$$\mathrm{Im}\, R_{11}^{00}(\omega^2) = \frac{G_{11}^{00}(\omega^2)}{\varepsilon M \frac{d}{d\omega^2}[\omega^2 G_{11}^{00}(\omega^2)]} \cdot \delta(\omega^2 - \omega_{\mathrm{loc}}^2)\,; \quad \omega^2 > \omega_m^2 \quad (23.49\,\mathrm{b})$$

$\omega_{\mathrm{loc}}^2$ is the localized frequency. The generalisation to other than isotopic defects is obvious. The explicit integration according to (23.45) can be done only numerically. *Dawber* and *Elliott* [63 D 5] have used a Debye-spectrum for the ideal lattice spectrum, and then evaluated the auto-correlation for the displacements and velocities (both with $t=0$). Some of their results are shown in Fig. 6.23. In the classical limit ($T > \theta$) the displacement-autocorrelation is not influenced by the isotopic defect, whereas the velocity-autocorrelation is changed by $M/M' = (1+\varepsilon)^{-1}$. At low temperatures there is a change in both correlations. The change in the velocity-correlation, if multiplied by $M'/M$, is nearly equal to the change in the displacement correlation. For isotopes lighter than the host lattice ions, the autocorrelation is enlarged, for heavier ones it is diminished.

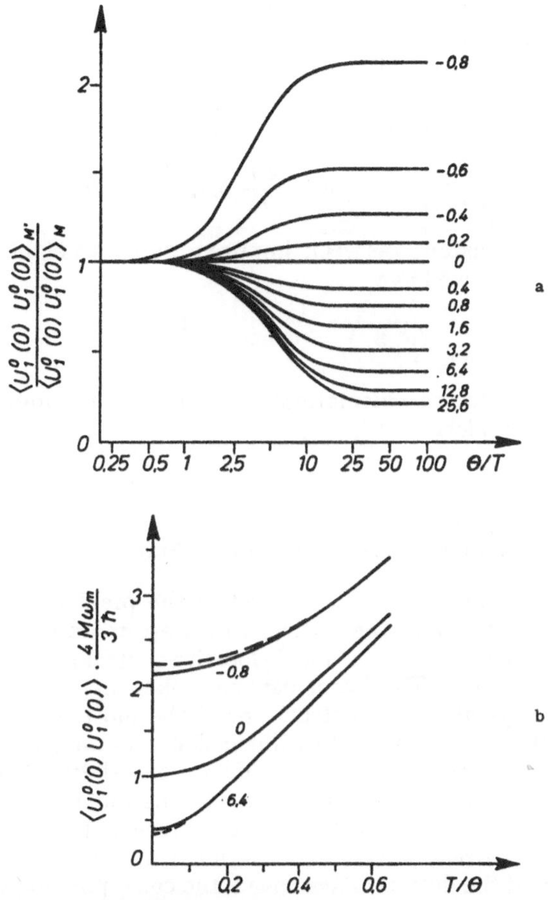


**Fig. 6.23 a and b.** The change in the displacement-autocorrelation for an isotopic defect, with the relative mass change $\varepsilon$ as a parameter. a) Relative to the ideal value, as function of temperature, b) The absolute dependence on temperature. According to [63 D 5]

The classical behavior can be estimated more easily by using classical averages. It is (Bravais-lattices, $t = 0$ for example)

$$\left\langle u_i^m u_j^n \right\rangle_T = \frac{\int \exp\{-\Phi'/kT\}\, u_i^m u_j^n \,\mathrm{d}\boldsymbol{u}^1 \dots \mathrm{d}\boldsymbol{u}^N}{\int \exp\{-\Phi'/kT\}\,\mathrm{d}\boldsymbol{u}^1 \dots \mathrm{d}\boldsymbol{u}^N} \tag{23.50a}$$

or

$$\left\langle \dot{u}_i^m \dot{u}_j^n \right\rangle_T = \frac{\int \exp\left\{-\sum_{mk} v_k^m v_k^m/2kT\right\} v_i^m v_j^n \,\mathrm{d}\boldsymbol{v}^1 \dots \mathrm{d}\boldsymbol{v}^N}{\sqrt{M_m M_n}\int \exp\left\{-\sum_{mk} v_k^m v_k^m/2kT\right\}\mathrm{d}\boldsymbol{v}^1 \dots \mathrm{d}\boldsymbol{v}^N} = \frac{kT}{M_m}\,\delta_{i\,j}^{mn} \tag{23.50b}$$

with $\boldsymbol{v}^m = \sqrt{M_m}\,\dot{\boldsymbol{u}}^m$; $M_m$ is the mass of ion $\boldsymbol{m}$, $\Phi'$ the harmonic potential energy, including defects. Because $\Phi'$ does not depend on any mass, it can be seen that only the velocity-correlation (23.50b) is influenced by a mass-change, whereas a change in force-constants affects only the displacement-correlation ($T \gtrsim \theta$!). Detailed calculations of this are more cumbersome; but the difficulties arise only from the occuring integrals. In a cubic lattice with central forces the autocorrelation for the impurity displacement is given essentially by

$$\left\langle \overset{0}{u}_i \overset{0}{u}_j \right\rangle_T = \mathrm{const} \cdot \frac{k\,T}{f'}\,\delta_{ij}; \quad (T \gtrsim \theta)\,. \tag{23.51}$$

$f'$ is the force constant between impurity and its neighbors, whereas in the ideal lattice we have

$$\left\langle \overset{0}{u}_i \overset{0}{u}_j \right\rangle_T = \mathrm{const} \cdot \frac{k\,T}{f}\,\delta_{ij}\,. \tag{23.51a}$$

For more quantitative calculations we refer to a number of papers using different models.

## 24. Extended Defects. Surfaces. Localized States

Besides the point defects, discussed in the previous sections, other kinds of defects occur in the lattice, which are more extended in size. Special kinds of such defects are dislocations, stacking faults, internal or even free surfaces. The latter can be looked upon as defects in an ideal infinite crystal. Those defects are of the most interest which are extended uniformly in one or two dimensions, forming straight lines or plane surfaces. At these defects, the lattice translational invariance in one or two dimensions, resp. is conserved. The immediate consequence for vibration states is that the wave-vector-components parallel to these directions are real; since the translational invariance perpendicular to the extension of the defects is destroyed, the corresponding wave-vector-components may become complex. Localized states and also scattering states occur. Of course, these states are not limited to defects in form of straight lines or plane surfaces, but in the following we limit ourselves to a discussion of the simple cases.

The procedure is investigated for defects in form of plane surfaces in the following. The application to straight lines and various other types of defects can be done simply. Also a large number of the results can be used for other types of defects with obvious changes. Lattice points in the surface are described by a vector $\boldsymbol{\varrho}^m$, lying in the surface, the direction perpendicular to the surface is the $z$-direction. The origin is chosen to be a point in the surface. Then the lattice points are given by

$$\mathbf{R}_\mu^m = \left\{\varrho_\mu^m;\quad z_\mu^m\right\}. \tag{24.1}$$

Further it is convenient to choose the basis-vectors of the lattice in such a way, that two of them are parallel to the surface, whereas the third one is not coplanar with the surface. Such a choice can always be made. This means, that we can split $\boldsymbol{m}$ into the components

$$\tilde{\boldsymbol{m}} = \{m_1, m_2\} \quad \text{and} \quad m_3; \qquad \boldsymbol{m} = \{\tilde{\boldsymbol{m}}\,; m_3\}\,. \tag{24.2}$$

Correspondingly, the wave vectors are composed of their components parallel to the surface and perpendicular to it:

$$\boldsymbol{q} = \{\tilde{\boldsymbol{q}}, q_3\}; \quad \tilde{\boldsymbol{q}} = \{q_1, q_2\}\,. \tag{24.3}$$

The equation of motion of the defect lattice can be written in the form (20.1) or (20.2, 3) again. But as the translational invariance in the surface is conserved, we have for the perturbation

$$I^{\tilde{m} m_3, \tilde{n}\, n_3}_{\mu\, i\ \nu\, j} = I^{\tilde{m} + \tilde{h} m_3, \tilde{n} + \tilde{h} n_3}_{\mu\, i\ \nu\, j} \tag{24.4}$$

Other relations, the $I$ have to satisfy, are given by the point group symmetries, and by the general invariances discussed in sect. 3. The eigensolutions of the equation of motion must have wave-character for the $\tilde{\boldsymbol{q}}$-components of the wave-vector $\boldsymbol{q}$. Therefore we put

$$u^{m}_{\mu\, i} = \frac{1}{N^{1/3}}\, e^{i\tilde{q}\,\varrho^m} \cdot v^{m_3}_{\mu\, i}(q_3\sigma)\,. \tag{24.5}$$

Here we have divided the surface into periodicity areas each containing $N^{2/3}$ "unit cells" of the surface. The formal solution of the defect equation of motion (20.7) is valid also in this case. Using (24.5) we have from (20.7)

$$v^{sm_3}_{\mu\, i} = e^{-i\tilde{q}\varrho^m}\, G^{mn}_{\mu\, i\ \nu\, j}\, I^{np}_{\nu\, j\ \varkappa\, k}\, e^{i\tilde{q}\varrho^p}\, v^{p_3}_{\varkappa\, k}$$

and with (21.21b) for $\mathbf{G}$ we obtain

$$\begin{aligned}\sum_{\tilde{m}} e^{-i\tilde{q}\varrho^m}\, G^{mn}_{\mu\, i\ \nu\, j} &= \frac{1}{sN\sqrt{M_\mu M_\nu}} \sum_{q_3\sigma} \frac{e^{\mu}_{i}\, e^{*\nu}_{j}(\tilde{q}, q_3\sigma)\, e^{iq_3(z^{m_3} - z^{n_3})}}{\omega^2(\tilde{q}, q_3\sigma) - \omega^2 - i\varepsilon}\, e^{-i\tilde{q}\varrho^n} \\ &= \tilde{G}^{m_3 n_3}_{\mu\, i\ \nu\, j}\, e^{-i\tilde{q}\varrho^n}\end{aligned} \tag{24.6}$$

which defines $\tilde{\mathbf{G}}$. It is a modified Greens function, containing only one summation (or integration) over the $q_3$-component. Therefore in general it is much more easier to calculate this Greens function instead of the whole Greens function $G$ used for point defects. It is a mixed representation, for it depends on $m_3 - n_3$ (space coordinate) and on $\tilde{\boldsymbol{q}}$ (momentum), and of course on the frequency $\omega$. We introduce further

$$\begin{aligned}\tilde{I}^{m_3 n_3}_{\mu\, i\ \nu\, j}(\tilde{q}) &= \frac{1}{N^{2/3}} \sum_{\tilde{m}\tilde{n}} e^{-i\tilde{q}\varrho^m}\, I^{mn}_{\mu\, i\ \nu\, j}\, e^{i\tilde{q}\varrho^n} \\ &= \sum_{\tilde{h}} I^{0 m_3, \tilde{h} n_3}_{\mu\, i\ \nu\, j}\, e^{i\tilde{q}\varrho^h}\end{aligned} \tag{24.7}$$

because of (24.4). Therefore

$$\overset{s}{v}{}^{m_3}_{\underset{i}{\mu}} = \tilde{G}^{m_3}_{\underset{i}{\mu}}{}^{n_3}_{\underset{j}{\nu}}(\tilde{q}, \omega)\, \tilde{I}^{n_3}_{\underset{j}{\nu}}{}^{p_3}_{\underset{k}{\varkappa}}(\tilde{q})\, v^{p_3}_{\underset{k}{\varkappa}}\,. \tag{24.8}$$

This equation can be handled just in the same way as discussed in sect. 20—23. It connects only the ionic layers at different distance from the surface, whereas the translational invariance has been used for the elimination of the ions in the surface.

Just as in the case of point defects, one can introduce the Greens function of the defect equation of motion $\mathbf{R}$ or $\tilde{\mathbf{R}}$ resp. and then one can use all the formulas of the preceding section to calculate spectral properties and correlations of surface ions.

Localized surface states can be calculated by using (24.5-7), if the $\boldsymbol{v}$ on the right side of this equation is replaced by $\overset{s}{\boldsymbol{v}}$, or

$$\|1 - \tilde{\mathbf{G}}(\tilde{q}\omega)\, \tilde{\mathbf{I}}(\tilde{q})\| = 0\,. \tag{24.9}$$

This secular equation gives the frequencies $\omega$ of localized surface states as a function of the wave vector $\tilde{\boldsymbol{q}}$ parallel to the surface, whereas the component of $\boldsymbol{q}$ perpendicular to the surface has to contain an imaginary part of such a kind, that the eigenstates have decreasing amplitudes perpendicular to the surface.

These states have been investigated in a number of papers [57 W 1, 59 W 1, 60 W 2, 62 G 1, 63 T 2, 64 L 5, 65 L 4], mainly for the simple cubic and the cubic face centered lattice, with different orientations of the surface (100, 110, 111). Localized states at free surfaces have to agree with the (generalized) Rayleigh-states of elastic continuum theory in the limit of long waves ($\boldsymbol{q} \to 0$, $\tilde{\boldsymbol{q}} \to 0$). In Section 6 it has been discussed that it is *necessary* to satisfy the condition of rotational invariance (2.9) to obtain the agreement between long wave lattice theory and elastic theory. If the continuum is described using a step function for the change of the elastic properties at the free surface, and using homogeneity inside the medium, the condition of rotational invariance is also sufficient, if similar assumptions on the coupling parameters are made.

But if the elastic medium contains a surface (transition) region with elastic constants being different from the "interior" of the crystal and also different from zero, appropriate choices for the coupling parameters have to be made in order to get agreement between lattice theory and elastic theory in the long-wave limit.

In a number of papers on surface modes the condition of rotational invariance (2.9) has been disregarded. Though the general features of surface states are not changed essentially by this inconsistency, the long wave limit does not give the correct elastic theory. The best model for a free surface seems to be that of *Wallis, Gazis, Herman* [60 W 2], which starts from an explicit expression for the potential energy, therefore (2.9) is satisfied.

It is often more convenient, not to start with the description according to (24.5—9), but with a variational procedure (Sect. 20) or with an appropriate Ansatz for the state-vectors. This avoids the calculation of the Greens function. We will use the last procedure for a simple discussion of possible localized states. As a model we use again the simple cubic lattice, the ideal lattice frequencies of which are given by

$$M\,\omega^2(\boldsymbol{q},\sigma=1)=4\,\alpha\sin^2 q_1 a/2+4\,\beta\sin^2 q_2 a/2+4\,\beta\sin^2 q_3 a/2$$
$$\sigma=2,3\quad\text{cyclic}\tag{24.10}$$

Suppose the surface of defects is a (001)-surface. Because of the translational invariance in the 1- and 2-direction, $q_1$ and $q_2$ have to be real, but $q_3$ may be complex: $q_3\Rightarrow q_3+i\varkappa$. Then

$$2\sin^2 q_3 a/2=1-\cos q_3 a\cosh\varkappa a+i\sin q_3 a\sinh\varkappa a\,.$$

But in eigenstates the squared frequencies have to be real and positive. Therefore either $\varkappa=0$, $q_3$ arbitrary or $q_3=\pi\nu/a$, $\varkappa$ arbitrary. The first case corresponds to wave like solutions, not showing localized behavior. The second case gives localized states. If the state vector is written as

$$u_i^{m}\sim e_i(\boldsymbol{q}\sigma)\,\mathrm{e}^{i\boldsymbol{q}\boldsymbol{R}^m}\,;\qquad \boldsymbol{R}^m=a\boldsymbol{m}\tag{24.11}$$

we receive

$$u_i^{m}\sim(-1)^{\nu m_3}\,e_i(\boldsymbol{q}\sigma)\,\mathrm{e}^{ia(q_1 m_1+q_2 m_2)}\,\mathrm{e}^{-\varkappa a m_3}\,.\tag{24.12}$$

For physical reasons*, the states must be decreasing with increasing $m_3$, therefore we have to take $\varkappa\gtrless 0$ for $m_3\gtrless 0$ or to write $\exp\{-\varkappa a\,|m_3|\}$ with $\varkappa>0$. From (24.10) and (24.12) it follows, that we can limit $\nu$ to 0 and 1: There are two different kinds of localized states. With $\nu=0$, all the displacements in the long wave limit have amplitudes equal in sign. These states are called *acoustical* localized states. With $\nu=1$, neighboring layers ($m_3$) have antiparallel displacements, the states are *optical localized states*. The frequencies are

$$M\,\omega^2(\boldsymbol{q},\sigma=1)=4\,\alpha\sin^2 q_1 a/2+4\,\beta\sin^2 q_2 a/2\begin{cases}-4\,\beta\sinh^2\varkappa\,a/2, & \nu=0\\ +4\,\beta\cosh^2\varkappa\,a/2, & \nu=1\end{cases}$$
$$\sigma=2,3\quad\text{cyclic}\tag{24.13}$$

Because $\sinh\varkappa a/2$ is not bounded, the frequencies of acoustical states can become zero. The region of acoustical frequencies is the low frequency region up to a certain limit; the optical frequencies lie at the top of the frequency region of the ideal lattice and there is no upper limit for the frequencies. These optical states correspond to the localized states at point defects. Acoustical states are not possible at point defects in ionic lattices.

Though we have discussed this with the simplest example, it is clear that the general features will occur at every surface. Acoustical and optical surface states are always possible, but whether they occur in practice depends on the parameters describing the surface (mass, coupling constants).

We consider first an "isotopic surface", i.e. a plane consisting of ions of mass $M'$, whereas the host lattice has mass $M$. Though this is

* The defect is supposed to be at $m_3\cong 0$.

a rather artificial model, it gives some insight. The isotopes form a (001)-plane at $m_3 = 0$. The equation of motion for the ions of this plane reads

$$
\begin{aligned}
M' \omega^2 u_1^{lm0} &= \alpha \left(2 u_1^{lm0} - u_1^{l+1\,m0} - u_1^{l-1\,m0}\right) \\
&+ \beta \left(2 u_1^{lm0} - u_1^{lm+1\,0} - u_1^{lm-1\,0}\right) \\
&+ \beta \left(2 u_1^{lm0} - u_1^{lm1} - n_1^{lm\bar{1}}\right)
\end{aligned}
$$

$$m = \{m_i\} = \{l, m, n\}; \qquad (24.14)$$

$i = 2, 3$ cyclic

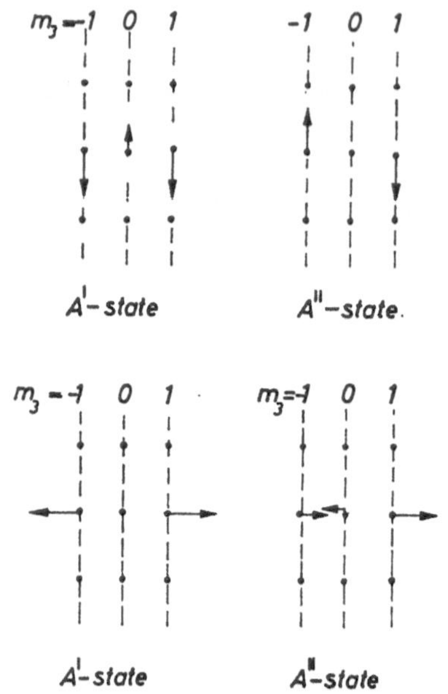


Fig. 6.24 a and b. Localized vibration states at planes of defects. a) displacements parallel to the surface, b) displacements perpendicular to the surface

The solution is simplified by symmetry considerations again (group theory). For displacements parallel to the surface ($i = 1, 2$) we have the two states shown in Fig. 6.24a, because the surface is a mirror plane. In this simple model, this state is twofold degenerated. For displacements perpendicular to the surface, we have Fig. 6.24b; by an "isotopic surface" only the states, where the $m_3 = 0$-amplitude is different from zero, are influenced and therefore give rise to localized states. Using this symmetry and (24.12) we obtain

$$
\begin{aligned}
M' \omega^2 (q, \sigma = 1) &= 4 \alpha \sin^2 q_1 a/2 + 4 \beta \sin^2 q_2 a/2 + 2 \beta [1 - (-1)^\nu e^{-\varkappa_1 a}] . \\
M' \omega^2 (q, \sigma = 2) &= 4 \beta \sin^2 q_1 a/2 + 4 \alpha \sin^2 q_2 a/2 + 2 \beta [1 - (-1)^\nu e^{-\varkappa_2 a}] . \\
M' \omega^2 (q, \sigma = 3) &= 4 \beta \sin^2 q_1 a/2 + 4 \beta \sin^2 q_2 a/2 + 2 \alpha [1 - (-1)^\nu e^{-\varkappa_3 a}] .
\end{aligned}
\qquad (24.15)
$$

The frequencies defined by (24.13) and (24.15) must be equal. By comparison we obtain

$$\varepsilon = [2\beta + 4\alpha \sin^2 q_1 a/2 + 4\beta \sin^2 q_2 a/2] = (-1)^\nu \cdot 2\beta [\sinh \varkappa_1 a + \varepsilon \cosh \varkappa_1 a]$$

and cyclic (24.16)

which allows for a determination of $\varkappa_i$ as a function of $\varepsilon$ and $q_1, q_2$; thus the frequencies of the localized states are completely determined by the defect parameter $\varepsilon = (M' - M)/M$ and by the wave-vector-components $q_1, q_2$ in the plane of translational invariance. It can be seen from (24.16), that localized acoustical states can occur only if $\varepsilon > 0$, whereas optical states occur for $\varepsilon < 0$. This corresponds to localized states at point defects ($\varepsilon < 0$), whereas there are no acoustical states for $\varepsilon > 0$ at point defects. Only resonance states occur if $\varepsilon$ is sufficiently

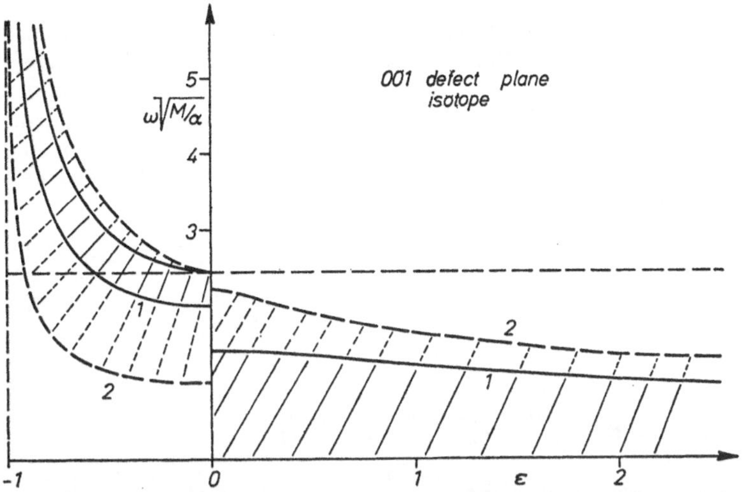


Fig. 6.25. Band of frequencies of localized states at a plane of isotopes, for $\varepsilon < 0$ there is an optical band of states, for $\varepsilon > 0$ an acoustical band. curve 1: displacements perpendicular, curve 2: displacements parallel to surface

large. The frequencies of the localized states form frequency bands, because there is a translational invariance conserved in two dimensions. Fig. 6.25 shows the position of the localized frequency bands and their widths as function of the change in mass $\varepsilon$. Fig. 6.26 shows the dispersion curves for special directions.

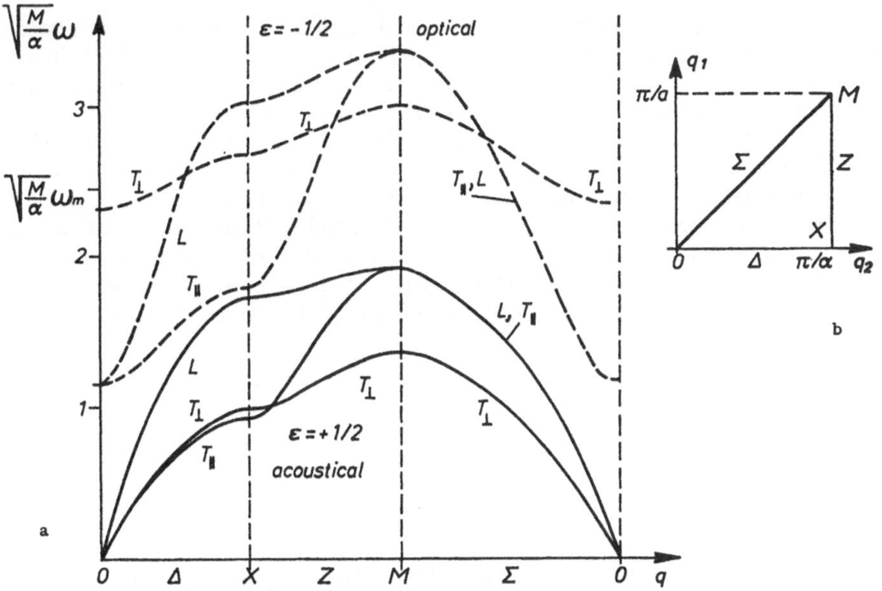


Fig. 6.26 a and b. a) Dispersion curves for localized states at a plane of isotopes. Acoustical states ($\varepsilon > 0$) full lines, optical states ($\varepsilon < 0$) broken lines. b) defines the directions of surface propagation

A somewhat more realistic model is that of a free surface, say a (001)-surface again. We have mentioned already that the condition of rotational invariance (2.9) has to be satisfied in order to have a consistent model. According to Sect. 5, equ. (5.7–9) this implies the use of non-diagonal coupling matrices at the surface. Then the simple features of the simple cubic lattice vanish, because these are related to diagonal coupling-matrices. The equations of motion for ions in the surface layer now read ($\boldsymbol{m} = \{l, m, n\}$)

$$
\begin{aligned}
M\omega^2 u_1^{lm0} &= \alpha\left(2u_1^{lm0} - u_1^{l+1\,m0} - u_1^{l-1\,m0}\right) - \delta\left(u_3^{l+1\,m0} - u_3^{l-1\,m0}\right)\\
&\quad + \beta\left(2u_1^{lm0} - u_1^{lm+1\,0} - u_1^{lm-1\,0}\right)\\
&\quad + \beta\left(u_1^{lm0} - u_1^{lm1}\right)\\
M\omega^2 u_2^{lm0} &= \beta\left(2u_2^{lm0} - u_2^{l+1\,m0} - u_2^{l-1\,m0}\right)\\
&\quad + \alpha\left(2u_2^{lm0} - u_2^{lm+1\,0} - u_2^{lm-1\,0}\right) - \delta\left(u_3^{lm+1\,0} - u_3^{lm-1\,0}\right)\\
&\quad + \beta\left(u_2^{lm0} - u_2^{lm1}\right)\\
M\omega^2 u_3^{lm0} &= \beta\left(2u_3^{lm0} - u_3^{l+1\,m0} - u_3^{l-1\,m0}\right) + \delta\left(u_1^{l+1\,m0} - u_1^{l-1\,m0}\right)\\
&\quad + \beta\left(2u_3^{lm0} - u_3^{lm+1\,0} - u_3^{lm-1\,0}\right) + \delta\left(u_2^{lm+1\,0} - u_2^{lm-1\,0}\right)\\
&\quad + \alpha\left(u_3^{lm0} - u_3^{lm1}\right)
\end{aligned}
\tag{24.17}
$$

whereas in layers with $m_3 \geqq 1$ the ideal lattice equation of motion holds. (2.9) demands $2\delta = \beta$. With (24.12, 13) the equations are solved in the "interior" of the crystal, and also the equation of the ions at the surface would be solved by (24.12, 13) *if it were the unperturbed equation.* To solve (24.17) simultaneously with (24.13) we have to insert (24.12). But now we have to realize, that there might be different "damping constants" $\varkappa_\sigma$ for different states $\sigma$ or in other words, $\varkappa$ might depend on $\sigma$. This does not influence the above discussed isotopic case, because all the different states $\sigma$ were separated and no mixing of states was caused by the defect. Now (24.17) mixes states with different $\sigma$ and we have to allow for $\varkappa_\sigma$. But $e_i(\boldsymbol{q}\sigma)$ is again equal to $\delta_{i\sigma}$, therefore the most general solution is

$$
\begin{aligned}
u_i^{\boldsymbol{m}} &= \sum_\sigma b_\sigma (-1)^{\nu m_3}\,\delta_{i\sigma}\, e^{ia(q_1 m_1 + q_2 m_2)}\, e^{-a m_3 \varkappa_\sigma}\\
&= b_i(-1)^{\nu m_3}\, e^{ia(q_1 m_1 + q_2 m_2)}\, t_i^{m_3}; \quad t_i = e^{-a\varkappa_i}.
\end{aligned}
\tag{24.18}
$$

Inserting this into (24.17) and using (24.13) we obtain

$$
\begin{Bmatrix}
\beta(1 \mp 1/t_1) & 0 & 2i\delta\sin q_1 a\\
0 & \beta(1 \mp 1/t_2) & 2i\delta\sin q_2 a\\
-2i\delta\sin q_1 a & -2i\delta\sin q_2 a & \alpha(1 \mp 1/t_3)
\end{Bmatrix}
\begin{Bmatrix} b_1\\ b_2\\ b_3\end{Bmatrix} = 0 \tag{24.19a}
$$

or

$$\frac{\alpha\beta}{4\delta^2}(1 \mp t_1)(1 \mp t_2)(1 \mp t_3) = t_1 t_3 (1 \mp t_2)\sin^2 q_1 a + t_2 t_3 (1 \mp t_1)\sin^2 q_2 a\,. \tag{24.19b}$$

The upper sign holds for acoustical states, the lower sign for optical states. To get the complete solution, $t_i$ has to be eliminated from (24.19) by using (24.13) for each $i = \sigma$ with $\varkappa_i$ and the resulting equation has to be solved for $\omega^2$. Even in this simplest case this can be done only by large numerical work. What can be seen easily is, that there are no optical localized states at free surfaces. This is clear already by physical reasons; only acoustical waves can be identical with the solutions of the elastic theory for long waves*. This limiting case can be obtained easily from (24.19). We discuss two different cases.

i) $q_2 = 0$; $q_1 a \ll 1$.

We obtain two solutions for localized surface states:

$$\begin{aligned}
\omega^2 &= q_1^2 \frac{a^2}{2M}\left\{\alpha + \beta \overset{(+)}{-} \sqrt{(\alpha-\beta)^2 + 64\delta^4/\alpha\beta}\right\}\\
\varkappa_1^2 &= q_1^2 \frac{1}{2\beta}\left\{\alpha - \beta \overset{(-)}{+} \sqrt{(\alpha-\beta)^2 + 64\delta^4/\alpha\beta}\right\}\\
\varkappa_3^2 &= q_1^2 \frac{1}{2\alpha}\left\{-\alpha + \beta \overset{(-)}{+} \sqrt{(\alpha-\beta)^2 + 64\delta^4/\alpha\beta}\right\}\\
b_3/b_1 &= -i\,\beta\varkappa_1/2\delta q_1\,.
\end{aligned} \tag{24.20}$$

Here only the lower sign is reasonable, because the damping-constants $\varkappa_i$ have to be real and positive. The „third solution" is $t_2 = 1$, $b_2$ arbitrary and represents no localized state. Thus there is only one localized surface state in this case, and it can be shown that this state is identical with the surface state of the corresponding elastic theory if the condition of rotational invariance (2.9) is satisfied ($2\delta = \beta$!) [64 L 6]. Numerically we have for $\alpha = 4\beta$

$$\omega = 0.424 q_1 a \sqrt{\alpha/M}\,; \quad \varkappa_1 = 1.75 q_1; \quad \varkappa_3 = 0.45 q_1$$
$$b_3/b_1 = -i \cdot 1.75\,.$$

This means, that the displacement of the surface ions has the main component perpendicular to the surface and only a smaller component parallel to the direction of propagation. The transverse component ($b_3$) is less damped with increasing distance from the surface. The frequencies are somewhat smaller than the frequencies of corresponding modes in the interior of the crystal.

ii) $q_1 a = q_2 a \ll 1$.

---

* In non-primitive lattices, however, there might be optical free surface states.

Similarly we have

$$\begin{aligned}
\omega^2 &= q_1^2 \frac{a^2}{2M} \left\{\alpha + 3\beta \overset{(+)}{-} \sqrt{(\alpha-\beta)^2 + 256\,\delta^4/\alpha\,\beta}\right\} \\
\varkappa_1^2 &= q_1^2 \frac{1}{2\beta} \left\{\alpha - \beta \overset{(-)}{+} \sqrt{(\alpha-\beta)^2 + 256\,\delta^4/\alpha\,\beta}\right\} \\
\varkappa_3^2 &= q_1^2 \frac{1}{2\alpha} \left\{-\alpha + \beta \overset{(-)}{+} \sqrt{(\alpha-\beta)^2 + 256\,\delta^4/\alpha\,\beta}\right\} \\
b_1 &= b_2; \quad b_3/b_1 = -i\,\beta\varkappa_1/2\,\delta q_1\,.
\end{aligned} \tag{24.21}$$

The remarks to case i) hold also in this case. With the same assumptions we have

$$\omega = 0.652 q_1 a \sqrt{\alpha/M}\,; \quad \varkappa_1 = 1.82 q_1; \quad \varkappa_3 = 0.27 q_1; \quad b_3/b_1 = -i \cdot 1.82\,,$$

which gives a similar behavior as for the waves in $q_1$-direction. Fig. 6.27 shows the complete dispersion curves for symmetric directions in the Brillouin zone of the surface. The general shape of the surface dispersion curves is very similar to that for plane waves in the interior of a crystal. Differences occur only in quantitative respect.

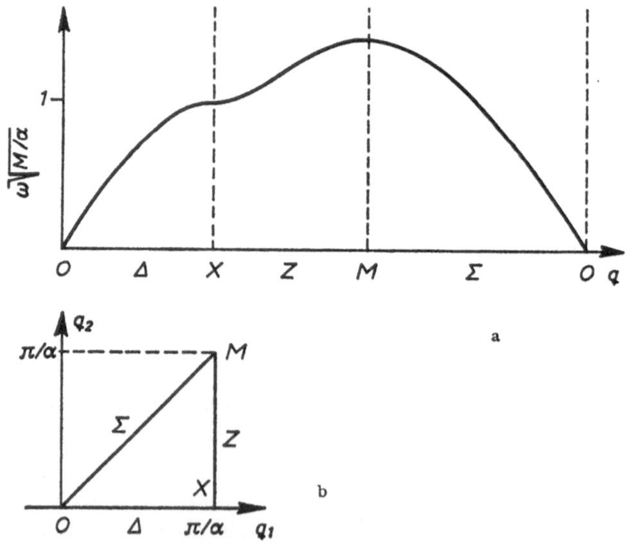


Fig. 6.27 a and b. a) Dispersion curves for localized states at a free surface. (The directions of propagation are defined in b)

The results derived with this model hold qualitatively also in other, more realistic models. Only the numerical behavior differs from model to model. But there is another, more general kind of surface states, which can be discussed with the help of a (011)-plane in a simple cubic crystal.

Because only the $\boldsymbol{q}$-component perpendicular to the surface can become complex, it is convenient to transform to a coordinate system which has one of its axis (3) perpendicular to the (011)-plane. To make

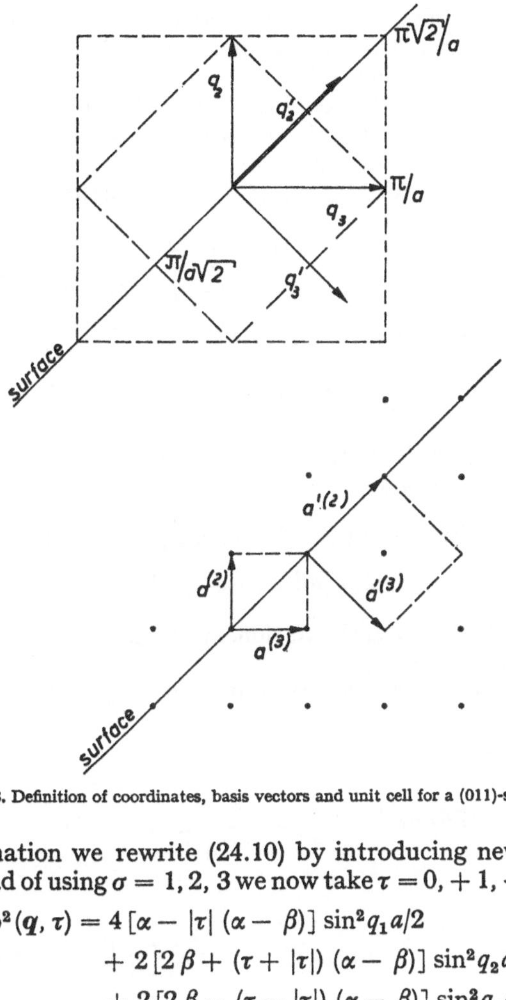

Fig. 6.28. Definition of coordinates, basis vectors and unit cell for a (011)-surface

this transformation we rewrite (24.10) by introducing new polarization indices. Instead of using $\sigma = 1, 2, 3$ we now take $\tau = 0, +1, -1$ and obtain

$$\begin{aligned} M\,\omega^2(\boldsymbol{q}, \tau) = {} & 4\,[\alpha - |\tau|\,(\alpha - \beta)]\,\sin^2 q_1 a/2 \\ & + 2\,[2\,\beta + (\tau + |\tau|)\,(\alpha - \beta)]\,\sin^2 q_2 a/2 + \\ & + 2\,[2\,\beta - (\tau - |\tau|)\,(\alpha - \beta)]\,\sin^2 q_3 a/2\,. \end{aligned} \tag{24.10a}$$

Transforming this according to

$$q_2' = \frac{1}{\sqrt{2}}\,(q_3 + q_2)\,; \quad q_3' = \frac{1}{\sqrt{2}}\,(q_3 - q_2) \tag{24.22}$$

we have

$$\begin{aligned} M\,\omega^2(\boldsymbol{q}, \tau) = {} & 4\,[\alpha - |\tau|\,(\alpha - \beta)]\,\sin^2 q_1 a/2 + \\ & + 2\,[2\,\beta + |\tau|\,(\alpha - \beta)] \cdot [1 - \cos q_2' a/\sqrt{2}\,\cos q_3' a/\sqrt{2}\,] - \\ & - 2\tau(\alpha - \beta)\,\sin q_2' a/\sqrt{2}\,\sin q_3' a/\sqrt{2} \end{aligned} \tag{24.23}$$

$q_2$ and $q_3$ run from $-\pi/a$ to $+\pi/a$; the corresponding Brillouin zone is indicated in Fig. 6.28a. There results a more complicated zone for $q_2'$ and $q_3'$. However, we can limit $q_2'$ and $q_3'$ to the range $-\pi/a\sqrt{2}$ to $\pi/a\sqrt{2}$; in the rest of the zone the cos-terms in (24.23) change sign; this can be described by introducing a second "optical" branch of frequencies with reduced wave vectors, lying in the first new Brillouin zone. We describe this by a new polarization index $\sigma'$, taking the value $-1$("acoustical") and $+1$ ("optical"). In the coordinate space this means a non-primitive description of the simple cubic lattice with new basis-vectors, as indicated in Fig. 6.28b. The new Brillouin zone is just appropriate for the description of surface states. The frequencies are now written as

$$\begin{aligned} M\omega^2(\boldsymbol{q},\tau,\sigma') = & 4[\alpha-|\tau|(\alpha+\beta)]\sin^2 q_1 a/2 + \\ & + 2[2\beta+|\tau|(\alpha-\beta)]\cdot[1+\sigma'\cos q_2' a/\sqrt{2}\cos q_3' a/\sqrt{2}] - \\ & - 2\tau(\alpha-\beta)\sin q_2' a/\sqrt{2}\sin q_3' a/\sqrt{2} \end{aligned} \tag{24.24}$$

$$-\frac{\pi}{a} < q_1 \leqq \frac{\pi}{a} \quad ; -\frac{\pi}{a\sqrt{2}} < q_2', q_3' \leqq \frac{\pi}{a\sqrt{2}}; \quad \tau = 0, \pm 1, \ \sigma' = \pm 1 .$$

At defect surfaces we have to allow for complex $q_3' \to q_3' + i\varkappa$. The imaginary part of $\omega^2$ becomes

$$\begin{aligned} \operatorname{Im}\omega^2 = & -2\sigma'[2\beta+|\tau|(\alpha-\beta)]\cos q_2' a/\sqrt{2}\sin q_3' a/\sqrt{2}\sinh\varkappa a/\sqrt{2} - \\ & -2\tau(\alpha-\beta)\sin q_2' a/\sqrt{2}\cos q_3' a/\sqrt{2}\sinh\varkappa a/\sqrt{2} . \end{aligned} \tag{24.25}$$

In order to make this vanish (stationary states) we can choose

i) $\varkappa = 0$, $q_3'$ arbitrary.

These are no localized states, but scattering states; they shall not be discussed in the moment.

ii) $\varkappa$ arbitrary.

Then

$$\operatorname{tg} q_3' a/\sqrt{2} = -\frac{\tau\sigma'(\alpha-\beta)}{2\beta+|\tau|(\alpha-\beta)}\operatorname{tg} q_2' a/\sqrt{2} \tag{24.26}$$

or, if

$$\tau = 0; \quad q_3' = 0 \tag{24.26a}$$

and if

$$\tau = \pm 1; \quad \operatorname{tg} q_3' a/\sqrt{2} = -\tau\sigma'\frac{\alpha-\beta}{\alpha+\beta}\operatorname{tg} q_2' a/\sqrt{2} \tag{24.26b}$$

$\tau = 0$ corresponds to waves were the polarization vector is parallel to the surface in the direction, which is not affected by the above transformation. In this case the possible surface waves are similar to those discussed for the (001)-plane. A new type of surface waves arises from eq. (24.26b), which corresponds to waves with polarization vectors in the $2'-3'$-plane (drawing plane in Fig. 6.28). Every surface wave having a wave vector with non-zero $q_2$-component has a oscillating behavior in the direction perpendicular to the surface, which is superposed on the decreasing amplitude determined by $\varkappa$. The wavelength of these oscillations is strongly connected with $q_2$. The displacements are then given by

$$u^{m}_{\mu i} = \sum_{\sigma,\tau} b_{\sigma\tau} e^{\mu}_{i}(\sigma,\tau)\, e^{i q_1 a m_1 + i a \sqrt{2} q_2' m_2 + i a \sqrt{2} q_3' m_3} \cdot e^{-a\sqrt{2} m_3 \varkappa(\sigma,\tau)} \tag{24.27}$$

The index $\mu$ now arises from the non-primitive description of the lattice. The oscillating behavior in the 3′-direction holds of course also in the elastic limit, as can be seen immediately from (24.26b). Elastic surface waves of this type are called generalized Rayleigh-waves.

If the surfaces have higher indices, or in crystals with more complicated structure and force-constants, $q_3$ will be a function of $q_1$ and $q_2$. This is the most general case. A number of models and different orientations of surfaces have been investigated. Qualitatively we always find a similar behavior of surface waves, but quantitatively very different aspects occur. At the surface of a given crystal there might be surface states for a certain orientation of this surface and no surface states for another orientation. It might also be, that at a given surface waves are allowed to propagate in a certain direction, whereas the propagation in

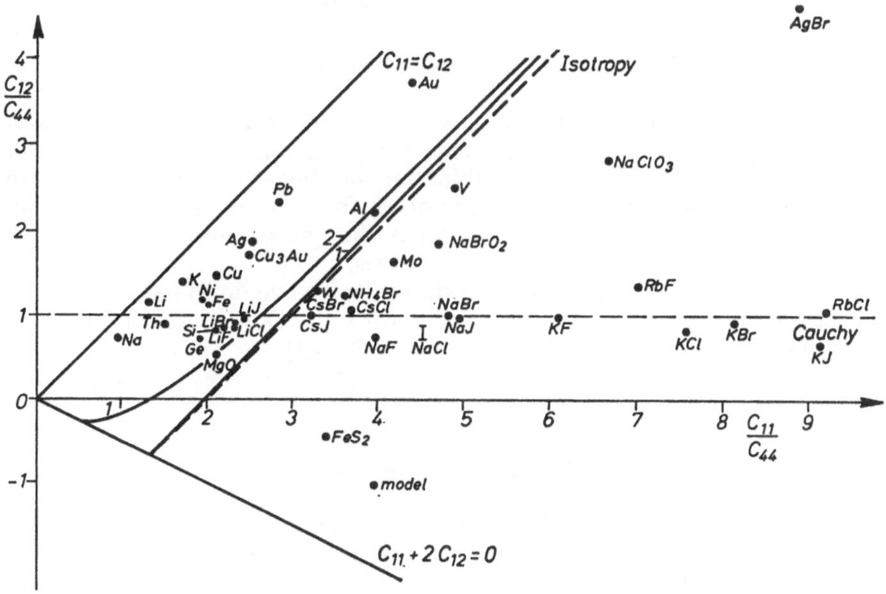


Fig. 6.29. Positions of various crystals in a map which shows the possibility of surface states. The details are explained in the text

other directions is not allowed for surface waves. Also the types of the surface waves may differ from direction to direction of the propagation and from orientation to orientation of the surface, as is shown already by our simple example.

For a (001)-surface of a cubic crystal, *Gazis*, *Herman* and *Wallis* [60 W 2] have given the possible wave forms in the elastic limit as a function of the elastic constants of the medium. Fig. 6.29 shows their result. The straight lines $c_{11} = c_{12}$ and $c_{11} + 2c_{12} = 0$ define the region of lattice stability. The dashed lines give the condition of isotropy and Cauchy relations.

Between the curve $c_{11} + 2c_{12} = 0$ and the curve marked with 1 Rayleigh-waves at the (001)-surface are always possible, for all the directions of propagation. These Rayleigh-waves have purely exponentially decreasing amplitudes, when the distance from the surface increases.

Between curve 1 and 2 generalized Rayleigh-waves are possible, again for all directions of propagation. Generalized Rayleigh-waves are the waves mentioned in (24.26b), which have an oscillating behavior superposed over the exponential decrease.

Between curve 2 and curve $c_{11} = c_{12}$ again generalized Rayleigh-waves occur, but only for certain directions of propagation; there are always some directions or a cone of directions, where no surface waves are possible [60 W 2].

In other surfaces the situation is similar, but it might be even more complex. It should be mentioned further, that sometimes there is a distinction between so-called Love- and Rayleigh-waves. In Rayleigh-waves the displacements of the ions lie in a plane which is defined by the normal of the surface and the direction of propagation; the waves have a longitudinal and a transverse (perpendicular to the surface) component. Love-waves are purely transverse waves with the displacements parallel to the surface. They can occur only in very special situations [60 W 2]. A linear combination of these two wave-types is the most general one, and it makes not much sense to distinguish strongly between the different types, because they all are only special aspects of the general surface or Rayleigh waves.

The elastic theory of surface states (long wave limit) has been treated in a number of papers, beginning with the famous one of Rayleigh [85 R 1, 55 S 3, 57 D 1, 57 S 2, 59 M 2, 63 B 3].

## 25. Scattering at Surfaces. Reflection and Transmission

Extended defects, like surfaces, give rise to a scattering of plane waves. From the geometry of the problem it is obvious, that plane waves remain plane waves, if the surface is really plane. Therefore the scattering of incoming plane waves can be described by a reflection coefficient $\mathscr{R}$ and, if the surface is imbedded in a crystal, by a transmission coefficient $\mathscr{T}$. Because there are the polarization states of waves in a threedimensional (Bravais) crystal lattice, an incoming wave of a defined polarization will be scattered in three different polarization states, if there is a coupling between different polarizations. This is determined by the form of the coupling-matrices. The frequency remains constant during the scattering process; thus different polarizations will have different $\boldsymbol{q}$-vectors in general. However, the translational invariance of the surface, say e.g. in the 1- and 2-direction, needs equal $q_1$- and $q_2$-components for incoming and scattered waves. Different polarizations will differ in the $q_3$-component alone, after scattering. This means, that an incoming plane wave of given polarization $\sigma$ suffers triple refraction and triple reflection into polarizations $\sigma'$ in general. In certain cases this will „degenerate" to double or single reflection and refraction. In non-primitive lattices even more transitions to other polarization states might occur, if optical and acoustical branches of frequencies overlap in such a way, that transitions are possible without violating the conservation of energy (frequency).

From these considerations it follows, that the reflection coefficient is a matrix $\mathscr{R}_{\sigma\sigma'}(\boldsymbol{q})$ and also the transmission coefficient $\mathscr{T}_{\sigma\sigma'}(\boldsymbol{q})$ at interior surfaces. The conservation of energy demands

$$\sum_{\sigma'} \{\mathscr{R}_{\sigma\sigma'}(\boldsymbol{q}) + \mathscr{T}_{\sigma\sigma'}(\boldsymbol{q})\} = 1 . \tag{25.1}$$

Therefore all the coefficients must be smaller than one. This implies that there cannot be strong resonances as in the scattering amplitudes at point defects; however, there will be maxima and minima or even zeros in the $\mathscr{R}$- and $\mathscr{T}$-coefficients, which mean a sort of a resonance [67 L 1]. In any case, they are not very pronounced and the description used for point defect scattering fails.

Because the scattering of waves at *plane* surfaces is a very definite scattering, this scattering does not contribute to thermal resistance. To explain the effect of surface scattering on thermal resistance, one has to assume rough surfaces, causing a diffuse scattering at the surface. Then the energy is distributed into a large variety of scattered waves. This can be described by assuming a statistical distribution of surface orientations and averaging over the waves scattered by different oriented surface elements. For the details of the surface contribution to thermal resistance we refer to the papers mentioned already [38 C 1, 59 C 2, 61 C 2, 62 B 6]. In the following we will discuss the principles of surface scattering using simple examples. Because plane waves remain plane waves in surface scattering, the description of this process is easier than the discussion of point defect scattering in many cases.

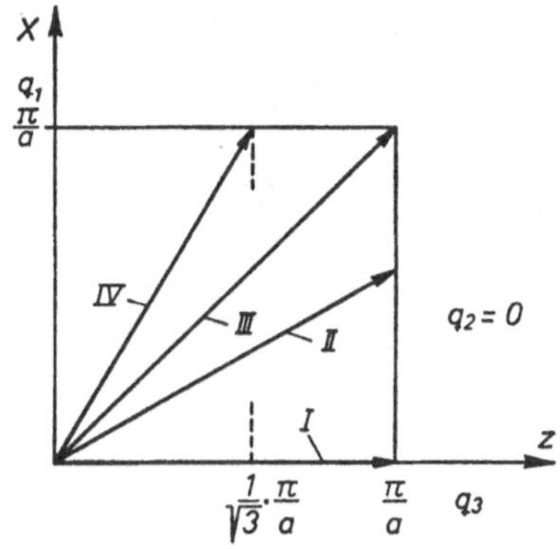


Fig. 6.30. Definition of directions of wave propagation used in the following figures

First we use the example of Sect. 24: A simple cubic lattice with a (001)-plane of isotopes. In this case the coupling matrices are diagonal and there is no interaction of different polarization states. $\mathscr{R}$ and $\mathscr{T}$

are diagonal. The wave may come from the left ($m_3 < 0$) of the isotopic plane ($m_3 = 0$). We then have*

$$u_i^m = e_i(\boldsymbol{q}\sigma)\, e^{iq_1 a m_1 + i q_2 a m_2} \{A_\sigma\, e^{i q_3 a m_3} + B_\sigma\, e^{-i q_3 a m_3}\}; \quad m_3 < 0 \tag{25.2a}$$

$$u_i^m = e_i(\boldsymbol{q}\sigma)\, e^{iq_1 a m_1 + i q_2 a m_2} C_\sigma\, e^{i q_3 a m_3}; \quad m_3 > 0;\; e_i(\boldsymbol{q}\sigma) = \delta_{i\sigma}\,. \tag{25.2b}$$

With this ansatz we have to satisfy the equation of motion for the $m_3 = 0$-plane (24.14). For $m_3 = 0$ (25.2a) and (25.2b) have to agree because of the continuity of the displacements. Therefore

$$A_\sigma + B_\sigma = C_\sigma\,. \tag{25.3}$$

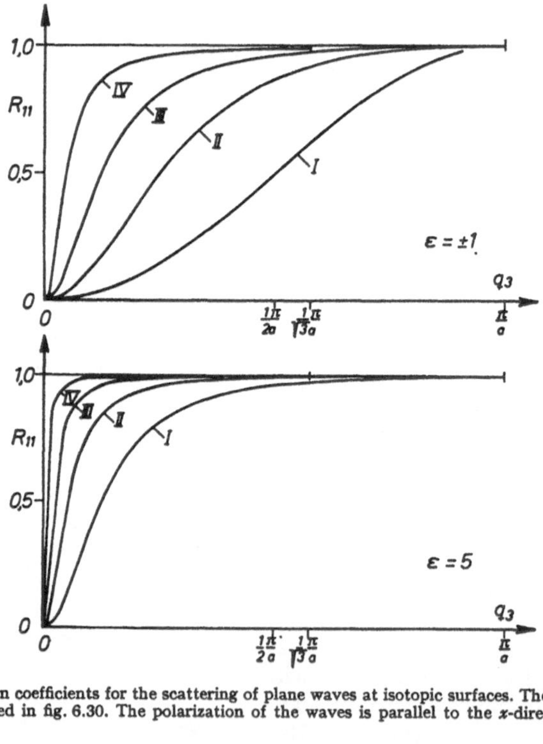


**Fig. 6.31.** Reflection coefficients for the scattering of plane waves at isotopic surfaces. The directions of propagation are defined in fig. 6.30. The polarization of the waves is parallel to the $x$-direction

Inserting (25.2) into (24.14) we obtain, using (24.10)

$$\frac{B_\sigma}{A_\sigma} = -\frac{2\,\varepsilon\, r_\sigma}{2\,\varepsilon\, r_\sigma + i \sin q_3 a}; \quad \frac{C_\sigma}{A_\sigma} = \frac{i \sin q_3 a}{2\,\varepsilon\, r_\sigma + i \sin q_3 a}$$

$$\mathscr{R}_{\sigma\sigma} = \left|\frac{B_\sigma}{A_\sigma}\right|^2; \quad \mathscr{T}_{\sigma\sigma} = \left|\frac{C_\sigma}{A_\sigma}\right|^2; \; \mathscr{R}_{\sigma\sigma} + \mathscr{T}_{\sigma\sigma} = 1; \; \varepsilon = (M' - M)/M$$

* This is the only Ansatz compatible with the requirement that the frequencies for each $\sigma$ remain constant.

$$r_1 = \alpha^* \sin^2 q_1 a/2 + \sin^2 q_2 a/2 + \sin^2 q_3 a/2; \quad \alpha^* = \alpha/\beta$$
$$r_2 = \sin^2 q_1 a/2 + \alpha^* \sin^2 q_2 a/2 + \sin^2 q_3 a/2$$
$$r_3 = \frac{1}{\alpha^*} \sin^2 q_1 a/2 + \frac{1}{\alpha^*} \sin^2 q_2 a/2 + \sin^2 q_3 a/2 \,. \tag{25.4}$$

The same result can be gotten easily by using the description of eq. (24.5); the Greens function method corresponds to the procedure used

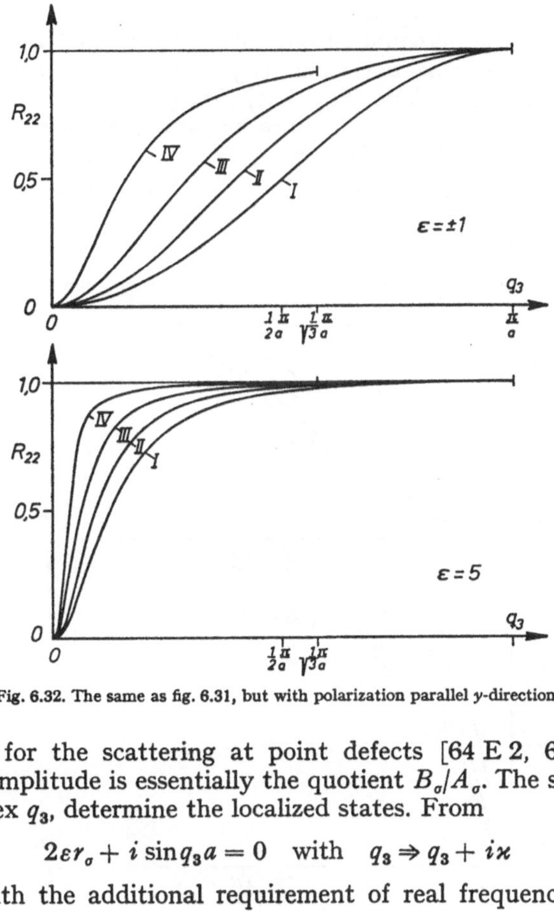


Fig. 6.32. The same as fig. 6.31, but with polarization parallel $y$-direction

in Sect. 22 for the scattering at point defects [64 E 2, 65 T 2]. The scattering amplitude is essentially the quotient $B_\sigma/A_\sigma$. The singularities, with complex $q_3$, determine the localized states. From

$$2\varepsilon r_\sigma + i \sin q_3 a = 0 \quad \text{with} \quad q_3 \Rightarrow q_3 + i\varkappa \tag{25.5}$$

we have with the additional requirement of real frequencies, e.g. for $\sigma = 1$

$$\varepsilon \left\{ \alpha^* \sin^2 q_1 a/2 + \sin^2 q_2 a/2 \begin{matrix} -\sinh^2 \varkappa a/2 \\ +\cosh^2 \varkappa a/2 \end{matrix} \right\} = \pm \sinh \varkappa a/2 \cosh \varkappa a/2 \tag{25.6}$$

which is identical with the use of (24.15) and (24.16).

In Fig. 6.31 to 6.33 the reflection coefficient for several incoming directions (defined in Fig. 6.30) and the three polarizations is shown;

it depends only on the absolute value of $\varepsilon$, therefore it is equal for a plane of lighter or heavier isotopes. With increasing $\varepsilon$, the reflection becomes larger for all the wave-lengths; in the limit $\varepsilon \to \infty$ $\mathcal{R}_{\sigma\sigma} = 1$ over the whole range. This is obvious, because a plane of infinite masses has no transparency for phonons.

Similar calculations can be done with a plane of impurities. This can be described by coupling parameters $\alpha' = (1 + \zeta')\,\alpha$ between ions in the impurity plane and $\alpha'' = (1 + \zeta'')\,\alpha$ between ions of the plane and those in neighboring planes. The non-central coupling parameters $\beta$ can be looked upon as equal between all ions [rotational invariance (2.9)]. For polarizations parallel to the impurity plane there is a behavior similar to that of an isotopic plane (Fig. 6.34). For $\zeta' \to \infty$, i.e. when the force-constants in the plane have infinite stiffness, the plane is again not transparent for phonons. For incoming waves with polarization perpendicular to the plane, the behavior for $\zeta'' \gg 1$ is equal (Fig. 6.35). For small

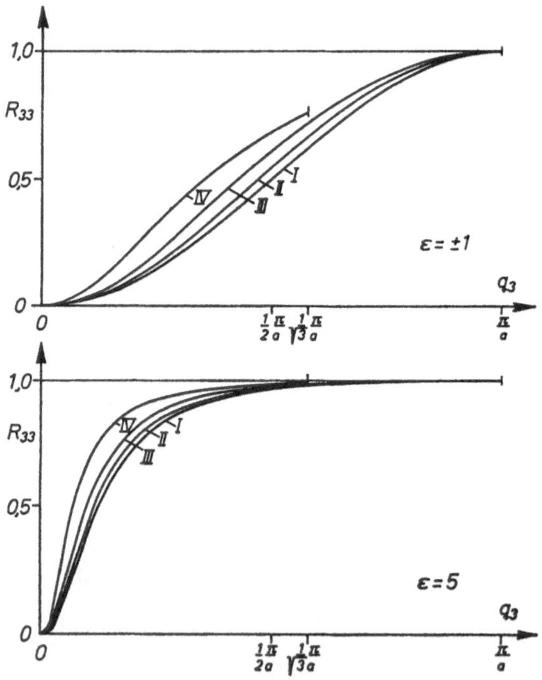


Fig. 6.33. The same as fig. 6.31, but with polarization parallel z-direction

positive changes a zero in the reflection coefficient occurs. $\mathcal{R}_{33}$ has a maximum at a certain frequency. This behavior resembles a resonance behavior. But it is not strongly pronounced, and the characteristic properties of point defect resonances are missing. For $\zeta'' = 1$, the zero is just at the Brillouin zone-boundary. This is connected to localized states; they occur if $\zeta'' \geqq 1$. The condition for the zero to be at $q_3 = \pi/a$ can be shown to be just the condition for the first occurence of localized modes [see 64 L 1, 67 L 1].

Another case, perhaps of somewhat more interest, is the reflection of waves at the boundary of two different crystals (Fig. 6.36). Both

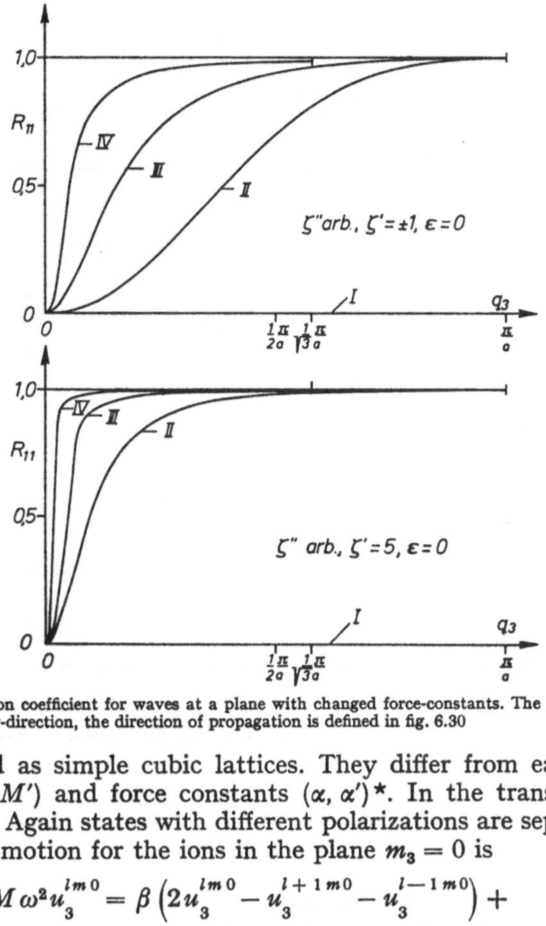


Fig. 6.34. Reflection coefficient for waves at a plane with changed force-constants. The polarization of the waves is parallel $x$-direction, the direction of propagation is defined in fig. 6.30

are assumed as simple cubic lattices. They differ from each other in masses ($M$, $M'$) and force constants ($\alpha$, $\alpha'$)*. In the transition region we have $\alpha''$. Again states with different polarizations are separated. The equation of motion for the ions in the plane $m_3 = 0$ is

$$\begin{aligned} M\omega^2 u_3^{lm0} &= \beta\left(2u_3^{lm0} - u_3^{l+1\,m0} - u_3^{l-1\,m0}\right) + \\ &+ \beta\left(2u_3^{lm0} - u_3^{lm+1\,0} - u_3^{lm-1\,0}\right) + \qquad (25.7\text{a}) \\ &+ \alpha\left(u_3^{lm0} - u_3^{lm\bar{1}}\right) + \alpha''\left(u_3^{lm0} - u_3^{lm1}\right). \end{aligned}$$

The equation for $m_3 = 1$ has $m_3$ replaced by $m_3 + 1$ and

$$M \text{ by } M',\quad \alpha \text{ by } \alpha'',\quad \alpha'' \text{ by } \alpha'. \qquad (25.7\text{b})$$

* The condition of rotational invariance (2.9) requires in this case all the non-central constants $\beta$ between layers with different $m_3$ to be equal. It is therefore reasonable, to assume also the non-central-constants in the layers with $m_3$ constant to be equal. All the characteristic features which might occur can be seen with this simple model, though in realistic cases the relations are more complicated.

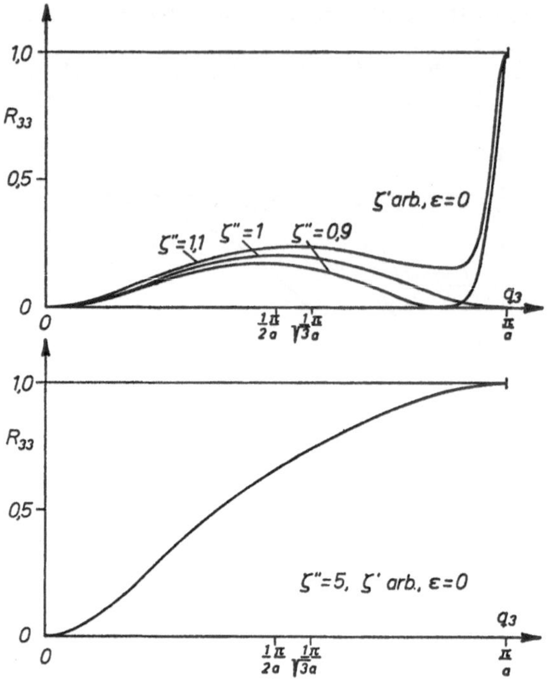


Fig. 6.35. The same as fig. 6.34, but with polarization in z-direction

The frequencies are given by

left: $M\omega^2(\boldsymbol{q}\sigma = 3) = 4\beta\sin^2 q_1 a/2 + 4\beta\sin^2 q_2 a/2 + 4\alpha\sin^2 q_3 a/2$

(25.8)

right: $M'\omega^2(\boldsymbol{q}'\sigma = 3) = 4\beta\sin^2 q_1 a/2 + 4\beta\sin^2 q_2 a/2 + 4\alpha'\sin^2 q_3 a/2$

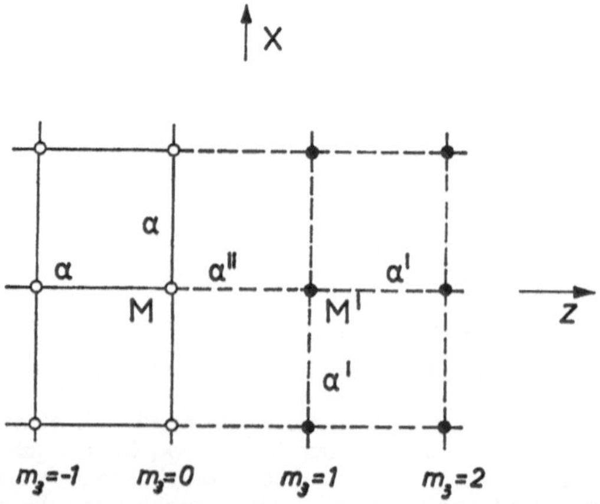


Fig. 6.36. Definition of force-constants and masses in a model, which contains a free surface as a limit, otherwise it is the boundary between two media

Use has been made of the translational invariance in the 1-, 2-directions. The frequencies have to be equal, therefore

$$\alpha(1+\zeta')\sin^2 q_3' a/2 = \beta\varepsilon(\sin^2 q_1 a/2 + \sin^2 q_2 a/2) + \alpha(1+\varepsilon)\sin^2 q_3 a/2 \tag{25.9}$$

$$\varepsilon = (M'-M)/M;\quad \zeta' = (\alpha'-\alpha)/\alpha;\quad \zeta'' = (\alpha''-\alpha)/\alpha. \tag{25.10}$$

The incoming wave is assumed to come from the left:

$$u_i^m = e^{iq_1am_1+iq_2am_2}\{A_i e^{iq_3am_3} + B_i e^{-iq_3am_3}\};\quad m_3 \leqq 0 \tag{25.11}$$

$$u_i^m = e^{iq_1am_1+iq_2am_2}\{C_i e^{iq_3'a(m_3-1)}\}\quad m_3 \geqq 1.$$

Inserting this into (25.7) we obtain with

$$r = \zeta'' + e^{-iq_3a};\quad r' = \zeta'' - \zeta' + (1+\zeta')e^{-iq_3'a} \tag{25.12}$$

$$\frac{B_3}{A_3} = -\frac{r'r^* - (1+\zeta'')^2}{rr' - (1+\zeta'')^2};\ \frac{C_3}{A_3} = -i\frac{2(1+\zeta'')\sin q_3 a}{rr' - (1+\zeta'')^2}$$

Since the group-velocity and the flux is different in both semi-spaces, to get the transmission-coefficient we have to multiply the square of $C_3/A_3$ by the quotient of the fluxes right and left of the boundary, or more precisely, by the quotient of the 3-components*. The flux is proportional to

$$j_i \sim 2M\omega\frac{\partial\omega}{\partial q_i} = M\frac{\partial\omega^2}{\partial q_i} = 2M\omega\cdot v_i \tag{25.13}$$

$v_i$ is the group velocity. With (25.8) we have

$$\frac{j_3'}{j_3} = \frac{M'v_3'}{Mv_3} = \frac{\alpha'\sin q_3' a}{\alpha\sin q_3 a}. \tag{25.14}$$

Thus

$$\mathcal{R}_{33} = \left|\frac{B_3}{A_3}\right|^2;\quad \mathcal{T}_{33} = \frac{(1+\zeta')\sin q_3' a}{\sin q_3 a}\left|\frac{C_3}{A_3}\right|^2 \tag{25.15}$$

which can be proved with the preceding relations. The discussion gives for different cases:

i) $\varepsilon > 0,\quad \zeta' > -1.$

Both sides of eq. (25.9) are positive. For real $q_3'$, outgoing plane waves, the $\sin^2 q_3' a/2$ is limited to the interval 0 to 1. This means a limitation for the values of $\varepsilon$, $\zeta$ and $\boldsymbol{q}$, for which a transmission is possible. The boundary is intransparent if $\boldsymbol{q}$ exceeds a certain value, i.e. at large frequencies. If $\boldsymbol{q}$ is larger than the limiting value, the phonons suffer total-reflection. $q_3'$ becomes complex; this means that there are excited localized states in the vicinity of the boundary (exponentially decreasing amplitudes) just as in the case of electro-magnetic total-reflection. This enables the possibility of a *tunnel-effect for phonons*. If there is a medium divided in two parts by a film of another medium, the parameters of which are chosen appropriately, in the film there are excited exponentially decreasing amplitudes, which allow for waves in the medium of the other

* This can be seen by considering the energy in a volume element near the boundary.

side of the film, with a smaller intensity. The quantitative results depend sensitively on the parameters involved. Fig. 6.37 shows the reflection coefficient for $\zeta' = -1/4$, $\zeta'' = -1/8$, $\varepsilon = 0$ and $q_1 = q_2 = 0$. For small wavelength, it is practically zero, whereas at larger $q_3$ there is an abrupt change to total-reflection. No total-reflection for all the possible $\boldsymbol{q}$-values occurs, if $\zeta' > \varepsilon(1 + 2\beta/\alpha)$. But the coefficient increases with increasing $\zeta'$, and it is again equal to one for $\zeta' \to \infty$. This is clear, because waves cannot propagate if $\zeta' = \infty$. For $\zeta' \to -1$, the region of total-reflection becomes larger. It covers the whole $\boldsymbol{q}$-range if $\zeta' = -1$ (no wave propagation is possible in the right crystal). But this limiting case makes no sense, because then the semi-crystal becomes unstable in this model.

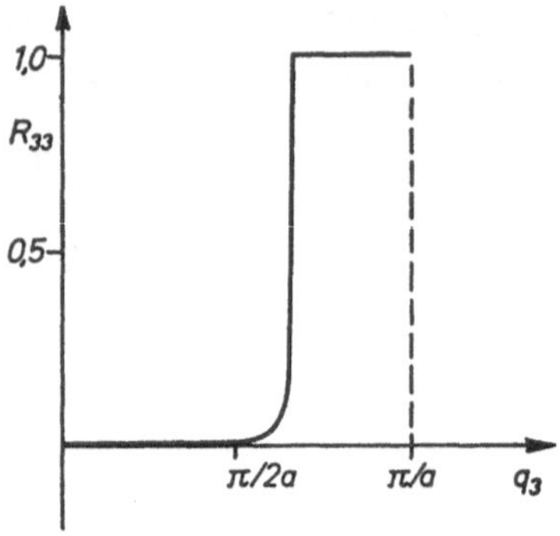


Fig. 6.37. Reflection coefficient for the scattering of waves at the boundary of two media. The parameters are explained in the text

ii) $\varepsilon < 0$.

Qualitatively the situation is similar. But there occurs the possibility of total-reflection at small wave-lengths. In the elastic limit the condition for the occurence of total-reflection is

$$\frac{q_1^2 + q_2^2}{q_3^2} = \mathrm{tg}^2\varphi > -\frac{\alpha(1+\varepsilon)}{\beta\varepsilon} \tag{25.16}$$

which defines a sort of a Brewster-angle. For larger wave-lengths the $q_i$-components have to be replaced by $\sin q_i a/2$ in (25.16). This gives another possibility of a tunnel-effect for acoustical phonons. Fig. 6.38 shows the reflection coefficient for $\zeta' = \zeta'' = 0$, $\varepsilon = -0.55$; $q_1 = 2q_3$, $q_2 = 0$, $\alpha = 4\beta$. If the wave-lengths are out of the region of total-reflection, $\mathscr{R}_{33}$ decreases rapidly and increases again near the maximum value of $q_3$. In the limiting case $q_3 = \pi/a$ we have complete reflection.

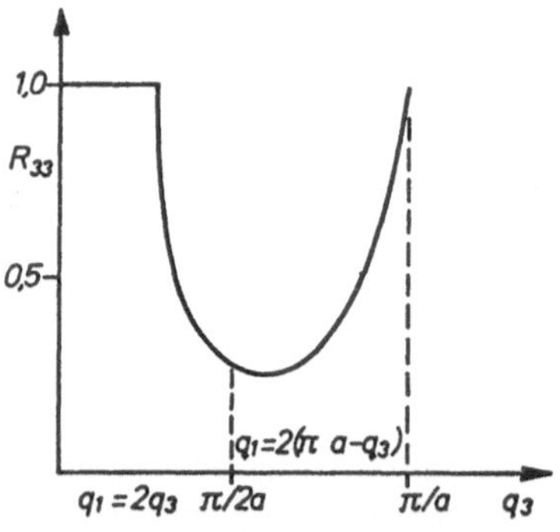


Fig. 6.38. Reflection coefficient for the scattering of waves at the boundary of two media

A similar calculation can be done for incoming waves of polarization parallel to the surface ($\sigma = 1, 2$). For $\sigma = 1$ we obtain

$$\beta \sin^2 q_3' a/2 = \alpha(\varepsilon - \zeta') \sin^2 q_1 a/2 + \beta\varepsilon \sin^2 q_2 a/2 + \beta(1+\varepsilon)\sin^2 q_3 a/2 \tag{25.17}$$

whereas for $\sigma = 2$ $q_1$ and $q_2$ have to be interchanged. The reflection and transmission coefficients satisfy the same formula:

$$\mathscr{R}_{11} = \mathscr{R}_{22} = \frac{1 - \cos a(q_3 - q_3')}{1 - \cos a(q_3 + q_3')}; \quad \mathscr{T}_{11} = \mathscr{T}_{22} = \frac{2 \sin q_3 a \sin q_3' a}{1 - \cos a(q_3 + q_3')}. \tag{25.18}$$

Again we have the possibility of total-reflection as well for high frequencies as for low frequencies (elastic limit). The details depend on the choice of the parameters $\alpha$, $\beta$, $\varepsilon$, $\zeta'$. In the elastic limit total-reflection occurs if (for $\sigma = 1$)

$$\frac{q_1^2}{q_3^2} + \frac{\beta\,\varepsilon}{\alpha\,(\varepsilon - \zeta')}\,\frac{q_2^2}{q_3^2} > -\frac{\beta\,(1+\varepsilon)}{\alpha\,(\varepsilon - \zeta')} \tag{25.19}$$

which is satisfied again mainly with $\varepsilon < 0$.

From this model we could extrapolate the case of a free surface by equating the force-constants between the layers $m_3 = 0$ and $m_3 = 1$ to zero. Then all the $\mathcal{R}_{\sigma\sigma}$ become equal to one, which, of course, must result because in this model there is no coupling between different polarizations. But it is a poor model for a free surface; for the condition of rotational invariance allows for only one $\beta$ between layers with different $m_3$ and for a free surface we have $\beta = 0$ and the crystal becomes unstable.

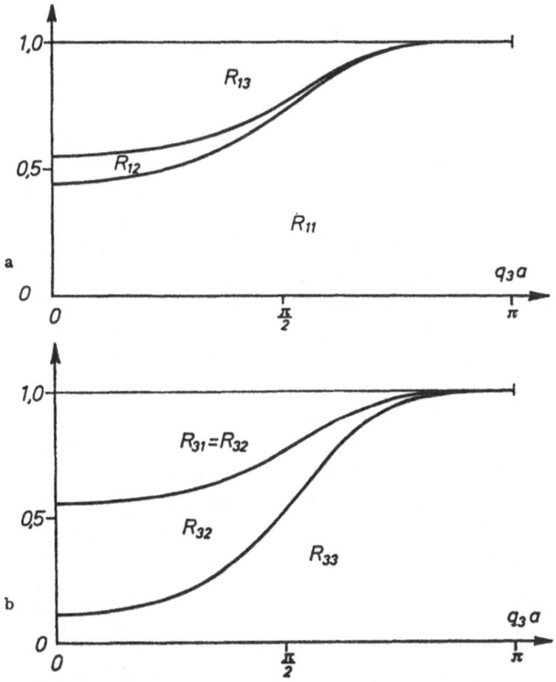

a
b

Fig. 6.39a and b. Reflection coefficient for the scattering of waves at a free surface in a simple cubic lattice. The polarization changes during scattering; the different contributions are shown one above the other, the sum being one. a) incoming polarization parallel to the surface, b) incoming polarization perpendicular to the surface. The direction of propagation is $q_1 = q_2 = q_3$

A somewhat better model is that of the preceding section [eq. (24.17)]. This is rotational invariant and because the coupling matrices are non-diagonal, the interaction of different polarizations can be studied also with this model.

If the incoming wave has polarization $\sigma$ and wave-vector $\boldsymbol{q}$, the appropriate ansatz for the scattering problem is

$$u_i^{\boldsymbol{m}} = e^{iq_1am_1 + iq_2am_2}\{A_\sigma e_i(\boldsymbol{q}\sigma)\, e^{iq_3am_3} + \sum_{\sigma'} B_{\sigma\sigma'} e_i(\boldsymbol{q}'\sigma')\, e^{iq_3(\sigma')am_3}\} \tag{25.20}$$

because a transmission of phonons through a free surface is not possible. There is no conservation law for the $q_3'$-component; therefore in general it can and will be different for each of the three polarizations (triple reflection). But all the frequencies involved have to be equal. From this the three possible values $q_3'(\sigma')$ can be determined as a function of $\sigma'$ and the incoming $\boldsymbol{q}$, $\sigma$. The $B_{\sigma\sigma'}$ will be calculated from inserting (25.20) into(24.17). The reflection coefficients are not directly given by $B_{\sigma\sigma'}/A_\sigma$, because the outgoing waves $\sigma'$ have different group-velocities $v_i(\sigma')$. The masses are always equal, therefore

$$\mathcal{R}_{\sigma\sigma'} = \frac{v_3(\sigma')}{v_3(\sigma)} \left|\frac{B_{\sigma\sigma'}}{A_\sigma}\right|^2; \quad v_i(\sigma) = \frac{\partial\omega(\boldsymbol{q}\sigma)}{\partial q_i}; \quad \sum_{\sigma'} \mathcal{R}_{\sigma\sigma'} = 1\,. \tag{25.21}$$

The resulting formulas are very long and will not be given here [67 L 1]. Instead in Fig. 6.39 the reflection-coefficients are shown, for incident waves with $q_1 = q_2 = q_3$ and polarization parallel and perpendicular to the surface, the three coefficients are shown one above the other, the sum being one.

Similar phenomena arise in more realistic and complicated models, and at other planes of reflection [67 L 1], but the principle features are the same. Often detailed calculations can be avoided, for example, if one is interested in total-reflection. Then a discussion of the frequencies as function of $\boldsymbol{q}$ is sufficient. If the coupling-matrices are non-diagonal, the polarizations are coupled. There are different critical conditions for total-reflection for different polarizations.

It should be mentioned, that the investigations can be done also using the method of Greens function or a variational method. In many cases it is further sufficient to use the elastic theory to calculate reflection and transmission in the long wave limit. Also the effect of surfaces on correlations and thermodynamic functions can be calculated with the same methods as indicated for point defects. If one uses Greens functions it is convenient to use them in the way as defined in (24.6) and (24.7), which takes into account the translational invariance. The correlation is completely determined if the Greens function of the defect problem $\mathbf{R}$ is known (23.45). The Greens function $\tilde{\mathbf{R}}$ can be determined from

$$(\tilde{\mathbf{1}} - \tilde{\mathbf{G}}\tilde{\mathbf{l}})\,\tilde{\mathbf{R}} = \tilde{\mathbf{G}} \tag{25.22}$$

which is analogous to (23.8). The original $\mathbf{R}$, which enters (23.35) is then

$$R_{i\ j}^{\boldsymbol{m}\boldsymbol{n}}(\omega) = \frac{1}{N^{2/3}} \sum_{\tilde{\boldsymbol{q}}} e^{i\tilde{\boldsymbol{q}}(\boldsymbol{\varrho}^m - \boldsymbol{\varrho}^n)}\, \tilde{R}_{i\ j}^{m_3 n_3}(\tilde{\boldsymbol{q}}; \omega)\,. \tag{25.23}$$

For the auto-correlation we only need

$$R_{i\ j}^{m\,m}(\omega) = R_{i\ j}^{0 m_3 0 n_3} = \frac{1}{N^{2/3}} \sum_{\tilde{q}} \tilde{R}_{i\ j}^{m_3 n_3}(\tilde{q}, \omega) \tag{25.24}$$

which is a trace in the two-dimensional surface-space.

We can also use (23.45) for a calculation of correlations. The translational invariance can be taken into account in the same way as in (25.23, 24). The main difficulty lies in the inversion of (25.22) or of $D$ in (23.45), resp. Already the simplest models for surface correlations need a large amount of numerical work. Two models have been discussed in some detail. *Maradudin* and *Melngailis* [64 M 4] considered a simple cubic lattice with central forces between nearest and next nearest neighbors. They essentially used the Greens function method to calculate the displacement correlations of ions in the vicinity of a (100)-surface. *Clark, Herman* and *Wallis* [66 C 1] discussed the correlations of surface ions in a face-centered cubic crystal with nearest neighbor central force interaction. They proceed with the method of (23.45), corrected for translational invariance in the surfaces, which are assumed to be (100)-, (110)-, and (111)-surfaces. The crystals are supposed to consist of 20 layers parallel to the surface, and for these 20 layers the $\tilde{D}$-matrix is inversed. The results are shown in Fig. 6.40. Whereas in the interior of cubic crystals the displacement-autocorrelation satisfies

$$\left\langle u_i^m(0)\, u_j^m(0)\right\rangle \sim \text{const.}\ \delta_{ij}\,,$$

this is not true for surface atoms. The correlations are different for components parallel and perpendicular to the surface, and even the two

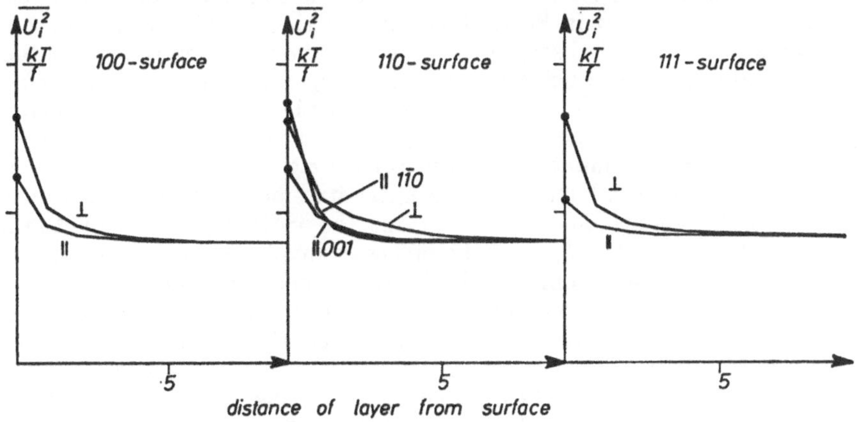


Fig. 6.40. Displacement-autocorrelation of ions near a free surface, as function of the distance from the surface. According to [66 C 1]

parallel components might be different because of the anisotropy of the surface (110-surface with a twofold axis only). The displacement correlations for surface ions are larger than those for interior ions because of the missing force-constants at the surface. In general the correlation-components perpendicular to the surface are larger than those parallel to the surface because the perpendicular ones are more affected by the missing force-constants. However, in very complicated structures and surfaces this might be different, namely if the missing forces are such as to affect the parallel components stronger than the perpendicular ones. It should be mentioned further that the autocorrelation matrix $\left\langle u_i^m(0)\, u_j^m(0)\right\rangle$ is no longer diagonal for surface ions, even in cubic crystals, because the symmetry of the ion-positions is lowered in the surface. The correlations approach the interior values with increasing distance from the surface. The decrease is roughly exponential. This holds also for other cases. It is easy to see the principles of surface-ion autocorrelations, but quantitative statements involve large numerical work even for the simplest models.

## 26. Anharmonic Effects on Defect Lattice Modes

Anharmonic effects manifest themselves in a number of phenomena related to the frequency spectrum of the defect crystal. Here we will give a short account on the possible features. It should be mentioned, that not all of the phenomena have been understood very well.

Perhaps the most striking effect is the finite life-time of localized modes, which is connected with cubic and also higher anharmonic terms in the potential energy of the defect lattice. This leads to a finite width of the discrete frequencies of localized states. Using the anharmonic Hamiltonian for the defect lattice, the calculation can be done in a very similar way to that in Sect. 27 and 30 for the width of neutron-cross-sections and optical absorption. There we have discussed the calculation for ideal lattices. But using the corresponding Hamiltonian for the defect lattice, and using the eigenstates of the defect lattice instead of the plane-wave-eigenstates of the ideal lattice, the theory can be done in the same way. The anharmonic constants in (27.72) e.g. have to be replaced by those of the defect lattice. The essential statements can be taken over. If the cubic anharmonicity is important, the width is proportional to the temperature, for $T > \theta_D$. But the processes involved in cubic terms give a finite contribution only, if the localized mode frequency $\omega_{loc}$ is not larger than twice the maximum frequency $\omega_m$ of the ideal lattice frequency band. This follows from the conservation of energy involved in (27.72). If $2\omega_m < \omega_{loc} < 3\omega_m$, the first nonvanishing term is from quartic anharmonicity. It can be shown by similar arguments as in Sect. 27, that this term is proportional to $T^2$ for $T > \theta_D$. Thus we have different temperature dependences of the widths of localized states depending on the magnitude of $\omega_{loc}$ relative to $\omega_m$. This seems to be in accordance with experimental results [61 K 1, 61 K 2, 63 K 1, 63 D 4, 64 M 1, 64 M 5, 64 V 1]. Further, there is a shift of the frequencies of localized states due to cubic and quartic anharmonicity, being linear in $T$ for $T > \theta_D$.

Another effect is the occurence of side-bands of the localized states, which can be seen for example in the $U$-center absorption [65 F 1, 65 F 2, 66 H 1] (Fig. 6.41). These side-bands can be understood also with the formula (27.72b) for the widths; for $\Gamma_J$ is a function of frequency. If

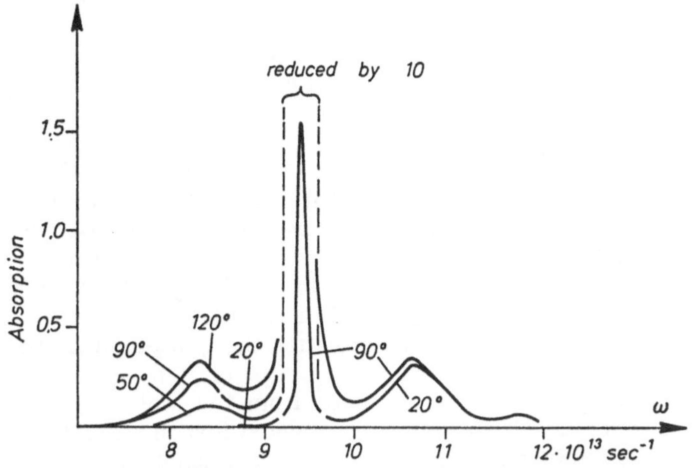


Fig. 6.41. Sidebands of the absorption peak of a localized $U$-center mode in KCl, according to [65 F 1]

one of the frequencies in $\Gamma_J$, say $\omega_{J_1}$, is a localized state frequency, and $\omega_{J_2}$ an inband-frequency, we have absorption for all the frequencies left and right of $\omega_{J_1}$ in the range of $\omega_{J_2}$. This corresponds to the sum- and difference-processes discussed in Sect. 30. Because of the different temperature factors in (27.72), the low-frequency side-bands vanish for $T \Rightarrow 0$. At high temperatures, both side-bands vary proportional to $T$, provided the cubic anharmonicity is sufficient to explain the behavior. These effects are also in agreement with experiments.

Other possible effects, which may lead to sidebands are higher order dipole-moments (Sect. 30), which we will not discuss in detail. Also overtones of localized frequencies may occur. A great help in the investigation of possible processes is the use of group theory. This gives an answer, which processes may occur at all [64 T 2, 66 B 3, 64 L 7].

If anharmonicity is present, which of course is always the case, all other effects of defects on lattice vibrations have to be reinvestigated. The effects are most striking in connection with localized modes, but they manifest themselves also in further interaction processes in thermal conductivity, in connection with resonance states and so on. Detailed investigations, experimental as well as theoretical, are very difficult, because in the latter cases anharmonicity gives only additional effects, which have to be separated from the "large" harmonic effects, whereas in the effects related to localized states anharmonicity leads to striking effects.

# VII. Interaction of Phonons with Particles and Radiation

The investigations of phonons and lattice properties are often done by using the interactions of the phonons with particles and radiation. For many purposes these methods are the most powerful ones. We will give a comprehension of the theory of phonon-interactions which are treated generally by quantum-mechanical perturbation methods. These methods we assume to be known in the following (e.g. *Born*'s and *Dirac*'s approximations). We suppose further, that the description and the properties of the free systems of particles or radiation (neutrons, photons, electrons, . . .) are known. The essential starting point thus is the discussion of the transition-probabilities of the system under the influence of external radiation. The measurement of these transition probabilities or the cross-sections derived from them gives an information on the properties of the phonons. It is mainly in connection with interaction processes that the concept of phonons in the theory of lattice vibrations shows its importance. These interactions are governed by the conservation of energy and the selection rules for the quasi-momentum of the phonon.

Most information can be obtained from the scattering of thermal neutrons at crystals. This shall be considered first in some detail. Many of the results can then be used also in the discussion of other interactions.

## 27. Interaction of Phonons with Thermal Neutrons

A room-temperature neutron has an energy of about $2.6 \cdot 10^{-2}$ eV, an angular frequency of $3.9 \cdot 10^{13}$ sec$^{-1}$ and a wavelength of $1.8 \cdot 10^{-8}$ cm. These data are comparable with the interionic distances, with the ionic displacements and the lattice frequencies of crystals. Thus it is clear that by scattering of thermal neutrons at crystals we can get an information on the lattice properties, provided there is a sufficiently strong interaction between neutrons and crystal ions. Such an interaction occurs via the nuclear potential between the neutron and the nuclei of the ions. For small energies of the incident neutron, the scattering of the neutron at a free nucleus can be described by a scattering length, which in general depends on the relative orientations of the spins of neutron and nucleus. Since the incoming neutrons are unpolarized in general, one has to take an average over the relative orientations*.

In crystals the nuclei are not free, but bound to their equilibrium positions by (harmonic) restoring forces. It is then questionable whether the description of the neutron scattering by a scattering length is sufficient or whether we have to introduce a more detailed knowledge on the interaction between neutron and a nucleus, bound to a crystal. This

* See any book on nuclear theory, e. g. *J. M. Blatt, V. F. Weisskopf*, Theoretical Nuclear Physics, Wiley and Sons, New York, London, *R. G. Sachs*, Nuclear Theory, Addison-Wesley, Cambridge, Mass., *G. Eder*, Kernkräfte, Braun, Karlsruhe.

problem has been dealt with by *Fermi* [36 F 2]; a refined treatment has been given later by *Breit* [47 B 1] and *Lippman* and *Schwinger* [50 L 2]. It seems that the description by a scattering length holds as long as the range of the nuclear forces is small compared to the vibration amplitudes of the nuclei and the neutron wave-length is large compared to these amplitudes. But a strong proof of its validity or its limitations seems to be missing. At any case, it is justified only, if the first Born or Dirac approximation is sufficient in calculating the scattering cross-section. In the following we assume this Fermi scattering theory to be sufficient.

The interaction between the neutron and nuclei of a crystal can then be written as

$$W(\boldsymbol{r}) = \frac{2\pi\hbar^2}{m}\sum_n a_n\,\delta(\boldsymbol{r}-\boldsymbol{r}^n)\,. \tag{27.1}$$

$a_n$ is the scattering-length of the $n$-th nucleus, at position $\boldsymbol{r}^n$. $\boldsymbol{r}$ is the coordinate of the neutron, $m$ its mass. The Hamiltonian of the crystal is given by

$$\mathscr{H}_{\mathrm{c}} = \sum_n \frac{(\boldsymbol{p}_n)^2}{2M_n} + \Phi(\ldots\boldsymbol{r}^n\ldots)\,. \tag{27.2}$$

The crystal is in state $|\alpha\rangle$ with energy $E_\alpha$ before the interaction with the neutron and in state $|\varepsilon\rangle$ with energy $E_\varepsilon$ afterwards. The free neutron is described by a wavefunction

$$\varphi_a = \frac{1}{\sqrt{V}}\,\mathrm{e}^{i\boldsymbol{k}_a\boldsymbol{r}} \text{ and } \varphi_e = \frac{1}{\sqrt{V}}\,\mathrm{e}^{i\boldsymbol{k}_e\boldsymbol{r}} \tag{27.3}$$

for incoming or outgoing (scattered) neutron, resp. $V$ is an arbitrary normalizing volume, which does not enter the result. The change in energy and momentum by the scattering process is

$$\begin{gathered}\hbar\omega = \frac{\hbar^2}{2m}(k_a^2 - k_e^2)\,; \quad \boldsymbol{K} = \boldsymbol{k}_a - \boldsymbol{k}_e\,;\\ K^2 = 2k_a^2\left\{1 - \Delta/2 - \sqrt{1-\Delta}\cdot\cos\vartheta\right\}; \quad \Delta = \frac{2m\omega}{\hbar k_a^2}\end{gathered} \tag{27.4}$$

$\vartheta$ is the scattering angle. In elastic scattering processes, $\omega = \Delta = 0$ and $K = 2k_a\cdot\sin\vartheta/2$.

We assume an incoming stream of neutrons with fixed energy and momentum, i.e. $\boldsymbol{k}_a = \text{const}$. The transition probability from the initial state $|\alpha a\rangle$ of crystal and neutron to a final state $|\varepsilon e\rangle$ with neutron momentum in the interval $\boldsymbol{k}_e$, $\boldsymbol{k}_e + \mathrm{d}\boldsymbol{k}_e$ may be denoted by $w_{\alpha a\to\varepsilon e}\,\mathrm{d}\boldsymbol{k}_e$. From (27.4) we have

$$\mathrm{d}\boldsymbol{k}_e = k_e^2\,\mathrm{d}k_e\,\mathrm{d}\Omega = \frac{m}{\hbar}\,k_e\,\mathrm{d}\omega\,\mathrm{d}\Omega\,. \tag{27.5}$$

$\mathrm{d}\Omega$ is the spherical angle, in which the neutron is scattered. The cross-section for this process is related to the transition probability by

$$\mathrm{d}\sigma_{\alpha a\to\varepsilon e} = \frac{mV}{\hbar k_a}\,w_{\alpha a\to\varepsilon e}\,\mathrm{d}\boldsymbol{k}_e = \frac{\mathrm{d}\sigma_{\alpha a\to\varepsilon e}}{\mathrm{d}\omega\,\mathrm{d}\Omega}\cdot\mathrm{d}\omega\,\mathrm{d}\Omega\,. \tag{27.6}$$

The transition probability follows with the usual procedure

$$w_{\alpha a\to\varepsilon e}\,\mathrm{d}\boldsymbol{k}_e=\frac{2\pi}{\hbar}\left|\left\langle\varepsilon\left|\frac{1}{V}\int \mathrm{e}^{-i\boldsymbol{k}_e\boldsymbol{r}}\,W\mathrm{e}^{i\boldsymbol{k}_a\boldsymbol{r}}\,\mathrm{d}\boldsymbol{r}\right|\alpha\right\rangle\right|^2\delta(\hbar\omega+E_\alpha-E_\varepsilon)\times$$
$$\times\frac{V}{(2\pi)^3}\,\mathrm{d}\boldsymbol{k}_e\,.$$

Using (27.1) and (27.6) we obtain

$$\frac{\mathrm{d}\sigma_{\alpha a\to\varepsilon e}}{\mathrm{d}\omega\,\mathrm{d}\Omega}=\frac{k_e}{k_a}\sum_{m,n}a_m^*a_n\left\langle\alpha\left|\mathrm{e}^{-i\boldsymbol{K}\boldsymbol{r}^m}\right|\varepsilon\right\rangle\left\langle\varepsilon\left|\mathrm{e}^{i\boldsymbol{K}\boldsymbol{r}^n}\right|\alpha\right\rangle\delta\left(\omega+\frac{E_\alpha-E_\varepsilon}{\hbar}\right).\tag{27.7}$$

But this is the cross-section only, if we have a definite state at the beginning of the process. Now the crystal is in thermal equilibrium before scattering, therefore there is a thermal mixture of states, the probability of state $\alpha$ being

$$p_\alpha=\mathrm{e}^{-E\alpha/kT}\Big/\sum_\alpha\mathrm{e}^{-E\alpha/kT}\,.\tag{27.8}$$

To get the true scattering cross-section, we have to sum over all the states $\alpha$ with their probabilities. Further, we are not interested in the final states of the crystal, but only in that of the neutron; this means an additional sum over $\varepsilon$. We have finally

$$\frac{\mathrm{d}\sigma}{\mathrm{d}\omega\,\mathrm{d}\Omega}=\frac{k_e}{k_a}\sum_{m,n}a_m^*a_n\,s^{mn}(\boldsymbol{K},\omega)\tag{27.9}$$

with

$$s^{mn}(\boldsymbol{K},\omega)=\sum_{\alpha,\varepsilon}p_\alpha\langle\alpha\,|\,\mathrm{e}^{-i\boldsymbol{K}\boldsymbol{r}^m}\,|\,\varepsilon\rangle\,\langle\varepsilon\,|\,\mathrm{e}^{i\boldsymbol{K}\boldsymbol{r}}\,|\,\alpha\rangle\,\delta\left(\omega+\frac{E_\alpha-E_\varepsilon}{\hbar}\right).\tag{27.10}$$

For a discussion of this expression we use a method of *van Hove* [54 H 2]. Instead of discussing (27.10) we introduce Fourier-transformed functions defined by

$$\chi^{mn}(\boldsymbol{K},t)=\int s^{mn}(\boldsymbol{K},\omega)\,\mathrm{e}^{i\omega t}\,\mathrm{d}\omega\,,$$
$$g^{mn}(\boldsymbol{r},t)=\frac{1}{(2\pi)^3}\int\chi^{mn}(\boldsymbol{K},t)\,\mathrm{e}^{-i\boldsymbol{K}\boldsymbol{r}}\,\mathrm{d}\boldsymbol{K}\tag{27.11}$$
$$=\frac{1}{(2\pi)^3}\int s^{mn}(\boldsymbol{K},\omega)\,\mathrm{e}^{-i(\boldsymbol{K}\boldsymbol{r}-\omega t)}\,\mathrm{d}\boldsymbol{K}\,\mathrm{d}\omega\,.$$

With

$$\langle\varepsilon\,|\mathrm{e}^{itE_\varepsilon/\hbar}\,\mathrm{e}^{i\boldsymbol{K}\boldsymbol{r}^n}\,\mathrm{e}^{-itE_\alpha/\hbar}|\,\alpha\rangle=\langle\varepsilon\,|\mathrm{e}^{it\mathscr{H}_c/\hbar}\,\mathrm{e}^{i\boldsymbol{K}\boldsymbol{r}^n}\,\mathrm{e}^{-it\mathscr{H}_c/\hbar}|\,\alpha\rangle\times$$
$$=\langle\varepsilon\,|\mathrm{e}^{i\boldsymbol{K}\boldsymbol{r}^n(t)}|\,\alpha\rangle$$

and

$$\sum_\varepsilon|\varepsilon\rangle\,\langle\varepsilon|=1\quad\text{(completeness)}$$

we obtain

$$\chi^{mn}(\boldsymbol{K},t)=\langle\mathrm{e}^{-i\boldsymbol{K}\boldsymbol{r}^m(0)}\cdot\mathrm{e}^{i\boldsymbol{K}\boldsymbol{r}^n(t)}\rangle_{\mathrm{T}}\tag{27.12}$$

where $\langle\ \rangle_T$ denotes the thermal average. Correspondingly we have (note that $\boldsymbol{r}^m(0)$ and $\boldsymbol{r}^n(t)$ do not commute!)

$$g^{mn}(\boldsymbol{r}, t) = \int \mathrm{d}\boldsymbol{r}' \langle \delta(\boldsymbol{r}' - \boldsymbol{r}^m(0)) \cdot \delta(\boldsymbol{r}' + \boldsymbol{r} - \boldsymbol{r}^n(t))\rangle_T . \tag{27.13}$$

If one of these functions is known completely, the neutron cross-section can be determined immediately. In experiments, essentially $s^{mn}(\boldsymbol{K}, \omega)$ is measured. The easiest method of calculation starts with (27.12). From (27.13) it can be seen, that $g^{mn}(\boldsymbol{r}, t)$ represents a correlation: it gives the probability that if ion $m$ is in $\boldsymbol{r}^m$ at $t = 0$, ion $n$ is a distance $\boldsymbol{r}$ apart (between $\boldsymbol{r}$ and $\boldsymbol{r} + \mathrm{d}\boldsymbol{r}$) at time $t$ (between $t$ and $t + \mathrm{d}t$). In deriving (27.9–13) no assumptions on the crystal have been made. It includes an anharmonic crystal as well as a defect crystal, or even any system of interacting particles (gases, fluids, crystals).

Before going to calculate (27.12), we give some simplifications of (27.9). The scattering-length of a nucleus in general consists of a spin-dependent and a spin-independent part; it is different for different relative orientations of nuclear and neutron spin. Further, the scattering length is different for different isotopes of a nucleus, in the spin-dependent as well as in the spin-independent part. If there is no spin-dependent part in the scattering length, the neutron spin is not changed during the scattering process and the scattered neutron waves interfere. Otherwise the neutron spin may change during the scattering process (spin-flip). If we make the assumption, that the ions with different scattering lengths $a_m$ are distributed statistically and independently over the crystal, we obtain by averaging (27.9)

$$\frac{\mathrm{d}\sigma}{\mathrm{d}\omega\,\mathrm{d}\Omega} = \frac{k_e}{k_a}\left\{\sum_m \overline{a^2}\, s^{mm}(\boldsymbol{K}, \omega) + \sum_{m \neq n} \bar{a}^2\, s^{mn}(\boldsymbol{K}, \omega)\right\}$$

or with

$$\left.\begin{aligned} S(\boldsymbol{K}, \omega) &= \sum_{m,n} s^{mn}(\boldsymbol{K}, \omega) = N \cdot \sum_{m-n} s^{m-n}(\boldsymbol{K}, \omega) \\ S_s(\boldsymbol{K}, \omega) &= \frac{1}{N}\sum_m s^{mm}(\boldsymbol{K}, \omega) = s^{mm}(\boldsymbol{K}, \omega) \end{aligned}\right\} \begin{array}{l}\text{in homogeneous}\\ \text{systems}\\ \text{of } N \text{ particles}\end{array}$$

$$\frac{\mathrm{d}\sigma}{\mathrm{d}\omega\,\mathrm{d}\Omega} = \frac{k_e}{k_a}\{(\overline{a^2} - \bar{a}^2)\, N S_s(\boldsymbol{K}, \omega) + \bar{a}^2\, S(\boldsymbol{K}, \omega)\} . \tag{27.14}$$

This holds if the crystal is a mixture of different isotopes or different spin-states. If all the ions have the same scattering length $a = \bar{a} = \sqrt{\overline{a^2}}$ (equal isotopes with spin-independent $a$), only the second term is different from zero *(coherent scattering)*. If $\bar{a} = 0$, we have only *incoherent scattering*.

Correspondingly we can define the Fourier-transformed functions:

$$\begin{aligned} \chi_s(\boldsymbol{K}, t) &= \langle \mathrm{e}^{-i\boldsymbol{K}\boldsymbol{r}^n(0)} \cdot \mathrm{e}^{i\boldsymbol{K}\boldsymbol{r}^n(t)}\rangle_T \quad \text{in homogeneous systems} \\ \chi(\boldsymbol{K}, t) &= \sum_{m,n} \langle \mathrm{e}^{-i\boldsymbol{K}\boldsymbol{r}^m(0)} \cdot \mathrm{e}^{i\boldsymbol{K}\boldsymbol{r}^n(t)}\rangle_T = \langle \varrho_{-\boldsymbol{K}}(0) \cdot \varrho_{\boldsymbol{K}}(t)\rangle_T . \end{aligned} \tag{27.15}$$

$\varrho_{\boldsymbol{K}} = \sum_n \mathrm{e}^{i\boldsymbol{K}\boldsymbol{r}^n}$ is the Fourier-transformed particle-density

$$\varrho(\boldsymbol{r}) = \sum_n \delta(\boldsymbol{r} - \boldsymbol{r}^n) .$$

The correlation-functions in normal space are finally

$$g(\boldsymbol{r}, t) = \frac{1}{N} \sum_{m,n} g^{mn}(\boldsymbol{r}, t) = \frac{1}{N} \int \mathrm{d}\boldsymbol{r}' \langle \varrho(\boldsymbol{r}', 0) \cdot \varrho(\boldsymbol{r}' + \boldsymbol{r}, t) \rangle_{\mathrm{T}} . \tag{27.16}$$

In homogeneous systems (ideal lattice) this can be simplified to

$$\begin{aligned} g(\boldsymbol{r}, t) &= \frac{V}{N} \langle \varrho(0, 0)\, \varrho(\boldsymbol{r}, t) \rangle_{\mathrm{T}} \\ &= \frac{1}{\varrho_0} \sum_{m,n} \langle \delta(\boldsymbol{r}^m(0)) \cdot \delta(\boldsymbol{r} - \boldsymbol{r}^n(t)) \rangle_{\mathrm{T}} , \end{aligned} \tag{27.17}$$

because the thermal average cannot depend on $\boldsymbol{r}'$. $N$ is the number of ions in volume $V$, $\varrho_0$ the mean particle density. $g(\boldsymbol{r}, t)$ gives the probability of finding a particle at $\boldsymbol{r}, t$ (between $\boldsymbol{r}$ and $\boldsymbol{r} + \mathrm{d}\boldsymbol{r}$, $t$ and $t + \mathrm{d}t$), if there was a particle at $\boldsymbol{r} = 0$ when $t = 0$. The density-autocorrelation in homogeneous systems is

$$g_{\mathrm{s}}(\boldsymbol{r}, t) = V \langle \delta(\boldsymbol{r}^m(0)) \cdot \delta(\boldsymbol{r} - \boldsymbol{r}^m(t)) \rangle_{\mathrm{T}} . \tag{27.18}$$

Besides these two functions sometimes a pair-correlation defined by

$$g_{\mathrm{D}}(\boldsymbol{r}, t) = g(\boldsymbol{r}, t) - g_{\mathrm{s}}(\boldsymbol{r}, t) \tag{27.19}$$

is introduced.

We have to calculate (27.12) for the situation of a crystal; this means we have to replace the arbitrary index $m$ by $\overset{m}{\mu}$ now, using the description of the lattice as indicated in (1.3). The displacement of the ions from the equilibrium positions $\boldsymbol{R}_{\mu}^{m}$ are small:

$$\boldsymbol{r}_{\mu}^{m} = \boldsymbol{R}_{\mu}^{m} + \boldsymbol{u}_{\mu}^{m} . \tag{27.20}$$

With (27.20) we have from (27.12)

$$\chi_{\mu\,\nu}^{m\,n}(\boldsymbol{K}, t) = \mathrm{e}^{i\boldsymbol{K}\left(\boldsymbol{R}_{\nu}^{n} - \boldsymbol{R}_{\mu}^{m}\right)} \left\langle \mathrm{e}^{-i\boldsymbol{K}\boldsymbol{u}_{\mu}^{m}(0)} \cdot \mathrm{e}^{i\boldsymbol{K}\boldsymbol{u}_{\nu}^{n}(t)} \right\rangle_{\mathrm{T}} , \tag{27.21}$$

because the equilibrium positions are time-independent and commute with each other [54 H 2].

*Harmonic crystal*

The simplest case of a harmonic crystal can be evaluated directly. The displacements are completely determined by their representation in terms of normal-coordinates (16.9, 21.2, 23.42). The commutator between $\boldsymbol{u}_{\mu}^{m}(0)$ and $\boldsymbol{u}_{\nu}^{n}(t)$ is a $c$-number, as can be seen from (21.8) for ideal lattices. For defect lattices it can be proved similarly. For two operators $A$ and $B$, the commutator of which is a $c$-number, we have

$$\mathrm{e}^{A} \cdot \mathrm{e}^{B} = \exp\{A + B + 1/2\,[A, B]_{-}\} . \tag{27.22}$$

Because the thermal average of a $c$-number is the $c$-number itself, we have for this average in (27.21)

$$\langle\ \rangle_{\mathrm{T}} = \left\langle \exp\left\{-i\boldsymbol{K}\left[\boldsymbol{u}_{\mu}^{m}(0) - \boldsymbol{u}_{\nu}^{n}(t)\right]\right\}\right\rangle_{\mathrm{T}} \times \exp\left\{\frac{1}{2}\left[\boldsymbol{K}\boldsymbol{u}_{\mu}^{m}(0), \boldsymbol{K}\boldsymbol{u}_{\nu}^{n}(t)\right]_{-}\right\} . \tag{27.23}$$

The displacements of a harmonic vibration have a Gaussian thermal distribution [I]. For such a distribution holds*

$$\langle e^{\alpha u}\rangle_T = e^{\alpha^2 \langle u^2\rangle_T/2}\,. \tag{27.24}$$

Then we have

$$\langle\ \rangle_T = \exp\left\{-\frac{1}{2}\left\langle\left(\boldsymbol{K}\boldsymbol{u}^m_\mu(0)\right)^2\right\rangle_T - \frac{1}{2}\left\langle\left(\boldsymbol{K}\boldsymbol{u}^n_\nu(t)\right)^2\right\rangle_T + \right.$$
$$\left. + \left\langle \boldsymbol{K}\boldsymbol{u}^m_\mu(0)\cdot\boldsymbol{K}\boldsymbol{u}^n_\nu(t)\right\rangle_T\right\}.$$

If the Hamiltonian $\mathscr{H}_c$ does not depend explicitly on time, $\left\langle\left(\boldsymbol{K}\boldsymbol{u}^n_\nu(t)\right)^2\right\rangle_T$ is independent of $t$ and we are left with

$$\chi^{m\,n}_{\mu\,\nu}(\boldsymbol{K},t) = e^{-i\boldsymbol{K}(\boldsymbol{R}^m_\mu - \boldsymbol{R}^n_\nu)}\cdot e^{-\frac{1}{2}\langle(\boldsymbol{K}\boldsymbol{u}^m_\mu)^2\rangle_T - \frac{1}{2}\langle(\boldsymbol{K}\boldsymbol{u}^n_\nu)^2\rangle_T}\times$$
$$\times\, e^{+\langle \boldsymbol{K}\boldsymbol{u}^m_\mu(0)\cdot\boldsymbol{K}\boldsymbol{u}^n_\nu(t)\rangle_T}. \tag{27.25}$$

This holds for any harmonic crystal, also for those with defects. The displacement-correlations which enter this expression, $\left\langle u^m_{\mu\,i}(0)\,u^n_{\nu\,j}(t)\right\rangle_T$ can be calculated as indicated in Sect. 23, and can be expressed by the corresponding Greens-function (**G** for ideal, **R** for defect lattices). It becomes rather simple for *homogeneous (ideal) lattices*. Using (21.2) and taking the thermal averages of the products of creation and annihilation operators (23.43) we have the Debye-Waller-exponent

$$W_\mu = \frac{1}{2}\left\langle\left(\boldsymbol{K}\boldsymbol{u}^m_\mu\right)^2\right\rangle_T = \frac{1}{2s\,N\,M_\mu}\sum_{\boldsymbol{q}\,\sigma}\frac{|\boldsymbol{K}\boldsymbol{e}^\mu(\boldsymbol{q}\,\sigma)|^2}{\omega^2(\boldsymbol{q}\,\sigma)}\,\varepsilon(\omega,T) \tag{27.26}$$

and

$$\left\langle\boldsymbol{K}\boldsymbol{u}^m_\mu(0)\cdot\boldsymbol{K}\boldsymbol{u}^n_\nu(t)\right\rangle_T = \frac{\hbar}{2s\,N\sqrt{M_\mu M_\nu}}\sum_{\boldsymbol{q}\sigma}\frac{(\boldsymbol{K}\boldsymbol{e}^\mu)(\boldsymbol{K}\boldsymbol{e}^{*\nu})}{\omega(\boldsymbol{q}\,\sigma)}\,e^{i\boldsymbol{q}(\boldsymbol{R}^m-\boldsymbol{R}^n)}\times$$
$$\times\{\bar{n}(\boldsymbol{q}\sigma)\,e^{-i\omega(\boldsymbol{q}\sigma)t} + (\bar{n}(\boldsymbol{q}\sigma)+1)\,e^{i\omega(\boldsymbol{q}\sigma)t}\}\,.$$

In neutron-diffraction experiments the quantity measured is the cross-section. To see the quantities which can be obtained by such an investigation, we use the procedure of (27.14). Assuming a statistical and independent distribution of isotopes or spin-states in every sublattice, we obtain similarly**

$$\frac{d\sigma}{d\omega\,d\Omega} = \frac{k_e}{k_a}\left\{\sum_{m\mu}(\overline{a^2_\mu} - \bar{a}^2_\mu)\,s^{m\,m}_{\mu\,\mu} + \sum_{\substack{m n\\ \mu\,\nu}}\bar{a}_\mu\,\bar{a}_\nu\,s^{m\,n}_{\mu\,\nu}\right\}. \tag{27.27}$$

---

* A simple but not very elegant proof is the following: $\langle e^{\alpha u}\rangle_T = \sum_{\nu=0}^{\infty}\frac{1}{(2\nu)!}\times$ $\times\langle(\alpha u)^{2\nu}\rangle_T$ since the odd powers vanish in a Gaussian distribution. Now $\langle u^{2\nu}\rangle_T = \frac{(2\nu)!}{\nu!}\left[\frac{1}{2}\langle u^2\rangle_T\right]^\nu$ in a Gaussian distribution, therefore (27.24).

** Here, as well as already in (27.14) we have tacitly assumed that the average over the scattering length can be taken independently from those over the masses of the isotopes which enter in $S^{m\,n}_{\mu\,\nu}$ by means of (27.26).

$s_{\mu\ \nu}^{m\,n}$ can be obtained from (27.25) by using the back-transformation of (27.11a). We make the further assumption, that the exponent (27.26) is „small", so that the exponential function can be expanded. This means, that the displacement correlations are not too large. The limits of this expansion will become more clear, when the terms of the expansion are discussed. At any case, there are restrictions on the mean thermal occupation numbers $\bar{n}(\boldsymbol{q}\sigma)$, on the momentum transfer $\boldsymbol{K}$, and the other quantities. Expanding (27.25) with respect to the correlation $\left\langle \boldsymbol{K}\boldsymbol{u}_\mu^m(0)\cdot \boldsymbol{K}\boldsymbol{u}_\nu^n(t)\right\rangle_T$, introducing it into (27.27), summing over $\boldsymbol{m}$ and $\boldsymbol{n}$ and using the properties of the ideal lattice we obtain for the incoherent part

$$\sum_{m\,\mu} (\overline{a_\mu^2} - \bar{a}_\mu^2)\, s_{\mu\ \mu}^{mm} = N \sum_{\mu} (\overline{a_\mu^2} - \bar{a}_\mu^2)\, \mathrm{e}^{-2W_\mu} \times$$
$$\times \left\{\delta(\omega) + \frac{\hbar}{2s\,NM_\mu} \sum_{\boldsymbol{q}\sigma} \frac{|\boldsymbol{K}\,\boldsymbol{e}^\mu|^2}{\omega(\boldsymbol{q}\,\sigma)} [\bar{n}\,\delta(\omega + \omega(\boldsymbol{q}\sigma)) + \right.$$
$$\left. + (\bar{n}+1)\,\delta\,(\omega\,(\boldsymbol{q}\sigma) - \omega)] + \ldots\right\} \tag{27.28}$$

and for the coherent part

$$\sum_{\substack{m\,n\\ \mu\ \nu}} \bar{a}_\mu\, \bar{a}_\nu\, s_{\mu\ \nu}^{m\,n} = N^2 \sum_{\mu\nu} \bar{a}_\mu\, \mathrm{e}^{-W_\mu - i\boldsymbol{K}\boldsymbol{R}_\mu} \cdot \bar{a}_\nu\, \mathrm{e}^{-W_\nu + i\boldsymbol{K}\boldsymbol{R}_\nu} \times$$
$$\times \sum_{\boldsymbol{h}} \left\{\delta(\omega)\,\delta(\boldsymbol{K} - 2\pi\,B\,\boldsymbol{h}) + \frac{\hbar}{2s\,N\sqrt{M_\mu\,M_\nu}} \sum_{\boldsymbol{q}\sigma} \frac{(\boldsymbol{K}\boldsymbol{e}^\mu)\,(\boldsymbol{K}\boldsymbol{e}^{*\nu})}{\omega(\boldsymbol{q}\sigma)} \times \right. \tag{27.29}$$
$$\left. \times\ \delta(\boldsymbol{K} - \boldsymbol{q} - 2\pi B\boldsymbol{h})\ [\bar{n}\,\delta(\omega(\boldsymbol{q}\sigma) + \omega) + (\bar{n}+1)\,\delta(\omega(\boldsymbol{q}\sigma) - \omega)] + \cdots\right\}$$

$B\boldsymbol{h}$ is a vector in the reciprocal lattice. It is obvious that the next order contribution contains factors like $\delta(\boldsymbol{K} - \boldsymbol{q} - \boldsymbol{q}' - 2\pi\, B\boldsymbol{h})$, $\delta(\omega(\boldsymbol{q}\sigma) + \omega(\boldsymbol{q}'\sigma') + \omega)$ and so on.

The temperature-dependence of the scattered intensity, coherent as well as incoherent, is essentially determined by the Debye-Waller-exponent $W_\mu$. It can be calculated, if the polarization and the frequencies are known. In cubic crystals it can be expressed by the spectral frequency distribution; for then

$$\sum_{\boldsymbol{q}\sigma} \frac{e_i^\mu\, e_j^{*\mu}(\boldsymbol{q}\sigma)}{\omega^2(\boldsymbol{q}\,\sigma)}\, \varepsilon(\omega, T) = \frac{1}{3}\,\delta_{ij} \sum_{\boldsymbol{q}\sigma} \frac{f_\mu(\boldsymbol{q}\,\sigma)}{\omega^2(\boldsymbol{q}\,\sigma)}\, \varepsilon(\omega, T) \tag{27.30}$$

because a second rank tensor in cubic crystals is proportional to the unit tensor. In Bravais-crystals $f_\mu(\boldsymbol{q}\sigma) = 1$ and we have*

$$W_\mu = \frac{K^2}{2\,M_\mu} \int \frac{z(\omega)}{\omega^2}\, \varepsilon(\omega, T)\, \mathrm{d}\omega\,. \tag{27.31}$$

* The index $\mu$ can be dropped then.

For a rough estimate, we use the Debye-spectrum to give the value in the limits $T \to 0$ and $T \gg \Theta_D$, where $k\Theta_D = \hbar\omega_D$ defines the Debye-temperature:

$$W_\mu = \frac{3k\,\Theta_D}{8M_\mu} \cdot \left(\frac{\hbar K}{k\,\Theta_D}\right)^2 = \frac{3P_\mu}{4k\,\Theta_D}\,; \quad T \to 0 \tag{27.31a}$$

$$W_\mu = \frac{3kT}{2M_\mu}\left(\frac{\hbar K}{k\,\Theta_D}\right)^2 = \frac{3P_\mu kT}{(k\,\Theta_D)^2} = ; \quad T \gtrsim \Theta_D\,. \tag{27.31b}$$

$P_\mu = \frac{\hbar^2 K^2}{2M_\mu}$ is the repulsion energy of ion $\mu$, if the whole momentum-transfer of the neutron is given to this ion.

The meaning of the different order terms in (27.28, 29) can be seen immediately:

i) The first term describes the *elastic scattering* of the neutron by the crystal. The energy $\omega$ of the neutron is not changed: $k_a = k_e$, $K = 2k_a \sin\vartheta/2$. The incoherent *zero-phonon cross section* is apart from the factor $\exp\{-W_\mu(K^2)\}$ independent of the scattering angle $\vartheta$ and gives no essential statement.

In coherent scattering, the further factor $\delta(\boldsymbol{K} - 2\pi B\boldsymbol{h})$ gives the possibility of investigating the structure of a lattice. If the momentum transfer $\boldsymbol{K}$ of the neutron is just equal to a reciprocal lattice vector, and

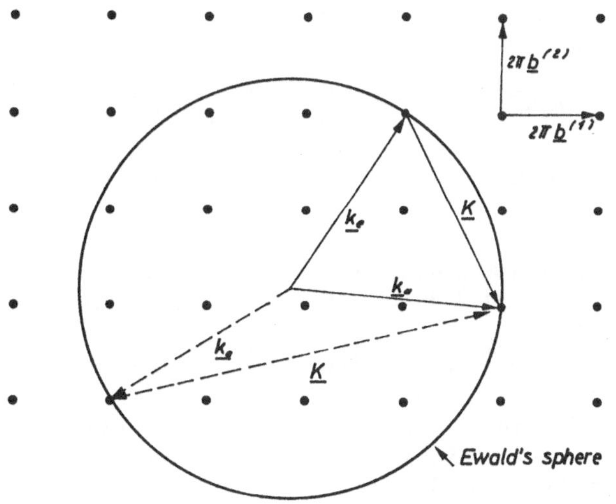


Fig. 7.1. Reciprocal lattice and Ewald's sphere, defining the elastic scattering processes in neutron spectroscopy

$k_a = k_e$, there is a peak in the cross-section. The situation is discussed in Fig. 7.1, which shows the *Ewald-sphere.* In the reciprocal lattice the incoming wave-vector is drawn in such a way, that its top coincides with a reciprocal lattice point. The outgoing wave-vector has its top on a sphere with radius $k_a$. For all those lattice points which lie on the sphere, the two conditions $\omega = 0$, $\boldsymbol{K} = 2\pi B\boldsymbol{h}$ are satisfied*. We observe intensity in the corresponding directions. The factors before, called the cell-form-factors, $\sum_\mu \bar{a}_\mu \mathrm{e}^{-W_\mu - i\boldsymbol{K}\boldsymbol{R}_\mu}$ determine the intensity. They can cause that some of the peaks determined by the Ewald-sphere vanish or have smaller intensity; this depends on the magnitudes of the $\bar{a}_\mu$ and on the $\boldsymbol{R}_\mu$. So $\boldsymbol{K} = 2\pi B\boldsymbol{h}$ determines the Bravais-lattice-structure, the cell-form-factor the ionic positions in the unit cell.

ii) The second terms represent processes, in which a neutron interacts with one phonon with $\boldsymbol{q}$, $\sigma$, $\omega(\boldsymbol{q}\sigma)$; therefore these terms are called the incoherent or coherent *one-phonon-cross-section*. $\bar{n}(\boldsymbol{q}\sigma)\,\delta(\omega(\boldsymbol{q}\sigma) + \omega)$ means the absorption of a phonon by the neutron; this process is not possible at $T = 0$. Correspondingly $(\bar{n}(\boldsymbol{q}\sigma) + 1)\,\delta(\omega(\boldsymbol{q}\sigma) - \omega)$ is the emission of a phonon by the neutron; at $T = 0$ we have only the spontaneous emission.

Apart from the factor $(\boldsymbol{K}\boldsymbol{e}^\mu)$, in incoherent scattering this process is independent of the momentum-transfer. To see the quantity measured somewhat more clearly let us restrict to a cubic Bravais crystal. By similar arguments as in (27.30) we obtain for the phonon-*emission*

$$\frac{\mathrm{d}\sigma_1^{\mathrm{inc}}}{\mathrm{d}\omega\,\mathrm{d}\Omega} = \frac{k_e \hbar K^2}{k_a 6NM} \cdot N(\overline{a^2} - \bar{a}^2)\,\mathrm{e}^{-2W} \sum_{\boldsymbol{q}\sigma} \frac{1}{\omega(\boldsymbol{q}\,\sigma)}\,\delta(\omega(\boldsymbol{q}\sigma) - \omega)\,(\bar{n}(\boldsymbol{q}\sigma) + 1)\,.$$

Replacing the sum by an integral over the spectral distribution we have finally for the second term in (27.28)

$$\begin{aligned}\frac{\mathrm{d}\sigma_1^{\mathrm{inc}}}{\mathrm{d}\omega\,\mathrm{d}\Omega} &= \frac{k_e}{k_a} N(\overline{a^2} - \bar{a}^2)\,\mathrm{e}^{-2W}\,\frac{\hbar K^2}{2M} \int \frac{z(\omega')}{\omega'}\,(\bar{n}(\omega') + 1)\,\delta(\omega - \omega')\,\mathrm{d}\omega' \\ &= \frac{k_e}{k_a} N(\overline{a^2} - \bar{a}^2)\,\mathrm{e}^{-2W}\,\frac{\hbar K^2}{2M} \cdot \frac{z(\omega)}{\omega}\,[\bar{n}(\omega) + 1]\,. \end{aligned} \tag{27.32a}$$

Thus in cubic Bravais-crystals the one-phonon-incoherent cross-section measures the spectral distribution of phonon frequencies multiplied by a known function of $\omega$. The absorption process measures

$$\frac{\mathrm{d}\sigma_1^{\mathrm{inc}}}{\mathrm{d}\omega\,\mathrm{d}\Omega} = -\frac{k_e}{k_a} N(\overline{a^2} - \bar{a}^2)\,\mathrm{e}^{-2W}\,\frac{\hbar K^2}{2M} \cdot \frac{z(-\omega)}{\omega}\,\bar{n}(-\omega)\,. \tag{27.32b}$$

In other lattices the situation is somewhat more complicated, but by combining *appropriate measurements* it is also possible to determine the spectral distribution of phonons.

In coherent scattering we have a second conservation law (selection rule) for the momentum transfer: $\boldsymbol{K} = \boldsymbol{q} + 2\pi B\boldsymbol{h}$. This enables us by analyzing the peaks in the neutron-cross-section to determine frequency $\omega(\boldsymbol{q}\sigma)$ and pseudomomentum $\boldsymbol{q}$ of the phonon involved. Because of the

* The Ewald sphere is not a sharp sphere, because there are always uncertainties. $\boldsymbol{k}_e$ has an uncertainty $\mathrm{d}\boldsymbol{k}_e$ and also $\boldsymbol{k}_a$ is not sharp. The sphere is smeared out over a small region, giving the possibility of "seeing" more points than indicated.

periodicity of $\omega(\boldsymbol{q}\sigma)$ and $\boldsymbol{e}^{\mu}(\boldsymbol{q}\sigma)$ in the reciprocal lattice we obtain

$$\frac{d\sigma_1^{coh}}{d\omega\, d\Omega} = \frac{k_e}{k_a}\frac{N\hbar}{2s}\sum_{\mu,\nu}\bar{a}_\mu\, e^{-W_\mu - i\boldsymbol{K}\boldsymbol{R}_\mu}\,\bar{a}_\nu\, e^{-W_\nu + i\boldsymbol{K}\boldsymbol{R}_\nu} \times$$
$$\times \sum_\sigma \frac{(\boldsymbol{K}\boldsymbol{e}^{\mu}(\boldsymbol{K}\sigma))\,(\boldsymbol{K}\boldsymbol{e}^{*\nu}(\boldsymbol{K}\sigma))}{\sqrt{M_\mu M_\nu}\cdot\omega(\boldsymbol{K}\sigma)} \times \tag{27.33}$$
$$\times\left[\bar{n}\,\delta(\omega(\boldsymbol{K}\sigma)+\omega) + (\bar{n}+1)\,\delta(\omega(\boldsymbol{K}\sigma)-\omega)\right].$$

Fig. 7.2 gives a graphical description of the scattering process. If $k_e < k_a$ we have phonon-emission, if $k_e > k_a$ phonon-absorption. The two equations (in absorption e.g.)

$$-\omega = \omega(\boldsymbol{K}\sigma)\,; \quad \boldsymbol{q} + 2\pi B\boldsymbol{h} = \boldsymbol{K} \tag{27.34}$$

define for any of the polarizations $\sigma$ a surface in the reciprocal lattice, called the scattering surface. Only in those directions of $\boldsymbol{k}_e$ which are determined by a point on the scattering surface, scattered intensity can be observed. But we must take into account again, that none of the

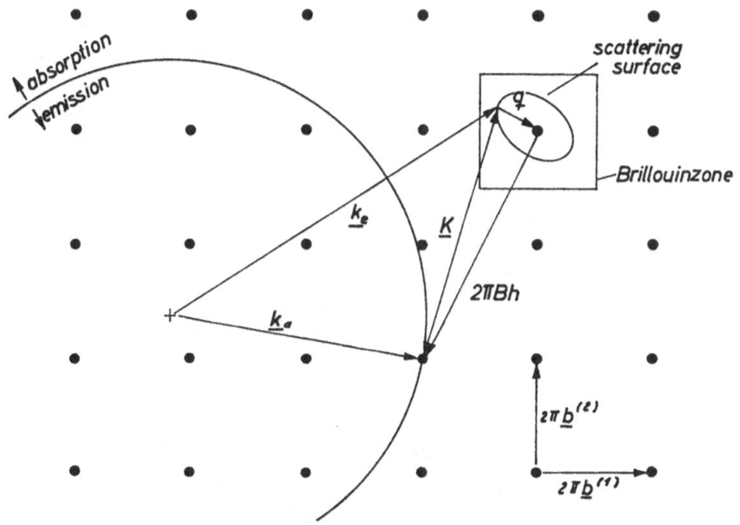


Fig. 7.2. Scattering surface in the reciprocal lattice, defining the possible inelastic one-phonon-processes

quantities involved ($\boldsymbol{k}_a, \boldsymbol{k}_e, \ldots$) is sharp. Therefore the scattering surface in general is smeared out over a region, which is large compared to the distance between different $\boldsymbol{q}$-points*. We can find always $\boldsymbol{q}$-points on the broadened scattering surface. Thus in coherent scattering we can determine the full dispersion curves $\omega(\boldsymbol{q}\sigma)$ by measuring the change of the neutron's energy $\omega$ and its scattering angle. For $\boldsymbol{K}$ being in the first

* Because the crystal is finite, the possible $\boldsymbol{q}$-values are not distributed continuously, but have small distances.

Brillouinzone ($B\boldsymbol{h} = 0$), we have $\boldsymbol{K} = \boldsymbol{q}$ and $(\boldsymbol{K}\boldsymbol{e}^{\mu}) = (\boldsymbol{q}\boldsymbol{e}^{\mu})$ which is different from zero only for longitudinal waves or waves with longitudinal components. For the determination of transverse components $\boldsymbol{K}$ must be a vector in Brillouinzones with $B\boldsymbol{h} \neq 0$.

To give a short idea of the scattering surface, we consider an elastic isotropic continuum, with $B\boldsymbol{h} = 0$. We have, for longitudinal waves

$$\omega = c_l q = c_l K .$$

With $(\boldsymbol{k}_a \boldsymbol{q}) = k_a q \cos\varphi$ we have from (27.34), omitting elastic scattering,

$$q = K = 2k_a \cos\varphi \pm \frac{2m}{\hbar} c_l \begin{cases} + : \text{absorption} \\ - : \text{emission} . \end{cases}$$

The emission of phonons by neutrons is possible only if

$$\cos\varphi > \frac{m c_l}{\hbar k_a} = \frac{c_l}{v_a} = \cos\varphi_0 ;$$

$v_a$: velocity of incoming neutron, which implies $v_a > c_l$. The neutron can create phonons only, if its velocity is larger than sound velocity. Then it produces phonons in a way, which corresponds to Čerenkov-radiation. The scattering surface is the sphere indicated in Fig. 7.3.

According to this outline, measurements of dispersion curves and spectral distributions of phonons by neutron-spectroscopy have been done in a number of cases [58 B 2, 60 W 1, 62 B 2, 3, 7, 63 C 1, 63 W 1, 3 65 D 3, 65 W 1, 65 M 1]. At the present time this method seems to be the best one in the investigation of lattice vibrations.

In this harmonic theory the absorption or emission peaks in (27.28, 29) have the form of $\delta$-functions. But this is valid only if all anharmonicities are neglected (see below). What is measured really are broadened peaks with a finite width, being a measure of anharmonicity.

iii) The higher terms in (27.28, 29) describe two-, three-, . . . phonon processes. They can be treated in a similar, but more complicated way, and give a correction to the first order process. By choosing appropriate conditions it is often possible to suppress higher order processes. The expansion in (27.28, 29) can be justified afterwards if the probability of many-phonon-processes is small compared to that of one- or zero-phonon-processes. It is not reasonable to use the expansion if many-phonon-processes are likely to occur.

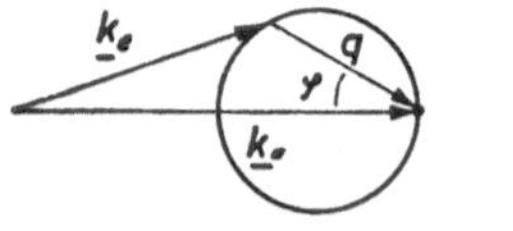

Fig. 7.3. Scattering surface for an isotropic elastic medium, considering longitudinal waves only

We will add some remarks on the properties of the scattering functions and the correlation functions. Because

$$s^{mn}(\boldsymbol{K}, \omega) = \frac{1}{2\pi} \int \chi^{mn}(\boldsymbol{K}, t)\, e^{-i\omega t}\, dt$$

we receive

$$\int_{-\infty}^{+\infty} s^{mn}(\boldsymbol{K}, \omega)\, d\omega = \chi^{mn}(\boldsymbol{K}, 0) . \tag{27.35}$$

Similarly we can derive expressions for the moments, e.g. the first moment, of $\omega$,

$$\int_{-\infty}^{+\infty} \omega\, s^{mn}(\boldsymbol{K},\omega)\,\mathrm{d}\omega = -i\left[\frac{\partial}{\partial t}\chi^{mn}(\boldsymbol{K},t)\right]_{t=0} \tag{27.36a}$$

Using (27.12) we obtain

$$\left[\frac{\partial}{\partial t}\chi^{mn}(\boldsymbol{K},t)\right]_{t=0} = \frac{i}{\hbar}\langle \mathrm{e}^{-i\boldsymbol{K}\boldsymbol{r}^m}[\mathscr{H}_{\mathrm{c}}, \mathrm{e}^{i\boldsymbol{K}\boldsymbol{r}^n}]_-\rangle_{\mathrm{T}}\,.$$

The Hamiltonian $\mathscr{H}_{\mathrm{c}}$ contains a potential energy part, which depends on the displacements only and gives no contribution to the commutator. The kinetic energy part gives a non-vanishing commutator, which can be determined immediately; thus we have

$$\int_{-\infty}^{+\infty} \hbar\,\omega\, s^{mn}(\boldsymbol{K},\omega)\,\mathrm{d}\omega = \frac{\hbar^2 K^2}{2M_n}\chi^{mn}(\boldsymbol{K},0) + \frac{\hbar}{M_n}\langle \mathrm{e}^{-i\boldsymbol{K}\boldsymbol{r}^m}\mathrm{e}^{i\boldsymbol{K}\boldsymbol{r}^n}(\boldsymbol{K}\boldsymbol{p}^n)\rangle_{\mathrm{T}}\,. \tag{27.36b}$$

This expression simplifies further in homogeneous systems (27.14). Then

$$\overline{\hbar\,\omega} = \int_{-\infty}^{+\infty} \hbar\,\omega\, S(\boldsymbol{K},\omega)\,\mathrm{d}\omega \Bigg/ \int_{-\infty}^{+\infty} S(\boldsymbol{K},\omega)\,\mathrm{d}\omega = \frac{\hbar^2 K^2}{2M} \tag{27.36c}$$

where $M$ is the mass of the atoms in the homogeneous system. The mean energy transfer of the neutron is equal to the repulsive energy of the ion with mass $M$. In inhomogeneous systems $M$ has to be replaced by a certain weighted average, which can be calculated from (27.36b). This holds in inhomogeneous systems (non-Bravais-crystals), if they have inversion symmetry or if they are harmonic.

The second moment can be determined from

$$\int_{-\infty}^{+\infty} \omega^2\, s^{mn}(\boldsymbol{K},\omega)\,\mathrm{d}\omega = -\left[\frac{\partial^2}{\partial t^2}\chi^{mn}(\boldsymbol{K},t)\right]_{t=0} \tag{27.36d}$$

(similar relations hold for higher moments).

It contains, as can be seen simply, a twofold commutator and for this reason it depends on more detailed properties of the system:

$$-\left[\frac{\partial^2}{\partial t^2}\chi^{mn}(\boldsymbol{K},t)\right]_{t=0} = \frac{1}{\hbar^2}\langle \mathrm{e}^{-i\boldsymbol{K}\boldsymbol{r}^m}[\mathscr{H}_{\mathrm{c}}, [\mathscr{H}_{\mathrm{c}}, \mathrm{e}^{i\boldsymbol{K}\boldsymbol{r}^n}]_-]_-\rangle_{\mathrm{T}}\,.$$

For example, in homogeneous, isotropic harmonic systems we obtain in the high-temperature (classical limit), similarly as in (27.36c)

$$\overline{(\hbar\,\omega)^2} = \frac{\hbar^2 K^2}{2M}\left(\frac{\hbar^2 K^2}{2M} + 2kT\right).$$

Other forms of these expressions are possible, using special representations of the functions involved. From (27.10) we have

$$\begin{aligned} s^{mn}(-\boldsymbol{K},-\omega) &= \sum_{\alpha\varepsilon} p_\alpha \langle\alpha\,|\,\mathrm{e}^{i\boldsymbol{K}\boldsymbol{r}^m}\,|\,\varepsilon\rangle\langle\varepsilon\,|\,\mathrm{e}^{-i\boldsymbol{K}\boldsymbol{r}^n}\,|\,\alpha\rangle\,\delta\left(-\omega + \frac{E_\alpha - E_\varepsilon}{\hbar}\right) \\ &= \sum_{\alpha\varepsilon} p_\varepsilon \langle\varepsilon\,|\,\mathrm{e}^{i\boldsymbol{K}\boldsymbol{r}^m}\,|\,\alpha\rangle\langle\alpha\,|\,\mathrm{e}^{-i\boldsymbol{K}\boldsymbol{r}^n}\,|\,\varepsilon\rangle\,\delta\left(\omega + \frac{E_\alpha - E_\varepsilon}{\hbar}\right) \end{aligned}$$

by interchanging the summation indices for the crystal states: $\alpha \Leftrightarrow \varepsilon$. Because of the $\delta$ function it is

$$p_\varepsilon = p_\alpha \, e^{-\hbar\omega/kT}$$

and therefore

$$s^{mn}(-\boldsymbol{K}, -\omega) = e^{-\hbar\omega/kT} \cdot s^{nm}(\boldsymbol{K}, \omega) . \qquad (27.37)$$

This relation allows for a derivation of the energy loss from the energy gain and vice versa (principle of detailed equilibrium). It follows further from (27.10), that

$$s^{*mn}(\boldsymbol{K}, \omega) = s^{nm}(\boldsymbol{K}, \omega) . \qquad (27.38)$$

From (27.11)

$$s^{mn}(\boldsymbol{K}, \omega) = \frac{1}{2\pi} \int e^{i(\boldsymbol{K}\boldsymbol{r} - \omega t)} g^{mn}(\boldsymbol{r}, t) \, d\boldsymbol{r} \, dt$$

and with (27.38)

$$s^{nm}(-\boldsymbol{K}, -\omega) = \frac{1}{2\pi} \int e^{i(\boldsymbol{K}\boldsymbol{r} - \omega t)} g^{*mn}(\boldsymbol{r}, t) \, d\boldsymbol{r} \, dt .$$

Combining these two equations according to (27.37) we obtain

$$\int e^{i(\boldsymbol{K}\boldsymbol{r} - \omega t)} \operatorname{Im} g^{mn}(\boldsymbol{r}, t) \, d\boldsymbol{r} \, dt$$
$$= -i \, (\operatorname{tgh} \hbar\omega/2kT) \int e^{i(\boldsymbol{K}\boldsymbol{r} - \omega t)} \operatorname{Re} g^{mn}(\boldsymbol{r}, t) \, d\boldsymbol{r} \, dt .$$

Taking the Fourier-transform of this equation we have

$$\operatorname{Im} g^{mn}(\boldsymbol{r}, t) = -\operatorname{tg}\left(\frac{\hbar}{2kT}\frac{\partial}{\partial t}\right) \operatorname{Re} g^{mn}(\boldsymbol{r}, t) \qquad (27.39)$$

which connects real and imaginary part of the van Hove-correlation-function in a similar way as does (21.25) for the phonon-Greens-function. Equations like these can be derived for all correlation functions. (27.39) clearly shows, that the correlation function contains an imaginary part in general. We have to discuss the meaning of this imaginary part in the correlation-function. *Van Hove* [54 H 2] has shown that the imaginary part is a function which describes the propagation of a momentaneous local perturbation in the system. It is that part of the change in the particle density at $\boldsymbol{r}$ and $t$, which arises from the fact, that the neutron was at $t = 0$ in $\boldsymbol{r} = 0$ and has thus influenced the system. In other words, the imaginary part describes the linear response of the system to an external perturbation (dissipation). For a detailed proof of this statement we refer to van Hove's paper. The real part describes the density fluctuations. Therefore (27.39) is called the dissipation-fluctuation theorem. Similar relations can be derived for $g(\boldsymbol{r}, t)$ and $g_s(\boldsymbol{r}, t)$

So far we have discussed only the scattering cross section for an ideal harmonic crystal and some of the properties of the functions involved. In non-homogeneous lattices, e.g. lattices with defects we cannot use (21.2) but rather (23.42) for a calculation of the displacement correlations involved in (27.25). Using (23.42) in an obvious slight modification we obtain

$$\left\langle\left(\boldsymbol{K}\boldsymbol{u}_\mu^m(0) \cdot \boldsymbol{K}\boldsymbol{u}_\nu^n(t)\right)\right\rangle_T = \frac{1}{2} \sum_\alpha A_{\mu\nu}^{mn}(\alpha, T) \left[e^{-\frac{\hbar\omega_\alpha}{2kT} - i\omega_\alpha t} + e^{+\frac{\hbar\omega_\alpha}{2kT} + i\omega_\alpha t}\right] \qquad (27.40)$$

with
$$A_{\mu\ \nu}^{mn}(\alpha, T) = \frac{\hbar (\boldsymbol{K} \overset{0}{\boldsymbol{u}}{}_\mu^m(\alpha)) (\boldsymbol{K} \overset{0}{\boldsymbol{u}}{}_\nu^{*n}(\alpha))}{\sqrt{M_\mu M_\nu}\, \omega_\alpha \cdot 2 \sinh(\hbar \omega_\alpha / 2kT)}\,.$$

Using the generating functions for the Bessel-functions or expanding both sides in a series, it is easy to prove that

$$e^{\frac{1}{2} A (z + 1/z)} = \sum_{n=-\infty}^{+\infty} z^n I_n(A) = I_0(A) + \sum_{n=1}^{\infty} (z^n + 1/z^n) I_n(A)\,,$$

where $I_n(A) = (-i)^n J_n(iA)$ is the modified, $J_n(iA)$ the normal Bessel-function of the first kind. It is

$$I_n(A) = \sum_{\nu=0}^{\infty} \frac{1}{\nu!(n+\nu)!} (A/2)^{n+2\nu}\,. \tag{27.41}$$

With these relations we have, $z_\alpha = \exp[-\hbar\omega_\alpha/2kT - i\omega_\alpha t]$

$$\begin{aligned} \exp\left\langle \boldsymbol{K}\boldsymbol{u}_\mu^m(0) \cdot \boldsymbol{K}\boldsymbol{u}_\nu^n(t) \right\rangle_T &= \prod_\alpha \left\{ I_0\left(A_{\mu\ \nu}^{mn}(\alpha)\right) + \sum_{n=1}^{\infty} I_n\left(A_{\mu\ \nu}^{mn}(\alpha)\right)\left(z_\alpha^n + 1/z_\alpha^n\right) \right\} \\ &= \prod_\alpha \{ I_0(A_\alpha) + (z_\alpha + 1/z_\alpha) I_1(A_\alpha) + \cdots \} \\ &\approx \prod_\alpha I_0(A_\alpha) + \sum_\alpha (z_\alpha + 1/z_\alpha) I_1(A_\alpha) \times \\ &\quad \times \prod_{\beta \neq \alpha} I_0(A_\beta) + \cdots \end{aligned} \tag{27.42}$$

where we have limited the expansion to zero- and one-phonon-processes. (27.42) looks somewhat different and more complicated compared to the corresponding ideal lattice expression (27.28, 29), mainly because of the Bessel-functions. This complication is related to the possible occurence of localized vibration states in defect lattices. We have to distinguish between two different sorts of states.

i) Scattering states, i.e. vibration states with frequencies in the region of ideal lattice frequencies (in lattices with point defects), or more precisely: states which have an oscillating character without an (exponential) decrease of amplitudes with increasing distance from the defect. These states can be normalized only in a box (periodicity volume). Their normalization factor contains a term $1/\sqrt{N}$, if $N$ is the number of unit cells in the periodicity volume. For these states $A_{\mu\ \nu}^{mn}(\alpha, T)$ is proportional to $1/N$; the expansion (27.41) then is an expansion with respect to $1/N$ and it is sufficient to take the lowest order terms of the Bessel-function-expansion, because $N$ can be assumed of the order of magnitude of $10^{23}$. For the contribution of these scattering states to the neutron cross-section we obtain, using (27.9, 11, 25, 40, 42)

$$\begin{aligned} \left(\frac{d\sigma}{d\omega\, d\Omega}\right)_{sc} &= \frac{k_e}{k_a} \sum_{\substack{mn \\ \mu\nu}} a_\mu^{*m} e^{-i\boldsymbol{K}\boldsymbol{R}_\mu^m - W_\mu^m} \cdot a_\nu^n e^{i\boldsymbol{K}\boldsymbol{R}_\nu^n - W_\nu^n} \times \\ &\times \left\{ \delta(\omega) + \frac{\hbar}{2\sqrt{M_\mu M_\nu}} \sum_\alpha \frac{(\boldsymbol{K} \overset{0}{\boldsymbol{u}}{}_\mu^m(\alpha)) (\boldsymbol{K} \overset{0}{\boldsymbol{u}}{}_\nu^{*n}(\alpha))}{\omega_\alpha} \times \right. \\ &\times \left. [(\bar{n}_\alpha + 1)\, \delta(\omega_\alpha - \omega) + \bar{n}_\alpha\, \delta(\omega_\alpha + \omega)] \right\} + \cdots \end{aligned} \tag{27.43}$$

— apart from factors, which arise from possible localized states and which contain Besselfunctions (see ii), —
which is an obvious generalization of (27.28, 29). Of course, the Debye-Waller-exponent $W_{\mu}^{m}$ has to be calculated from (27.40) or (23.45) rather than from (27.26).

The question arises where the higher order terms (in $1/N$), neglected in the derivation of (27.43), are hidden in (27.29). The second order processes, not written down in (27.29) contain a cross term of the form

$$\delta(\boldsymbol{q}+\boldsymbol{q}'+2\pi\, B\boldsymbol{h}-\boldsymbol{K})\;\delta(\omega(\boldsymbol{q}\sigma)-\omega(\boldsymbol{q}'\sigma')-\omega)$$

and similar ones. This term involves contributions from $\boldsymbol{q}=-\boldsymbol{q}'$, or $\delta(\boldsymbol{K}-2\pi\, B\boldsymbol{h})\;\delta(\omega)$ which is just a zero-order process. But the processes with $\boldsymbol{q}=-\boldsymbol{q}'$ are a factor $N$ less than those with $\boldsymbol{q}\neq-\boldsymbol{q}'$, and because the fore-factor is proportional to $1/N^2$, they are in effect of the order of $1/N$ and can be neglected. In this way all the Bessel-function terms can be found also in (27.29).

If the concentration of defects is very small, this part of the scattering cross-section is not very different from that for the scattering of neutrons at ideal lattices. For higher concentrations there are characteristic features, but we will not go into details here [61 K 1, 62 K 1, 63 K 2, 63 W 2, 64 I 1, 64 D 2, 65 E 1, 66 M 2].

The more interesting case is the following one.

ii) Localized vibration states. These have exponentially decreasing amplitudes. For such states the normalization factor is of the order 1 (compared to $1/\sqrt{N}$)*, if we limit the consideration to point defects; for localized states at lines or surfaces there is again a factor $1/\sqrt{N_s}$, if $N_s$ is the number of ions in a periodicity length in the line, or periodicity area in the surface, resp. The normalization can be seen from the simple example of a single mass ($M'$) defect in a cubic Bravais-crystal, which we use as an example in the following. According to (20.19, 20) the localized states follow from

$$1=\varepsilon\, M\omega^2\cdot G_{1\,1}^{00}(\omega^2);\quad \varepsilon=\frac{M'-M}{M}<0\,, \tag{27.44a}$$

which gives three degenerated states $\alpha$. The amplitude of the defect ion (in the origin!) can be choosen as

$$\overset{s}{u}(\alpha)=c\cdot\delta_{i\alpha}\,. \tag{27.44b}$$

The components of the eigenstates follow from

$$\overset{sm}{u}_i(\alpha)=\varepsilon M\omega^2\, G_{i\,j}^{m0}\cdot\overset{s0}{u}_j(\alpha)=\varepsilon\, M\,\omega^2\, G_{i\,\alpha}^{m0}\cdot c \tag{27.44c}$$

where $c$ can be determined from normalization

$$\sum_{mi}\overset{sm}{u}_i(\alpha)\cdot\overset{sm}{u}_i(\alpha)=1\,.$$

* See for example the normalization of the bound states of an electron in a Coulomb-field, from where it is known, that the normalization factor is not small.

A simple calculation gives

$$1 = c^2 \cdot (\varepsilon M \omega^2)^2 \cdot \frac{\partial G_{11}^{00}}{\partial M \omega^2}; \tag{27.44d}$$

$$\frac{\partial G_{11}^{00}}{\partial M \omega^2} = \frac{1}{N M^2} \sum_{q\sigma} \frac{e_1(q\sigma)\, e_1^*(q\sigma)}{[\omega^2(q\sigma) - \omega^2]^2} = \frac{1}{3 N M^2} \sum_{q\sigma} \frac{1}{[\omega^2(q\sigma) - \omega^2]^2}$$

showing that $c$ contains no factor $1/\sqrt{N}$, because all quantities involved are independent of $N$.

$$\dot{u}_i^m(\alpha) = -\left|\frac{\partial G_{11}^{00}}{\partial M \omega^2}\right|_{\omega=\omega_\alpha}^{-1/2} \cdot G_{i\ \alpha}^{m0}. \tag{27.44e}$$

In the determination of the scattering cross section we cannot restrict ourselves to the first term in the expansion of the Bessel-function and we obtain with (27.9, 11, 25, 40, 42)

$$\left(\frac{\mathrm{d}\sigma}{\mathrm{d}\omega\,\mathrm{d}\Omega}\right)_{\mathrm{loc}} = \frac{k_e}{k_a} \sum_{mn} a^{*m}\, \mathrm{e}^{-iKR^m - W_m} \cdot a^n\, \mathrm{e}^{iKR^n - W_n} \times \tag{27.45}$$

$$\times \Bigg\{ \prod_{\alpha(\mathrm{loc})} I_0(A_\alpha) \cdot \delta(\omega) +$$

$$+ 2 \sum_{\alpha(\mathrm{loc})} \sinh(\hbar\omega_\alpha/2kT) \cdot I_1(A_\alpha)\, [(\overline{n}_\alpha + 1)\, \delta(\omega_\alpha - \omega) + \overline{n}_\alpha\, \delta(\omega_\alpha + \omega)] \times$$

$$\times \prod_{\beta \neq \alpha(\mathrm{loc})} I_0(A_\beta) + \cdots \Bigg\}.$$

Here we can use the leading terms in the Bessel-function expansion only, if the $A_\alpha$ become small compared to one by other reasons than by $1/N$-arguments. It can be seen readily that at low temperatures ($T \to 0$) also $A_\alpha \to 0$. For higher temperatures it is a monotonic function, for which in the classical limit

$$|A^{mn}(\alpha)| < \frac{2P_0}{(\hbar\omega_\alpha)^2} kT\, (\dot{u}^0(\alpha), \dot{u}^0(\alpha)) < \frac{2P_0}{k\Theta_{\mathrm{D}}} \cdot \frac{T}{\Theta_{\mathrm{D}}}\, (\dot{u}^0(\alpha), \dot{u}^0(\alpha)) \ .$$

$P_0$ is the repulsive energy of the defect ion (page 239); the localized frequencies $\omega_\alpha$ have been replaced by the Debye-frequency in order to enlarge the right hand side. The most unfavorable case $(\dot{u}^0(\alpha), \dot{u}^0(\alpha)) = 1$ gives $|A^{mn}(\alpha)| < 1$ if $P_0 < k\Theta_{\mathrm{D}}$ and $T$ is not too large. A more detailed analysis for $(\dot{u}^0(\alpha), \dot{u}^0(\alpha))$ using a numerical value of the Greens-function, gives an even better estimate [66 M 2]. So in all cases of practical interest the Bessel-functions in (27.45) can be replaced by their leading terms also for localized states, though the argumentation proceeds in an other way as for scattering states.

For the one-phonon-cross-section of an isotopic (mass) defect we obtain, when we replace $I_1(A_\alpha)$ by its argument*

$$\left(\frac{\mathrm{d}\sigma_1}{\mathrm{d}\omega\,\mathrm{d}\Omega}\right)_{\mathrm{loc}} = \frac{k_e}{2k_a} \sum_\alpha \frac{\hbar}{M\omega_\alpha} \left|\frac{\partial G_{11}^{00}}{\partial M \omega^2}\right|_{\omega=\omega_\alpha}^{-1} [(\overline{n}_\alpha + 1)\, \delta(\omega_\alpha - \omega) +$$

$$+ \overline{n}_\alpha \cdot \delta(\omega_\alpha + \omega)] \times \left| \sum_m a^m\, \mathrm{e}^{-W_m}\, \mathrm{e}^{-iKR^m} \sum_i K_i\, G_{i\ \alpha}^{m0} \right|^2. \tag{27.46}$$

* The total one-phonon-cross-section is the sum of $(\mathrm{d}\sigma_1)_{\mathrm{sc}} + (\mathrm{d}\sigma_1)_{\mathrm{loc}}$

Here we have used eqs. (27.44) for the isotopic defect. The three localized frequencies $\omega_\alpha$ are equal because of the degeneracy in cubic crystals. The defect ion may be situated in the origin. We write

$$\tilde{a}^{m} = a^{m}\, \mathrm{e}^{-W m}$$

and make the further assumption that the Debye-Waller-factor is equal for all the ions apart from the isotope itself. This assumption is not justified very well, but it simplifies the calculation and the result obtained in this way shows the essential points. The Debye-Waller-factor can be used as derived on page 239. The properties related to the defect ion are indicated by a prime. Then we have

$$\sum_{m} \tilde{a}^{m}\, \mathrm{e}^{-i\boldsymbol{K}\boldsymbol{R}^{m}} \sum_{i} K_i\, G^{m0}_{i\ \alpha} = (\tilde{a}' - \tilde{a}) \sum_{i} K_i\, G^{00}_{i\alpha} + \sum_{mi} \tilde{a}\, \mathrm{e}^{-i\boldsymbol{K}\boldsymbol{R}^{m}}\, K_i\, G^{m0}_{i\ \alpha}$$

$$= \frac{1}{M} \sum_{\sigma} (\boldsymbol{K}\boldsymbol{e}(\boldsymbol{K}\sigma))\, e^{*}_{\alpha}(\boldsymbol{K}\sigma) \left\{\frac{\tilde{a}' - \tilde{a}}{\varepsilon\, \omega^2_\alpha} + \frac{\tilde{a}}{\omega^2(\boldsymbol{K}\sigma) - \omega^2_\alpha}\right\}.$$

In this derivation use has been made of (27.44a) and of the representation (21.21b) of the Greens function as well as of some other properties of the Greens function. Inserting this into (27.46) and summing over $\alpha$ (all frequencies $\omega_\alpha$ are equal!) we obtain finally

$$\left(\frac{\mathrm{d}\sigma_1}{\mathrm{d}\omega\, \mathrm{d}\Omega}\right)_{\mathrm{loc}} = + \frac{k_e}{2k_a} \cdot \frac{\hbar}{M^3\, \omega_\alpha} \left(\frac{\partial G^{00}_{11}}{\partial M \omega^2}\right)^{-1}_{\omega=\omega_\alpha} \cdot [(\bar{n}_\alpha + 1)\, \delta\, (\omega_\alpha - \omega) +$$

$$+ \bar{n}_\alpha\, \delta(\omega_\alpha + \omega)] \times pN \sum_{\sigma} |\boldsymbol{K}\boldsymbol{e}(\boldsymbol{K}\sigma)|^2 \left\{\frac{\tilde{a}' - \tilde{a}}{\varepsilon\, \omega^2_\alpha} + \frac{\tilde{a}}{\omega^2(\boldsymbol{K}\sigma) - \omega^2_\alpha}\right\}^2 \qquad (27.47)$$

for the one-phonon-cross-section of a localized mode ($\varepsilon < 0$). This equation holds for the localized states of a single mass defect ($Np = 1$) or for finite numbers of isotopes with particle concentration $p$, if the concentration is so small, that the interaction between different isotopes can be neglected ($p \lesssim 1\%$).

In case of other defects, a derivation of a corresponding formula is more complicated, but we can expect a similar result.

From measurements of these cross-sections we can get the following informations. First we can determine the localized mode frequencies from the change in energy of the neutron: $\omega_\alpha = \pm\omega$. This, of course, is a more or less trivial statement and obviously holds for any localized states at any point defects. But by measuring the angular-dependence (momentum transfer $\boldsymbol{K}$) of the scattered neutrons, we can even obtain information on the ideal lattice vibration states; for from (24.47) we see that $\omega^2(\boldsymbol{K}\sigma)$ enters the cross-section, $\omega^2(\boldsymbol{K}\sigma) - \omega^2_\alpha$ becoming large for high frequencies. For $\boldsymbol{K}$-values in the first Brillouinzone $\boldsymbol{K} = \boldsymbol{q}$, only longitudinal modes, or longitudinal components of mixed modes enter. But using $\boldsymbol{K}$ in a higher Brillouinzone $\boldsymbol{K} = \boldsymbol{q} + 2\pi B\boldsymbol{h}$, $\boldsymbol{h} \neq 0$ we have also information on transverse modes. A third possibility of information has been pointed out by *Maradudin* [66 M 2]. From (27.44e) we have the Fourier-transform

$$\sum_{mi} \mathrm{e}^{-i\boldsymbol{q}\boldsymbol{R}^{m}}\, e^{*}_{i}(\boldsymbol{q}\sigma)\, \overset{sm}{u_i}(\alpha) = - \left|\frac{\partial G^{00}_{11}}{\partial M \omega^2}\right|^{-1/2}_{\omega=\omega_\alpha} \cdot \frac{e^{*}_{\alpha}(\boldsymbol{q}\sigma)}{M[\omega^2(\boldsymbol{q}\sigma) - \omega^2_\alpha]} \qquad (27.48)$$

showing that (27.47) measures essentially the Fourier-transform of the localized state vector. It is the transform of a function like $\exp(-\alpha R)/R$ and just this function determines the spatial distribution of the amplitudes of localized states. The asymptotic expansion of a localized state function is [see 60 M 1]

$$u_i(\alpha, R) \approx \frac{1}{\hat{R}} \exp\left\{-\sqrt{2}\sqrt{\omega_\alpha^2 - \omega_m^2}\,\hat{R}\right\}, \tag{27.49}$$

where $\omega_m$ is the maximum frequency of the ideal lattice vibrations*. Expanding $\omega^2$ about this frequency we have

$$\omega^2(\boldsymbol{q}\sigma) = \omega_m^2 - \frac{1}{2}\sum_{ij} \Lambda_{ij}(\boldsymbol{q} - \boldsymbol{q}_m)_i (\boldsymbol{q} - \boldsymbol{q}_m)_j \ldots; \quad \Lambda_{ij} = \frac{\partial^2 \omega^2}{\partial q_i \partial q_j} \tag{27.50a}$$

$\boldsymbol{q}_m$ is the wave-vector belonging to $\omega_m$. $\hat{\boldsymbol{R}}$ is defined by

$$\hat{R}^2 = \sum_{ij} X_i X_j (\Lambda^{-1})_{ij}. \tag{27.50b}$$

Thus from measuring the localized-mode-one-phonon cross-section in the neighborhood of $\boldsymbol{K} = \boldsymbol{q}_m$ and determining the half-width and the height of the cross-section in this neighborhood, one can estimate $\Lambda_{ij}$ and therefore the spatial decrease of the localized mode amplitudes.

Investigations of more complicated defect models have been done also [64 D 2, 63 K 2, 63 W 2, 65 E 1]. The essential points are similar, but the functions involved have not the simple shape of those in (27.47). It should be mentioned further, that the interaction of defects at larger concentrations results in a frequency band for localized modes. In effect, this leads to a broadening of the energy change of the neutron. All the recent experiments have been done with larger concentrations [60 L 2, 60 W 1, 61 B 2, 63 W 3]. The maximum absorption or emission frequency, however, gives a good estimate of the localized frequencies, if different localized states do not overlap according to a strong interaction between different defects.

*Anharmonic crystals*

In harmonic theory the strong periodic dependence of the eigenstates on time, $\exp(-i\omega_\alpha t)$, or in other words, the infinite life-time of the phonons was the origin for the energy transfer of the neutrons being $\delta$-functions $\delta(\omega)$, $\delta(\omega_\alpha \pm \omega)$, ... i.e. peaks of zero width. Anharmonic terms in the potential energy of a crystal cause a finite life-time of phonons, and of course, a change in the phonon-frequencies. The absorption or emission peaks of the neutron-energy-transfer will have a finite width and will be shifted somewhat. In terms of the scattering amplitude of the neutrons we may say, that in harmonic theory the poles of the scattering amplitude lie on the real axis of the $\omega$-plane, whereas in anharmonic theory they

* The asymptotic expansion (27.49) is a good approximation only if $\omega_\alpha$ is only slightly larger than $\omega_m$. Otherwise the subsequent terms in the expansion give a bad convergence [see also *K. Dettmann*, Diplom-Thesis, Aachen 1963].

lie below the real axis. We expect the $\delta$-function in the scattering amplitude to be replaced by

$$\delta(\omega - \omega_\alpha) \Rightarrow \frac{1}{(\omega - \omega_\alpha - \Delta_\alpha) + i\,\Gamma_\alpha} \tag{27.51a}$$

where $\Delta_\alpha$ is the frequency shift and $\Gamma_\alpha$ the life-time of the phonons due to anharmonicity. The cross-section is expected to be

$$\frac{\mathrm{d}\sigma}{\mathrm{d}\omega\,\mathrm{d}\Omega} \sim \frac{\Gamma_\alpha/\pi}{(\omega - \omega_\alpha - \Delta_\alpha)^2 + \Gamma_\alpha^2} \tag{27.51b}$$

$\Delta_\alpha$ and $\Gamma_\alpha$ have to be calculated by a perturbation procedure for the anharmonic terms. It turns out further, that the exact cross-section does not have the symmetrical Lorentzian shape (27.51b) but that there has to be added an additional term which produces an unsymmetry in the emission-function [62 M 2, 62 K 2, 64 M 6].

The difficulties in the calculation of (27.21) result from the fact, that the commutator of $\boldsymbol{u}^m(0)$ and $\boldsymbol{u}^n(t)$ is no longer a $c$-number, and that the distribution for the $\boldsymbol{u}^m$ is no longer a Gaussian one. In the following we give an outline of the derivation, and refer to the appendix for more detailed calculations. We start with the relation

$$\begin{aligned}
\langle \mathrm{e}^A \cdot \mathrm{e}^B \rangle_T = \mathrm{e}^{-W_A - W_B} \cdot \exp\{&\langle A\,B\rangle - \langle A\rangle\langle B\rangle + \\
&+ \tfrac{1}{2}\langle A^2 B\rangle - \tfrac{1}{2}\langle A^2\rangle\langle B\rangle - \langle A\rangle\langle A\,B\rangle + \langle A\rangle^2\langle B\rangle + \\
&+ \tfrac{1}{2}\langle A\,B^2\rangle - \tfrac{1}{2}\langle A\rangle\langle B^2\rangle - \langle A\,B\rangle\langle B\rangle + \langle A\rangle\langle B\rangle^2 + \\
&+ \text{higher order terms}\}; \\
-W_A = \langle A\rangle + &\tfrac{1}{2}\{\langle A^2\rangle - \langle A\rangle^2\} + \\
+ &\tfrac{1}{6}\{\langle A^3\rangle - 3\langle A\rangle\langle A^2\rangle + 2\langle A\rangle^3\} + \cdots
\end{aligned} \tag{27.52}$$

given in (32.4), which holds for two arbitrary non-commuting operators $A$ and $B$. In our case

$$A = -\,i\boldsymbol{K}\boldsymbol{u}^m(0)\,; \quad B = B(t) = i\boldsymbol{K}\boldsymbol{u}^n(t)\,. \tag{27.53}$$

With this definition, $W$ is the anharmonic Debye-Waller-factor which shall be discussed later. In *harmonic* theory all odd averages $\langle A\rangle$, $\langle A^2 B\rangle, \ldots$ vanish. Also the terms $\langle A^4\rangle - 3\langle A^2\rangle^2$ and those of similar shape, occuring in higher order terms, vanish in a Gaussian distribution and we have the old result. Thus we expect all these terms to be directly proportional to anharmonic force constants in anharmonic theory. The leading term, also in anharmonic theory, will be $\langle A\,B\rangle$, which has to be averaged with the anharmonic distribution now.

It can be shown by simple symmetry arguments, that the linear averages $\langle A\rangle$, $\langle B\rangle$ vanish also in anharmonic theory, if each ion of the lattice is a center of inversion (parameterfree crystals). Because we will restrict to Bravais-crystals in the following, we drop these terms now. This is only for simplicity and they can be calculated just in the same way as the other terms.

Further we will limit ourselves to the discussion of the one phonon cross-section, as before, and therefore we expand the exponential in (27.52) in the same way as in harmonic theory, being left with

$$\langle e^A \cdot e^B \rangle = e^{-W_A - W_B} \{1 + \langle A\,B \rangle + \langle A^2 B \rangle + \text{higher terms}\}\,. \quad (27.54)$$

The first term in { }, 1, gives the zero-phonon-cross-section, the second term is essentially the one-phonon-cross-section; the third term*

$\frac{1}{2}\langle A^2 B \rangle + \frac{1}{2}\langle A\,B^2 \rangle = \langle A^2 B \rangle$, vanishing in harmonic theory, gives also a contribution to the one-phonon cross-section, and it is just this term which gives the unsymmetry of the Lorentzian (27.51b) mentioned above. It can be seen from (27.53) that it is proportional to $K^3$, whereas the „normal" cross-section is proportional to $K^2$. Thus it can be neglected for small momentum transfer. Some of the higher order terms in (27.54), e.g. terms as $\langle A^2 B^2 \rangle - \langle A^2 \rangle \langle B^2 \rangle - 2\langle A\,B \rangle^2$ give also contributions to the one-phonon-cross-section, but i) they are directly proportional to anharmonic force constants, ii) they are proportional to $K^4$ and higher powers, thus being essential only for large momentum transfer. As far as known, these terms have not been seen in experiments and we will not discuss them here.

We give a short summary of the calculation for $\langle A\,B \rangle$ and the result for the unsymmetry term $\langle A^2 B \rangle$ only. Using (27.53) the zero- and one-phonon-scattering function is given by

$$s_0^{mn}(\boldsymbol{K}, \omega) = e^{-2W} \cdot e^{i\boldsymbol{K}(\boldsymbol{R}^n - \boldsymbol{R}^m)} \cdot \delta(\omega)$$

$$s_1^{mn}(\boldsymbol{K}, \omega) = \frac{1}{2\pi} e^{-2W} \cdot e^{i\boldsymbol{K}(\boldsymbol{R}^n - \boldsymbol{R}^m)} \sum_{ij} K_i K_j \int dt\, e^{-i\omega t} \left\langle u_i^m(0)\, u_j^n(t) \right\rangle_T . \quad (27.55)$$

Apart from the Debye-Waller-factor (see eq. 27.77), the zero-phonon-cross-section is not changed by anharmonicity. The one-phonon-cross-section contains the anharmonicity via the averaging process in $\left\langle u_i^m(0)\, u_j^n(t) \right\rangle_T$, which has to be done with the anharmonic Hamiltonian (1.6), (16.1)

$$\mathscr{H}_c = \mathscr{H}_0 + \Phi_3 + \Phi_4 + \cdots. \quad (27.56)$$

We have seen in Sect. 16, that contributions from $\Phi_3^2$ are of the same order of magnitude as $\Phi_4$; higher order contributions, $\Phi_3^3$, $\Phi_3^4$, $\Phi_4^2$ and so on, would be comparable with contributions from $\Phi_5$, $\Phi_6$, .... For a first reasonable expansion with respect to anharmonic terms, it is therefore sufficient to consider terms up to $\Phi_3^2$ and $\Phi_4$. It has turned out further (see Sect. 16) that higher order contributions cannot be seen in most the experiments which have been done and would be essential only near the melting point.

* The equality $\langle A^2\,B \rangle = \langle A\,B^2 \rangle$ is not quite trivial and general, but it can be proved simply in case of (27.53).

A second point to be mentioned again is that we have taken the thermal equilibrium positions as the starting point for the expansion (27.56) [Sect. 15, 16]. That means that all the harmonic quantities involved in $\mathscr{H}_0$ depend on temperature via the mean equilibrium positions (quasiharmonic approximation). Formally, all the equations derived in harmonic theory have the same shape in the quasiharmonic approximation, only all the positions have to be replaced by the thermal equilibrium positions. All the averages with the harmonic distribution are replaced by averages with the quasiharmonic approximation ($\mathscr{H}_{\mathrm{qh}}$).

Let the eigenstates of the crystal Hamiltonian $\mathscr{H}_{\mathrm{c}}$ be denoted by $|N\rangle$ with eigenvalues $E_N$ and the matrix-elements of operators $A$ by $A_{MN}$. Then it can be proved simply that

$$\begin{aligned}
&\int_{-\infty}^{+\infty} \mathrm{d}t \cdot \mathrm{e}^{i\omega t} \langle B(t) A\rangle = 2\pi\,\varrho(\omega)\,,\\
&\varrho(\omega) = \frac{\hbar}{Z} \sum_{MN} \delta(E_M - E_N - \hbar\omega)\,\mathrm{e}^{-E_N/kT} A_{MN} B_{NM}\,,\\
&Z = \mathrm{Trace}\,\mathrm{e}^{-\mathscr{H}_{\mathrm{c}}/kT}\,, \qquad (27.57)\\
&\overset{mn}{s_1}(\boldsymbol{K}, \omega) = \mathrm{e}^{-2W} \cdot \mathrm{e}^{i\boldsymbol{K}(\boldsymbol{R}^n - \boldsymbol{R}^m)} \cdot \varrho(\omega)\,.
\end{aligned}$$

We then define a function $(-\beta\hbar < \tau < \beta\hbar)$

$$f_{BA}(\tau) = \langle T_{\mathrm{D}}\,\mathrm{e}^{\tau\mathscr{H}_{\mathrm{c}}/\hbar}\,B\,\mathrm{e}^{-\tau\mathscr{H}_{\mathrm{c}}/\hbar}\,A\rangle = \langle T_{\mathrm{D}}\,B(\tau)\,A\rangle \qquad (27.58)$$

with the „time"-ordering operator $T_{\mathrm{D}}$ defined by

$$\langle T_{\mathrm{D}} B(\tau_1)\,A(\tau_2)\rangle = \begin{cases} \langle B(\tau_1)\,A(\tau_2)\rangle & \text{if} \quad \tau_1 > \tau_2 \\ \langle A(\tau_2)\,B(\tau_1)\rangle & \text{if} \quad \tau_2 > \tau_1\,. \end{cases}$$

It is shown in the Appendix, that $f(\tau)$ can be represented by a Fourier series

$$f(\tau) = \sum_{\nu=-\infty}^{+\infty} a_\nu\,\mathrm{e}^{2\pi i kT\tau\nu/\hbar}$$

the coefficients of which can be continued into the complex plane with continuous argument:

$$a_\nu \Rightarrow a(z)\,; \quad z = 2\pi i\,kT\nu/\hbar$$

and that the function $\varrho(\omega)$ can be determined from the discontinuity of $a(z)$ on the real axis:

$$\varrho(\omega) = \frac{\hbar/kT}{\mathrm{e}^{-\hbar\omega/kT} - 1} \lim_{\epsilon\to 0+} \frac{a(-\omega + i\epsilon) - a(-\omega - i\epsilon)}{2\pi i}\,. \qquad (27.59)$$

Thus the problem is reduced to finding the Fourier-coefficients of (27.58).

We start with (27.56) using the displacements in the transformed representation (16.9, 21.2). The perturbation is then given by (16.13) with Fourier-transformed coupling parameters (16.14). They satisfy (16.15–17). The indices $\boldsymbol{q}\sigma$ are abbreviated as $J$, etc. in the following, where $-J$ means $-\boldsymbol{q}, \sigma$. We introduce further

$$B_J = b_J - b_{-J}^\dagger\,; \quad B_{-J} = b_{-J} - b_J^\dagger = -B_J^\dagger\,. \qquad (27.60)$$

In analogy to (27.58) we define

$$\begin{aligned} g_J(\tau)\cdot\delta_{JJ'} &= \langle T_D\,\tilde{B}_J(\tau)\,\widetilde{B_{J'}^{+}}(0)\rangle_{\mathrm{qh}} = \langle T_D\,\widetilde{B_J^{+}}(\tau)\,\tilde{B}_{J'}(0)\rangle_{\mathrm{qh}} \\ &= -\langle T_D\,\tilde{B}_J(\tau)\,\widetilde{B}_{-J'}(0)\rangle_{\mathrm{qh}} = -\langle T_D\,\widetilde{B_{J'}^{+}}(\tau)\,\widetilde{B_{-J}^{+}}(0)\rangle_{\mathrm{qh}} \quad (27.61)\\ &= \delta_{JJ'}\{\bar{n}_J\,\mathrm{e}^{|\tau|\,\omega_J} + (\bar{n}_J+1)\,\mathrm{e}^{-|\tau|\,\omega_J}\}\,, \qquad \text{for every } J, J', \end{aligned}$$

where

$$\tilde{B}_J(\tau) = \mathrm{e}^{\tau\mathscr{H}_{\mathrm{qh}}/\hbar}\,B_J\,\mathrm{e}^{-\tau\mathscr{H}_{\mathrm{qh}}/\hbar} \quad\text{or}\quad \tilde{W}(\tau) = \mathrm{e}^{\tau\mathscr{H}_{\mathrm{qh}}/\hbar}\,W\,\mathrm{e}^{-\tau\mathscr{H}_{\mathrm{qh}}/\hbar}$$

is constructed with the *quasiharmonic* Hamiltonian $\mathscr{H}_{\mathrm{qh}}$. Correspondingly $\omega_J$ and $\bar{n}_J$ are the frequency and the mean occupation numbers in the *quasiharmonic* approximation. Obviously the function $g_J(\tau)$ is a special kind of $f(\tau)$ in (27.58), having the same properties and satisfying additionally

$$g_J(\tau) = g_J(-\tau)\,. \qquad (27.62)$$

We introduce Fourier-transformed functions

$$\begin{aligned} g_J(\tau) &= \sum_{-\infty}^{+\infty} h_J^{\nu}\,\mathrm{e}^{2\pi i k T\tau\nu/\hbar};\quad \hbar\omega_\nu = 2\pi\,k T\nu \\ h_J^{\nu} &= h_J^{-\nu} = \frac{kT}{\hbar}\int_0^{\hbar/kT} \mathrm{d}\tau\, g_J(\tau)\,\mathrm{e}^{-2\pi i k T\tau\nu/\hbar} = \frac{2kT\,\omega_J}{\hbar}\cdot\frac{1}{\omega_J^2+\omega_\nu^2}\,. \end{aligned} \qquad (27.63)$$

What we are really concerned with is (27.55); we obtain (when in $\left\langle u_i^m(\tau)\,u_j^n(0)\right\rangle$ the $u_i^m$ are expressed by means of (16.9, 21.2)) for Bravais lattices now:

$$\begin{aligned} \left\langle u_i^m(\tau)\,u_j^n(0)\right\rangle &= \frac{\hbar}{2NM}\sum_{JJ'}\frac{e_i(J)\,e_j^*(J')}{\sqrt{\omega_J\,\omega_{J'}}}\,\mathrm{e}^{i\boldsymbol{q}\boldsymbol{R}^m - i\boldsymbol{q}'\boldsymbol{R}^n}\cdot f_{JJ'}(\tau) \\ f_{JJ'}(\tau) &= \langle T_{\mathrm{D}}\,\mathrm{e}^{\tau\mathscr{H}_{\mathrm{c}}/\hbar}\,B_J\,\mathrm{e}^{-\tau\mathscr{H}_{\mathrm{c}}/\hbar}\cdot B_{J'}^{+}\rangle\,, \end{aligned} \qquad (27.64)$$

with the complete crystal Hamiltonian $\mathscr{H}_{\mathrm{c}}$. Using the perturbation expansion (16.6) for $\mathscr{H}_{\mathrm{c}} = \mathscr{H}_{\mathrm{qh}} + W$ we have

$$f_{JJ'}(\tau) = \frac{\langle T_{\mathrm{D}}\,\tilde{B}_J(\tau)\,\tilde{B}_{J'}^{+}(0)\,S(\beta)\rangle_{\mathrm{qh}}}{\langle T_{\mathrm{D}}\,S(\beta)\rangle_{\mathrm{qh}}} \qquad (27.65)$$

with

$$S(\beta) = \sum_{n=0}^{\infty}\frac{(-1)^n}{n!}\int_0^{\beta}\mathrm{d}\tau_1\ldots\int\mathrm{d}\tau_n\,\tilde{W}(\tau_1)\ldots\tilde{W}(\tau_n)\,,$$

where the average now is meant with the *quasiharmonic* Hamiltonian. The „time ordering" operator $T_{\mathrm{D}}$ means, that the function with largest argument has to be written left to that with smaller argument.

The calculation of (27.65) can be illustrated by using a graphical description for the different terms in the expansion, being explained in some detail in the Appendix. Because the perturbation is essentially a function of $B_J$'s, the averaging process means a quasiharmonic average about powers (or products) of $B_J$. Because of the properties (27.61) only even products of $B_J$ contribute. There are two different types of terms involved in (27.65): those, in which the $\tilde{B}_J(\tau)\,\tilde{B}^\dagger_{J'}(0)$ are averaged independently of the rest, and those in which the averaging process connects a $\tilde{B}_J$ with any $\tilde{B}_{J''}$ contained in the $\tilde{W}(\tau)$. The former types of contributions can be represented by unconnected diagrams, the latter ones by connected diagrams. The effect of the unconnected diagrams is just to cancel the normalization factor in (27.65) (Appendix); thus if we restrict the averaging process to a sum over connected diagrams, we can drop the normalization factor:

$$f_{JJ'}(\tau) = \Big\langle T_D\,\tilde{B}_J(\tau)\,\tilde{B}^\dagger_{J'}(0)\sum_{n=0}^{\infty}\frac{(-1)^n}{n!}\times$$
$$\times\int_0^{1/kT} d\tau_1\ldots\int d\tau_n\,\tilde{W}(\tau_1)\ldots\tilde{W}(\tau_n)\Big\rangle_{\mathrm{qh,\,connected}} \qquad (27.65\,\mathrm{a})$$

As an example we calculate the contribution of a second order term in $\Phi_3$. Inserting $\Phi_3$ from (16.13) we have

$$\frac{1}{2}\,\frac{\hbar^3}{288N}\sum_{J_1\ldots J_6}\frac{\Phi_{J_1J_2J_3}\Phi_{J_4J_5J_6}}{(\omega_{J_1}\,\omega_{J_2}\,\omega_{J_3}\,\omega_{J_4}\,\omega_{J_5}\,\omega_{J_6})^{1/2}}\Big\langle T_D\,\tilde{B}_J(\tau)\,\tilde{B}^\dagger_{J'}(0)\int_0^{1/kT}\!\!\int d\tau_1\,d\tau_2\times$$
$$\times\,\tilde{B}_{J_1}(\tau_1)\,\tilde{B}_{J_2}(\tau_1)\,\tilde{B}_{J_3}(\tau_1)\,\tilde{B}_{J_4}(\tau_2)\,\tilde{B}_{J_5}(\tau_2)\,\tilde{B}_{J_6}(\tau_2)\Big\rangle\,,\,.$$

Disregarding terms which occur $1/N$ less, typical connected diagrams are shown in Fig. 7.4. The first type of diagrams gives no contribution,

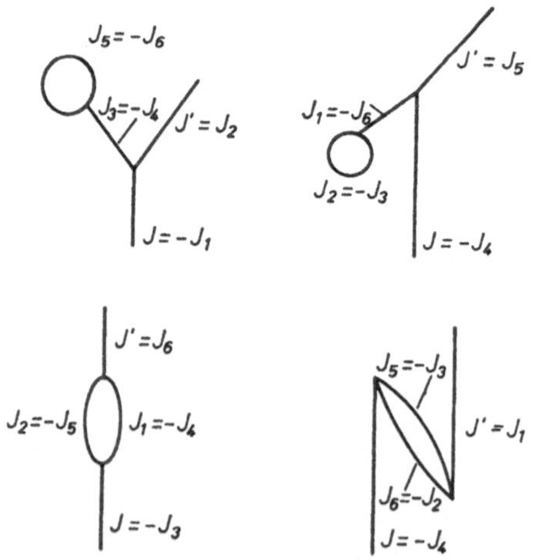


Fig. 7.4. Diagrams representing contributions of second order in $\Phi_3$; every line represents the average of two paired $B_J$, i. e. $g_J(\tau)$. Every vertex represents an interaction. Because we consider three phonon interactions, in every vertex there meet three phonon lines (see fig. 8.2.)

because in $\Phi_{J_4J_5J_6}$ the conservation of quasimomentum is involved, that means

$$\Delta(\boldsymbol{q}_4 + \boldsymbol{q}_5 + \boldsymbol{q}_6 - 2\pi\,B\boldsymbol{h}) = \Delta(\boldsymbol{q}_4 + \boldsymbol{q}_5 - \boldsymbol{q}_5 - 2\pi\,B\boldsymbol{h}) = \Delta(\boldsymbol{q}_4)$$

and $\Phi^{0\ \boldsymbol{q}_5 - \boldsymbol{q}_5}_{\sigma_4 \sigma_5\ \ \sigma_5}$ vanishes in Bravais-lattices and in lattices where every ion is a center of symmetry. This can be seen simply from its properties. The second type diagrams contribute, and it can be seen, that they give the same contribution. Alltogether there are 36 possibilities of pairing

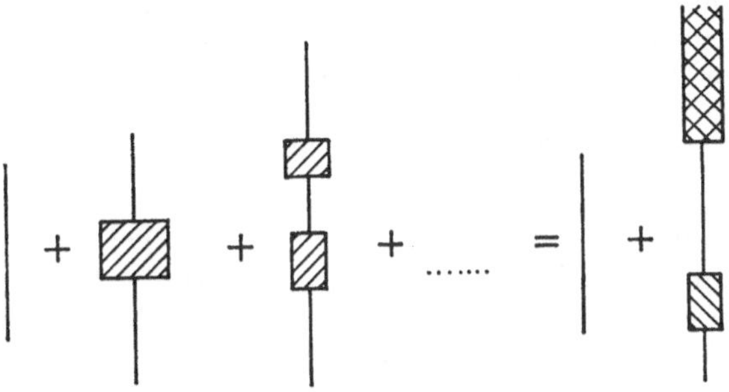

Fig. 7.5. Schematic representation of the Dyson-equation (27.67)

$J$'s corresponding to these diagrams. Doing this and using (27.61) we obtain

$$\frac{\hbar^3}{16N} \sum_{J_1 J_2} \frac{\Phi_{-J J_1 J_2} \Phi_{J' - J_1 - J_2}}{(\omega_J \omega_{J'})^{1/2} \omega_{J_1} \omega_{J_2}} \int\int_0^{1/kT} \mathrm{d}\tau_1 \, \mathrm{d}\tau_2 \times$$
$$\times \; g_J(\tau - \tau_1) \, g_{J'}(\tau_2) \, g_{J_1}(\tau_1 - \tau_2) \, g_{J_2}(\tau_1 - \tau_2);$$

with (27.63) we perform the integrations and have

$$\sum_\nu h^\nu_J \cdot \frac{\hbar^3}{16N(kT)^2} \sum_{J_1 J_2} \frac{\Phi_{-J J_1 J_2} \Phi_{J' - J_1 - J_2}}{(\omega_J \omega_{J'})^{1/2} \omega_{J_1} \omega_{J_2}} \sum_{\nu_1} h^{\nu_1}_{J_1} h^{\nu - \nu_1}_{J_2} h^\nu_{J'} \cdot e^{2\pi i k T \tau \nu/\hbar} \tag{27.66}$$

Similar expressions can be derived for higher order contributions and also for the contributions of $\Phi_4$. It can be seen, that all these contributions are proportional to $\exp\{2\pi i\, k T \tau \nu/\hbar\}$. Thus if we define the Fourier-transform of (27.65a) by

$$f_{JJ'}(\tau) = \sum_{\nu = -\infty}^{+\infty} F^\nu_{JJ'} \, e^{2\pi i k T \tau \nu/\hbar}$$

we can find the coefficients simply from (27.66). Because $\Phi_{-J J_1 J_2} \Phi_{J' - J_1 - J_2}$ contains Kronecker factors $\Delta(-\boldsymbol{q} + \boldsymbol{q}_1 + \boldsymbol{q}_2 - 2\pi B \boldsymbol{h}) \cdot \Delta(\boldsymbol{q}' - \boldsymbol{q}_1 - \boldsymbol{q}_2 - 2\pi B \boldsymbol{h})$ it follows, that $F^\nu_{JJ'}$, and therefore $f_{JJ'}(\tau)$ is proportional to $\Delta(\boldsymbol{q}' - \boldsymbol{q})$, but *not* to a factor $\delta_{\sigma\sigma'}$! The same statement holds for the other contributions.

To obtain the complete contribution of $\Phi_3$ and $\Phi_4$ we observe, that the sum of all the diagrams can be represented as shown in Fig. 7.5. The hatched square represents the sum of all diagrams, which cannot divided in two parts by cutting a single $g_J(\tau)$-line. These diagrams are called proper ones. The first type in Fig. 7.4 are improper diagrams, but they give no contribution in this special form. We denote the sum over all the proper diagrams by $\Sigma_{JJ'}$. They are diagonal with respect to $\boldsymbol{q}\boldsymbol{q}'$ (see above), but not to $\sigma\sigma'$. The sum represented by Fig. 7.5 is an integral-equation which can be written, in the Fourier-transformed quantities

$$F^{\nu}_{\sigma\sigma'} = h^{\nu}_{\sigma\sigma'} + h^{\nu}_{\sigma\sigma''}\, \Sigma^{\nu}_{\sigma''\sigma'''}\, F^{\nu}_{\sigma'''\sigma'} \,. \tag{27.67}$$

We have dropped the $\boldsymbol{q}$-index, because it is equal in all the factors. $h^{\nu}_{\sigma\sigma'}$ is diagonal also in $\sigma\sigma'$ (see 27.61, 63). If we would consider only the diagonal elements of $\Sigma^{\nu}_{\sigma\sigma'}$, the solution of (27.67) would be simple, because then the $F^{\nu}_{\sigma\sigma'}$ would be diagonal too:

$$F = \frac{1}{\frac{1}{h} - \Sigma} \text{ for all } \quad q, \sigma, \nu \,, \tag{27.68}$$

|| if $\Sigma$ is diagonal or if the non-diagonal elements are small.

In the following we will make this assumption*. If we would like to avoid this assumption, we could decompose $\Sigma^{\nu}_{\sigma\sigma'}$ into its diagonal and non-diagonal parts with respect to $\sigma\sigma'$.

$$\Sigma = \Sigma^{\mathrm{D}} + \Sigma^{\mathrm{N}}$$

and solve (27.67) by iteration with respect to $\Sigma^{\mathrm{N}}$:

$$F = \frac{1}{h^{-1} - \Sigma^{\mathrm{D}}} + \frac{1}{h^{-1} - \Sigma^{\mathrm{D}}}\, \Sigma^{\mathrm{N}} \frac{1}{h^{-1} - \Sigma^{\mathrm{D}}} + \cdots \tag{27.69}$$

provided the non-diagonal elements are small. One could have solved also (27.67) by iteration with respect to the total $\Sigma$, but in (27.69) we have at least all the diagonal elements of $\Sigma$ in the first approximation. We suppose that the diagonal elements give the main contribution and neglect all the non-diagonal elements in following.

To obtain $\Sigma^{\mathrm{D}}$, we have to sum all the proper diagrams in the expansion (27.65a), up to terms quadratic in $\Phi_3$ and linear in $\Phi_4$. The only term contributing from $\Phi_3$ is already given in (27.66), namely

$$\Sigma^{\mathrm{D}}_3 = \frac{\hbar^3}{16N(kT)^3} \sum_{J_1 J_2} \frac{\Phi_{-J J_1 J_2}\, \Phi_{J - J_1 - J_2}}{\omega_J\, \omega_{J_1}\, \omega_{J_2}} \sum_{\nu_1} h^{\nu_1}_{J_1}\, h^{\nu - \nu_1}_{J_2} \tag{27.70a}$$

The contribution from $\Phi_4$ can be calculated even simpler to give

$$\Sigma^{\mathrm{D}}_4 = -\frac{\hbar^2}{8N\,kT} \sum_{J_1 J_2} \frac{\Phi_{-J J - J_1 J_1}}{\omega_J\, \omega_{J_1}} \sum_{\nu_1} h^{\nu_1}_{J_2} \,. \tag{27.70b}$$

The sums in (27.70) can be determined readily by using the properties of $h^{\nu}_J$ and its Fourier-transformed periodic $g_J$.

$$\sum_{\nu_1}{}' h^{\nu_1}_{J_1} = g_{J_1}(0) = 2\bar{n}_{J_1} + 1 = \frac{2}{\hbar\,\omega_{J_1}}\, \varepsilon(J_1, T) = \frac{2}{\hbar\,\omega_1}\, \varepsilon_1$$

$$\sum_{\nu_1} h^{\nu_1}_{J_1}\, h^{\nu - \nu_1}_{J_2} = \frac{kT}{\hbar^2}\left(\frac{\varepsilon_1}{\omega_1} + \frac{\varepsilon_2}{\omega_2}\right)\left\{\frac{1}{i\,\omega_\nu + \omega_1 + \omega_2} - \frac{1}{i\,\omega_\nu - \omega_1 - \omega_2}\right\}$$
$$+ \frac{kT}{\hbar^2}\left(\frac{\varepsilon_1}{\omega_1} - \frac{\varepsilon_2}{\omega_2}\right)\left\{\frac{1}{i\,\omega_\nu - \omega_1 + \omega_2} - \frac{1}{i\,\omega_\nu + \omega_1 - \omega_2}\right\}$$

$$i\hbar\omega_\nu = 2\pi\, i\, kT\nu \Rightarrow -\omega \mp i\epsilon \,.$$

---

* For a discussion of the contribution of non-diagonal elements see *R. A. Cowley* [66 C 2]. The non-diagonal elements mean for example, that the cross-sections in Fig. 7.6 become more or less structured.

When inserting $F^{\nu}_{JJ'}$, which is the Fourier-transform of $f_{JJ}(\tau)$ into (27.59) we have to form limites with $\Sigma^{D}(-\omega \mp i\epsilon)$, $\epsilon \to 0^{+}$. We define

$$\lim_{\epsilon\to 0+} \Sigma^{D}(-\omega \pm i\epsilon) = -\frac{\hbar}{kT}\left\{\Delta_J(\omega) \pm i\,\Gamma_J(\omega)\right\} \tag{27.71}$$

and therefore from (27.70)

$$\begin{aligned}\Delta_J(\omega) &= \frac{1}{4N\omega_J}\sum_{J_1}\frac{\Phi_{-JJ-J_1J_1}}{\omega_{J_1}^2}\varepsilon_1 \\ &- \frac{1}{6N\omega_J}\sum_{J_1J_2}\frac{\Delta(-\boldsymbol{q}+\boldsymbol{q}_1+\boldsymbol{q}_2-2\pi B\boldsymbol{h})}{\omega_{J_1}\omega_{J_2}}|\Phi_{-JJ_1J_2}|^2\times \\ &\times P\left\{\left(\frac{\varepsilon_1}{\omega_1}+\frac{\varepsilon_2}{\omega_1}\right)\left(\frac{1}{\omega+\omega_1+\omega_2}-\frac{1}{\omega-\omega_1-\omega_2}\right)+\right. \\ &\left.+\left(\frac{\varepsilon_1}{\omega_1}-\frac{\varepsilon_2}{\omega_2}\right)\left(\frac{1}{\omega-\omega_1+\omega_2}-\frac{1}{\omega+\omega_1-\omega_2}\right)\right\}\end{aligned} \tag{27.72a}$$

$$\begin{aligned}\Gamma_J(\omega) &= \frac{\pi}{16N\omega_J}\sum_{J_1J_2}\frac{\Delta(-\boldsymbol{q}+\boldsymbol{q}_1+\boldsymbol{q}_2-2\pi B\boldsymbol{h})}{\omega_{J_1}\omega_{J_2}}|\Phi_{-JJ_1J_2}|^2\times \\ &\times\left\{\left(\frac{\varepsilon_1}{\omega_1}+\frac{\varepsilon_2}{\omega_2}\right)(\delta(\omega+\omega_1+\omega_2)-\delta(\omega-\omega_1-\omega_2))\right. \\ &\left.+\left(\frac{\varepsilon_1}{\omega_1}-\frac{\varepsilon_2}{\omega_2}\right)(\delta(\omega-\omega_1+\omega_2)-\delta(\omega+\omega_1-\omega_2))\right\}.\end{aligned} \tag{27.72b}$$

Using now (27.57, 53, 59, 64, 69, 71, 72) we obtain

$$\begin{aligned}\int_{-\infty}^{+\infty} dt\, e^{+i\omega t}\langle B(t)\,A\rangle &= -\frac{2\pi\hbar}{1-e^{-\hbar\omega/kT}}\cdot\frac{1}{2NM}\sum_J\frac{e_i e_j^*(J)}{\omega_J}e^{i\boldsymbol{q}(\boldsymbol{R}^m-\boldsymbol{R}^n)}\times \\ &\times\frac{1}{2\pi i}\lim_{\epsilon\to 0^+}\left\{\frac{1}{-\omega+\omega_J+i\epsilon-(kT/\hbar)\,\Sigma^{\nu}_{JJ}(-\omega+i\epsilon)}-\text{c.c.}+\right. \\ &\left.+\frac{1}{\omega+\omega_J-i\epsilon-(kT/\hbar)\,\Sigma^{\nu}_{JJ}(-\omega+i\epsilon)}-\text{c.c.}\right\}\end{aligned}$$

and finally

$$\begin{aligned}s_1^{mn}(\boldsymbol{K},\omega) &= \frac{\hbar e^{-2W}}{2\pi NM}\frac{1}{1-e^{-\hbar\omega/kT}}\sum_J e^{i(\boldsymbol{K}-\boldsymbol{q})(\boldsymbol{R}^m-\boldsymbol{R}^n)}\frac{(\boldsymbol{K}\,e(J))(\boldsymbol{K}\,e^*(J))}{\omega_J}\times \\ &\times\left\{\frac{\Gamma_J(\omega)}{[\omega+\omega_J+\Delta_J(\omega)]^2+\Gamma_J^2(\omega)}+\frac{\Gamma_J(\omega)}{[\omega-\omega_J-\Delta_J(\omega)]^2+\Gamma_J^2(\omega)}\right\}.\end{aligned} \tag{27.73}$$

This gives a Lorentzian form for the absorption or emission cross-section, whereas in harmonic theory these factors are $\delta$-functions. (27.73) clearly shows that $\Delta_J(\omega)$ is a frequency shift due to anharmonicity, whereas $\Gamma_J(\omega)$ is the width of the Lorentzian, or in other words, $\Gamma^{-1}$ is the lifetime of the phonons, which depends, in this approximation, only on $\Phi_3$. However, (27.73) is a strong Lorentzian only, if we replace the $\omega$ in $\Delta$ and $\Gamma$ as well as in $\exp(-\hbar\omega/kT)$ by $\omega_J$ or $-\omega_J$, resp. This can be done near the resonance, because we have assumed always that $\Delta$ and $\Gamma$ are small. If $\omega \approx \omega_J$ we have $\{1-\exp(-\hbar\omega/kT)\}^{-1} \approx \bar{n}_J + 1$, if $\omega \approx -\omega_J$ this is equal to $\approx \bar{n}_J$, which shows the similarity to (27.33) even more.

We have mentioned that the second term in (27.54) leads to an unsymmetry of the Lorentzian. This term can be determined along the same lines as indicated for the first term. The result can be expressed by replacing in the nominator of the Lorentzian

$$\Gamma_J(\omega) \Rightarrow \Gamma_J(\omega) + E_J(\omega) \cdot \frac{\omega^2 - [\omega_J + \Delta_J(\omega)]^2}{2\,\omega_J} \tag{27.74}$$

where

$$\begin{aligned} E_J(\omega) = {} & \frac{i\pi}{4N\sqrt{M}} \cdot \frac{1}{(\boldsymbol{K}\boldsymbol{e}_J)} \sum_{J_1 J_2} \frac{(\boldsymbol{K}\boldsymbol{e}_{J_1})(\boldsymbol{K}\boldsymbol{e}_{J_2})}{\omega_{J_1}\omega_{J_2}} \Delta(\boldsymbol{q} - \boldsymbol{q}_1 - \boldsymbol{q}_2 - 2\pi B\boldsymbol{h}) \times \\ & \times \Phi_{J-J_1-J_2} \cdot \left\{ \left[ \frac{\varepsilon_{J_1}}{\omega_{J_1}} + \frac{\varepsilon_{J_2}}{\omega_{J_2}} \right] [\delta(\omega + \omega_{J_1} + \omega_{J_2}) - \delta(\omega - \omega_{J_1} - \omega_{J_2})] + \right. \\ & \left. + \left[ \frac{\varepsilon_{J_1}}{\omega_{J_1}} - \frac{\varepsilon_{J_2}}{\omega_{J_2}} \right] [\delta(\omega - \omega_{J_1} + \omega_{J_2}) - \delta(\omega + \omega_{J_1} - \omega_{J_2})] \right\}. \end{aligned} \tag{27.75}$$

$E$ is real, because $\Phi_{J-J_1-J_2}$ is purely imaginary.

If we put

$$\Omega_J = \omega_J + \Delta_J(\omega_J)$$

we have approximately the following line shapes:

absorption of a phonon: $\dfrac{\Gamma - E(\omega + \Omega)}{(\omega + \Omega)^2 + \Gamma^2}$ ; emission of a phonon: $\dfrac{\Gamma + E(\omega - \Omega)}{(\omega - \Omega)^2 + \Gamma^2}$

which are shown in Fig. 7.6. From an experimental measurement we thus can determine frequency shift, lifetime and unsymmetry factor

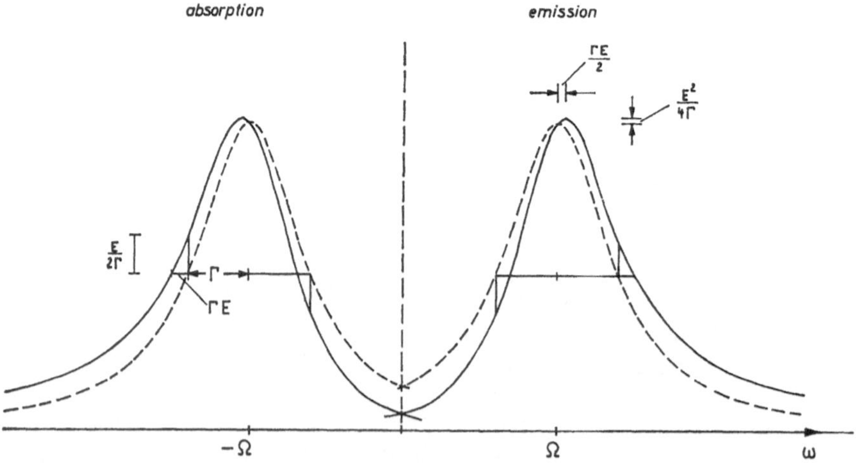


Fig. 7.6. One phonon absorption and emission peak, schematically to show the main influences of anharmonicity

and therefore the anharmonic parameters. As the line-unsymmetry contains a third power of the momentum transfer, it is very small and as far as known it has not been determined experimentally up till now.

We will add some remarks concerning the determination of harmonic values and the elastic limit.

At high temperatures ($T \gtrsim \Theta_D$) we obtain from (27.72)

$$\Delta_J(\omega) = \frac{kT}{4N\omega_J} \sum_{J_1} \frac{\Phi_{J-JJ_1-J_1}}{\omega_{J_1}^2} - \\ - \frac{kT}{8N\omega_J} \sum_{J_1 J_2} \frac{\Delta(-\boldsymbol{q} + \boldsymbol{q}_1 + \boldsymbol{q}_2 - 2\pi B \boldsymbol{h})}{\omega_{J1}^2 \omega_{J2}^2} |\Phi_{-JJ_1J_2}|^2 \times \\ \times P\left\{\frac{(\omega_1+\omega_2)^2}{(\omega_1+\omega_2)^2-\omega^2} + \frac{(\omega_1-\omega_2)^2}{(\omega_1-\omega_2)^2-\omega^2}\right\}. \tag{27.76}$$

The resonance frequency* is

$$\Omega_J \approx \omega_J + \Delta_J \quad \text{or} \quad \Omega_J^2 \approx \omega_J^2 + 2\omega_J \Delta_J$$

$\omega_J$ are the *quasiharmonic* frequencies, which are determined in chapter V. They consist of a *true harmonic* part and a part, which is also proportional to $T$ at high temperatures. Thus we have at high temperatures

$$\Omega_J = \omega_{J,\text{har}} + T \cdot \chi_J \quad \text{or} \quad \Omega_J^2 = \omega_{J,\text{har}}^2 + 2T\chi_J \cdot \omega_{J,\text{har}}.$$

Measuring the lattice frequencies by neutron diffraction at high temperatures ($T \gtrsim \Theta_D$) as a function of $T$ at constant $J$ and extrapolating this dependence to $T = 0$, we obtain the true harmonic frequencies and the true dispersion curves. From these dispersion curves we can determine the harmonic coupling parameters (force-constants), essentially by taking the Fourier-transform. *This is the only way of obtaining the harmonic coupling-parameters, and of making statements on the harmonic forces and their ranges.* In this approximation ($\Phi_3^2$, $\Phi_4$) it should be irrelevant, whether we extrapolate $\Omega_J$ or $\Omega_J^2$, though $\Omega_J^2$ seems to be the primary quantity. However, if there are very large differences in the behaviour of $\Omega_J$ and $\Omega_J^2$, this is a hint for other influences, e.g. higher anharmonic contributions. But then it is even worse to make statements on harmonic quantities and ranges of forces etc.

This holds as long as our assumption of mechanical potential forces applies. In semi-conductors with small band gaps, comparable with $kT$, the electronic state changes with temperature and so does the potential energy. In such cases the definition of temperature-independent coupling parameters is questionable; one has to reconsider the whole procedure for such effects.

A second remark concerns the elastic limit. In harmonic theory the long wave limit of $\omega^2(\boldsymbol{q}\sigma)$ with $\boldsymbol{q} \to 0$ essentially determines the elastic constants. The question arises whether this is true also in anharmonic theory, or more precisely: In anharmonic theory there are two kinds of elastic constants: isothermal and adiabatic ones. If at all, which of these two corresponds to the long wave limit of $\Omega^2(\boldsymbol{q}\sigma)$? In the anharmonic theory of X-ray diffraction (see below) it has been shown [61 H 1, 2],

* Sometimes the $\Omega_J$ are called quasiharmonic frequencies [63 C 2]. This is misleading, because they never can be determined by a harmonic procedure, whereas the $\omega_J$ can.

that the properly defined anharmonic frequencies correspond to the isothermal elastic constants, in the limit $\boldsymbol{q} \to 0$, at least in simple symmetry-directions for the $\boldsymbol{q}$-vector. Now the frequencies (27.72) are not identical with the X-ray anharmonic frequencies in general. But they seem to be identical with the X-ray-frequencies if $\boldsymbol{q} \to 0$, and then they should measure the isothermal elastic constants [61 H 1, 63 H 2].

The last task is to determine the anharmonic Debye-Waller-factor (27.52)

$$-W_A = \frac{1}{2}\langle A^2\rangle + \frac{1}{24}\{\langle A^4\rangle - 3\langle A^2\rangle^2\}, \tag{27.77}$$

where we have assumed already lattices where every ion is a center of symmetry. The second term is proportional to the fourth power of the momentum transfer $\boldsymbol{K}$, and we drop it here [63 M 5]. Then $\langle A^2\rangle$ has to be evaluated with the anharmonic distribution:

$$\begin{aligned} W_A &= \frac{1}{2}\sum_{ij} K_i K_j \left\langle U_i^m(0)\, U_j^m(0)\right\rangle = \\ &= \frac{\hbar}{4NM}\sum_{JJ'} \frac{(\boldsymbol{K}\boldsymbol{e}_J)(\boldsymbol{K}\boldsymbol{e}^*_{J'})}{\sqrt{\omega_J\,\omega_{J'}}}\, e^{i(\boldsymbol{q}-\boldsymbol{q}')\boldsymbol{R}^m} \cdot f_{JJ'}(0) \end{aligned} \tag{27.78}$$

where we have used (27.64). The calculation of $f_{JJ'}$ has been done in (27.66), where we have to add the quasiharmonic contribution and that from $\Phi_4$, which can be evaluated similarly. The sum $\sum_{\nu,\nu_1} h^{\nu}_J h^{\nu}_{J'} h^{\nu_1}_{J_1} h^{\nu-\nu_1}_{J_2}$ can be determined by a straightforward, though lengthy procedure. We only give the final result for Bravais-lattices, the generalization to non-primitive lattices is obvious ($J = \boldsymbol{q}, \sigma$!):

$$\begin{aligned} W &= \frac{1}{2NM}\sum_{JJ'} \frac{(\boldsymbol{K}\boldsymbol{e}_J)(\boldsymbol{K}\boldsymbol{e}^*_{J'})}{\omega_J\,\omega_{J'}}\,\delta_{\boldsymbol{q}\boldsymbol{q}'} \times \\ &\times \Bigg\{\varepsilon_J\,\delta_{\sigma\sigma'} - \frac{\hbar}{4N}\left[\frac{\bar n_J + \bar n_{J'}}{\omega_J + \omega_{J'}} - \frac{\bar n_J - \bar n_{J'}}{\omega_J - \omega_{J'}}\right]\sum_{J_1}\frac{\Phi_{-JJ'J_1-J_1}}{\omega^2_{J_1}}\,\varepsilon_{J_1} + \\ &+ \frac{\hbar^2}{32N}\sum_{J_1J_2}\frac{\Delta(\boldsymbol{q}-\boldsymbol{q}_1-\boldsymbol{q}_2-2\pi B\boldsymbol{h})}{\omega_{J_1}\,\omega_{J_2}}\,\Phi_{-JJ_1J_2}\,\Phi_{J'-J_1-J_2}\cdot P_{JJ'J_1J_2}\Bigg\} \end{aligned} \tag{27.79}$$

with

$$\begin{aligned} P_{JJ'J_1J_2} &= \frac{2\bar n_{J_1}\bar n_{J_2} + \bar n_{J_1} + \bar n_{J_2} + 1}{\bar n_{J_1}\bar n_{J_2}(\bar n_{J_1}+1)(\bar n_{J_2}+1)}\,Q_{JJ_1J_2}\,Q_{J'J_1J_2} + \\ &+ \frac{2\bar n_{J_1}\bar n_{J_2} + \bar n_{J_1} + \bar n_{J_2}}{\bar n_{J_1}\bar n_{J_2}(\bar n_{J_1}+1)(\bar n_{J_2}+1)}\,R_{JJ_1J_2}\,R_{J'J_1J_2} \end{aligned}$$

$$Q_{JJ_1J_2} = \frac{\bar n_J(\bar n_{J_1} + \bar n_{J_2} + 1) + \bar n_{J_1}\bar n_{J_2} + \bar n_{J_1} + \bar n_{J_2}}{\omega_J + \omega_{J_1} + \omega_{J_2}} + \frac{\bar n_{J_1}\bar n_{J_2} - \bar n_J(\bar n_{J_1} + \bar n_{J_2} + 1)}{\omega_J - \omega_{J_1} - \omega_{J_2}}$$

$$R_{JJ_1J_2} = \frac{\bar n_{J_1}(\bar n_J + \bar n_{J_2} + 1) - \bar n_J\bar n_{J_2}}{\omega_J - \omega_{J_1} + \omega_{J_2}} + \frac{\bar n_{J_2}(\bar n_J + \bar n_{J_1} + 1) - \bar n_J\bar n_{J_1}}{\omega_J + \omega_{J_1} - \omega_{J_2}}$$

For high temperatures $T \gtrsim \Theta_D$ it is

$$\left[\frac{\bar{n}_J + \bar{n}_{J'}}{\omega_J + \omega_{J'}} - \frac{\bar{n}_J - \bar{n}_{J'}}{\omega_J - \omega_{J'}}\right] \varepsilon_{J_1} \Rightarrow \frac{2(kT)^2}{\hbar\, \omega_J\, \omega_{J'}}$$

$$P_{JJ'J_1J_2} \Rightarrow \frac{16(kT)^2}{\hbar^2\, \omega_J \omega_{J'} \omega_{J_1} \omega_{J_2}}\,. \tag{27.80}$$

Then (27.79) agrees with the high temperature expressions given by *Maradudin* and *Flinn* [63 M 5]. At lower temperatures the different terms are rather complicated. On similar lines the higher order terms, being proportional to the fourth power of the momentum transfer can be given, but they have a reasonable shape only at $T \gtrsim \Theta_D$, then being proportional to $T^3$ [63 M 5].

## 28. Interaction of Phonons with Emitted or Absorbed γ-Rays (Mössbauer-Effect)

We now consider an ion bound in a crystal, the nucleus of which suffers a γ-transition from one nuclear level to another one. However, the transition discussed in the following, needs not be necessarily a nuclear transition, but it might be also an electronic transition in one of the interior shells of the electrons, provided the coupling of the ion to its neighbors is not changed essentially by the transition. An example is the transition between $f$-levels in rare-earth-ions. We will give the details only for the emission process; the changes which have to be made for the absorption process will be obvious.

As in section 27, the crystal state before emission may be $|\alpha\rangle$ with energy $E_\alpha$, the final state $|\varepsilon\rangle$ with $E_\varepsilon$. The nucleus (or ion) is in the state $|a\rangle$, $E_a$ before emission, in $|e\rangle$, $E_e$ afterwards. There is no photon before: $|0\rangle$, one photon with momentum $\hbar\boldsymbol{K}$ after emission $|\boldsymbol{K}\rangle$. The excited state of the nucleus has a finite life-time, or the transition has a half-width $\hbar\gamma$. The transition probability from the initial to the final state, taking into account the finite life-time, then is

$$w_{\alpha a 0 \to \varepsilon e \boldsymbol{K}}(t \to \infty) = \left|\frac{\langle \varepsilon\, e\, \boldsymbol{K}\, |W|\, \alpha\, a\, 0\rangle}{E_{\varepsilon e \boldsymbol{K}} - E_{\alpha a 0} + i\,\hbar\,\gamma/2}\right|^2 \tag{28.1}$$

where $W$ describes the interaction between the different systems: nucleus (ion), photons, crystal lattice. This interaction can be shown to be*

$$W = -\frac{e}{m\,c} \sum_\nu (\boldsymbol{A}\boldsymbol{p}_\nu)\,. \tag{28.2}$$

$\boldsymbol{A}$ is the vector potential of the electromagnetic field, $\boldsymbol{p}_\nu$ the momentum of the particle $\nu$ which makes the transition. $\boldsymbol{A}$ can be expressed by the creators $a^+_{\boldsymbol{K}\lambda}$ and annihilators $a_{\boldsymbol{K}\lambda}$ of photons, mainly by representing $\boldsymbol{A}$ as a Fourier-series:

$$\boldsymbol{A} = \frac{c}{\sqrt{V}} \sum_{\boldsymbol{K}\lambda} \sqrt{\frac{2\pi\hbar}{\omega_{\boldsymbol{K}}}}\; \boldsymbol{s}_{\boldsymbol{K}\lambda} \left[a_{\boldsymbol{K}\lambda}\, e^{i\boldsymbol{K}(\boldsymbol{r}+\boldsymbol{r}^m)} + a^+_{\boldsymbol{K}\lambda}\, e^{-i\boldsymbol{K}(\boldsymbol{r}+\boldsymbol{r}^m)}\right] \tag{28.2a}$$

* See any book on quantum theory.

$\omega_{\boldsymbol{K}}$ is the frequency of the photon with wave vector $\boldsymbol{K}$ and polarization $\lambda$, $\boldsymbol{s}_{\boldsymbol{K}\lambda}$ the polarization vector. $\boldsymbol{r}^m$ is the center of mass coordinate of the emitting system (nucleus) in the lattice, $\boldsymbol{r}$ the relative interior coordinate. $V$ is a suitable normalization volume. Therefore

$$W = -\frac{e}{m}\sqrt{\frac{2\pi\hbar}{V}} \sum_{\boldsymbol{K}\lambda,\nu} \frac{(\boldsymbol{p}_\nu\, \boldsymbol{s}_{\boldsymbol{K}\lambda})}{\sqrt{\omega_{\boldsymbol{K}}}} \left[a_{\boldsymbol{K}\lambda}\, e^{i\boldsymbol{K}(\boldsymbol{r}_\nu+\boldsymbol{r}^m)} + \overset{+}{a}_{\boldsymbol{K}\lambda}\, e^{-i\boldsymbol{K}(\boldsymbol{r}_\nu+\boldsymbol{r}^m)}\right]. \tag{28.3}$$

The only quantities, where the lattice enters, are the exponentials in (28.3). We consider the emission process, it is the *creation of one photon*: all terms do not contribute except just one $\overset{+}{a}_{\boldsymbol{K}\lambda}$. We obtain

$$|\langle \varepsilon e\boldsymbol{K}\,|W|\,\alpha a 0\rangle|^2 = \frac{1}{4}\hbar^2\gamma^2\sigma_0\,|\langle\varepsilon\,|e^{-i\boldsymbol{K}\boldsymbol{r}^m}|\,\alpha\rangle|^2$$

with (28.4)

$$\frac{1}{4}\hbar^2\gamma^2\sigma_0 = |\langle e\boldsymbol{K}| - \frac{e}{m}\sqrt{\frac{2\pi\hbar}{V\omega_{\boldsymbol{K}}}}\sum_\nu(\boldsymbol{p}_\nu\,\boldsymbol{s}_{\boldsymbol{K}\lambda})\,\overset{+}{a}_{\boldsymbol{K}\lambda}\,e^{-i\boldsymbol{K}\boldsymbol{r}_\nu}\,|a0\rangle|^2 .$$

Because we are interested only in the interaction between photon and phonon, we do not discuss $\sigma_0$ in the following. The lattice is in thermal mixture, before the emission takes place, and the final state is not of interest for us. Using (27.8) we thus have finally for the emission cross-section

$$\sigma_e(\boldsymbol{K},\omega) = \frac{1}{4}\gamma^2\sigma_0\sum_\alpha p_\alpha\sum_\varepsilon \frac{\langle\alpha\,|e^{+i\boldsymbol{K}\boldsymbol{r}^m}|\,\varepsilon\rangle\,\langle\varepsilon\,|e^{-i\boldsymbol{K}\boldsymbol{r}^m}|\,\alpha\rangle}{\left(\omega_0-\omega-\frac{E_\varepsilon-E_\alpha}{\hbar}\right)^2+\gamma^2/4} \tag{28.5}$$

with the abbreviations

$$\hbar\omega_0 = E_a - E_e;\quad \hbar\omega = \hbar\omega_{\boldsymbol{K}} = E_{\boldsymbol{K}}\,.$$

$\hbar\omega_0$ is the energy difference of the nuclear levels, $\omega$ the frequency of the emitted photon. (28.5) is equivalent to (27.10) for the neutron scattering. The difference in shape is connected mainly with the finite life-time $\gamma$ in (28.5). We make a similar transformation as in (27.11), using

$$\frac{\gamma}{\omega^2+\gamma^2/4} = \int_{-\infty}^{+\infty} e^{-\gamma|t|/2+i\omega t}\,dt \tag{28.6}$$

to obtain

$$\sigma_e(\boldsymbol{K},\omega) = \frac{1}{4}\gamma\,\sigma_0\int_{-\infty}^{+\infty} dt\, e^{-\gamma|t|/2+i(\omega-\omega_0)t}\cdot\chi^{mm}(-\boldsymbol{K},t) \tag{28.7}$$

where $\chi^{mm}(\boldsymbol{K},t)$ is identical with the definition in (27.12):

$$\chi^{mm}(-\boldsymbol{K},t) = \langle e^{i\boldsymbol{K}\boldsymbol{r}^m(0)}\cdot e^{-i\boldsymbol{K}\boldsymbol{r}^m(t)}\rangle_T\,. \tag{28.7a}$$

Thus the problem is completely reduced to that of neutron-scattering; it is even simpler, for there occurs only the autocorrelation for the emitting nucleus (ion). This is a special type of the functions of Sect. 27, and we take the results from there. Especially we have with (27.35)

$$\sigma_e(\boldsymbol{K}) = \int_{-\infty}^{+\infty} d\omega\, \sigma_e(\boldsymbol{K}, \omega) = \frac{\pi}{2}\gamma\sigma_0\, \chi^{mm}(-\boldsymbol{K}, 0) = \frac{\pi}{2}\gamma\sigma_0 \quad (28.8a)$$

for the total emission cross-section and with (27.36)

$$\int_{-\infty}^{+\infty} d\omega \cdot \hbar\omega \cdot \sigma_e(\boldsymbol{K}, \omega) = \frac{\pi}{2}\gamma\sigma_0 \left(\hbar\omega_0 - \frac{\hbar^2 K^2}{2M_m}\right)\chi^{mm}(-\boldsymbol{K}, 0) \quad (28.8b)$$

or

$$\overline{\hbar\omega} = \int_{-\infty}^{+\infty} d\omega \cdot \hbar\omega \cdot \sigma_e(\boldsymbol{K}, \omega) \Bigg/ \int_{-\infty}^{+\infty} d\omega\, \sigma_e(\boldsymbol{K}, \omega) = \hbar\omega_0 - \frac{\hbar^2 K^2}{2M_m} \quad (28.8c)$$

if the emitting nucleus is in a center of inversion. The mean energy transfer to the photon is not the energy difference of the nuclear levels, but rather that diminished by the repulsion energy given to the ion of mass $M_m$. In the absorption process the photon energy is larger by the amount $\hbar^2 K^2/2M_m$. We shall see however, that, inspite of this repulsion effect, the emitted $\gamma$-line is not broadened by the Doppler effect, if the conditions are choosen appropriately.

*Ideal harmonic crystal*

Using (27.25, 26) we obtain ($m \Rightarrow {m \atop \mu}$)

$$\chi^{{m \atop \mu}{m \atop \mu}}(-\boldsymbol{K}, t) = \chi^{{m \atop \mu}{m \atop \mu}}(\boldsymbol{K}, t) = e^{-2W_\mu} \times \quad (28.9)$$
$$\times \exp\left\{\frac{\hbar}{2sNM_\mu}\sum_{\boldsymbol{q}\sigma}\frac{|\boldsymbol{K}\boldsymbol{e}^\mu|^2}{\omega(\boldsymbol{q}\sigma)}\left[\bar{n}(\boldsymbol{q}\sigma)\, e^{-i\omega(\boldsymbol{q}\sigma)t} + (\bar{n}(\boldsymbol{q}\sigma)+1)\, e^{i\omega(\boldsymbol{q}\sigma)t}\right]\right\}.$$

We now make the same assumptions, or approximations, resp. as we have done for neutron scattering: we expand the exp{ . . . } in a series, insert (28.9) into (28.7) and discuss the different terms. We give the justification and the limits of this approximation below. Then we have

$$\sigma_e(\boldsymbol{K}, \omega) = \frac{1}{4}\gamma^2\sigma_0\, e^{-2W_\mu}\left\{\frac{1}{(\omega-\omega_0)^2+\gamma^2/4} + \frac{\hbar}{2sNM_\mu}\sum_{\boldsymbol{q}\sigma}\frac{|\boldsymbol{K}\boldsymbol{e}^\mu|^2}{\omega(\boldsymbol{q}\sigma)}\times\right.$$
$$\left.\times\left[\frac{\bar{n}(\boldsymbol{q}\sigma)}{[\omega-\omega_0-\omega(\boldsymbol{q}\sigma)]^2+\gamma^2/4} + \frac{\bar{n}(\boldsymbol{q}\sigma)+1}{[\omega-\omega_0+\omega(\boldsymbol{q}\sigma)]^2+\gamma^2/4}\right]+\cdots\right\}. \quad (28.10)$$

The first term represents a contribution, which involves no phonons (zero-phonon cross-section, or zero-phonon-line). It is the natural line shape without any broadening or shift by Doppler-effect or collision of atoms or any other source for a broadening. The second term involves interaction with one phonon, giving an addition of natural lines at frequencies $\omega(\boldsymbol{q}, \sigma)$ which leads to a continuously distributed emission spectrum. The next term represents essentially two-phonon processes, and so on. If $|\omega - \omega_0 \pm \omega(\boldsymbol{q}\sigma)| \gg \gamma/2$, we could neglect $\gamma^2/4$ in the denominator of the different Lorentzians in (28.10). Now the region of lattice frequencies is from zero to approximately the Debye-frequency $\omega_D$, i.e. up to $\approx 10^{13}\,\text{sec}^{-1}$, whereas the half-widths of nuclear levels are $\gamma \approx 10^7$ to $10^9\,\text{sec}^{-1}$, thus $\gamma \ll \omega_D$. Similar considerations hold for many of the transitions in the inner electron shells. Then we can neglect $\gamma$ in the denominator of the one-, two-, ... phonon processes and obtain

$$\begin{aligned}
\sigma_e(\boldsymbol{K},\omega) = \frac{1}{4}\gamma\,\sigma_0\,e^{-2W_\mu}\Bigg\{&\frac{\gamma}{(\omega-\omega_0)^2+\gamma^2/4} + \\
&+ \frac{\pi\hbar}{sNM_\mu}\sum_J \frac{|\boldsymbol{K}\boldsymbol{e}_J^\mu|^2}{\omega_J} \times \\
&\times [\bar{n}_J\,\delta(\omega-\omega_0-\omega_J) + (\bar{n}_J+1)\,\delta(\omega-\omega_0+\omega_J)] + \\
&+ \pi\left(\frac{\hbar}{2sNM_\mu}\right)^2 \sum_{JJ'} \frac{|\boldsymbol{K}\boldsymbol{e}_J^\mu|^2\,|\boldsymbol{K}\boldsymbol{e}_{J'}^\mu|^2}{\omega_J\,\omega_{J'}}\,[\bar{n}_J\bar{n}_{J'}\,\delta(\omega-\omega_0-\omega_J-\omega_{J'}) + \\
&+ \bar{n}_J(\bar{n}_{J'}+1)\,\delta(\omega-\omega_0-\omega_J+\omega_{J'}) + \\
&+ (\bar{n}_J+1)\,\bar{n}_{J'}\,\delta(\omega-\omega_0+\omega_J-\omega_{J'}) + \\
&+ (\bar{n}_J+1)(\bar{n}_{J'}+1)\,\delta(\omega-\omega_0+\omega_J+\omega_{J'})] + \cdots\Bigg\}. \qquad (28.11)
\end{aligned}$$

To give a little more quantitative insight, we restrict first to $T = 0$, i.e. $\bar{n}_J = 0$ and to cubic Bravais-crystals. Then with the arguments leading to (27.30) we have

$$\begin{aligned}
\sigma_e(\boldsymbol{K},\omega) = \frac{1}{4}\gamma\,\sigma_0\,e^{-2W}\Bigg\{&\frac{\gamma}{(\omega-\omega_0)^2+\gamma^2/4} + \frac{2\pi}{\hbar}P\,\frac{z(\omega_0-\omega)}{\omega_0-\omega} + \\
&+ \frac{\pi}{\hbar^2}P^2\iint d\omega'\,d\omega''\,\frac{z(\omega')\,z(\omega'')}{\omega'\,\omega''}\,\delta(\omega-\omega_0+\omega'+\omega'') + \cdots\Bigg\}. \qquad (28.12)
\end{aligned}$$

$P = \dfrac{\hbar^2 K^2}{2M}$ is the repulsive energy given by the emitting phonon to an ion of mass $M$. The result for the zero-, one- and two-phonon-emission cross-section is shown in Fig. 7.7. Of course, this behavior is not quite exact in the region $\omega_0 - \omega \approx \gamma$, i.e. the region where zero- and multi-phonon cross-section overlaps; this follows from the arguments given above. The one-phonon-cross-section represents essentially the spectral frequency distribution of the lattice, divided by the frequencies. From the figure it can be seen that there are two conditions which have to be satisfied in order this approximation to be a good one: i) the natural line width $\gamma$ has to be small compared to the Debye-frequency $\omega_D$:

$$\gamma \ll \omega_D\,; \quad 1/\gamma \gg 1/\omega_D\,. \qquad (28.13)$$

This can be read in the following way: $1/\omega_D$ is a measure for the periods of lattice vibrations. During the life-time of the nuclear state, i.e. during the time of emitting electromagnetic radiation the wave-front must have travelled a distance $r \approx c/\gamma \gg c/\omega_D \approx a$, ($a$: lattice constant, $c$: sound velocity). This means, that the repulsion energy is transferred to a large region, or in other words, the emitting ion behaves as tightly bound to its position. This is a qualitative argument only, as can be seen from the fact, that there is a second condition: ii) There must be enough intensity scattered into the zero-phonon-line to make it visible. A

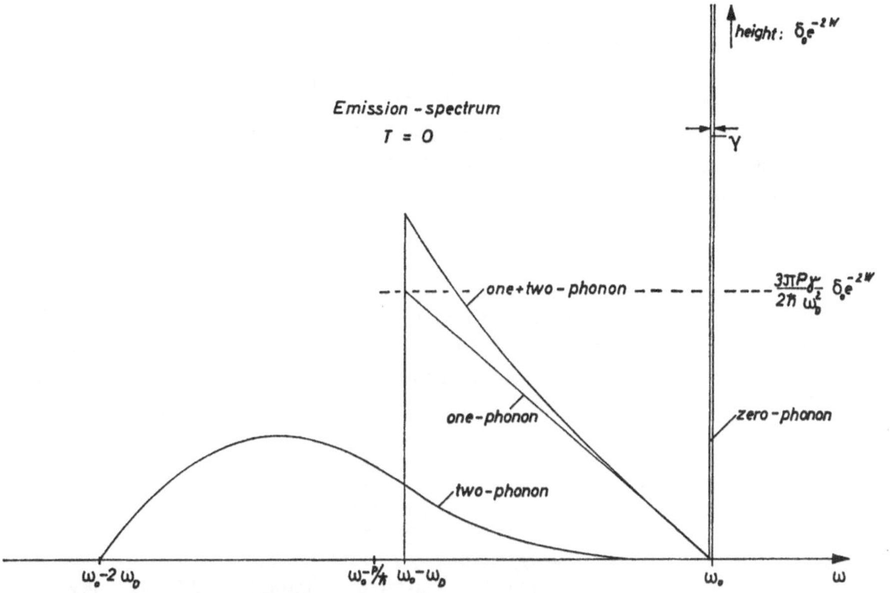


Fig. 7.7. Zero-, one-, and two-phonon-emission spectrum in the Mössbauer-resonance, calculated with a Debye-spectrum for the lattice frequencies

comparison of the intensities in the zero- ($I_0$), one- ($I_1$) and two- ($I_2$) phonon-cross-section gives

$$\frac{I_1}{I_0} = \frac{3}{2} \frac{P}{\hbar \omega_D} ; \quad \frac{I_2}{I_0} = \frac{7}{4} \left( \frac{P}{\hbar \omega_D} \right)^2 \tag{28.14}$$

thus $P/\hbar\omega_D$ may not be too large. It has not to be smaller than one in any case, for if the line-width $\gamma$ is very small, the height of the zero-line is $(4/\gamma) \cdot e^{-2W}$, whereas the maximum of the one-phonon-cross section is $(6\pi P/\hbar\omega_D^2) \cdot e^{-2W}$. The zero-phonon line can be seen also, if $P/\hbar\omega_D$ is larger than one, if $\gamma$ is sufficiently small. This holds at least for $T = 0$.

At higher temperatures we have to take into account the mean occupation numbers $\bar{n}_J$. This means, that there is not only intensity for $\omega < \omega_0$, but also for $\omega > \omega_0$; this intensity is less than that for $\omega < \omega_0$, even at $T \approx \Theta$, because for most of the frequencies $\bar{n}_J$ is not large compared to one at this temperature. The fraction of the recoilless emitted intensity (zero phonon line) can be given simply for all the temperatures. The total emitted intensity is given by (28.8a), the zero-phonon-intensity by integration of the first term of (28.11) over $\omega$. Then we have

$$\frac{\sigma_{e0}(\boldsymbol{K})}{\sigma_e(\boldsymbol{K})} = e^{-2W_\mu} . \tag{28.15}$$

The fraction of recoilless emitted $\gamma$-intensity is just given by the Debye-Waller-factor. Thus one would suppose that the zero-phonon-line can be observed, if most of the scattered intensity is in the zero-phonon-line, or roughly speaking, $e^{-2W_\mu} \gtrsim 0.5$. However, it can be observed even, if $e^{-2W_\mu}$ is much smaller, provided $\gamma$ is small compared to $\hbar\omega_D$ and $P$. Then the zero-phonon-cross-section is very small, and has a large peak, whereas the rest of the intensity is spread over a large region. At high temperatures, the width of the inelastic scattered intensity can be estimated to about $2(PkT/\hbar^2)^{1/2}$ at high temperatures, thus the maximum height of this intensity curve is about $\sqrt{\pi}\,\gamma\,\sigma_0\hbar/4(PkT)^{1/2}$ and the proportion of the heights of elastic scattered intensity to that of inelastic scattered intensity is

$$\frac{\sigma_{e0\max}}{\sigma_{e\max}} \approx \frac{\sqrt{PkT}}{\hbar\gamma}\, e^{-2W}, \quad T \gg \Theta_D. \tag{28.16}$$

Thus the zero-phonon-line exceeds the inelastic background even if $e^{-2W}$ is relatively small, provided $(PkT)^{1/2}/\hbar\gamma$ is large enough. In Table V some characteristic data are given for some representative nuclei. At $T = 0$ the zero-phonon-line can be observed in $Ir^{191}$ as well as in $Fe^{57}$, though it is much more pronounced in $Fe^{57}$. At $T \approx \Theta_D$,

Table V. *Some data for nuclei, which are used in Mössbauer-resonance experiments*

| | $Ir^{191}$ | $Au^{197}$ | $Ni^{61}$ | $Zn^{67}$ | $Fe^{57}$ | |
|---|---|---|---|---|---|---|
| $\hbar\omega_0$ | $1.29 \cdot 10^5$ | $7{,}7 \cdot 10^4$ | $7.1 \cdot 10^4$ | $9.3 \cdot 10^4$ | $1.44 \cdot 10^4$ | eV |
| $P$ | 0.0465 | 0.0163 | 0.0443 | 0.0691 | 0.00172 | eV |
| $\gamma$ | $5.56 \cdot 10^9$ | $3.65 \cdot 10^8$ | $1.33 \cdot 10^8$ | $7.37 \cdot 10^4$ | $1.395 \cdot 10^7$ | $sec^{-1}$ |
| $\omega_D$ | $3.75 \cdot 10^{13}$ | $2.12 \cdot 10^{13}$ | $6.2 \cdot 10^{13}$ | $4.16 \cdot 10^{13}$ | $6.15 \cdot 10^{13}$ | $sec^{-1}$ |
| $P/\hbar\omega_D$ | 1.89 | 1.16 | 1.08 | 2.52 | 0.0425 | |
| $2W(T=0)$ | 2.84 | 1.74 | 1.62 | 3.78 | 0.0638 | |
| $2W(T=\Theta_D)$ | 11.34 | 6.96 | 6.48 | 15.12 | 0.255 | |
| $e^{-2W}(T=0)$ | 0.058 | 0.173 | 0.198 | $2.28 \cdot 10^{-2}$ | 0.938 | |
| $e^{-2W}(T=\Theta_D)$ | $1.23 \cdot 10^{-5}$ | $9.5 \cdot 10^{-4}$ | $1.27 \cdot 10^{-7}$ | $2.71 \cdot 10^{-7}$ | 0.775 | |
| $\frac{\sigma_{0\max}}{\sigma_{\max}}(T=0)$ | $7.5 \cdot 10^2$ | $1.06 \cdot 10^4$ | $9.13 \cdot 10^4$ | $4.76 \cdot 10^7$ | $2.2 \cdot 10^7$ | |
| $\frac{\sigma_{0\max}}{\sigma_{\max}}(T=\Theta_D)$ | 0.201 | 59.4 | $6 \cdot 10^2$ | $2.42 \cdot 10^2$ | $1.2 \cdot 10^6$ | |
| $v(\gamma)$ | 0.85 | 0.094 | 0.037 | $1.56 \cdot 10^{-5}$ | 0.019 | cm/sec |
| $v(\omega_D)$ | 57 | 55 | 172 | 88 | 900 | m/sec |

that means essentially room-temperature, the zero phonon-line in $Ir^{191}$ has gone, whereas in $Fe^{57}$ it can be seen nearly as good as for $T \approx 0$.

To make the difference of the behavior of $\gamma$-emitting nuclei in gases somewhat more significant, we consider the high-temperature limit in somewhat detail. We replace $\bar{n}_J + 1/2$ by $kT/\hbar\omega_J$ and obtain from (28.9) for cubic Bravais-lattices having a Debye-spectrum

$$\chi^{mm}(\boldsymbol{K}, t) = e^{-2W} \cdot \exp\left\{\frac{PkT}{\hbar^2\omega_D^2} \cdot 6\,\frac{\sin\omega_D t}{\omega_D t} + i\,\frac{3P}{\hbar\omega_D} \cdot \frac{\sin\omega_D t - \omega_D t\cos\omega_D t}{(\omega_D t)^2}\right\}. \tag{28.17}$$

This has to be inserted into (28.7). The main contribution to the integral is from $|t| \lesssim 2/\gamma$. Thus if $\gamma \gtrsim \omega_D$, we can use the expansion of (28.17) with respect to $\omega_D t$ in the integration of (28.7). After some manipulations, taking into account (27.31) we have

$$\gamma \gtrsim \omega_D : \sigma_e(\boldsymbol{K}, \omega) = \frac{1}{2} \gamma \sigma_0 / y \ \operatorname{Re} e^{-\frac{x^2}{4y^2}} \int_0^\infty e^{-\left(z - i\frac{x}{2y}\right)^2} dz$$

$$x = \omega - \omega_0 + P/\hbar + i\gamma/2; \quad y^2 = P k T/\hbar^2 . \tag{28.18}$$

This is essentially the distribution of the emitted $\gamma$-energy if the emitting nucleus is free. From $\gamma \gtrsim \omega_D$ it can be seen that the emission in this case is so quick, that the repulsion energy is transferred mainly to the emitting ion itself. The intensity distribution is a slightly modified Gaussian, which is centered at the repulsion energy $P/\hbar$ (Doppler-shift) and has a width $2y = 2(P k T)^{1/2}/\hbar$ (Doppler-width).

If $\gamma \ll \omega_D$, we divide the integration into two parts: one from 0 to $1/\omega_D$ and one from $1/\omega_D$ to $\infty$. In the first part we can use the same procedure as in (28.18), whereas in the second part we may assume approximately, that $\sin \omega_D t$ and $\cos \omega_D t$ oscillate so rapidly that they can be replaced by zero. We then obtain

$$\gamma \ll \omega_D : \quad \sigma_e(\boldsymbol{K}, \omega) \approx \frac{\gamma \sigma_0}{2y} \operatorname{Re} e^{-\frac{x^2}{4y^2}} \int_0^{\sqrt{PkT}/\hbar\omega_D} e^{-(z - ix/2y)^2} dz +$$

$$+ \frac{1}{2} \gamma \sigma_0 e^{-2W} \cdot \frac{(\gamma/2)\cos(\omega - \omega_0)/\omega_D - (\omega - \omega_0)\sin(\omega - \omega_0)/\omega_D}{(\omega - \omega_0)^2 + \gamma^2/4} . \tag{28.19}$$

The emission spectrum contains a zero-phonon-line, which has the natural line width and which is a Lorentzian for $|\omega - \omega_0| \lesssim \gamma/2$; this is identical with the zero-phonon-line in (28.11, 12). For $|\omega - \omega_0| > \gamma/2$ the Lorentzian is modified, but this is uninteresting because then the main contribution is from the first part, which is a modified Gaussian again, and which represents the gas-like behaving multiphonon-background at high temperatures. The conditions for observing the zero-phonon-line are just those given above.

The absorption spectrum can be gotten from the emission spectrum by reflecting the latter at $\omega = \omega_0$. The zero-phonon-absorption line is identical with the emission line, whereas the multiphonon-background is now concentrated at $\omega > \omega_0$, whereas the emission background is mainly at $\omega < \omega_0$.

The experiments are done usually in the following way: The emitted intensity is sent through an absorber, which contains the same nuclei as the emitter, but in the ground state. The absorbing nuclei are excited to the level, which produce the $\gamma$-energy. But this only works, if the emitted energy corresponds to the energy of the absorption process. This is the case for the zero-phonon-line (if temperatures of emitter and absorber are equal, see below), but there is only small overlap between the emitter and absorber multiphonon-background. The re-absorption probability is proportional to $\sigma_e \cdot \sigma_a$. This explains the relatively easy possibility of observing the zero-phonon-line; up till now, experimentalists have not succeeded in measuring the phonon-background (Fig. 7.8). The zero-phonon-line offers a large number of possibilities in investigating nuclear levels, e.g. splitting of levels in electric and magnetic fields

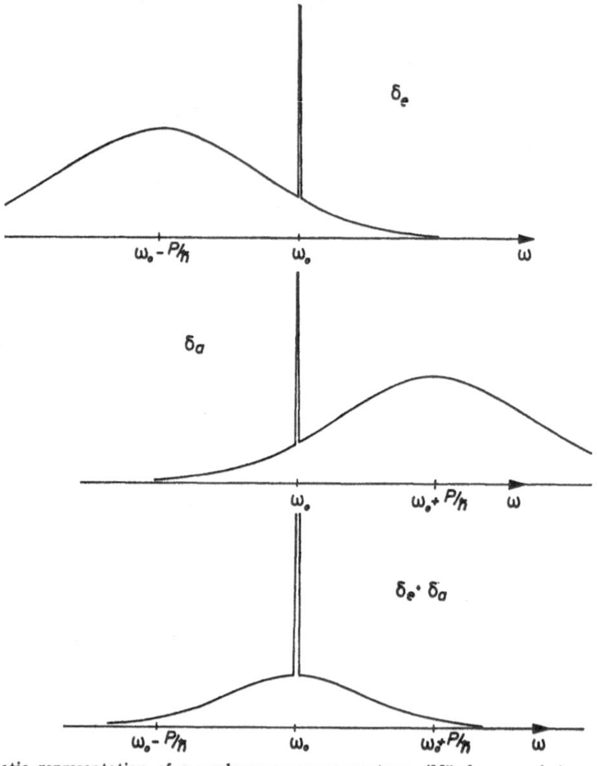


Fig. 7.8. Schematic representation of a nuclear resonance-spectrum (Mössbauer-emission and absorption spectrum)

in crystals, because the zero-phonon-part just gives the natural lines [see 62 F 3]. Also relativistic line shifts can be observed by means of the extremely sharp zero phonon-line [60 P 1, 60 J 1]. If the emitter is moved toward the absorber with a constant velocity $v$, the zero-phonon-line (and all the emission spectrum) suffers a Doppler shift $v/c$, there is a relative displacement between absorption and emission spectrum. Thus by choosing appropriate velocities it is possible to measure the complete line shape and the shape of the background, at least in principle. The velocity which is necessary to produce a shift in the order of the line-width is given by

$$v = \gamma c/\omega_0 .$$

We have mentioned already that the same considerations hold for electronic transitions in inner shells of ions, if conditions are appropriate

[62 M 3, 4 63 T 3, 64 S 2, 64 M 7]. It is possible also, to apply these formulae to the resonance absorption or emission of slow (thermal) neutrons by nuclei. This implies sufficiently small widths of the resonance levels; the repulsion energy $P$ in this case is of the order of $10^3$ to $10^5$ eV. It is just the neutron resonance capture, for which *W. E. Lamb* has developed the theory of resonance absorption or emission first [39 L 1]. Only a long time later *R. Mössbauer* has applied the theory to the resonance absorption of $\gamma$-rays [58 M 1]. Since then a number of papers has been devoted to this subject [62 F 3, 66 M 2].

*Impurities in crystals*

The theory as discussed so far applies to ideal lattices only. Now in most cases the emitting nucleus or the absorbing one is an impurity in the lattice. From (28.7) it can be seen, that the cross section is related to the autocorrelation $\chi^{mm}$ of the emitting ion, and thus it is clear, that the properties of the emitting impurity nucleus determine essentially the emission spectrum. We have to take into account the different mass and force-constants of the impurity, i.e. we have to use $\chi^{mm}(\boldsymbol{K}, t)$ as discussed on page 245. According to (27.25, 40–42, 28.7) we obtain for the zero- and one-phonon-cross-section

$$\sigma_e(\boldsymbol{K}, \omega) = \frac{1}{4}\gamma\,\sigma_0\,e^{-2W_\mu^m}\left\{\frac{\gamma}{(\omega-\omega_0)^2+\gamma^2/4}\cdot\prod_\alpha I_0(A_\alpha) + \right.$$
$$+ 4\pi\sum_\alpha\left[\bar{n}_\alpha\,\delta(\omega-\omega_0-\omega_\alpha) + (\bar{n}_\alpha+1)\,\delta(\omega-\omega_0+\omega_\alpha)\right]\sinh\hbar\omega_\alpha/2kT\times$$
$$\left.\times I_1(A_\alpha)\prod_{\beta(\neq\alpha)} I_0(A_\beta)\right\}$$
$$A_\alpha = \frac{\hbar\,|\boldsymbol{K}\,\overset{0}{\boldsymbol{u}}{}_\mu^m(\alpha)|^2}{2M_\mu\,\omega_\alpha\sinh\hbar\,\omega_\alpha/2kT}\,. \qquad (28.20)$$

In the one-phonon-term we have again neglected $\gamma$ [see (28.11)]. In Sect. 27 we have discussed, that in a certain approximation we can replace the Bessel-functions $I_0$, $I_1$ by their leading terms; then (28.20) agrees formally with (28.11):

$$\sigma_e(\boldsymbol{K}, \omega) = \frac{1}{4}\gamma\,\sigma_0\,e^{-2W_\mu^m}\left\{\frac{\gamma}{(\omega-\omega_0)^2+\gamma^2/4} + \right. \qquad (28.21)$$
$$\left. + \frac{\pi\hbar}{M_\mu}\sum_\alpha\frac{|\boldsymbol{K}\,\overset{0}{\boldsymbol{u}}{}_\mu^m(\alpha)|^2}{\omega_\alpha}\left[\bar{n}_\alpha\,\delta(\omega-\omega_0-\omega_\alpha) + (\bar{n}_\alpha+1)\,\delta(\omega-\omega_0+\omega_\alpha)\right]\ldots\right\}$$

We can use the same considerations as on page 264; the emission cross-section only depends on the vibrational properties of the emitting ion; the Debye-Waller-factor can be gotten from the formulae discussed in Sect. 27. In the one-phonon-cross-section the band-frequencies of the lattice contribute as well as the localized frequencies, if they are present. In Sect. 27 the appropriate formulae for simple cases can be found.

It can be seen further from (28.21) that for Mössbauer-ions near a free surface, or near other defects, the intensity of the zero-phonon-line depends sensitively on the appropriate Debye-Waller-factor. Because the displacement-autocorrelation is larger for surface atoms (Sect. 25) than for interior ones, the Debye-Waller-factor is appreciably smaller for such ions, the zero-phonon-line has less intensity.

The influence of lattice anharmonicity on the emission cross-section can also be calculated according to the lines indicated in Sect. 27. Most of the results can be taken over. Therefore we will not discuss the details.

We will finish this section with a short remark on the second order Doppler-effect. As already mentioned, the characteristic time for lattice vibrations, $1/\omega_D$, is much shorter than the lifetime of the nuclear level $1/\gamma$. This is the reason why there is no first order Doppler effect in the zero-phonon-line, or in other words, the linear Doppler term $\boldsymbol{v}/c$ averages out ($\boldsymbol{v}$: velocity of the emitting ion). But there is a second order Doppler effect, which causes a line shift $\delta\omega/\omega_0 \approx -v^2/2c^2$; the average of this quantity does not vanish in the cases considered. It is essentially the velocity-autocorrelation function as discussed in Sect. 23 too. The results can be taken over. At high temperatures ($T \gtrsim \Theta_D$) we can replace it by the classical average, obtaining

$$\delta\omega/\omega_0 = -3kT/2M_m c^2 \tag{28.22}$$

where $M_m$ is the mass of the emitting ion. For lower temperatures $kT$ has to be replaced by the mean thermal energy of an oscillator, and then $\delta\omega/\omega_0$ also depends on the coupling constants to the neighbors. It is clear that the term (28.22) cannot be observed, if emitter or absorber are at equal temperature, of equal mass and vicinity. Only if emitter and absorber differ in temperature, or in mass, or in the surrounding coupling, there is an observable effect.

## 29. Interaction of Phonons with X-rays. Thermal Diffuse Scattering

X-rays are scattered by the crystal ions by means of the electromagnetic interaction between X-rays and the electrons of the system. The incoming photon interacts with the electronic system and leaves the crystal with different momentum in general. There is no absorption or emission of photons in this kind of processes we will discuss now. Therefore the interaction cannot be described by (28.2); we have to take into account the second term in the interaction, i.e.

$$W = -\frac{e}{mc}\sum_s \boldsymbol{p}_s \boldsymbol{A}(\boldsymbol{r}_s) + \frac{e^2}{2mc^2}\sum_s A^2(\boldsymbol{r}_s)\,. \tag{29.1}$$

The sum is over all the electrons of the system. As discussed in Sect. 28, the first term describes absorption and emission of photons, only the second term contains direct scattering processes of photons. Using (28.2a) and neglecting terms which describe the creation and annihilation of one or two photons we have

$$W = \frac{\pi\hbar e^2}{mV}\sum_{\substack{\boldsymbol{k}\boldsymbol{k}'s\\ \lambda\lambda'}} \frac{(\boldsymbol{s}_{\boldsymbol{k}\lambda}\boldsymbol{s}_{\boldsymbol{k}'\lambda'})}{\sqrt{\omega_k\,\omega_{k'}}}\{a_{\boldsymbol{k}\lambda}\overset{+}{a}_{\boldsymbol{k}'\lambda'}\,e^{-i\boldsymbol{K}\boldsymbol{r}_s} + \overset{+}{a}_{\boldsymbol{k}\lambda}a_{\boldsymbol{k}'\lambda'}\,e^{+i\boldsymbol{K}\boldsymbol{r}_s}\}\,. \tag{29.2}$$

$\boldsymbol{K} = \boldsymbol{k}' - \boldsymbol{k}$ is again the momentum transfer We consider the following process: There is an incoming photon $\boldsymbol{k}'\lambda'$ and a scattered photon $\boldsymbol{k}\lambda$. No other photons shall be involved. The crystal is in state $|\alpha\rangle$ before and $|\varepsilon\rangle$ at the end. The transition probability is given by

$$w_{\boldsymbol{k}'\lambda'\alpha\to\boldsymbol{k}\lambda\varepsilon} = \frac{2\pi^3 e^4}{m^2 V^2} \frac{|\boldsymbol{s}_{\boldsymbol{k}\lambda}\boldsymbol{s}_{\boldsymbol{k}'\lambda'}|^2}{\omega_k \omega_{k'}} \times \\ \times |\langle\varepsilon| \sum_s \mathrm{e}^{i\boldsymbol{K}\boldsymbol{r}_s} |\alpha\rangle|^2 \,\delta(\omega_\varepsilon - \omega_\alpha + \omega_k - \omega_{k'}) . \tag{29.3}$$

The crystal is in thermal equilibrium before, thus we have to average over the thermal distribution of crystal state $|\alpha\rangle$, i.e. $p_\alpha$. Because we are not interested in the final state, we sum over $\varepsilon$. Further we assume, that the energy transfer of the photon, $\hbar\omega = \hbar(\omega_{k'} - \omega_k)$ is so small, that internal excitations in the electronic system do not occur. This is satisfied in the usual X-ray experiments. We divide the coordinates $\boldsymbol{r}_s$ of the electrons in the following way:
$\boldsymbol{r}^n$ is the center of mass coordinate of the $n$-th ion, $\boldsymbol{r}_\nu$ the relative-coordinates of the $\nu$-th electron belonging to the ion $n$. We define the atomic-form-factor of the ion $n$

$$f_n(\boldsymbol{K}) = \left\langle \sum_\nu \mathrm{e}^{i\boldsymbol{K}\boldsymbol{r}_\nu} \right\rangle = \left\langle \int \mathrm{e}^{i\boldsymbol{K}\boldsymbol{r}} \sum_\nu \delta(\boldsymbol{r} - \boldsymbol{r}_\nu)\, \mathrm{d}\boldsymbol{r} \right\rangle = \int \mathrm{e}^{i\boldsymbol{K}\boldsymbol{r}} \varrho_\mathrm{e}(\boldsymbol{r})\, \mathrm{d}\boldsymbol{r} . \tag{29.4}$$

The sum extends over all the electrons of the ion $n$; $\varrho_\mathrm{e}(\boldsymbol{r})$ is the average electronic density of the ion $n$. Using this and the functions defined in (27.11, 12), we obtain in a completely analogous way

$$w_{\boldsymbol{k}'\lambda'\to\boldsymbol{k}\lambda} = \frac{\pi^2 e^4}{m^2 V^2} \frac{(\boldsymbol{s}_{\boldsymbol{k}\lambda}\boldsymbol{s}_{\boldsymbol{k}'\lambda'})^2}{\omega_k \omega_{k'}} \sum_{mn} f_m^*(\boldsymbol{K})\, f_n(\boldsymbol{K}) \int \mathrm{d}t\, \chi^{mn}(\boldsymbol{K}, t)\, \mathrm{e}^{-i\omega t}$$

or of the cross-section (dividing by the incoming flux)

$$\frac{\mathrm{d}\sigma}{\mathrm{d}\omega\, \mathrm{d}\Omega} = \left(\frac{e^2}{m c^2}\right)^2 \frac{k}{k'} (\boldsymbol{s}_{\boldsymbol{k}\lambda}\boldsymbol{s}_{\boldsymbol{k}'\lambda'})^2 \sum_{mn} f_m^*(\boldsymbol{K})\, f_n(\boldsymbol{K}) \cdot \frac{1}{2\pi} \int \mathrm{d}t\, \chi^{mn}(\boldsymbol{K}, t)\, \mathrm{e}^{-i\omega t} . \tag{29.5}$$

This equation resembles completely the corresponding one in neutron diffraction (27.9). However, the time dependence in (29.5) can be neglected, as can be seen from the following consideration. The X-ray photon used in crystal investigations has a wavelength of approximately lattice constant, say $3 \cdot 10^{-8}$ cm; the corresponding frequency is $6.3 \cdot 10^{18}\,\mathrm{sec}^{-1}$ and the energy $\hbar\omega_k \approx 4.1 \cdot 10^3$ eV. This energy is large compared to the energy of the lattice vibrations ($\approx 10^{-2}$ eV). Therefore the energy loss of the photon by interaction with the lattice vibrations cannot be measured and it is reasonable to integrate (29.5) over the energy transfer $\omega$, being left with $\chi^{mn}(\boldsymbol{K}, 0)$. There is no time correlation; this can be seen from a comparison of the time, in which a photon travels a distance of $3 \cdot 10^{-8}$ cm ($10^{-18}$ sec) with that of a vibration period of a lattice ($\approx 10^{-13}$ sec). A neutron as described on page 232 needs $\approx 1.4 \cdot 10^{-13}$ sec, which is comparable with the lattice vibration time. Thus we have finally for X-ray scattering at crystals ($k/k' \approx 1$):

$$\frac{d\sigma}{d\Omega} = \left(\frac{e^2}{m c^2}\right)^2 \frac{k}{k'} (\mathbf{s}_{k\lambda}\, \mathbf{s}_{k'\lambda'})^2 \sum_{mn} f_m^*(\boldsymbol{K})\, f_n(\boldsymbol{K})\, \chi^{mn}(\boldsymbol{K}, 0)\,. \tag{29.6}$$

All the discussion of Sect. 27 can be used here also, but it is more convenient often, to begin with $\chi^{mn}(\boldsymbol{K}, 0)$ again because it contains now only commuting quantities:

$$\chi^{mn}(\boldsymbol{K}, 0) = e^{i\boldsymbol{K}(\boldsymbol{R}^m - \boldsymbol{R}^n)} \cdot \langle e^{+i\boldsymbol{K}(\boldsymbol{u}^m - \boldsymbol{u}^n)}\rangle_T\,. \tag{29.7}$$

Here the ion-coordinates have been separated in the equilibrium coordinates and the displacements from them. In harmonic (ideal) lattices we have $\left(m \Rightarrow {}^{m}_{\mu}\right)$

$$\chi^{mn}_{\mu\nu}(\boldsymbol{K}, 0) = e^{-W_\mu - W_\nu} \cdot e^{i\boldsymbol{K}(\boldsymbol{R}^m_\mu - \boldsymbol{R}^n_\nu)} \times$$

$$\times \exp\left\{\frac{1}{s N \sqrt{M_\mu M_\nu}} \sum_{\boldsymbol{q}\sigma} \frac{(\boldsymbol{K} \boldsymbol{e}^\mu)(\boldsymbol{K} \boldsymbol{e}^{*\nu})}{\omega^2(\boldsymbol{q}\sigma)}\, e^{i\boldsymbol{q}(\boldsymbol{R}^m - \boldsymbol{R}^n)} \cdot \varepsilon(\boldsymbol{q}\sigma, T)\right\}. \tag{29.8}$$

Inserting (29.8) into (29.6) and expanding the last exponential with respect to phonon-processes (zero-, one-, ...) we obtain

$$\begin{aligned} \frac{d\sigma}{d\Omega} = \left(\frac{e^2}{m c^2}\right)^2 (\mathbf{s}_{k\lambda}\, \mathbf{s}_{k'\lambda'})^2 N \sum_{\mu\nu} f_\mu^*\, e^{-W_\mu + i\boldsymbol{K}\boldsymbol{R}_\mu} \cdot f_\nu\, e^{-W_\nu - i\boldsymbol{K}\boldsymbol{R}_\nu} \times \\ \times \left\{\delta(\boldsymbol{K} - 2\pi B\boldsymbol{h}) + \frac{1}{s N \sqrt{M_\mu M_\nu}} \times \right. \\ \left. \times \sum_{\boldsymbol{q}\sigma} \frac{(\boldsymbol{K} \boldsymbol{e}^\mu)(\boldsymbol{K} \boldsymbol{e}^{*\nu})}{\omega^2(\boldsymbol{q}\sigma)}\, \varepsilon(\boldsymbol{q}\sigma, T)\, \delta(\boldsymbol{K} + \boldsymbol{q} - 2\pi B\boldsymbol{h}) + \cdots\right\}. \end{aligned} \tag{29.9}$$

According to the integration over the energy-transfer, there is no selection rule for the frequencies now. The first term, the zero-phonon-cross-section gives the possibility of the determination of the lattice structure from the reciprocal lattice $B$ and the cell-structure factors $\sum_\mu f_\mu\, e^{-W_\mu - i\boldsymbol{K}\boldsymbol{R}_\mu}$. We will not discuss the different methods with which this can be done*. The one-phonon-cross-section can be used in an investigation of phonon-dispersion curves [52 C 1, 55 J 1, 56 W 1], if the X-ray intensity is measured between two Laue-spots $\boldsymbol{K} = 2\pi B\boldsymbol{h}$. However, in general the contribution from higher order terms cannot be neglected in these investigations, and the measurements have to be corrected for the higher order contributions. In neutron scattering the multi-phonon-processes can be avoided more easily.

Anharmonic corrections to these expressions can be found by simple perturbation treatment; there are no difficulties arising from noncommuting displacements at different times [61 H 1, 2]. We will not proceed

* See the standard books on the of X-ray diffraction for crystal structure determination, e. g. *R. W. James*, The optical principles of the diffraction of X-rays. London 1958.

as done in [61 H 1, 2], because we can use part of the calculations of Sect. 27 now, if we limit ourselves to a consideration of the terms which are quadratic in the momentum transfer $\boldsymbol{K}$.

In [61 H 1, 2] also cubic and quartic terms have been calculated. The cubic term means an unsymmetry in the intensity again, though it is probably even more difficult, to analyze this term experimentally. Very careful analysis of the temperature dependence of scattered X-ray intensity between Lauespots is needed.

Up to quadratic terms we have from (27.52)

$$\chi^{mn}(\boldsymbol{K}, 0) = e^{i\boldsymbol{K}(\boldsymbol{R}^m - \boldsymbol{R}^n)}\, e^{-W_m - W_n} \cdot e^{\sum_{ij} K_i K_j \langle u_i^m u_j^n \rangle} \qquad (29.10)$$

and (again for Bravais-lattices)

$$\left\langle u_i^m u_j^n \right\rangle = \frac{\hbar}{2NM} \sum_{JJ'} \frac{e_i(J)\, e_j^*(J')}{\sqrt{\omega_J \omega_{J'}}}\, e^{i\boldsymbol{q}\boldsymbol{R}^m - i\boldsymbol{q}'\boldsymbol{R}^n} \cdot f_{JJ'}(0)\,. \qquad (29.11)$$

Use has been made of (27.64). $f_{JJ'}(0)$ occurs already in the Debye-Waller-factor (27.78) and we take the results from (27.79)

$$\begin{aligned}
\chi^{mn}(\boldsymbol{K}, 0) &= e^{i\boldsymbol{K}(\boldsymbol{R}^m - \boldsymbol{R}^n)} \cdot e^{-2W} \times \\
&\times \exp\left\{\frac{1}{NM} \sum_{JJ'} \frac{(\boldsymbol{K}\boldsymbol{e}_J)(\boldsymbol{K}\boldsymbol{e}_{J'}^*)}{\omega_J \omega_{J'}}\, \delta_{\boldsymbol{q}\boldsymbol{q}'}\, e^{i\boldsymbol{q}(\boldsymbol{R}^m - \boldsymbol{R}^n)} \times \right. \\
&\times \left[\varepsilon_J \delta_{\sigma\sigma'} - \frac{\hbar}{4N}\left(\frac{\bar{n}_J + \bar{n}_{J'}}{\omega_J + \omega_{J'}} - \frac{\bar{n}_J - \bar{n}_{J'}}{\omega_J - \omega_{J'}}\right) \sum_{J_1} \frac{\Phi_{-JJ'J_1-J_1}}{\omega_{J_1}^2}\, \varepsilon_{J_1} \right. \qquad (29.12) \\
&\left.\left. + \frac{\hbar^2}{32N} \sum_{J_1 J_2} \frac{\Delta(\boldsymbol{q} - \boldsymbol{q}_1 - \boldsymbol{q}_2 - 2\pi B\boldsymbol{h})}{\omega_{J_1}\omega_{J_2}}\, \Phi_{-JJ_1J_2}\, \Phi_{J'-J_1-J_2}\, P_{JJ'J_1J_2}\right]\right\}.
\end{aligned}$$

This result is, up to quadratic terms in $\boldsymbol{K}$, identical with that given in [61 H 1]. It should be emphasized again, that $\omega_J$ now are the quasi-harmonic frequencies, which depend on the mean thermal equilibrium positions, and therefore on temperature. The anharmonic contribution in (29.12) is diagonal with respect to $\boldsymbol{q}\boldsymbol{q}'$, but not to $\sigma\sigma'$. In (27.69) we have dropped these terms in a determination of effective neutron frequencies. We can do so here again and define a sort of effective phonon frequencies in X-ray experiments from the diagonal elements $JJ$ by writing the essential term in (29.12)

$$\left\{\frac{1}{NM} \sum_J \frac{|\boldsymbol{K}\boldsymbol{e}_J|^2}{\Omega_J^2}\, e^{i\boldsymbol{q}(\boldsymbol{R}^m - \boldsymbol{R}^n)} \cdot \varepsilon_J\right\}. \qquad (29.12a)$$

From a comparison of (29.12) and (29.12a) we obtain for high temperatures ($T \gtrsim \Theta$), using (27.80)

$$\begin{aligned}
\Omega_J^2 = \omega_J^2 \left\{1 + \frac{kT}{2N\omega_J^2}\left[\sum_{J_1} \frac{\Phi_{-JJ-J_1J_1}}{\omega_{J_1}^2} - \right.\right. \\
\left.\left. - \sum_{J_1J_2} \frac{\Delta(\boldsymbol{q} - \boldsymbol{q}_1 - \boldsymbol{q}_2 - 2\pi B\boldsymbol{h})}{\omega_{J_1}^2 \omega_{J_2}^2} |\Phi_{-JJ_1J_2}|^2\right]\right\}.
\end{aligned} \qquad (29.13)$$

For intermediate temperatures the expression can be gotten similarly, but it is much more complicated. The terms with $J \neq J'$, i.e. $\sigma \neq \sigma'$

must be considered separately*. (29.13) is not identical with the effective frequencies determined from neutron diffraction, but it becomes equal to those frequencies in the elastic limit $\Omega_J$ or $\omega_J \to 0$. At high temperatures the anharmonic frequencies are proportional to $T$; also the quasiharmonic ones are proportional to $T$. From linear extrapolation of the high temperature behavior to $T = 0$ the true harmonic frequencies can be determined, as has been pointed out sometimes [61 L 1, 65 L 5, 64 L 2]. The extrapolation procedure is explained in Fig. 7.9. The extrapolation for $\Omega(\boldsymbol{q}\sigma, T)$ has to be done with $\boldsymbol{q}a = \text{const.}$, where $a = a(T)$ is the lattice constant at that temperature, at which $\Omega(\boldsymbol{q}\sigma, T)$ has been measured.

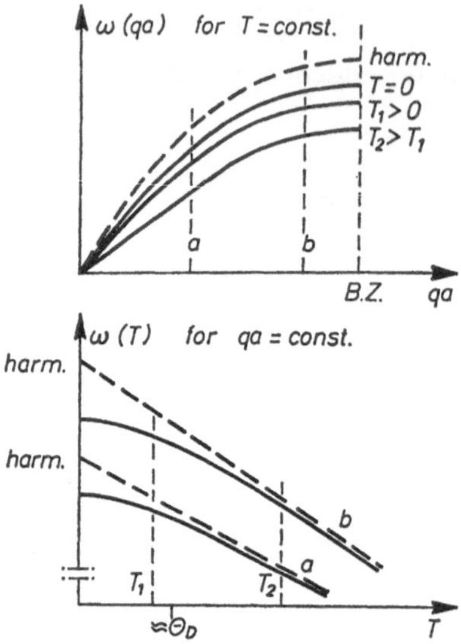


Fig. 7.9 a and b. Determination of harmonic values from measurements of the dispersion curves at different temperatures. For fixed $qa$, one has to extrapolate the frequencies linearly from high temperatures to $T = 0$ (fig. b); this gives the harmonic dispersion curves (fig. a). Only the harmonic curves should be used on statements about force-constants and the ranges of forces

The contributions to the X-ray scattering cross section arising from terms with $K^3$, $K^4$ show, at high temperatures, another temperature dependence than those with $K^2$. This enables one to find the one-phonon-cross-section in X-ray-scattering experiments; for the multi-phonon-terms have also a higher temperature-dependence.

From a thorough investigation of the temperature behavior one can find the true harmonic frequencies of the lattice. The details are explained in some length in [61 L 1, 65 L 5], to which papers we refer.

Eq. (29.5) can be used also in the investigation of the scattering of visible photons at defects. These have a wavelength of approximately $5 \cdot 10^{-5}$ cm, and a frequency of $3.8 \cdot 10^{15}\ \text{sec}^{-1}$. The corresponding photon energy is 2.4 eV, which is no longer very large compared to the energy of the phonons. (29.5) cannot be averaged about the frequencies; the complete $\chi^{mn}(\boldsymbol{K}, t)$ has to be used. In homogeneous harmonic lattices we have with (27.25, 26)

$$\sigma_{\boldsymbol{k}'\lambda' \to \boldsymbol{k}\lambda} = \left(\frac{e^2}{m c^2}\right)^2 \cdot \frac{k}{k'} (\boldsymbol{s}_{\boldsymbol{k}\lambda} \boldsymbol{s}_{\boldsymbol{k}'\lambda'})^2 |f(\boldsymbol{K})|^2 e^{-2W} \times \tag{29.14}$$

$$\times N^2 \Big\{ \delta(\omega)\, \delta(\boldsymbol{K}) + \frac{\hbar}{2NM} \sum_{\substack{\boldsymbol{q}\sigma \\ \boldsymbol{h}}} \frac{|\boldsymbol{K}\boldsymbol{e}|^2}{\omega(\boldsymbol{q}\sigma)}\, \delta(\boldsymbol{K} + \boldsymbol{q} - 2\pi B \boldsymbol{h}) \times$$

$$\times [\bar{n}(\boldsymbol{q}\sigma)\, \delta(\omega(\boldsymbol{q}\sigma) + \omega) + (\bar{n}(\boldsymbol{q}\sigma) + 1)\, \delta(\omega(\boldsymbol{q}\sigma) - \omega)] \Big\} + \cdots$$

* It can be shown that they vanish if $\boldsymbol{q}$ is in a symmetry direction of the lattice.

Here we have restricted to Bravais-lattices again. Further we have neglected the reciprocal lattice vectors $2\pi B\boldsymbol{h}$ in the momentum-selection-rule $\delta(\boldsymbol{K})$ for the following reason: Visible photons have a wave-length of $5 \cdot 10^{-5}$ cm, or the maximal momentum transfer is of the order $K \approx 10^5\,\mathrm{cm}^{-1}$; this holds even if there is inelastic scattering by the phonons because they change the energy of the photon not appreciably. Therefore $\boldsymbol{q}$ in the one phonon-cross-section of (29.14) is nearly equal to one of the reciprocal lattice points, i.e. the corresponding frequencies $\omega(\boldsymbol{q}\sigma)$ are those of the elastic limit, i.e. they are determined by the elastic constants and the mass density of the crystal. The one-phonon-cross-section describes the emission or absorption of an elastic phonon by the photon. Besides the elastic scattered photons ($\omega = 0$) there are two photon intensity maxima at

$$\omega = \pm\omega(\boldsymbol{q}\sigma); \quad 2\pi B\boldsymbol{h} - \boldsymbol{q} = \boldsymbol{K}. \tag{29.15}$$

If $\boldsymbol{h} = 0$, only longitudinal components of the phonons are involved because of the fore-factor $(\boldsymbol{K}\boldsymbol{e}(\boldsymbol{K}, \sigma))$. Let us assume an isotropic medium for simplicity. Then $\omega_l(\boldsymbol{q}\sigma) = c_l \cdot q$, where $c_l$ is the longitudinal sound velocity. Then*

$$\omega_{k'} - \omega_k = \omega = \pm c_l q = \pm c_l K \approx \pm c_l \cdot \frac{4\pi}{\lambda} \sin\vartheta/2 = \pm 2\omega_k \cdot \frac{c_l}{c} \sin\vartheta/2\,. \tag{29.16}$$

$\vartheta$ is the scattering angle of the photon, $c$ the velocity of light. This kind of scattering is generally referred to as Brillouin-scattering. The two intensity maxima defined by (29.15, 16) are called the Brillouin-doublet. It is clear, that it will occur also in liquids, if conditions are appropriate. Further it can be investigated also by neutron diffraction, if the neutrons have a sufficiently large wave-length, say $10^{-6}$ cm (subthermal neutrons, $10^{-5}$ eV). In anisotropic media the situation is somewhat more involved, but the occuring phenomena are essentially the same.

## 30. Interaction of Phonons with Infrared Light. Infrared Absorption

The infrared absorption in crystals is a special kind of the interaction of electromagnetic radiation with phonons, some aspects of which are already considered in sections 28 and 29. This interaction is connected with the dipole-moment of the optical vibrations of the crystals; the frequencies are of the order of magnitude of about $\omega \approx 10^{13} - 10^{14}\,\mathrm{sec}^{-1}$. This corresponds to wave numbers $k \approx 0.3 \cdot 10^3 - 0.3 \cdot 10^4\,\mathrm{cm}^{-1}$, which is small compared to the maximum wave numbers of the lattices ($q \approx 10^8\,\mathrm{cm}^{-1}$). Thus the optical absorption is related mainly to the $q \approx 0$ optical phonons (in parameterfree crystals).

In this connection we will discuss only the relation between the optical absorption and the Greens-function or the correlations of the ions, resp.

* We neglect the difference in the absolute values of $\boldsymbol{k}$ and $\boldsymbol{k}'$ for the photons, because its change is small (below 1%).

This has been discussed first in a number of papers [62 V 1, 63 M 6, 66 W 1, 66 B 1]. The Hamiltonian can be written again as

$$\mathscr{H} = \mathscr{H}_{\text{cryst}} + \mathscr{H}_{\text{el.m.f.}} + W = \mathscr{H}_0 + W \tag{30.1}$$

where the interaction term generally is assumed to be*

$$W = -\boldsymbol{M} \cdot \boldsymbol{E} \cdot e^{-i\omega t + \epsilon t}, \ \epsilon \to 0+ . \tag{30.2}$$

$\boldsymbol{M}$ is the dipole-moment of the crystal in a periodicity volume, $\boldsymbol{E}$ is the amplitude of the effective field which is assumed to be monochromatic with frequency $\omega$. The factor $\exp\{\epsilon t\}$ makes the interaction vanish for $t \to -\infty$ (adiabatic switching on of the field). When we calculate the thermal average of the dipole-moment $\langle M_i \rangle_\beta$ as a function of the effective field $\boldsymbol{E}$, the linear term in $\boldsymbol{E}$ defines the electric susceptibility

$$P_i = \frac{\langle M_i \rangle_\beta}{V} = \sum_j \chi_{ij}(\omega)\, E_j . \tag{30.3}$$

$E_j$ is the amplitude of the monochromatic wave with frequency $\omega$; $V$ is the periodicity volume. The imaginary part of the dielectric constant

$$\varepsilon_{ij} = \delta_{ij} + 4\pi\, \chi_{ij} \tag{30.4}$$

describes the absorption. What we have to calculate thus is the linear response of the dipoles to an electric field.

The averaging process of the dipole-moment can be done with the density matrix

$$\varrho(t) = \frac{1}{Z} e^{-\beta \mathscr{H}} = \frac{1}{Z} e^{-\beta(\mathscr{H}_0 + W)} \tag{30.5}$$

with the complete Hamiltonian $\mathscr{H}$. $\varrho(t)$ satisfies the Liouville-equation

$$i\hbar\dot{\varrho} = [\mathscr{H}_0 + W, \varrho] \tag{30.6}$$

$\mathscr{H}_0$ is independent of time. We put

$$\varrho(t) = \varrho_0 + \Delta\varrho(t); \quad [\varrho_0, \mathscr{H}_0] = 0 \tag{30.7}$$

and obtain up to linear terms in the perturbation

$$i\hbar(\Delta\dot{\varrho}) = [\mathscr{H}_0, \Delta\varrho] + [W, \varrho_0] \tag{30.8}$$

with the solution

$$\Delta\varrho = \frac{i}{\hbar} e^{-i\mathscr{H}_0 t/\hbar} \int_{-\infty}^{t} [\varrho_0, \tilde{W}(t')]\, dt' \cdot e^{+i\mathscr{H}_0 t/\hbar} \tag{30.9}$$

$$\tilde{W}(t') = e^{i\mathscr{H}_0 t'/\hbar}\, W(t')\, e^{-i\mathscr{H}_0 t'/\hbar} .$$

The linear response of the polarization to the field then is, according to (30.3)

$$P_i = \frac{1}{V} \operatorname{Trace}(\Delta\varrho \cdot M_i) \tag{30.10}$$

* We will not discuss the question whether this term contains all the possible interactions or not. For large wavelengths (infrared) it is correct, called dipole-approximation. $\mathscr{H}_{\text{el.m.f.}}$ does not enter in the following, we can use $\mathscr{H}_0 = \mathscr{H}_{\text{cryst}}$.

and with (30.9) and (30.2), $M_i(t) = e^{i\mathcal{H}_0 t/\hbar} M_i e^{-i\mathcal{H}_0 t/\hbar}$,

$$P_i = \sum_j \text{Trace} - \frac{i}{\hbar V} M_i(t) \int_{-\infty}^{t} [\varrho_0, M_j(t')] e^{-i\omega t' + \epsilon t'} dt' \cdot E_j$$

$$= \sum_j \text{Trace}\, \varrho_0 \cdot \frac{i}{\hbar V} \cdot \int_{-\infty} [M_i(t), M_j(t')] e^{-i\omega t' + \epsilon t'} dt' \cdot E_j .$$

Using the step-function $\Theta(t)$ and writing the trace as a thermal average we obtain

$$\chi_{ij}(\omega) = \frac{i}{\hbar V} \int_{-\infty}^{\infty} \Theta(t) \langle [M_i(t), M_j(0)] \rangle_{\beta,0} e^{i\omega t - \epsilon t} dt . \quad (30.11)$$

It has been used that the average depends only on $t - t'$. The explicit calculation of $\chi_{ij}$ depends on the knowledge of the dipole-moment. The general assumption is that it can be expanded with respect to the ionic displacements

$$M_k = \sum_{m \mu i} M_{k,\,\substack{m\\ \mu\\ i}}\, u^{\substack{m\\ \mu\\ i}} + \sum_{\substack{m \mu i\\ n \nu j}} M_{k,\,\substack{m n\\ \mu\, \nu\\ i\, j}}\, u^{\substack{m\\ \mu\\ i}}\, u^{\substack{n\\ \nu\\ j}} + \cdots . \quad (30.12)$$

For rigid ions, $M_{k,\,\substack{m\\ \mu\\ i}} = Q^{\substack{m\\ \mu}} \cdot \delta_{ki}$ represents the charge of the ions. The higher terms represent higher order dipole-moments and are essential in covalent crystals with vanishing linear moment. In *ideal* lattices because of the translational symmetry $M_{k,\,\substack{m\\ \mu\\ i}}$ is independent of $\boldsymbol{m}$, and $M_{k,\,\substack{m n\\ \mu\, \nu\\ i\, j}}$ depends only on $\boldsymbol{m} - \boldsymbol{n}$.

Inserting (30.12) into (30.11) we obtain

$$\chi_{ij}(\omega) = \frac{1}{V} \sum_{\substack{m \mu k\\ n \nu l}} M_{i,\,\substack{m\\ \mu\\ k}} M_{j,\,\substack{n\\ \nu\\ l}} \cdot \int_{-\infty}^{+\infty} G_{\mathrm{R}\,\substack{m n\\ \mu\, \nu\\ k\, l}}(t)\, e^{i\omega t - \epsilon t} dt + \cdots \quad (30.13)$$

with

$$G_{\mathrm{R}\,\substack{m n\\ \mu\, \nu\\ k\, l}} = \frac{i}{\hbar} \Theta(t) \left\langle \left[ u^{\substack{m\\ \mu\\ k}}(t), u^{\substack{n\\ \nu\\ l}}(0) \right] \right\rangle_{\beta,0} . \quad (30.13\text{a})$$

$G_{\mathrm{R}}$ is the retarded temperature-Greens-function, defined in (21.22) for zero-temperature (see appendix). It describes the correlations between displacements for positive times. Higher order dipole-moments lead to higher order correlation functions, which we will not consider here [66 W 1].

Using the normal mode representation (21.2) of the displacements, we can express (30.13) by Greens-functions for the $b$-, $b^+$-operators. We will discuss several simple cases only.

i) ideal harmonic crystal.

With the displacements given in (21.11) we obtain ($J = \boldsymbol{q}, \sigma$)

$$G_{\mathrm{R}\,\substack{m n\\ \mu\, \nu\\ k\, l}} = \frac{1}{s N \sqrt{M_\mu M_\nu}} \sum_J \frac{1}{\omega_J} e_{\substack{\mu\\ k}}(J)\, e^*_{\substack{\nu\\ l}}(J)\, e^{i\boldsymbol{q}(\boldsymbol{R}^m - \boldsymbol{R}^n)} \cdot \sin \omega_J t \cdot \Theta(t) \quad (30.14)$$

and

$$\chi_{ij}(\omega) = \frac{1}{V_z} \sum_\sigma{}' M_i(0\,\sigma)\, M_j^*(0\,\sigma) \left\{ \frac{1}{\omega_{0\sigma} - \omega - i\epsilon} + \frac{1}{\omega_{0\sigma} + \omega + i\epsilon} \right\} \tag{30.15}$$

with

$$M_i(\boldsymbol{q}\sigma) = \frac{1}{N} \sum_{\boldsymbol{m}\mu k} \frac{M_{i\mu}^{\boldsymbol{m}}{}_{k}\, e^{\mu}_{k}(\boldsymbol{q}\,\sigma)}{\sqrt{2s\, M_\mu\, \omega(\boldsymbol{q}\,\sigma)}}\, e^{i\boldsymbol{q}\boldsymbol{R}^{\boldsymbol{m}}} = \sum_{\mu k}{}' \frac{M_{i\mu}^{\boldsymbol{m}}{}_{k}\, e^{\mu}_{k}(0,\sigma)}{\sqrt{2s\, M_\mu\, \omega(0,\sigma)}}\, \delta^P(\boldsymbol{q})\,,$$

because $M_i$ is independent of $\boldsymbol{m}$ (ideal lattice). The prime at the summation sign indicates, that the center-of-mass motion (acoustical branches with $q = 0$) have to be omitted.

The imaginary part is then (considering only positive $\omega$)*

$$\operatorname{Im} \chi_{ij}(\omega) = \frac{\pi}{V_z} \sum_\sigma{}' M_i(0,\sigma)\, M_j(0,\sigma)\, \delta(\omega_{0\sigma} - \omega)\,. \tag{30.16}$$

The absorption has a sharp peak at the frequencies of the optical $q = 0$-vibrations (dispersion-oscillators), which is temperature-independent.

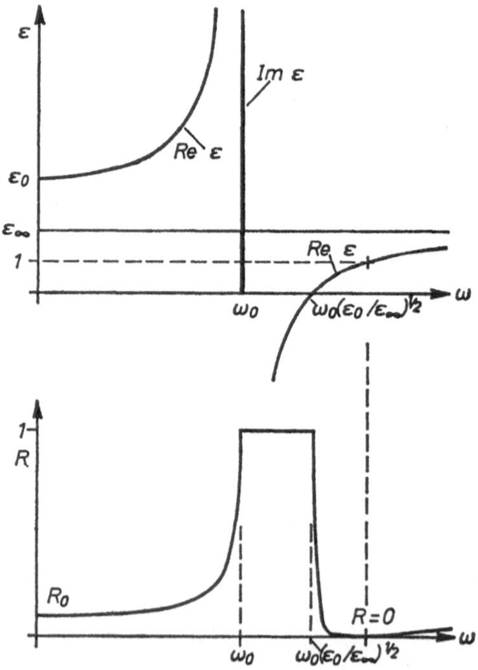


Fig. 7.10. Dielectric constant as function of frequency, schematically, in a harmonic approximation, together with the reflection coefficient

Only those oscillators contribute, which have non-vanishing dipole-moments. At the frequencies $\omega(0,\sigma)$ the wave vector $\boldsymbol{k}$ of the infrared-radiaton is nearly zero, so that we have the selection rule $\boldsymbol{k} = \boldsymbol{q} = 0$

* $M$ is real for $\boldsymbol{q} = 0$.

which is related to the translation symmetry of the lattice (conservation of quasi-momentum).

For a single dispersion oscillator $M(0, \sigma) = M_0$ in a cubic crystal (30.15) can be written as

$$\operatorname{Re}\chi(\omega) = \frac{M_0^2}{V_z} \cdot \frac{2\omega_0}{\omega_0^2 - \omega^2}; \quad \operatorname{Im}\chi(\omega) = \frac{\pi M_0^2}{V_z}\,\delta(\omega_0 - \omega)\,. \tag{30.15a}$$

In this case $\chi(\omega \to \infty) \Rightarrow 0$, or $\varepsilon(\omega \to \infty) \Rightarrow 1$. Now there are other dispersion mechanisms (electronic polarizabilities) which are essential for larger $\omega$, and contribute for $\omega > \omega_0$. The resulting dielectric constant we denote by $\varepsilon_\infty$. Introducing further

$$\varepsilon_0 - \varepsilon_\infty = \frac{8\pi M_0^2}{V_z \omega_0}$$

we obtain

$$\operatorname{Re}\varepsilon(\omega) = \varepsilon_\infty + \frac{\varepsilon_0 - \varepsilon_\infty}{1 - (\omega/\omega_0)^2}; \quad \operatorname{Im}\varepsilon(\omega) = \frac{1}{2}\pi\omega_0(\varepsilon_0 - \varepsilon_\infty)\,\delta(\omega_0 - \omega)\,.$$

The reflection coefficient for the reflection of infrared radiation at the surface of a crystal is given by

$$R(\omega) = \frac{(n-1)^2 + \varkappa^2}{(n+1)^2 + \varkappa^2}; \quad \varepsilon(\omega) = n^2 - \varkappa^2 + 2in\varkappa\,.$$

$\varepsilon(\omega)$ and $R(\omega)$ are shown in Fig. 7.10.

ii) Harmonic crystal with defects (especially point defects). In this case there is no translation symmetry of the crystal. (30.14) and (30.15) remain essentially valid, but the eigenstates are no longer plane waves. We have scattering states and localized states. Denoting the vectors by $\overset{0M}{u_i}(J)$ we have $\left(\overset{m}{\mu} \to M\right)$

$$M_i(J) = \frac{1}{\sqrt{2N}} \sum_{M,k} \frac{M_{i,\overset{M}{k}}}{\sqrt{M_M\,\omega(J)}}\,\overset{0M}{u_k}(J) \tag{30.17}$$

and instead of (30.16)*

$$\operatorname{Im}\chi_{ij}(\omega) = \frac{\pi}{V_z} \sum_J{}' \operatorname{Re}\{M_i(J)\,M_j^*(J)\}\,\delta(\omega_J - \omega)\,. \tag{30.18}$$

If there are localized states with non-vanishing dipole-moment (infrared-active), these contribute single absorption lines to the absorption spectrum. The contribution of the in-band-modes is more complicated. Assuming $M_i(J)$ independent of $J$, we have

$$\begin{aligned}\operatorname{Im}\chi_{ij}(\omega) &= \frac{\pi}{V_z}\operatorname{Re}(M_i M_j^*) \sum_J \delta(\omega_J - \omega) = \\ &= \operatorname{Re}(M_i M_j^*)\,\frac{3\pi s N}{V_z} \int z(\omega_J)\,\delta(\omega_J - \omega)\,\mathrm{d}\omega_J = \\ &= \frac{3\pi s N}{V_z}\operatorname{Re}(M_i M_j^*) \cdot z(\omega)\,,\end{aligned} \tag{30.19}$$

* $\operatorname{Im}\{M_i(J)\,M_j^*(J)\}$ gives no contribution.

where $z(\omega)$ is the frequency spectrum of the (defect) lattice, normalized to one. If the spectrum is not influenced very much by defects, as it is in most cases, with the optical absorption we measure the ideal lattice spectrum. If there are resonances, we measure these too. But in most

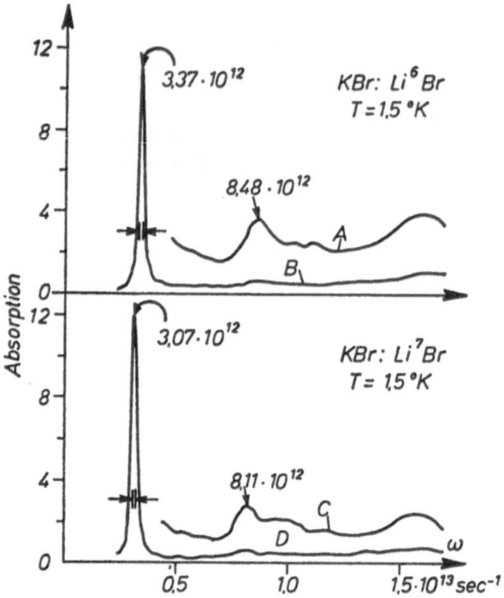


Fig. 7.11. Resonance absorption in a KBr-crystal with LiBr-impurities [65 S 3]. Due to the mass-difference between $Li^6$ and $Li^7$ there is an isotopic shift. The concentration is in curve *A:* 0,14 mol-% $Li^6Br$; *B:* 2,1 $10^{-3}$ mol-% $Li^6Br$; *C:* 0,17 mol-% $Li^7Br$; *D:* 1,8 $10^{-3}$ mol-% $Li^7Br$. The second absorption peak may be due to a second overtone of the low frequency absorption or to another infrared active resonance frequency. Debye-frequency is $\omega_D = 2{,}32\ 10^{13}\ \mathrm{sec}^{-1}$ (see also fig. 6.16)

cases, $M_i(J)$ will not be independent of $J$, so that we will measure a rather complicated absorption function. However, if there are strong resonances, these will be seen in the absorption spectrum at any case, provided the dipole-moment does not vanish. Fig. 7.11 shows some experimental results.

iii) Anharmonic crystal.

In the case of an ideal crystal, which we will consider only, we have to use the anharmonic Greens function, which is essentially that of the neutron-scattering at crystals. A similar procedure leads to the result

$$\operatorname{Im}\chi_{ij}(\omega) = \frac{1}{V_z}\sum_{\sigma}{}' M_i(0,\sigma)\, M_j(0,\sigma)\left\{\frac{\Gamma_{0\sigma}(\omega)}{[\omega+\omega(0\,\sigma)+\Delta_{0\sigma}(\omega)]^2+\Gamma_{0\sigma}^2(\omega)}+\right.$$
$$\left.+\frac{\Gamma_{0\sigma}(\omega)}{[\omega-\omega(0\,\sigma)-\Delta_{0\sigma}(\omega)]^2+\Gamma_{0\sigma}^2(\omega)}\right\}. \tag{30.20}$$

$\Gamma_{0\sigma}(\omega)$ and $\Delta_{0\sigma}(\omega)$ have the same meaning as in (27.72), now for the dispersion oscillators*. They are functions of temperature, being linear in $T$ for $T \gtrsim \Theta_D$ in the first anharmonic approximation.

* If we consider, as usual, positive frequencies, the first term in (27.72b) does not contribute. Further, the last two terms give the same contribution.

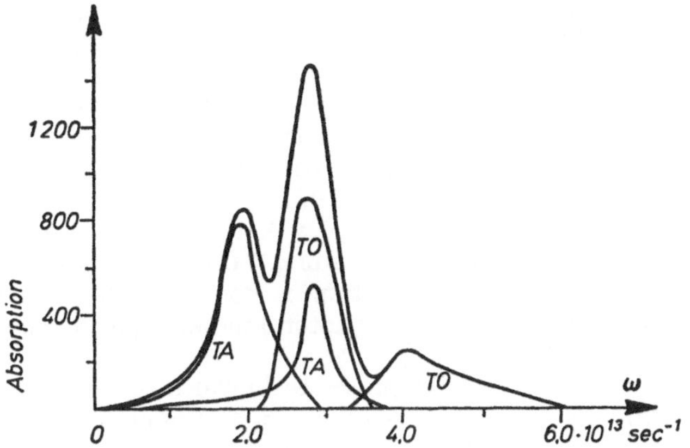


Fig. 7.12. Side-bands on the absorption spectrum of a NaCl-crystal, due to anharmonic processes in the absorption

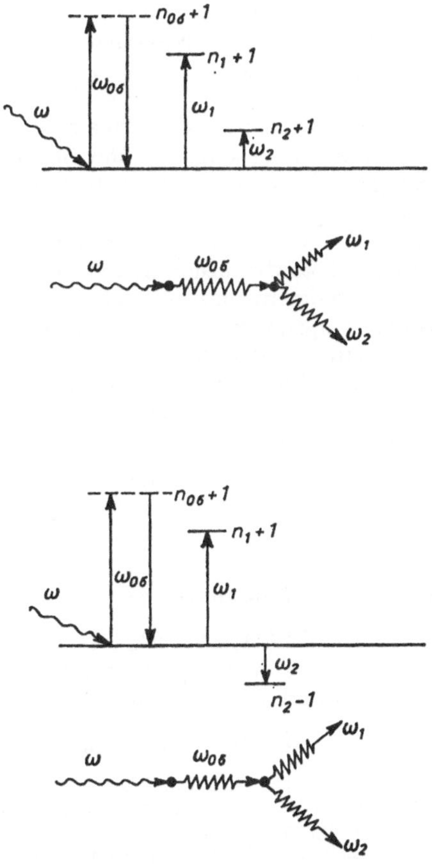

Fig. 7.13. Summation and difference processes in the optical absorption of crystals

The anharmonicity leads to a shift and a broadening of the absorption lines. There occur side-bands and further peaks (Fig. 7.12). The infrared-radiation has a direct interaction only with the dispersion oscillators $\boldsymbol{q} = 0$, but because of the anharmonicity these are coupled to other phonons, what can be seen directly from (27.72). The width $\Gamma_{0\sigma}$ is the reciprocal life-time of the dispersion oscillators. The possible processes contributing are the following (Fig. 7.13):

The second term in (27.72b), for $\omega > 0$, describes processes, in which the virtually excited dispersion-oscillator $\boldsymbol{q} = 0, \sigma$ decays into two phonons $\boldsymbol{q}_1 \sigma_1$ and $\boldsymbol{q}_2 \sigma_2$; the quasimomentum is conserved (photon $\boldsymbol{k} \approx 0 = \boldsymbol{q} = \boldsymbol{q}_1 + \boldsymbol{q}_2$). The frequency of the photon follows from the conservation of energy, $\omega = \omega_1 + \omega_2$.

The last two terms describe processes, in which the dispersion oscillator decays under creation of a phonon $\boldsymbol{q}_1\sigma_1$ and annihilation of another one $\boldsymbol{q}_2 = -\boldsymbol{q}_1, \sigma_2$. The frequency is given by $\omega = \omega_1 - \omega_2$. These difference processes vanish for $T = 0$, because then there is no phonon which can be annihilated.

Apart from all the frequency dependent factors, the frequency-dependence of $\Gamma_{0\sigma}(\omega)$ and of $\operatorname{Im} \chi_{ij}(\omega)$ is given by the combination spectra

$$z_{\pm}(\omega) = \sum_{J_1 J_2} \delta(\omega - (\omega_{J_1} \pm \omega_{J_2})) \,. \qquad (30.21)$$

Therefore the side-bands in ideal crystals give some information about the life-time of the dispersion oscillators and of the lattice-spectrum.

In defect lattices the anharmonicity leads to similar statements. This will not be discussed here [see Sect. 26].

The contribution of the higher order dipole-moments in (30.12) is the first non-vanishing contribution in covalent crystals (diamond, Si, Ge) where the linear moment vanishes. The infrared light is coupled directly to two phonons $q_1\sigma_1, q_2\sigma_2$, without any $\boldsymbol{q} = 0$-oscillator. Also in this case there are sum- and difference-processes, which determine the combination-spectra in the absorption. There is a similar structure as in the side-bands of anharmonic ionic crystals. For a very detailed discussion of the optical absorption we refer to [67 B 1]*.

# VIII. Appendix

## 31. Thermodynamic Greens Functions

[60 Z 1, 62 K 3, 62 B 8, 63 A 1, 64 N 2].

In the following we shall shortly discuss the thermodynamic Greens functions as far as they are related to the displacements or the operators

$$B_{q\sigma}(t) = b_{q\sigma} - b^{+}_{-q\sigma}\,; \quad B^{+}_{q\sigma}(t) = b^{+}_{q\sigma} - b_{-q\sigma} = -B_{-q\sigma} \qquad (31.1)$$

* In [67 B 1] Cowley refers to the anharmonic frequencies $\omega + \Delta$ in eq. (30.20) as well as for the frequencies in neutron scattering (27.73) as "quasiharmonic frequencies". This is misleading, for they cannot be obtained by any quasiharmonic procedure, see footnote p. 259. In this paper, the term quasiharmonic is used in connection with the procedure in sect. 15.

resp. The displacements are

$$u_{\mu i}^{m}(t)=\sqrt{\frac{\hbar}{2sNM_\mu}}\sum_{q\sigma}{}'\frac{1}{\sqrt{\omega(q\sigma)}}e_i^\mu(q\sigma)\cdot e^{iqR^m}\cdot B_{q\sigma}(t)\,. \tag{31.2}$$

The Greens function formed in this way is the only one we need in connection with phonons. Therefore we shall not discuss the function formed with the creation- and annihilation-operators $b_{q\sigma}^+$, $b_{q\sigma}$ alone, though some relations become simpler for the functions with $b_{q\sigma}^+$, $b_{q\sigma}$ instead of $B_{q\sigma}$, $B_{q\sigma}^+=-B_{-q\sigma}$.

The obvious generalization of (21.13) to excited states is*

$$G_{i\,j}^{mn}(t,t';n_{q\sigma})=\frac{i}{\hbar}\left\langle\ldots n_{q\sigma}\ldots\left|T_D u_i^m(t)\,u_j^n(t')\right|\ldots n_{q\sigma}\ldots\right\rangle. \tag{31.3}$$

This holds for a fixed phonon-state; but in general we have to deal with an average over a distribution of states determined by the temperature $T=1/k\beta$:

$$\begin{aligned}G_{i\,j}^{mn}(t,t';\beta)&=\frac{i}{\hbar}\cdot\frac{1}{Z}\sum_{\ldots n_{q\sigma}\ldots}e^{-\beta E(\ldots n_{q\sigma}\ldots)}\times\\&\quad\times\left\langle\ldots n_{q\sigma}\ldots\left|T_D u_i^m(t)\,u_j^n(t')\right|\ldots n_{q\sigma}\ldots\right\rangle\\&=\frac{i}{\hbar}\cdot\frac{1}{Z}\mathrm{Trace}\left\{e^{-\beta\mathcal{H}}\,T_D u_i^m(t)\,u_j^n(t')\right\}\\&=\frac{i}{\hbar}\left\langle T_D u_i^m(t)\,u_j^n(t')\right\rangle_\beta\end{aligned} \tag{31.4}$$

with

$$\langle A\rangle_\beta=\frac{1}{Z}\mathrm{Trace}\,\{e^{-\beta\mathcal{H}}A\};\quad Z=\mathrm{Trace}\,\{e^{-\beta\mathcal{H}}\}\,. \tag{31.5}$$

Correspondingly we define ($J=q\sigma$)

$$G(J,t;J't';\beta)=\frac{i}{\hbar}\langle T_D\,B_J(t)\,B_{J'}^+(t')\rangle_\beta\,. \tag{31.6}$$

The relation between (31.4) and (31.6) can be obtained immediately by using (31.2).

*Harmonic approximation*

First we give some explicit formulae for the simple harmonic case, as an example. We use

$$\mathcal{H}=\mathcal{H}_0=\sum_J\hbar\omega_J(b_J^+b_J+1/2)\,. \tag{31.7}$$

A straight-forward calculation gives

$$G(J,J';t-t')=\frac{i}{\hbar}\delta_{JJ'}\{(\bar n_J+1)\,e^{-i\omega_J|t-t'|}+\bar n_J e^{i\omega_J|t-t'|}\}=G(J;t-t')\cdot\delta_{JJ'} \tag{31.8}$$

* We drop the $\mu$, $\nu$-indices now, because all the following relations can be immediately used also if $\mu$, $\nu$ are present.

showing that the system is homogeneous in the lattice and in time. Using (31.2) we have

$$G_{i\,j}^{mn}(t-t') = \frac{i}{2NM}\sum_{q\sigma}\frac{1}{\omega(q\sigma)}\, e_i(q\sigma)\, e_j^*(q\sigma)\, e^{iq(R^m-R^n)}\cdot G(q\sigma; t-t')\,. \tag{31.9}$$

The Fourier-transform of (31.8) defined by

$$G(J, J'; \omega) = \int_{-\infty}^{+\infty} G(J, J'; t)\, e^{i\omega t-\epsilon|t|}\, dt, \quad \epsilon > 0 \tag{31.10}$$

is

$$\begin{aligned} G(J, J'; \omega) &= \frac{2\omega_J}{\hbar}\,\delta_{JJ'}\left\{\frac{(\bar n_J+1)}{\omega_J^2-\omega^2-i\epsilon} - \frac{\bar n_J}{\omega_J^2-\omega^2+i\epsilon}\right\} \\ &= \frac{2\omega_J}{\hbar}\,\delta_{JJ'}\left\{\mathrm{P}\frac{1}{\omega_J^2-\omega^2} + i\pi\,\mathrm{ctgh}(\beta\hbar\omega_J/2)\,\delta(\omega_J^2-\omega^2)\right\}. \end{aligned} \tag{31.11}$$

Dividing $G$ in its real and imaginary part

$$G = G' + iG'' \tag{31.12}$$

we have

$$G'(J, J'; \omega) = \frac{2\omega_J}{\hbar}\,\delta_{JJ'}\,\mathrm{P}\frac{1}{\omega_J^2-\omega^2} \tag{31.13a}$$

$$G''(J, J'; \omega) = \frac{2\pi\,\omega_J}{\hbar}\,\delta_{JJ'}\,\mathrm{ctgh}(\beta\hbar\,|\omega|/2)\,\delta(\omega_J^2-\omega^2)\,. \tag{31.13b}$$

From these definitions it can be seen easily, that the following "dispersion relation" holds:

$$G'(J, J'; \omega) = \frac{1}{\pi}\,\mathrm{P}\int_{-\infty}^{+\infty}\frac{G''(J, J'; x)}{x-\omega}\cdot \mathrm{tgh}(\beta\hbar x/2)\cdot \mathrm{d}x\,. \tag{31.14}$$

It is sometimes convenient, not to use $\omega$ but $\lambda = \omega^2$ as the variable. Then (31.14) can be rewritten into

$$G'(J, J'; \lambda) = \frac{1}{\pi}\,\mathrm{P}\int_0^{\infty}\mathrm{d}y\,\frac{G''(J, J'; y)}{y-\lambda}\,\mathrm{tgh}\left(\beta\hbar\sqrt{y}/2\right). \tag{31.14a}$$

$G(\omega)$ is not an analytic function in the complex $\omega$-plane. It can be seen, however, that the following function has analytic properties in one half-plane.

$$\begin{aligned} G_{\mathrm{R,A}}(J, J'; \omega) &= G'(J, J'; \omega) \pm iG''(J, J'; \omega)\,\mathrm{tgh}(\beta\hbar\omega/2) \\ &= \frac{2\omega_J}{\hbar}\left\{\mathrm{P}\frac{1}{\omega_J^2-\omega^2} \pm i\pi\,\mathrm{sgn}\,\omega\;\delta(\omega_J^2-\omega^2)\right\}\delta_{JJ'} \\ &= \frac{2\omega_J}{\hbar}\cdot\frac{\delta_{JJ'}}{\omega_J^2-(\omega\pm i\epsilon)^2}\,. \end{aligned} \tag{31.15}$$

We have

$$\mathrm{Re}\,G_{\mathrm{R,A}}(\omega) = \pm\frac{1}{\pi}\,\mathrm{P}\int_{-\infty}^{+\infty}\frac{\mathrm{Im}\,G_{\mathrm{R,A}}(x)}{x-\omega}\,\mathrm{d}x\,. \tag{31.16}$$

The time-dependent function $G_{R,A}(t)$ can be obtained with the inversion of (31.10):

$$G_{R,A}(J, J'; t) = \pm \frac{2}{\hbar} \sin\omega_J t \cdot \Theta(\pm t) \cdot \delta_{JJ'} . \qquad (31.17)$$

It can be shown easily, that the functions $G_R$ and $G_A$ are nothing else than the retarded and advanced harmonic Greens functions, defined alternatively by

$$G_{R,A}(J, J'; t) = \pm \frac{i}{\hbar} \Theta(\pm t) \langle [B_J(t), B_{J'}^{+}(0)]_{-} \rangle_\beta . \qquad (31.18)$$

Now we will continue the general discussion.

*General Remarks*

Using the cyclic properties of the trace in (31.4, 5) we can show that $G(t, t')$ depends only on $t - t'$, if the system is homogeneous in time:

$$G(t, t') = G(t - t') . \qquad (31.19)$$

In the following we put $t' = 0$. Consider the expression $\frac{i}{\hbar} \langle B_J(t)\, B_{J'}^{+}(0)\rangle_\beta$, which is just (31.6) *but with dropping the time-ordering operator* $T_D$ and let $M$, $N$ denote the eigenstates of the complete Hamiltonian $\mathscr{H}$ of the system. Then with $B_J(t) \to B(t)$, $B_{J'}^{+}(0) \to A$, we have

$$\begin{aligned} \frac{i}{\hbar} \langle B(t)\, A\rangle_\beta &= \mathrm{Trace} \frac{i}{\hbar Z} \{ e^{-\beta \mathscr{H}} e^{i\mathscr{H}t/\hbar} B\, e^{-i\mathscr{H}t/\hbar} A \} \\ &= \frac{i}{\hbar Z} \sum_N e^{-\beta E_N} \langle N \mid e^{i\mathscr{H}t/\hbar} B\, e^{-i\mathscr{H}t/\hbar} A \mid N \rangle \qquad (31.20) \\ &= \frac{i}{\hbar Z} \sum_{N,M} e^{-\beta E_N} e^{i(E_N - E_M)t/\hbar} B_{NM} A_{MN} . \end{aligned}$$

With (31.20) it is

$$\begin{aligned} &\int_{-\infty}^{+\infty} \langle B(t)\, A\rangle_\beta\, e^{i\omega t}\, \mathrm{d}t = 2\pi\, \varrho(\omega) , \\ &\varrho(\omega) = \frac{\hbar}{Z} \sum_{N,M} B_{NM} A_{MN} e^{-\beta E_N} \delta(E_M - E_N - \hbar\omega) . \end{aligned} \qquad (31.21)$$

Thus the Fourier-transform of the spectral representation $\varrho(\omega)$ just gives the correlation between $B(t)$ and $A$, i.e. $B_J(t)$ and $B_{J'}^{+}(0)$ or $u_i^m(t)$ and $u_j^n(0)$ and so on.

We next define a continuation of the Green function to imaginary (or complex) times: $it \Rightarrow \tau$:

$$f_{BA}(\tau) = \langle T_D\, e^{\tau\mathscr{H}/\hbar} B\, e^{-\tau\mathscr{H}/\hbar} A\rangle_\beta = \langle T_D\, B(\tau)\, A\rangle_\beta \qquad (31.22)$$

with

$$\langle T_D\, B(\tau_1)\, A(\tau_2)\rangle = \begin{cases} \langle B(\tau_1)\, A(\tau_2)\rangle & \text{for} \quad \tau_1 > \tau_2 \\ \langle A(\tau_2)\, B(\tau_1)\rangle & \text{for} \quad \tau_2 > \tau_1 . \end{cases}$$

$\tau$ will be limited to $-\beta\hbar < \tau < \beta\hbar$ now. From the definition (31.22) and the properties of $T_D$ and the trace it follows

$$\text{i)}\quad f_{BA}(\tau+\beta) = f_{BA}(\tau), \quad \text{if} \quad -\beta\hbar < \tau < 0\,, \tag{31.23a}$$

$$\text{ii)}\quad f_{BA}(\tau) = f_{AB}(-\tau)\,, \tag{31.23b}$$

$$\text{iii)}\quad f_{BA}(0^+) - f_{BA}(0^-) = \langle [B, A]_- \rangle_\beta = \frac{1}{Z}\,\text{Trace}\,\{e^{-\beta\mathscr{H}}[BA - AB]\} \tag{31.23c}$$

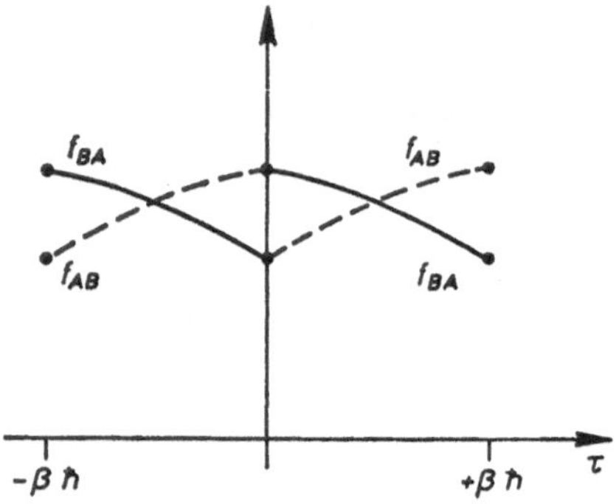


Fig. 8.1. Schematic representation of the properties of the thermodynamic Greens function

(see Fig. 8.1). $f_{BA}$ will be needed only in the interval $-\beta\hbar < \tau < \beta\hbar$. Therefore outside this interval we can take a periodic continuation of it and expand this continued function in a Fourier-series

$$f_{BA}(\tau) = \sum_{\nu=-\infty}^{+\infty} a_\nu\, e^{2\pi i\tau\nu/\beta\hbar}; \quad 2\pi\nu/\beta\hbar = \omega_\nu \tag{31.24}$$

with the coefficients

$$\begin{aligned} a_\nu &= \frac{1}{\beta\hbar}\int_0^{\beta\hbar} f_{BA}(\tau)\, e^{-i\tau\omega_\nu}\, d\tau \\ &= \frac{1}{\beta Z}\sum_{N,M} B_{NM} A_{MN} \frac{e^{-\beta E_M} - e^{-\beta E_N}}{E_N - E_M - i\hbar\omega_\nu}\,. \end{aligned} \tag{31.25}$$

The coefficients $a_\nu$, defined on discrete points $\nu$, can be continued into the complex plane

$$a_\nu \Rightarrow a(z) \quad \text{with} \quad i\omega_\nu = i\,2\pi\nu/\beta\hbar \Rightarrow z\,. \tag{31.26}$$

The function $a(z)$ is regular apart from a cut on the real axis, and it is further $\lim_{R\to\infty} a(x + iR) \to 0$ for every $x = \text{const}$. The discontinuity of $a(z)$ when passing the real axis is obtained from

$$\begin{aligned} a(x \pm i\epsilon) &= \frac{1}{\beta Z}\sum_{N,M} B_{NM} A_{MN} \frac{e^{-\beta E_M} - e^{-\beta E_N}}{E_N - E_M - \hbar x \mp i\hbar\epsilon} \\ &= \frac{1}{\beta Z}\sum_{NM} B_{NM} A_{MN} \{e^{-\beta E_M} - e^{-\beta E_N}\}\Big\{P\,\frac{1}{E_N - E_M - \hbar x} \pm \\ &\quad \pm i\pi\delta(E_N - E_M - \hbar x)\Big\} \end{aligned}$$

and therefore

$$a(x+i\epsilon)-a(x-i\epsilon)$$
$$=\frac{2\pi i}{\beta Z}\left(e^{\beta\hbar x}-1\right)\sum_{NM} e^{-\beta E_N} B_{NM} A_{MN}\,\delta(E_N-E_M-\hbar x)\,.$$

Comparison with (31.21) gives

$$\varrho(\omega)=\frac{\hbar\beta}{e^{-\beta\hbar\omega}-1}\lim_{\epsilon\to 0+}\frac{a(-\omega+i\epsilon)-a(-\omega-i\epsilon)}{2\pi i}\,. \tag{31.27}$$

Thus from $f_{BA}(\tau)$ we can calculate the $a_\nu$ and the $a(z)$ by Fourier-transformation, from their discontinuity on the real axis we have the representation $\varrho(\omega)$ and from the Fourier-transformation of $\varrho(\omega)$ we have the correlation function $\langle B(t)\,A\rangle_\beta$, which enters in many problems. In the same way as in (30.20, 21) we can show that

$$\int_{-\infty}^{+\infty}\langle A\,B(t)\rangle_\beta\, e^{i\omega t}\,dt=2\pi\,\varrho(\omega)\,e^{-\beta\hbar\omega} \tag{31.28}$$

and therefore

$$\langle B(t)\,A\rangle_\beta=\int_{-\infty}^{+\infty}\varrho(\omega)\,e^{-i\omega t}\,d\omega\,, \tag{31.29a}$$

$$\langle A\,B(t)\rangle_\beta=\int_{-\infty}^{+\infty}e^{-\beta\hbar\omega}\,\varrho(\omega)\,e^{-i\omega t}\,d\omega\,, \tag{31.29b}$$

from which it can be seen that

$$\langle B(t)\,A\rangle_\beta=\langle A\,B(t+i\beta\hbar)\rangle_\beta\,. \tag{31.30}$$

Next we look for a relation between the spectral representation $\varrho(\omega)$ and the retarded and advanced Greens-functions. According to (31.18) we define them by

$$G_{R,A}(t)=\pm\frac{i}{\hbar}\,\Theta(\pm t)\,\langle B(t)\,A-A\,B(t)\rangle_\beta \tag{31.31}$$

and their Fourier-transforms by

$$G_{R,A}(\omega)=\int_{-\infty}^{+\infty}G_{R,A}(t)\,e^{i\omega t-\epsilon|t|}\,dt\,,\quad \epsilon>0\,. \tag{31.31a}$$

Introducing (31.31) and (31.29) into (31.31a) we obtain

$$G_{R,A}(\omega)=\pm\frac{i}{\hbar}\int_{-\infty}^{+\infty}\Theta(\pm t)\,e^{i\omega t-\epsilon|t|}\{\langle B(t)\,A\rangle-\langle A\,B(t)\rangle\}\,dt$$

$$=\pm\frac{i}{\hbar}\int_{-\infty}^{\infty}dx\,\varrho(x)\,(1-e^{-\beta\hbar x})\int_{-\infty}^{+\infty}dt\,e^{i(\omega-x)t-\epsilon|t|}\,\Theta(\pm t)$$

$$G_{R,A}(\omega)=\frac{1}{\hbar}\int_{-\infty}^{+\infty}dx\,\varrho(x)\,(e^{-\beta\hbar x}-1)\,\frac{1}{\omega-x\pm i\epsilon} \tag{31.32}$$

$$=\frac{1}{\hbar}\,P\int_{-\infty}^{+\infty}\frac{(e^{-\beta\hbar x}-1)\,\varrho(x)\,dx}{\omega-x}\mp\frac{i\pi}{\hbar}\,(e^{-\beta\hbar\omega}-1)\,\varrho(\omega)\,.$$

From (31.29a) we have with $it = \tau > 0$

$$f_{BA}(\tau) = \int_{-\infty}^{+\infty} \varrho(\omega)\, e^{-\omega\tau}\, d\omega$$

and with (31.25)

$$a_\nu = -\frac{1}{\beta\hbar} \int_{-\infty}^{+\infty} dx\, \varrho(x)\, [e^{-\beta\hbar x} - 1] \frac{1}{x + i\omega_\nu}.$$

Using the continuation (31.26) and comparing the last equation with (31.32) we obtain

$$G_{R,A}(\omega) = \beta \cdot a(-\omega \mp i\epsilon). \tag{31.32a}$$

Similarly we obtain for the (causal) Greensfunction

$$G(t) = \frac{i}{\hbar} \langle T_D B(t) A \rangle_\beta \tag{31.33}$$

by Fourier-transformation and use of (31.29)

$$G(\omega) = \frac{1}{\hbar} \int_{-\infty}^{+\infty} dx\, \varrho(x)\, (e^{-\beta\hbar x} - 1) \left\{ P \frac{1}{\omega - x} - i\pi\, \mathrm{ctgh}\, (\beta\hbar x/2)\, \delta(\omega - x) \right\} \tag{31.34}$$

or

$$G'(\omega) = \mathrm{Re}\, G(\omega) = \frac{1}{\hbar} P \int_{-\infty}^{+\infty} \frac{\varrho(x)\, (e^{-\beta\hbar x} - 1)}{\omega - x}\, dx \tag{31.34a}$$

$$G''(\omega) = \mathrm{Im}\, G(\omega) = -\frac{\pi}{\hbar} \varrho(\omega)\, (e^{-\beta\hbar\omega} - 1)\, \mathrm{ctgh}\, (\beta\hbar\omega/2) \tag{31.34b}$$

therefore

$$\mathrm{Re}\, G(\omega) = \frac{1}{\pi} P \int_{-\infty}^{+\infty} \frac{\mathrm{Im}\, G(x)}{x - \omega}\, \mathrm{tgh}\, (\beta\hbar x/2) \cdot dx \tag{31.35}$$

which is identical with (31.14) derived in the harmonic approximation. $G(\omega)$ cannot be continued analytically into the complex $\omega$-plane; but $G_{R,A}(\omega)$ can. Let $\omega$ be complex; then (31.32) can be rewritten as

$$G_{ra}(\omega) = \frac{1}{\hbar} \int_{-\infty}^{+\infty} dx \frac{\varrho(x)\, (e^{-\beta\hbar x} - 1)}{\omega - x} = \begin{cases} G_R(\omega) & \text{if} \quad \mathrm{Im}\, \omega > 0 \\ G_A(\omega) & \text{if} \quad \mathrm{Im}\, \omega < 0. \end{cases} \tag{31.36}$$

Thus $G_{R,A}(\omega)$ may be considered as one analytical function $G_{ra}$ in the complex $\omega$-plane with a singularity on the real axis. The convergence of the integral in (31.31a) and of its derivatives with respect to $\omega$ are guaranteed by the cut-off factor

$$e^{-\epsilon|t|} \quad \text{with} \quad \epsilon > 0$$

or equivalently

$$e^{-\epsilon t} \quad \text{with} \quad \epsilon > 0 \quad \text{for} \quad t > 0, \epsilon < 0 \quad \text{for} \quad t < 0 .$$

$G_{ra}(\omega)$ consists of two branches, one defined in the upper half-plane ($G_R$) and one defined in the lower half-plane ($G_A$). The discontinuity of $G_{ra}(\omega)$ when passing the real axis determines again the spectral representation. Using (31.36) we have immediately

$$G_{ra}(\omega + i\epsilon) - G_{ra}(\omega - i\epsilon) = -\frac{2\pi i}{\hbar}(e^{-\beta\hbar\omega} - 1)\,\varrho(\omega) . \tag{31.37}$$

We can express the correlation functions (31.29) by using (31.37)

$$\langle B(t)\,A\rangle_\beta = \frac{i\hbar}{2\pi}\int \frac{G_{ra}(\omega + i\epsilon) - G_{ra}(\omega - i\epsilon)}{e^{-\beta\hbar\omega} - 1}\cdot e^{-i\omega t}\,d\omega \tag{31.38a}$$

$$\langle A\,B(t)\rangle_\beta = \frac{i\hbar}{2\pi}\int \frac{G_{ra}(\omega + i\epsilon) - G_{ra}(\omega - i\epsilon)}{1 - e^{\beta\hbar\omega}}\cdot e^{-i\omega t}\,d\omega . \tag{31.38b}$$

For the retarded and advanced Greens functions we have from (31.32) the dispersion relation

$$\operatorname{Re} G_{R,A}(\omega) = \pm\frac{1}{\pi}\,P\int \frac{\operatorname{Im} G_{R,A}(x)}{x - \omega}\,dx \tag{31.39}$$

in agreement with (31.16) for the harmonic approximation. So we can obtain all the correlation-functions and Greens functions from the spectral representation $\varrho(\omega)$, which in turn can be obtained from the function $f_{BA}(\tau)$ defined in (31.22). The main problem is the calculation of $f_{BA}(\tau)$, which is indicated for one example in Section 27.

Finally we give the equation of motion for the Greens functions; they can be derived from (31.31, 33) by derivation with respect to time:

$$-i\hbar\frac{\partial G(t)}{\partial t} = +\,\delta(t)\,\langle[B(t), A]_-\rangle_\beta + \left\langle T_D\frac{dB}{dt}A\right\rangle_\beta \tag{31.40a}$$

$$-i\hbar\frac{\partial G_{R,A}(t)}{\partial t} = +\,\delta(t)\,\langle[B(t), A]_-\rangle_\beta \pm \Theta(\pm t)\left\langle\left[\frac{dB}{dt}, A\right]_-\right\rangle_\beta . \tag{31.40b}$$

## 32. Remarks about some Theorems used in Sect. 27

[62 K 3, 62 B 8, 63 A 1, 64 N 2].

Also here we will give not a detailed derivation of the theorems in question, but merely indicate their statements and their use. The proofs have been given in many papers. The expression we have to calculate in (27.65) is

$$f_{JJ'}(\tau) = \frac{\langle T_D\{\tilde{B}_J(\tau)\,\tilde{B}_{J'}^+(0)\,S(\beta)\}\rangle_{qh}}{\langle T\,S(\beta)\rangle_{qh}} \tag{32.1}$$

with $\beta = 1/kT$

$$S(\beta) = \sum_{n=0}^{\infty}\frac{(-1)^n}{n!}\int_0^\beta d\tau_1 \ldots \int d\tau_n\,\tilde{W}(\tau_1)\ldots\tilde{W}(\tau_n) . \tag{32.1a}$$

The perturbation $\tilde{W}(\tau)$ contains operators $\tilde{B}_J, \tilde{B}_J^+$, in our case products of three and four of these operators. The averaging process in (32.1) is over all the occurring products of $\tilde{B}_J$-operators, taking into account the time-ordering $T_D$.

*Wick's theorem*

This theorem states, that the average occurring in (32.1) can be divided into products of averages over pairs of $\tilde{B}_J$– or $b_J, b_J^+$–operators, where all the possible combinations of operators to pairs have to be taken into account additively. The number of $b_J$ has to be even, otherwise the average vanishes. We illustrate this with the example of four $b$, $b^+$-operators. $a_j$ may be a representative for either $b_J$ or $b_J^+$. Assuming that $a_1 \ldots a_4$ have the correct "time" order, we obtain by successive commutations (all minus-commutators)

$$\langle a_1 a_2 a_3 a_4\rangle_{qh} = \langle [a_1, a_2]\, a_3 a_4\rangle + \langle a_2 [a_1, a_3]\, a_4\rangle + \\ + \langle a_2 a_3 [a_1, a_4]\rangle + \langle a_2 a_3 a_4 a_1\rangle .$$

With $\varrho = \exp\{-\beta \mathscr{H}_{qh}\}/Z$; $\langle A\rangle_{qh} = \mathrm{Trace}(\varrho A)$ we have for the difference of the left side and the right side (last term)

$$\mathrm{Trace}\{[\varrho, a_1]\, a_2 a_3 a_4\} = [a_1, a_2]\,\langle a_3 a_4\rangle + \\ + [a_1, a_3]\,\langle a_2 a_4\rangle + \\ + [a_1, a_4]\,\langle a_2 a_3\rangle ,$$

because the $[a_1, a_J]$ are numbers. It is

$$[\varrho, b_J] = (1 - \mathrm{e}^{-\beta\hbar\omega_J})\, \varrho b_J = c(-\omega_J)\, \varrho b_J ,$$

$$[\varrho, b_J^+] = (1 - \mathrm{e}^{\beta\hbar\omega_J})\, \varrho b_J^+ = c(+\omega_J)\, \varrho b_J^+ ,$$

$$[a_1, a_J] = \mathrm{Trace}\, \varrho\, [a_1, a_J] = \mathrm{Trace}\, [\varrho, a_1]\, a_J = c(\mp\omega_1)\, \langle a_1 a_J\rangle$$

therefore

$$\langle a_1 a_2 a_3 a_4\rangle_{qh} = \langle a_1 a_2\rangle \langle a_3 a_4\rangle + \langle a_1 a_3\rangle \langle a_2 a_4\rangle + \langle a_1 a_4\rangle \langle a_2 a_3\rangle . \quad (31.2)$$

In this simple example this is all. If there would be more than four $a_J$-operators, we would have to continue the procedure. For an average over a product of $2n$ operators we obtain $\frac{(2n)!}{2^n \cdot n!}$ terms. In the pairs, the time-ordering is conserved, so that we can put in the $T_D$-symbol in every bracket $\langle\ \rangle$. Starting with a product, which is not in a time-ordered form, it is easy to see, that the same result is valid. Also, starting with the $B_J = b_J - b_J^+$ one can show that a corresponding formula occurs. The relation is especially simple, because here only minus-commutators are involved. (31.2) and corresponding relations for higher order averages are used in Sect. 27.

*Graphical representation*

In the calculation of (32.1) there enter terms of cubic anharmonicity and of quartic anharmonicity, in our approximation. The cubic term is connected with $3B_f$-operators, the quartic term with 4. As only even terms enter, the lowest order term of cubic anharmonicity is of second

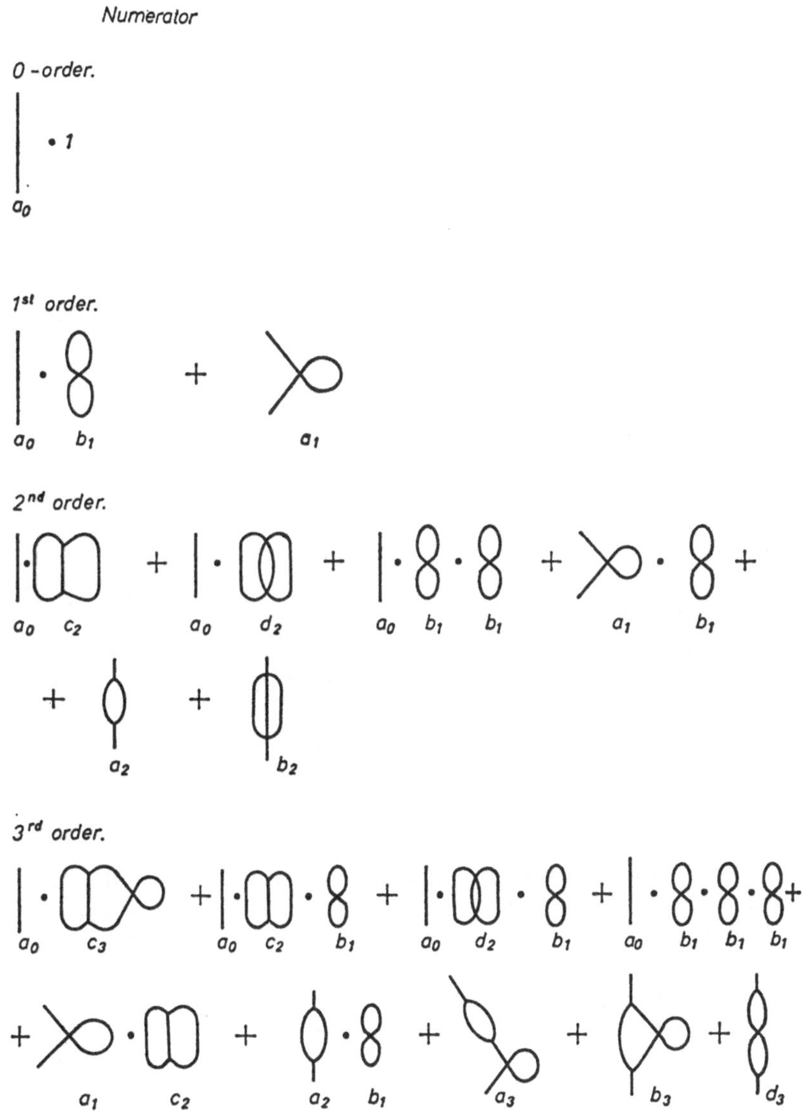


Fig. 8.2a

**Fig. 8.2a and b. Different diagrams which give contributions to thermal averages in $S(\beta)$; shown are only some representatives.**

order (6 $B_J$), that of quartic anharmonicity of first order (4 $B_J$). We represent the cubic interaction, three phonons involved, by a vertex with three outgoing lines, the quartic interaction (four phonons) by a vertex with four outgoing lines. The averages of two $B_J$-operators,

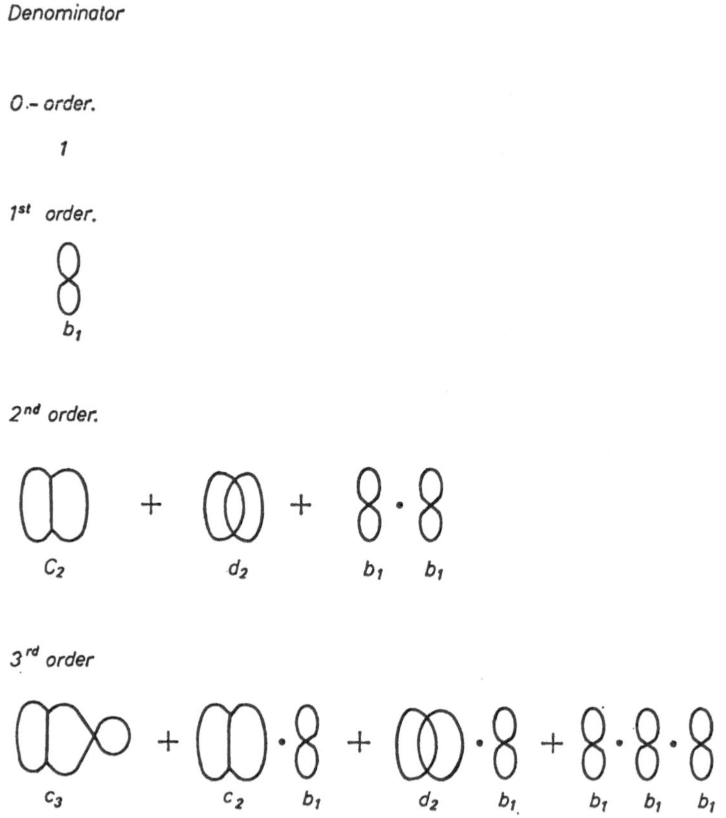


Fig. 8.2 b

defined in (27.61) represent a phonon and are denoted by a line. "Time" continues from bottom to top. In the numerator diagrams occur, where the two $B_J$ in front of the $S$ are averaged separately, and diagrams, where they are connected with some $B_J$'s from $S(\beta)$. In Fig. (8.2) some of the diagrams are represented, for the numerator (8.2a) as well as for the denominator. The "time"-ordering has not been taken into account, and equivalent diagrams arising from different pairings of the $B_J$ are not shown. Disconnected diagrams are for example $a_0b_1$; $a_0c_2$; $a_0d_2$; $a_0b_1b_1$; $a_1b_1$; $a_0c_3$; $a_0c_2b_1$; $a_0d_2b_1$; $a_0b_1b_1b_1$; $a_1c_2$; $a_2b_1$. The rest is connected, in Fig. (8.2a). In calculating (32.1) the sum has to be taken over all the diagrams, including equivalent ones from different pairings.

*Linked-cluster-theorem*

The sum of the numerator gives, indicated schematically

$$\begin{aligned}\mathcal{N} &= a_0\{1 + b_1 + c_2 + d_2 + b_1^2 + c_3 + c_2 b_1 + d_2 b_1 + b_1^3 + \cdots\} + \\ &\quad + a_1\{1 + b_1 + c_2 + \cdots\} + \\ &\quad + a_2\{1 + b_1 + \cdots\} + \\ &\quad + b_2\{1 + b_1 + \cdots\} + a_3\{1 + b_1 + \cdots\} + b_3\{1 + b_1 + \cdots\} + \cdots \\ &= \{a_0 + a_1 + a_2 + b_2 + a_3 + b_3 + \cdots\}\{1 + b_1 + c_2 + d_2 + b_1^4 + \cdots\}\end{aligned}$$

and it can be seen that this is correct also if the different factors arising from all possible pairings are taken into account. Similarly the denominator gives

$$\mathcal{D} = \{1 + b_1 + c_2 + d_2 + b_1^2 + \cdots\},$$

also including the correct factors from the pairings. Therefore

$$\begin{aligned}\frac{\mathcal{N}}{\mathcal{D}} &= a_0 + a_1 + a_2 + b_2 + a_3 + b_3 + \cdots \\ &= \text{sum over all the connected diagrams of } \mathcal{N}\end{aligned}$$

which includes again the factors arising form different pairings. This is the linked-cluster-theorem

$$f_{JJ'}(\tau) = \langle T_{\mathrm{D}}\{\hat{B}_J(\tau)\, \widetilde{B_{J'}^+}(0)\, S(\beta)\rangle_{\mathrm{qh,\,connected}}\,. \tag{32.3}$$

It can be shown to be valid for more difficult interactions; we have illustrated it only for the simple case we need in Sect. 27.

Finally we will derive the expansion (27.52). Expanding with respect to the exponents, multiplying and averaging we have

$$\begin{aligned}\langle e^{A-\langle A\rangle} \cdot e^{B-\langle B\rangle}\rangle &= 1 + \frac{1}{2}\{\langle A^2\rangle - \langle A\rangle^2 + \langle B^2\rangle - \langle B\rangle^2\} + \\ &\quad + \frac{1}{6}\{\langle A^3\rangle - 3\langle A^2\rangle\langle A\rangle + 2\langle A\rangle^3 + \langle B^3\rangle - \\ &\quad - 3\langle B^2\rangle\langle B\rangle + 2\langle B\rangle^3\} + \langle A B\rangle - \langle A\rangle\langle B\rangle + \\ &\quad + \frac{1}{2}\{\langle A^2 B\rangle - \langle A^2\rangle\langle B\rangle - 2\langle A\rangle\langle A B\rangle + \\ &\quad + 2\langle A\rangle^2\langle B\rangle + \langle A B^2\rangle - \langle A\rangle\langle B^2\rangle - \\ &\quad - 2\langle A B\rangle\langle B\rangle + 2\langle A\rangle\langle B\rangle^2\} + \\ &\quad + \text{terms of fourth and higher order}\,.\end{aligned}$$

This can be rewritten as

$$\begin{aligned}&\langle e^A \cdot e^B\rangle \\ &= \exp\left\{\langle A\rangle + \frac{1}{2}[\langle A^2\rangle - \langle A\rangle^2] + \frac{1}{6}[\langle A^3\rangle - 3\langle A^2\rangle\langle A\rangle + 2\langle A\rangle^3]\right\} \times \\ &\times \exp\left\{\langle B\rangle + \frac{1}{2}[\langle B^2\rangle - \langle B\rangle^2] + \frac{1}{6}[\langle B^3\rangle - 3\langle B^2\rangle\langle B\rangle + 2\langle B\rangle^3]\right\} \times \\ &\qquad \times \exp\left\{\langle A B\rangle - \langle A\rangle\langle B\rangle + \right.\end{aligned}$$

$$+\frac{1}{2}[\langle A^2B\rangle - \langle A^2\rangle\langle B\rangle + \langle AB^2\rangle - \langle A\rangle\langle B^2\rangle] -$$
$$- [\langle A\rangle\langle AB\rangle - \langle A\rangle^2\langle B\rangle + \langle AB\rangle\langle B\rangle - \langle A\rangle\langle B\rangle^2]\Big\} \times$$
$$\times \{1 + \text{terms of fourth and higher order}\}. \qquad (32.4)$$

The indicated procedure can now be used to obtain in a similar manner by an iteration process the higher order terms in the exponent, which represent just the higher order fluctuations. (32.4) is already correct up to third order which is used only in the discussion of Sect. 27.

## References

85 R 1. *Rayleigh, Lord (J. W. Strutt):* Proc. Math. Soc. London **17**, **4** (1885)
10 V 1. *Voigt, W.:* Lehrbuch der Kristallphysik. Leipzig und Berlin: Teubner 1910. Nachdruck 1966
12 D 1. *Debye, P.:* Ann. Physik **39**, **789** (1912)
12 G 1. *Grüneisen, E.:* Ann. Physik, **39**, **257** (1912)
14 D 1. *Debye, P.:* Vorträge über kinetische Theorie der Materie und Elektrizität Berlin 1914
22 B 1. *Born, M., u. E. Brody:* Z. Physik **11**, **327** (1922)
22 S 1. *Schrödinger, E.:* Z. Physik **11**, 170 (1922)
27 B 1. *Born, M., u. R. Oppenheimer:* Ann. Physik [4] **84**, **457** (1927)
33 B 1. — *u. M. Blackman:* Z. Physik **82**, **551** (1933)
33 B 2. *Blackman, M.:* Z. Physik **86**, 421 (1933)
35 F 1. *Fuchs, K.:* Proc. Roy. Soc. (London) Ser. A **153**, **622** (1935)
36 F 1. — Proc. Roy. Soc. (London) Ser. A **157**, **444** (1936)
36 F 2. *Fermi, E.:* Ric. Sci. **7**, **13** (1936)
37 M 1. *Murnaghan, F. C.:* Am .J. Math. **49**, 235 (1937), and: Finite deformation of an elastic solid. New York 1951
38 C 1. *Casimir, H. B. G.:* Physica **5**, **495** (1938)
39 L 1. *Lamb jr., W. E.:* Phys. Rev. **55**, 190 (1939)
39 M 1. *Margenau, H.:* Revs. Mod. Phys. **11**. 1 (1939)
40 K 1. *Kellermann, E. W.:* Phil. Trans. Roy. Soc. (London) Ser. A **238**, **513** (1940)
43 B 1. *Born, M.:* Rept. Progr. Phys. **9**, **294** (1943)
43 B 2. *Born, M.:* Proc. Cambridge Phil. Soc. **39**, 100 (1943)
45 B 1. — Revs. Mod. Phys. **17**, **245** (1945)
47 B 1. *Breit, G.:* Phys. Rev. **71**, **215** (1947)
47 B 2. *Birch, F.:* Phys. Rev. **71**, **809** (1947)
48 L 1. *Löwdin, P. O.:* Thesis. Upsala: Almquist and Wiksell 1948
49 A 1. *Axilrod, B. M.:* J. Chem. Phys. **17**, 1349 (1949)
50 H 1. *Huang, K.:* Proc. Roy. Soc. (London) Ser. A **203**, **178** (1950)
50 L 1. *Löwdin, P. O.:* J. Chem. Phys. **18**, **365** (1950)
50 L 2. *Lippmann, B. A., and J. Schwinger:* Phys. Rev. **79**, **469** (1950)
50 T 1. *Tolpygo, K. B.:* Zh. Eksperim i. Teor. Fiz. **20**, **497** (1950)
51 A 1. *Axilrod, B. M.:* J. Chem. Phys. **19**, 719, 724 (1951)
51 B 1. *Born, M.:* Nachr. Akad. Wiss. Göttingen, Math.-Physik. Kl. IIa, No. 6 (1951)
51 L 1. *Laval, J.:* C. R. Acad. Sci. (Paris) **232**, **1947** (1951)
51 R 1. *Reitz, J. R., and J. L. Gammel:* J. Chem. Phys. **19**, **894** (1951)
52 B 1. *Brauer, P.:* Z. Naturforsch. **7** A, **372** (1952)
52 C 1. *Curien, H.:* Bull. Soc. Franc. Mineral. **75**, **197** (1952)
— Acta Cryst. **5**, **393** (1952)
52 L 1. *Lundquist, S. O.:* Arkiv Fysik **6**, **25** (1952)
52 L 2. *Lax, M.:* J. Chem. Phys. **20**, **1752** (1952)

53 D 1. *Dyson, F. J.:* Phys. Rev. **92**, 1331 (1953)
53 H 1. *Herpin, A.:* Phys. Radium **14**, 611 (1953)
53 K 1. *Kröner, E.:* Z. Physik **136**, 402 (1953)
53 T 1. *Toya, T.:* Busseiron Kenkyu **64**, 1 (1953)
54 B 1. *Born, M., u. K. Huang:* Dynamical theory of crystal lattices. Oxford: University Press 1954
54 H 1. *Haasen, P., u. G. Leibfried:* Fortschr. Physik **2**, 73 (1954)
54 H 2. *Hove, L. v.:* Phys. Rev. **98**, 1189 (1954)
54 K 1. *Koster, G. F., and J. C. Slater:* Phys. Rev. **95**, 1167 (1954)
54 K 2. — Phys. Rev. **95**, 1436 (1954)
54 K 3. *Kerner, E. H.:* Phys. Rev. **95**, 687 (1954)
54 L 1. *Lax, M.:* Phys. Rev. **94**, 1391 (1954)
54 L 2. *Laval, J.:* C. R. Acad. Sci. (Paris) **238**, 1773 (1954)
54 P 1. *Prigogine, J., R. Bingen, and J. Jeener:* Physica **20**, 383 (1954)
55 B 1. *Barron, T. H. K.:* Phil. Mag. **46**, 720 (1955)
55 B 2. *Brenig, W.:* Z. Physik **143**, 168 (1955)
55 B 3. *Blackman, M.:* Hdb. der Physik, 7/I, 325—382. Berlin, Göttingen, Heidelberg: Springer 1955
55 B 4. *Bauer, H.:* Staatsexamensarbeit Göttingen (1955)
55 J 1. *Jakobsen, E. H.:* Phys. Rev. **97**, 654 (1955)
55 J 2. *Jansen, J., and L. M. Dawson:* J. Chem. Phys. **23**, 482 (1955)
55 K 1. *Kröner, E.:* Z. Physik **141**, 386 (1955)
55 L 1. *Leibfried, G.:* Hdb. der Physik, 7/I, 104—324. Berlin, Göttingen, Heidelberg: Springer 1955
55 L 2. *Lundquist, S. O.:* Arkiv Fysik **9**, 435 (1955)
55 M 1. *Montroll, E. W., and R. B. Potts:* Phys. Rev. **100**, 525 (1955)
55 N 1. *Nakajima, S.:* Advan. Phys. **4**, 363 (1955)
55 R 1. *Raman, C. V., and K. S. Viswanathan:* Proc. Indian Acad. Sci. A **42**, 1, 51 (1955)
—, *and D. Krishnamurti:* Proc. Indian Acad. Sci. A **42**, 111 (1955)
55 S 1. *Shostak, A.:* J. Chem. Phys. **23**, 1808 (1955)
55 S 2. *Seeger, A.:* Hdb. der Physik. VII/1, 2. Berlin, Göttingen, Heidelberg: Springer 1955
55 S 3. *Stoneley, R.:* Proc. Roy. Soc. (London) Ser. A **232**, 447 (1955)
55 T 1. *Toya, T.:* Busseiron Kenkyu **99**, 33 (1955)
55 Y 1. *Yamashita, J., and T. Kurosawa:* J. Phys. Soc. Japan **19**, 610 (1955)
56 L 1. *Löwdin, P. O.:* Phil. Mag. Suppl. **5**, 1—171 (1956)
56 L 2. *Lifshitz, I. M.:* Nuovo Cimento **3**, 716 (1956)
56 M 1. *Montroll, E. W., and R. B. Potts:* Phys. Rev. **102**, 72 (1956)
56 M 2. *McGinnes, R. T., and L. Jansen:* Phys. Rev. **101**. 1301 (1956)
56 M 3. *Mashkevich, V. S., i K. B. Tolpygo:* Dokl. Akad. Nauk. SSSR **111**, 575 (1956)
56 S 1. *Süssmann, G.:* Z. Naturforsch. **11a**, 1 (1956)
56 W 1. *Walker, C. B.:* Phys. Rev. **103**, 547 (1956)
57 B 1. *Barron, T. H. K.:* Ann. Phys. (N. Y.) **1**, 77 (1957)
57 B 2. *Blackman, M.:* Proc. Phys. Soc. (London) B **70**, 827 (1957)
57 D 1. *Deresiewicz, H., and R. D. Mindlin:* J. Appl. Phys. **28**, 669 (1957)
57 E 1. *Eshelby, J. D.:* Solid State Phys. **3**, 79—144 (1957)
57 F 1. *Fumi, F. G., and M. P. Tosi:* Discussions Faraday Soc. **23**, 92 (1957)
57 H 1. *Hall, G. L.:* J. Phys. Chem. Solids **3**, 210 (1957)
57 H 2. *Hori, J., and T. Asahi:* Progr. Theoret. Phys. (Kyoto) **17**, 523 (1957). — *Hori, J.*: Progr. Theoret. Phys. (Kyoto) **18**, 367 (1957)
57 K 1. *Kanzaki, H.:* J. Phys. Chem. Solids **2**, 24 (1957)
57 L 1. *Lundquist, S. O.:* Arkiv Fysik **12**, 263 (1957)
57 M 1. *Mashkevich, V. S., i K. B. Tolpygo:* Soviet Phys. JETP **5**, 435 (1957). — *Mashkevich, V. S.:* Soviet Phys. JETP **5**, 707 (1957)
57 S 1. *Schmidt, H.:* Phys. Rev. **105**, 425 (1957)
57 S 2. *Synge, J. L.:* J. Math. Phys. **35**, 323 (1957)
57 W 1. *Wallis, R. F.:* Phys. Rev. **105**, 540 (1957)
58 B 1. *Blackman, M.:* Phil. Mag. [8], **3**, 831 (1958)

58 B 2. *Brockhouse, B. N., and P. K. Iyengar:* Phys. Rev. **111**, 747 (1958)
58 D 1. *Dick, B. G., and A. W. Overhauser:* Phys. Rev. **112**, 90 (1958)
58 E 1. *Englman, R.:* Nuovo Cimento, **10**, 615 (1958)
58 H 1. *Haussühl, S.:* Acta Cryst. **11**, 58 (1958)
58 K 1. *Kröner, E.:* Kontinuumstheorie der Versetzungen und Eigenspannungen. Berlin, Göttingen, Heidelberg: Springer 1958
58 K 2. — Z. Physik **151**, 504 (1958)
58 K 3. *Klemens, P. G.:* Solid State Phys. **7**, 1 (1958)
58 L 1. *Ludwig, W.:* J. Phys. Chem. Solids **4**, 283 (1958)
58 L 2. *Leibfried, G., and H. Hahn:* Z. Physik **150**, 497 (1958)
58 M 1. *Mössbauer, R. L.:* Z. Physik **151**, 124 (1958)
58 S 1. *Süssmann, G.:* Z. Naturforsch. **13a**, 1 (1958)
58 S 1. *Smith, Ch. S.:* Solid State Phys. **6**, 175—249 (1958)
58 T 1. *Toya, T.:* Progr. Theoret. Phys. (Kyoto) **20**, 973 (1958); — J. Res. Inst. Catalysis, Hokkaido Univ. **6**, 161, 183 (1958)
59 B 1. *Blackman, M.:* Proc. Phys. Soc. (London) **74**, 17 (1959)
59 C 1. *Cochran, W.:* Phys. Rev. Letters **2**, 495 (1959); — Proc. Roy. Soc. (London) Ser. A **253**, 260 (1959); — Phil. Mag. **4**, 1082 (1959)
59 C 2. *Callaway, J.:* Phys. Rev. **113**, 1046 (1959)
59 F 1. *Fischer, K.:* Z. Physik **155**, 59 (1959); **157**, 198 (1959)
59 H 1. *Hanlon, J. E., and A. W. Lawson:* Phys. Rev. **113**, 472 (1959)
59 K 1. *Kohn, W.:* Phys. Rev. Letters **2**, 393 (1959)
59 L 1. *Lundquist, S. O., V. Lundström, E. Tenerz och I. Waller:* Arkiv Fysik **15**, 193 (1959)
59 M 1. *Mashkevich, V. S.:* Soviet Phys. JETP **9**, 76, 1237 (1959)
59 M 2. *Musgrave, M. J. P.:* Rep. Progr. Phys. **22**, 74 (1959)
59 W 1. *Wallis, R. F.:* Phys. Rev. **116**, 302 (1959)
60 B 1. *Bennemann, K. H., u. L. Tewordt:* Z. Naturforsch. **15a**, 772 (1960)
60 D 1. *Dean, P.:* Proc. Roy. Soc. (London) Ser. A **254**, 507 (1960). — *Dean, P., and J. L. Martin:* Proc. Roy. Soc. (London) Ser. A **259**, 409 (1960)
60 H 1. *Hedin, L. T.:* Arkiv Fysik **18**, 369 (1960)
60 H 2. *Havinga, E. E.:* Phys. Rev. **119**, 1193 (1960)
60 H 3. *Hori, J.:* Progr. Theoret. Phys. (Kyoto) **23**, 475 (1960)
60 J 1. *Josephson, B. D.:* Phys. Rev. Letters **4**, 341 (1960)
60 L 1. *Leibfried, G., u. W. Ludwig:* Z. Physik **160**, 80 (1960)
60 L 2. *Larsson, K. E.:* Arkiv Fysik **17**, 369 (1960)
60 M 1. *Maradudin, A. A., E. W. Montroll, G. H. Weiss, R. Herman et H. W. Milnes:* Mém. Acad. Roy. Belg. Cl. Sci., Coll. 4°, Ser. II, Vol. 14, nr. 7 (1960)
60 M 2. *Mahanty, J., A. A. Maradudin, and G. H. Weiss:* Progr. Theoret. Phys. (Kyoto) **24**, 648 (1960)
60 P 1. *Pound, R. V., and G. A. Rebka:* Phys. Rev. Letters **4**, 274, 337, 397 (1960)
60 S 1. *Seeger, A., u. O. Buck:* Z. Naturforsch. **15a**, 1056 (1960)
60 S 2. *—, and E. Mann:* J. Phys. Chem. Solids **12**, 326 (1960)
60 S 3. *Schaefer, G.:* J. Phys. Chem. Solids **12**, 233 (1960)
60 W 1. *Woods, A. D. B., W. Cochran, and B. N. Brockhouse:* Phys. Rev. **119**, 980 (1960)
60 W 2. *Wallis, R. F., D. C. Gazis, and R. Herman:* Phys. Rev. **119**, 533 (1960)
60 Z 1. *Zubarev, D. N.:* Soviet Phys. Usp. **3**, 320 (1960)
61 A 1. *Abrahamson, A. A., R. D. Hatcher, and G. H. Vineyard:* Phys. Rev. **121**, 159 (1961)
61 B 1. *Bennemann, K. H.:* Z. Physik **165**, 445 (1961)
61 B 2. *Brockhouse, B. N.* et al.: Inelastic scattering of neutrons. p. 531 I. A. E. A. Vienna 1961
61 B 3. *Barron, T. H. K.:* Phys. Rev. **123**, 1995 (1961)
61 C 1. *Chester, G. V.:* Advan. Phys. **10**, 357 (1961)
61 C 2. *Carruthers, P.:* Rev. Mod. Phys. **33**, 92 (1961)
61 D 1. *Dean, P.:* Proc. Roy. Soc. (London) Ser. A **260**, 263 (1961). — *Martin, J. L.:* Proc. Roy. Soc. (London) Ser. A **260**, 139 (1961)
61 H 1. *Hahn, H., u. W. Ludwig:* Z. Physik **161**, 404 (1961)

61 H 2. — Z. Physik **165**, 569 (1961)
61 K 1. *Krivoglaz, M. A.:* Soviet Phys. JETP **13**, 397 (1961)
61 K 2. *Klemens, P. G.:* Phys, Rev. **122**, 443 (1961)
61 L 1. *Leibfried, G., and W. Ludwig:* Solid State Phys. **12**, 275 (1961)
61 L 2. *Lundquist, S. O.:* Arkiv Fysik **19**, 113 (1961)
61 L 3. *Litzmann, O., and J. Cely:* Czech. J. Phys. B **11**, 320 (1961)
61 L 4. *Langer, J. S.:* J. Math. Phys. **2**, 584 (1961)
61 L 5. *Leibfried, G., u. W. Ludwig:* Z. Physik **161**, 475 (1961)
61 L 6. *Ludwig, W.:* Z. Physik **164**, 490 (1961)
61 M 1. *Maradudin, A. A., P. A. Flinn, and R. A. Coldwell-Horsfall:* Ann. Phys. (N.Y.) **15**, 337 (1961)
61 M 2. —, —, — Ann. Phys. (N.Y.) **15**, 360 (1961)
61 M 3. —, *and G. H. Weiss:* Phys. Rev. **123**, 1968 (1961)
61 M 4. *Mashkevich, V. S.:* Soviet Phys. Solid State **2**, 2345 (1961)
61 M 5. *Mann, E., R. von Jan, u. A. Seeger:* phys. stat. solidi **1**, 17 (1961)
61 R 1. *Rosenstock, H. B.:* Phys. Rev. **121**, 416 (1961)
61 S 1. *Swalin, R. A.:* J. Phys. Chem. Solids **18**, 290 (1961)
61 T 1. *Tolpygo, K. B.:* Soviet Phys. Solid State **3**, 685, 2482 (1961); **2**, 2367 (1961)
61 T 2. *Toya, T.:* J. Res. Inst. Catalysis, Hokkaido Univ. **9**, 178 (1961)
62 B 1. *Barron, T. H. K., and M. L. Klein:* Phys. Rev. **127**, 1997 (1962)
62 B 2. *Brockhouse, B. N., K. R. Rao, and A. D. B. Woods:* Phys. Rev. Letters **7**, 93 (1962)
62 B 3. —, *A. D. B. Woods, R. H. March, A. T. Stewart, and R. Bowers:* Phys.Rev. **128**, 1112 (1962)
62 B 4. *Brout, R., and W. M. Vissher:* Phys. Rev. Letters **9**, 54 (1962)
62 B 5. *Bacon, M. D., P. Dean, and J. L. Martin:* Proc. Phys. Soc. (London) **80**, 174 (1962)
62 B 6. *Bross, H.:* Phys. stat. solidi **2**, 481 (1962)
62 B 7. *Brockhouse, B. N.*, et al.: Phys. Rev. **128**, 1099 (1962)
62 B 8. *Bonch-Bruevich, V. L., and S. V. Tyablikov:* The Green-function method in statistical mechanics. Amsterdam: North-Holland P. C. 1962
62 C 1. *Cowley, R. A.:* Proc. Roy. Soc. (London) Ser. A **268**, 109, 121 (1962)
62 D 1. *Dantl, G.:* Z. Physik **166**, 115 (1962)
62 F 1. *Flinn, P. A., and A. A. Maradudin:* Ann. Phys. (N.Y.) **18**, 81 (1962)
62 F 2. *Fritz, B.:* J. Phys. Chem. Solids **23**, 375 (1962)
62 F 3. *Frauenfelder, H.:* The Mössbauer-effect. New-York: Benjamin Inc. 1962
62 G 1. *Gazis, D. C., and R. F. Wallis:* J. math. Phys. **3**, 190 (1962)
62 H 1. *Hardy, J. R.:* Phil. Mag. **7**, 315 (1962)
62 J 1. *Jansen, L.:* Phys. Rev. **125**, 1798 (1962)
62 K 1. *Krivoglaz, M. A.:* Soviet Phys. Solid State **3**, 2671 (1962)
62 K 2. *Kockedee, J. J. J.:* Physica **28**, 374 (1962)
62 K 3. *Kadanoff, L. P., and G. Baym:* Quantum statistical mechanics. New York: Benjamin Inc. 1962
62 K 4. *Kagan, Yu., and Ya. A. Iosilewskii:* Soviet Phys. JEPT **15**, 182 (1962)
62 K 5. *Keller, J. M., and D. C. Wallace:* Phys. Rev. **126**, 1275 (1962)
62 L 1. *Litzman, O., F. Klvana:* Phys. stat. solidi **2**, 42 (1962)
62 M 1. *Mozer, B., K. Otnes, and V. W. Myers:* Phys. Rev. Letters **8**, 278 (1962)
62 M 2. *Maradudin, A. A., A. E. Fein:* Phys. Rev. **128**, 2589 (1962)
62 M 3. —, *and P. A. Flinn:* Phys. Rev. **126**, 2059 (1962)
62 M 4. — — *and S. Ruby:* Phys. Rev. **126**, 9 (1962)
62 M 5. *Maradudin, A. A.:* Phys. status solidi **2**, 1493 (1962)
62 N 1. *Nardelli, G. F., and N. Tettamanzi:* Phys. Rev. **126**, 1283 (1962)
62 N 2. *Ninio, F.:* Phys. Rev. **126**, 962 (1962)
62 S 1. *Sandor, E.:* Acta Cryst. **15**, 463 (1962)
62 S 2. *Seeger, A., E. Mann, and R. von Jan:* J. Phys. Chem. Solids **23**, 639 (1962)
62 T 1. *Takeno, S.:* Progr. Theoret. Phys. (Kyoto) **28**, 33 (1962)
62 V 1. *Vinogradov, V. S.:* Soviet Phys. Solid State **4**, 519 (1962)
62 W 1. *Woll jr., E. J., and W. Kohn:* Phys. Rev. **126**, 1693 (1962)

63 A 1. *Abrikosov, A. A., L. P. Gorkov, and I. E. Dzyaloskinshi:* Methods of quantum field theory in statistical physics. Englewood Cliffs: Prentice Hall 1963
63 B 1. *Biem, W., H. Hahn:* Phys. stat. solidi **3**, 1911 (1963)
63 B 2. — Phys. stat. solidi **3**, 1927 (1963)
63 B 3. *Buchwald, V. T., and A. Davis:* Quart. J. Mech. Appl. Math. **16/3**, 283 (1963)
63 B 4. *Bernstein, B. T.:* Phys. Rev. **132**, 50 (1963)
63 C 1. *Cochran, W.:* Rept. Progr. Phys. **26**, 1 (1963)
63 C 2. *Cowley, R. A.:* Advan. Phys. **12**, 421 (1963)
63 D 1. *Dick, B. G.:* Phys. Rev. **129**, 1583 (1963)
63 D 2. *Damask, A. C., and G. J. Dienes:* Point defects in metals. New York: Gordon and Breach 1963
63 D 3. *Davies, R. W., and J. S. Langer:* Phys. Rev. **131**, 163 (1963)
63 D 4. *Dawber, P. G., and R. J. Elliott:* Proc. Phys. Soc. (London) **81**, 453 (1963)
63 D 5. — — Proc. Roy. Soc. (London) Ser. A **273**, 222 (1963)
63 F 1. *Fischer, K., u. H. Hahn:* Z. Physik **172**, 172 (1963)
63 F 2. *Fukai, Y.:* J. Phys. Soc. Japan **18**, 1413 (1963)
63 F 3. *Foreman, A. J. E., and A. B. Lidiard:* Phil. Mag. **8**, 97 (1963)
63 H 1. *Harrison, W. A.:* Phys. Rev. **129**, 2503, 2512 (1963); **131**, 2433 (1963)
63 H 2. *Hahn, H.:* Inelastic scattering of neutrons. p. 37. Vienna: I.A.E.A. 1963
63 K 1. *Kagan, Yu. M., i Ya. A. Iosilevskii:* Soviet Phys. JETP **17**, 195 (1963)
63 K 2. — — Soviet Phys. JETP **17**, 925 (1963)
63 K 3. *Klein, M. V.:* Phys. Rev. **131**, 1500 (1963)
63 L 1. *Lengeler, B., u. W. Ludwig:* Z. Physik **171**, 273 (1963)
63 L 2. *Litzman, O.:* Czech. J. Phys. **13**, 558 (1963)
63 L 3. *Lifshitz, I. M.:* Soviet Phys. JETP **17**, 1159 (1963)
63 L 4. *Lehman, G. W., and R. E. de Wames:* Phys. Rev. **131**, 1008 (1963)
63 L 5. *Lloyd, P., and J. J. O'Dwyer:* Australian J. Phys. **16**, 193 (1963)
63 M 1. *Maradudin, A. A., E. W. Montroll, and G. H. Weiss:* Sol. State Phys. Suppl. **3**, (1963)
63 M 2. — In: Astrophysics and the many body problem. p. 109. New York: Benjamin (1963)
63 M 3. *Mozer, B., and A. A. Maradudin:* Bull. Am. Phys. Soc. II, **8**, 193 (1963)
63 M 4. *Mozer, B.:* Bull. Am. Phys. Soc. II, **8**, 593 (1963)
63 M 5. *Maradudin, A. A., and P. A. Flinn:* Phys. Rev. **129**, 2529 (1963)
63 M 6. *Mitskevich, V. V.:* Soviet Phys. Solid State **4**, 2224 (1963)
63 M 7. *Melngailis, J., A. A. Maradudin, and A. Seeger:* Phys. Rev. **131**, 1972 (1963)
63 P 1. *Pohl, R. O., and C. T. Walker:* Phys. Rev. **131**, 1433 (1963)
63 P 2. *Pautamo, Y,:* Ann. Acad. Sci. Fennicae A VI, No 129 (1963)
63 S 1. *Silverman, B. D.:* Phys. Rev. **131**, 2478 (1963)
63 T 1. *Taylor, P. L.:* Phys. Rev. **131**, 1995 (1963)
63 T 2. *Takeno, Sh.:* Progr. Theoret. Phys. (Kyoto) **29**, 191, 328 (1963); **30**, 1, 144 (1963)
63 T 3. *Trifonov, E. D.:* Soviet. Phys. Doklady **7**, 1105 (1963)
63 V 1. *Vissher, W. M.:* Phys. Rev. **129**, 28 (1963)
63 W 1. *Woods, A. D. B., B. N. Brockhouse, R. A. Cowley, and W. Cochran:* Phys. Rev. **131**, 1025 (1963)
63 W 2. *Waller, I.:* Crystallography and crystal perfection. p. 189. London, New York: Academic Press 1963
63 W 3. *Woods, A. D. B.:* Inelastic scattering of neutrons. Vol. II, p. 3. Vienna: I.A.E.A. 1963
64 B 1. *Bak, T., (ed):* Phonons and phonon interactions New York: Benjamin 1964
64 C 1. *Cochran, W., and G. S. Pawley:* Proc. Roy. Soc. (London) Ser. A **280**, 1 (1964)
64 D 1. *Dettmann, K., u. W. Ludwig:* Phys. kondens. Materie, **2**, 241 (1964)
64 D 2. *Dzyub, J. P.:* Soviet Phys. Solid State **6**, 1469 (1964)
64 E 1. *Einspruch, N. G., and R. J. Manning:* J. Appl. Phys. **35**, 560 (1964)
64 E 2. *Elliott, R. J., and D. W. Taylor:* Proc. Phys. Soc. (London) **83**, 189 (1964)
64 F 1. *Friedel, J.:* Dislocations. Oxford: Pergamon 1964
64 H 1. *Harrison, W. A.:* Phys. Rev. **136**, A 1107 (1964)

64 I 1. *Ivanov, M. A., i M. A. Krivoglaz:* Soviet Phys. Solid State, **6**, 159 (1964)
64 J 1. *Jansen, L.:* Phys. Rev. **135**, A 1292 (1964). —, *and E. Lombardi:* Phys. Rev. **136**, A 1011 (1964)
64 J 2. *Johnson, R. A.:* Phys. Rev. **134**, A 1329 (1964). —, *G. J. Dienes, and A. C. Damask:* Acta Met. **12**, 1215 (1964)
64 K 1. *Kagle, B. J., and A. A. Maradudin:* Westinghouse Research Memo 64—929—442—M1—(1964)
64 L 1. *Ludwig, W.:* Ergebn. d. exakt. Naturwissenschaften **35**, 1—102 (1964)
64 L 2. *Ludwig, W.:* In: Phonons and phonon interactions. p. 23. New York: Benjamin 1964
64 L 3. *Lloyd, P.:* Australian J. Phys. **17**, 269 (1964)
64 L 4. *Laval, J.:* Ann. Inst. Henri Poincaré, Vol. I, n° **4**, 329 (1964)
64 L 5. *Lengeler, B., u. W. Ludwig:* Phys. stat. solidi **7**, 463 (1964)
64 L 6. *Ludwig, W., and B. Lengeler:* Solid State Commun. **2**, 83 (1964)
64 L 7. *Loudon, R.:* Proc. Phys. Soc. **84**, 379 (1964)
64 M 1. *Maradudin, A. A.:* In: Phonons and phonon interactions. p. 424. New York: Benjamin 1964
64 M 2. — Rev. Mod. Phys. **36**, 417 (1964)
64 M 3. *Mirlin, D. N., i I. I. Reshina:* Soviet Phys. Solid State **6**, 728 (1964)
64 M 4. *Maradudin, A. A., and J. Melngailis:* Phys. Rev. **133**, A 1188 (1964)
64 M 5. — Ann. Phys. (N.Y.) **30**, 371 (1964)
64 M 6. —, *and A. Ambegoakar:* Phys. Rev. **135**, A 1071 (1964)
64 M 7. *McCumber, D.:* J. Math. Phys. **5**, 221, 508 (1964)
64 M 8. *Maradudin, A. A., P. A. Flinn, and J. M. Radcliff:* Ann. Phys. **26**, 81 (1964)
64 N 1. *Narayanamurti, V.:* Phys. Rev. Letters **13**, 693 (1964)
64 N 2. *Nozières, P.:* Theory of interacting Fermi systems. New York: Benjamin Inc. (1964)
64 S 1. *Stekhanov, A. I., i M. B. Eliashberg:* Soviet Phys. Solid State **5**, 2185 (1964)
64 S 2. *Silsbee, R. H., and D. F. Fitchen:* Rev. Mod. Phys. **36**, 432 (1964)
64 T 1. *Timusk, T., and W. Staude:* Phys. Rev. Letters **13**, 373 (1964)
64 T 2. *Tinkham, M.:* Group theory and quantum mechanics. New York: McGraw-Hill 1964
64 V 1. *Vissher, W. M.:* Phys. Rev. **134**, A 965 (1964)
64 W 1. *Wallace, D. C.:* Phys. Rev. **133**, A 153 (1964)
64 W 2. *Weiss, R. J.:* Proc. Phys. Soc. (London) **83**, 1021 (1964)
65 B 1. *Bicknese, V.:* Physica **31**, 1473 (1965)
65 B 2. *Barron, T. H. K.:* Phys. Rev. **137**, A 487 (1965)
65 B 3. *Bernstein, B. T.:* Phys. Rev. **137**, A 1404 (1965)
65 C 1. *Cowley, R. A.:* Phil. Mag. **11**, 673 (1965)
65 C 2. —, *and E. R. Cowley:* Proc. Roy. Soc. (London) Ser. A **287**, 259 (1965)
65 D 1. *Diehl, J.:* In: Moderne Probleme der Metallphysik. p. 227. Berlin: Springer 1965
65 D 2. *Dettmann, K., u. W. Ludwig:* Phys. stat. solidi **10**, 689 (1965)
65 D 3. *Dolling, G., and A. D. B. Woods:* Proc. Intern Conf. on Inelastic Scattering of Neutrons, p. 373. I. A. E. A. Vienna 1965
65 E 1. *Elliott, R. J., and A. A. Maradudin:* Proc. Intern. Conf. on Inelastic Neutron Scattering. Bombay 1965
65 E 2. — et al. Proc. Roy. Soc. (London) Ser. A **289**, 1 (1965)
65 F 1. *Fritz, B.:* J. Phys. Chem. Solids, Suppl. **1**, 485 (1965)
65 F 2. —, *U. Gross u. D. Bäuerle:* Phys. stat. solidi **11**, 231 (1965). —, *F. Lüty u. G. Rausch:* Phys. stat. solidi **11**, 635 (1965)
65 F 3. *Fussgaenger, K., W. Martiensen u. H. Bilz:* Phys. stat. solidi **12**, 383 (1965)
65 G 1. *Grindlay, J.:* Canad. J. Phys. **43**, 1604 (1965)
65 G 2. *Gazis, D. C., and R. F. Wallis:* Acta Mech. **1**, 253 (1965)
65 H 1. *Hahn, H.:* Proc. Intern. Conf. on Inelastic Scattering of Neutrons, p. 279, I.A.E.A. Vienna 1965
65 H 2. *Harrison, W. A.:* J. Phys. Chem. Solids, Suppl. **1**, 85 (1965)
65 H 3. *Högberg, T.:* Arkiv Fysik **29**, 519 (1965); **30**, 563 (1965)

65 H 4. *Hayes, W., H. F. MacDonald, and R. J. Elliott:* Phys. Rev. Letters **15**, 961 (1965)
65 J 1. *Johnson, R. A.:* J. Phys. Chem. Solids **26**, 75 (1965). — Acta Met. **13**, 1259 (1965)
65 K 1. *Krebs, K.:* Phys. Rev. **138**, A 143 (1965)
65 K 2. *Krumhansl, J. A.:* J. Phys. Chem. Solids, Suppl. **1**, 627 (1965)
65 K 3. *Kunc, K.:* Czech. J. Phys. B **15**, 883 (1965)
65 L 1. *Lombardi, E., and L. Jansen:* Phys. Rev. **140**, A 275 (1965)
65 L 2. *Lax, M.:* J. Phys. Chem. Solids, Suppl. **1**, 583 (1965)
65 L 3. — J. Phys. Chem. Solids, Suppl. **1**, 179 (1965)
65 L 4. *Lengeler, B., and W. Ludwig:* J. Phys. Chem. Solids, Suppl. **1**, 439 (1965)
65 L 5. *Leibfried, G.:* J. Phys. Chem. Solids, Suppl. **1**, 237 (1965)
65 L 6. *Litzman, O., and K. Kunc:* J. Phys. Chem. Sol. **26**, 1825 (1965)
65 M 1. *Møller, H. B., and A. R. Mackintosh:* Phys. Rev. Letters **15**, 623 (1965) — et al. Inelastic scattering of neutrons, p. 95. Vienna: I.A.E.A. 1965
65 M 2. *Mahanty, J., and M. Yussouff:* Proc. Phys. Soc. (London) **85**, 1223 (1965)
65 O 1. *Overton, W. C.:* Sol. State Commun. **3**, 397 (1965)
65 P 1. *Pathak, K. N.:* Phys. Rev. **139**, A 1569 (1965)
65 P 2. *Perlin, Yu. E., i I. Ya. Ogurtsov:* Soviet Phys. Solid State **7**, 1180 (1965)
65 R 1. *Renk, K. F.:* Phys. Rev. Lett. **14**, 281 (1965)
65 S 1. *Sham, L. J.:* Proc. Roy. Soc. (London) Ser. A **283**, 33 (1965)
65 S 2. *Sievers, A. J., and C. D. Lytle:* Phys. Rev. Letters **14**, 271 (1965). —, *A. A. Maradudin, and S. S. Jaswal:* Phys. Rev. **138**, A 272 (1965)
65 S 3. —, *and S. Takeno:* Phys. Rev. Letters **15**, 1020 (1965). — — Phys. Rev. **140**, A 1030 (1965)
65 S 4. *Srivastava, P. L., and B. Dayal:* Phys. Rev. **140**, A 1014 (1965)
65 S 5. *Sham, L. J.:* Phys. Rev. **139**, A 1189 (1965)
65 S 6. *Swartz, K. D., and A. V. Granato:* J. Acoust. Soc. Am. **38**, 824 (1965)
65 S 7. *Singh, D. N.:* Indian J. Pure Appl. Phys. **39**, 555 (1965)
65 S 8. *Seward, W. D.:* Proc. 9th Intern. Conf. Low Temperat. Physics. New York (1965)
65 S 9. *Stockmeyer, R.:* Kernforschungsanlage Jülich, Bericht Jül-292-RW (1965)
65 T 1. *Toya, T.:* J. Phys. Chem. Solids, Suppl. **1**, 91 (1965)
65 T 2. *Thoma, K., u. W. Ludwig:* Phys. stat. solidi **8**, 487 (1965)
65 V 1. *Vosko, S. H., R. Taylor, and G. H. Keech:* Can. J. Phys. **43**, 1187 (1965)
65 W 1. *Wallace, D. C.:* Rev. Mod. Phys. **37**, 57 (1965)
65 W 2. *Weertman, J., and J. R. Weertman:* Elementary dislocation theory. New York: Mac Millan 1965.
65 W 3. *Warren, J. L.:* Inelastic scattering of neutrons, p. 361. I.A.E.A. Vienna (1965)
65 W 4. *Wilcox, R. M.:* Phys. Rev. **139**, A 1281 (1965)
65 *W* 5. *Wallace, D. C.:* Phys. Rev. **139**, A 877 (1965)
66 B 1. *Bilz, H.:* Lectures on infrared absorption. Summerschool Aberdeen, 1966
66 B 2. *Barron, T. H. K.:* Acta Cryst. **20**, 125 (1966)
66 B 3. *Birman, J. L.:* Phys. Rev. **150**, 771 (1966)
66 C 1. *Clark, C. B., R. Herman, R. F. Wallis:* Phys. Rev. **139**, A 860 (1965)
66 C 2. *Cowley, E. R., and R. A. Cowley:* Proc. Roy. Soc. (London) Ser. A **292**, 209 (1966)
66 C 3. *Carruthers, P., and K. S. Dy:* Phys. Rev. **147**, 214 (1966)
66 D 1. *Deprez, G., et R. Fouret:* J. Physique Radium **27**, 147 (1966)
66 D 2. *Doyama, R. M. J. Cotterill:* Intern. Conf. on Electron Diffraction . . . Melbourne, II-H-3, Pergamon (1966)
66 F 1. *Filser, J., u. Ludwig Thoma:* Z. Physik **193**, 384 (1966)
66 H 1. *Hayes, W., and J. W. Hodby;* Proc. Roy. Soc. (London) Ser. A **294**, 359 (1966)
66 H 1. *Hiki, Y., and A. V. Granato:* Phys. Rev. **144**, 411 (1966)
66 J 1. *Jansen, L., u. E. Lombardi:* Z. Physik **190**, 161 (1966)
66 J 2. *Johnson, R. A.:* Intern. Conf. on Electron Diffraction . . . Melbourne II-H-4, Pergamon (1966)

66 J 3. *Jones, E. R., J. T. McKinney, M. B. Webb:* Phys. Rev. **151**, 476 (1966)
66 K 1. *Kunc, K.:* Phys. stat. solidi **15**, 683 (1966). —, *J. Miklosko:* Czech. J. Physics B **17**, (1967) to be published
66 K 2. *Krumhansl, J. A.:* In: Elementary excitations and their interactions in solids. VII. Cortina: Nato Adv. Study Inst. 1966
66 K 3. *Kratochvil, J.:* In: Theory of crystal defects. p. 17. Prague: Academia 1966
66 K 4. *Klein, M. V.:* Phys. Rev. **141**, 716 (1966)
66 K 5. *Keating, P. N.:* Phys. Rev. **145**, 637; **149**, 674; **152**, 774 (1966)
66 L 1. *Ludwig, W.:* In: Theory of crystal defects. p. 57. Prague: Academia 1966
66 L 2. *Litzman, O.:* Phys. stat. solidi **13**, 71 (1966)
66 L 3. *Ludwig, W.:* In: Elementary excitations and their interactions in solids. III. Cortina: Nato Adv. Study Inst. 1966
66 M 1. *Mahesh, P. S., and B. Dayal:* Phys. Rev. **143**, 443 (1966)
66 M 2. *Maradudin, A. A.:* Solid State Phys. **18**, 273—420; **19**, 1—134 (1966)
66 M 3. *Mozer, B.:* Private Communication
66 M 4. *Mahanty, J.:* Proc. Phys. Soc. (London) **88**, 1011 (1966)
66 M 5. *Maradudin, A. A., and R. F. Wallis:* Phys. Rev. **148**, 945 (1966)
66 N 1. *Nardelli, G. F., (ed.):* Elementary excitations and their interactions in solids. Cortina: Nato Adv. Study Inst. 1966
66 O 1. *Overton, W. C.:* J. Chem. Phys. **44**, 934 (1966)
66 P 1. *Pick, H.:* Springer tracts in modern physics **38**, 1—83 (1966)
66 P 2. *Pohl, R. O.:* In: Elementary excitations and their interactions in solids. VIII, Cortina: Nato Adv. Study Inst. 1966
66 P 3. —, *R. Rollefson:* to be published
66 P. 4 —, *F. C. Baumann:* to be published
66 R 1. *Rejler, O., S. Wernberg, och O. Beckmann:* Arkiv Fysik **32**, 509 (1966)
66 S 1. *Schröder, U.:* Solid State Commun. **4**, 347 (1966)
66 S 2. *Seeger, A.:* In: Theory of crystal defects. p. 37. Prague: Academia 1966
66 S 3. *Seward, W. D., V. Narayanamurti:* Phys. Rev. **148**, 463 (1966). —, — *and R. O. Pohl:* Phys. Rev. **148**, 481 (1966)
66 S 4. *Sievers, A. J.:* In: Elementary excitations and their interactions in solids. VI, Cortina: Nato Adv. Study Inst. 1966
66 S 5. —, *R. W. Alexander, and S. Takeno:* Solid State Commun. **4**, 483 (1966)
66 W 1. *Wehner, R.:* Phys. stat. solidi **15**, 725 (1966)
66 W 2. *Wallace, D. C.:* Phys. Rev. **152**, 247, 261 (1966)
66 W 3. *Wallis, R. F., and A. A. Maradudin:* Phys. Rev. **148**, 962 (1966)
66 W 4. —, *I. P. Ipatova i A. A. Maradudin:* Soviet Phys. Solid State **8**, 850 (1966)
66 W 5. *Walton, D.:* Phys. Rev. **151**, 627 (1966)
66 Y 1. *Yussouff, M., and J. Mahanty:* Proc. Phys. Soc. (London) **87**, 689 (1966)
67 B 1. *Bilz, H., u. L. Genzel:* (editor) Encyclop. Physics, **25/2**. Berlin, Heidelberg, New York: Springer 1967
67 B 2. *Biem, W.:* Habilitationsschrift Aachen (1967)
67 C 1. *Choquard, Ph.:* In preparation
67 G 1. *Goetze, W., and K. H. Michel:* In press
67 H 1. *Horner, H.:* Z. Physik, in press
67 L 1. *Ludwig, W., and H. Pick:* Phys. status solidi **19**, 313 (1967)
67 L 2. *Ludwig, W.:* Nat. Bureau of Standards, Proc. of the Conf. Shennandoah, (1967) p. to be published
67 T 1. *Thoma, K.:* Dissertation Aachen (1967), to be published in Phys. status solidi

Professor Dr. *W. Ludwig*

Institut für Theoretische Physik III der Universität
63 Gießen, Leihgesterner Weg 108